Machine Trades Printreading

third edition

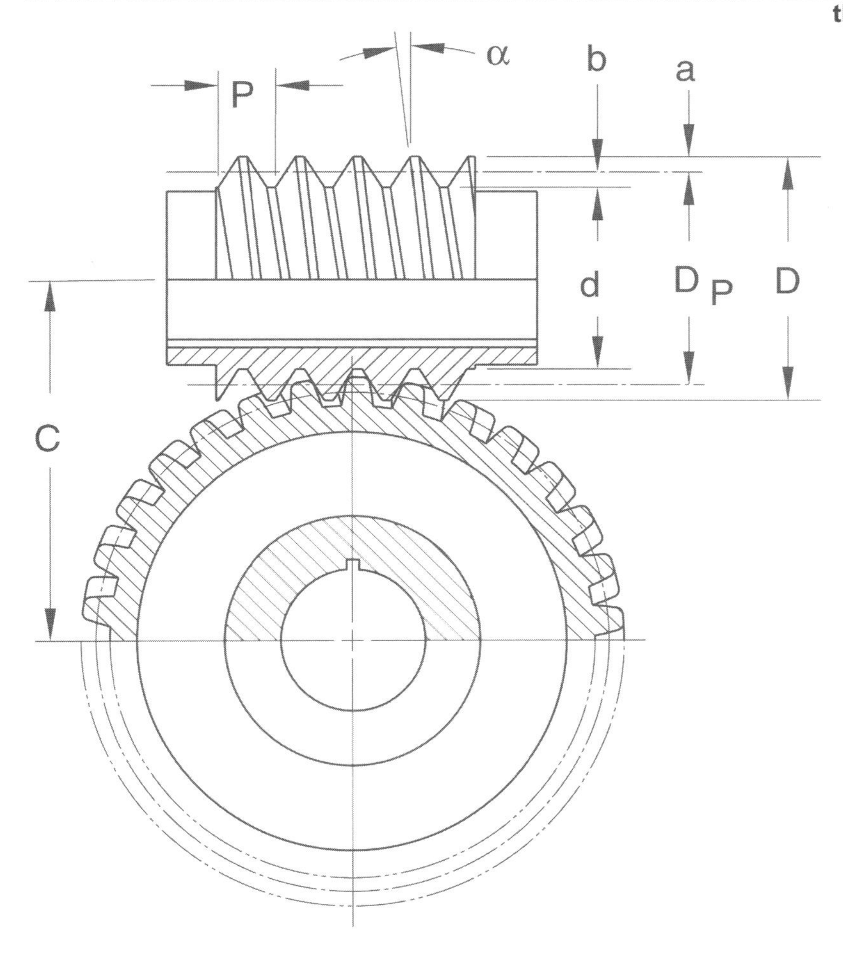

Thomas E. Proctor
J. David Holloway
Jonathan F. Gosse

atp AMERICAN TECHNICAL PUBLISHERS
ORLAND PARK, ILLINOIS 60467-5756

Machine Trades Printreading, 3rd edition, contains procedures commonly practiced in industry and the trade. Specific procedures vary with each task and must be performed by a qualified person. For maximum safety, always refer to specific manufacturer recommendations, insurance regulations, specific job site and plant procedures, applicable federal, state, and local regulations, and any authority having jurisdiction. The material contained is intended to be an educational resource for the user. American Technical Publishers, Inc. assumes no responsibility or liability in connection with this material or its use by any individual or organization.

American Technical Publishers, Inc., Editorial Staff

Editor in Chief:
Jonathan F. Gosse
Vice President—Production:
Peter A. Zurlis
Art Manager:
James M. Clarke
Multimedia Manager:
Carl R. Hansen
Technical Editor:
Julie M. Welch
Copy Editor:
Jeana M. Platz
Cover Design:
Melanie G. Doornbos
Illustration/Layout:
Melanie G. Doornbos
Thomas E. Zabinski
Jennifer M. Hines
CD-ROM Development:
Gretje Dahl
Nicole S. Polak

3 4 5 6 7 8 9 – 11 – 9 8 7 6 5 4

Printed in the United States of America

ISBN 978-0-8269-1881-9

 This book is printed on recycled paper.

ACKNOWLEDGMENTS

The authors and publisher are grateful for the technical information and assistance provided by the following companies, organizations, and individuals.

Autodesk, Inc.
Boston Gear
Carrier Corporation
EZ Loader Boat Trailers, Inc.
The Falk Corporation
General Electric Co. U. S. A.
Girtz Industries, Inc.
Hewlett-Packard Company
I. T. W. Devcon
L. S. Starrett Company
Miller Electric Manufacturing Company
Randall Machine Company
Skil Corporation
Staedtler, Inc.
Stratasys, Inc.
Toledo Scale Masstron Products
Worthington Industries, Inc.

Joe Gladkowski
Mechanical Production Technology Program
Joliet Junior College

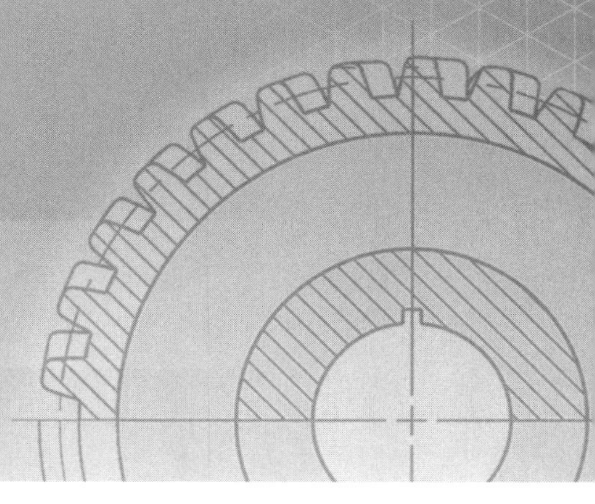

CONTENTS

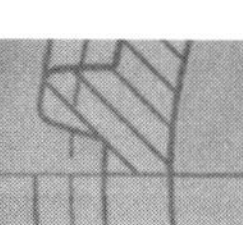

INTERACTIVE CD-ROM CONTENTS

- Using This CD-ROM
- Quick Quizzes®
- Illustrated Glossary
- Alphabet of Lines Tests
- Printreading Symbols Tests
- Printreading Tests
- Flash Cards
- Prints
- Reference Material

Machine Trades Printreading, 3rd Edition, presents a thorough foundation for understanding the drawings and prints used for manufacturing and machining parts. This text begins with print production and the various techniques for representing three-dimensional objects in two-dimensional drawings. Next is a review of the associated math and measurement tools that machinists frequently encounter, followed by dimensioning and tolerancing methods. Then, the text presents a discussion of various materials and their properties, special machined features, and other components, along with how each is specified on prints. The text concludes with an overview of numerical control machining and associated print information.

This new edition has been updated to conform to the latest ANSI standards for printreading symbols, view representation, and geometric dimensioning and tolerancing. It also features new and expanded topics, including the following:

- projection systems
- significant digits
- measurement systems
- units of measurement
- precision measurement tools
- tolerance stack
- standard tolerances
- engineering fits

Each chapter concludes with a set of review questions and a trade competency test, which includes questions based on actual manufacturing prints. These questions and tests are designed to reinforce the printreading skills and knowledge covered in each chapter. Selected chapters also include sketching exercises for practicing techniques for object representation. The final chapter is a section of 10 general trade competency tests and their associated prints. Large-format foldout prints are included for more complex manufacturing drawings.

The interactive CD-ROM in the back of the book includes Quick Quizzes® for each chapter, Alphabet of Lines Tests, Printreading Symbols Tests, Printreading Tests, an Illustrated Glossary, Flash Cards, Prints, and related Reference Material. Information about using the *Machine Trades Printreading* CD-ROM is included on the back inside cover of the book. Answers to all review questions, sketching exercises, and trade competency tests are included in the Answer Key, which is available separately.

Machine Trades Printreading, 3rd Edition, is one of many high-quality training products available from American Technical Publishers, Inc. To obtain information about related training products, visit the American Tech web site at www.go2atp.com.

The Publisher

FEATURES

Machine Trades PRINTREADING

Gears and Cams 10

Gears are toothed wheels used in pairs to transmit power or motion from one shaft to another. The American Gear Manufacturers Association (AGMA) provides specifications and tolerances for gear manufacture and applications. Cams are used with cam followers to transfer loads, change direction, or change the speed of a machine. Assemblies using cams must include cam prints because cams must be custom made for their particular applications.

GEARS

A *gear* is a toothed wheel that is used with other gears in order to transmit rotational power from one shaft to another. The meshing of gear teeth prevents slippage, which can otherwise occur with belt and pulley systems. Slippage is lost motion, which also creates friction heat, reducing the efficiency of the power transmission. However, gear shafts must be in close proximity. Longer distances require the use of a chain and sprockets.

Meshing gears rotate in opposite directions. A *pinion gear* is the smallest of meshing gears. With two meshing gears, one gear rotating counterclockwise causes the other gear to rotate clockwise. **See Figure 10-1.** An added third gear rotates in the same direction as the first gear.

Gear Ratio

Meshed gears of equal diameter rotate at the same rate. Since the circumferences are also equal, both return to the same position after each rotation. One complete revolution of one gear produces one complete revolution of the other gear.

If gears have different diameters, they turn at different rates. *Gear ratio* is the relationship between the sizes of two meshing gears. The gear ratio can be calculated by dividing the diameter of the driving gear by the diameter of the driven gear. **See Figure 10-2.** Another method of calculating gear ratio is to divide the number of teeth on the respective gears. Gear ratio is important to determining how rotational speed and forces change as they are transmitted through gears.

Boston Gear

Gears are made in many different sizes, shapes, and materials.

Chapter Introductions provide an overview of key content in the chapter.

Factoids include additional information about the history, application, or importance of printreading topics.

Two-color illustrations clearly convey printreading concepts and practices through examples.

Photographs supplement printreading topics and practices.

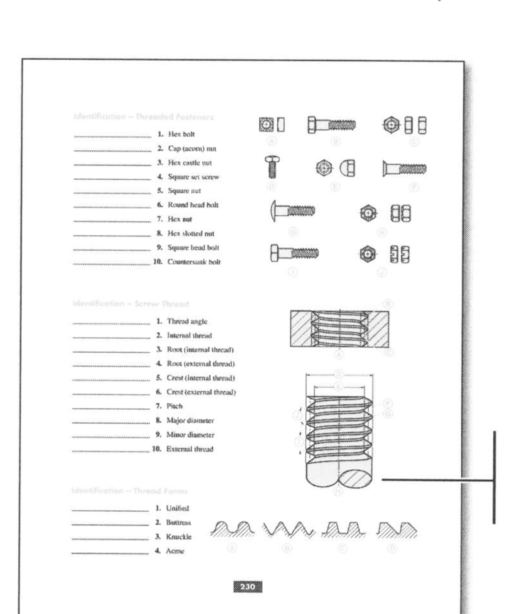

Identification – Threaded Fasteners

1. Hex bolt
2. Cap (acorn) nut
3. Hex castle nut
4. Square set screw
5. Square nut
6. Round head bolt
7. Hex nut
8. Hex slotted nut
9. Square head bolt
10. Countersunk bolt

Identification – Screw Thread

1. Thread angle
2. Internal thread
3. Root (internal thread)
4. Root (external thread)
5. Crest (internal thread)
6. Crest (external thread)
7. Pitch
8. Major diameter
9. Minor diameter
10. External thread

Identification – Thread Forms

1. Unified
2. Buttress
3. Knuckle
4. Acme

230

Review questions test for comprehension of content covered.

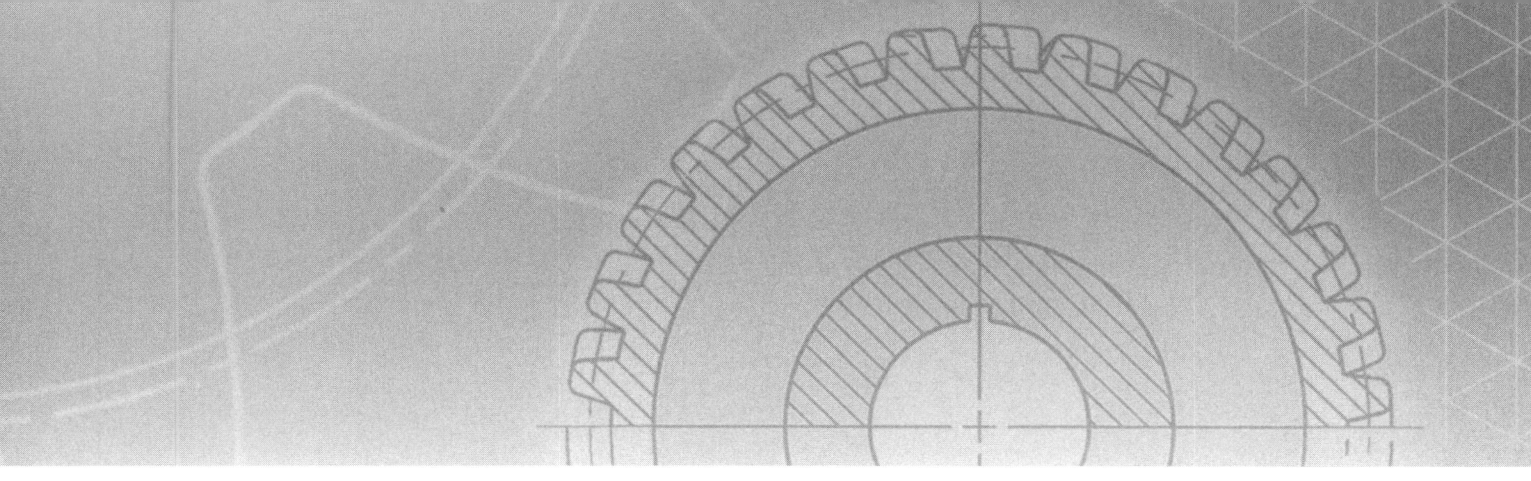

Included CD-ROM offers Quick Quizzes®, interactive tests on symbols, lines, and printreading skills, an Illustrated Glossary, and Reference Material.

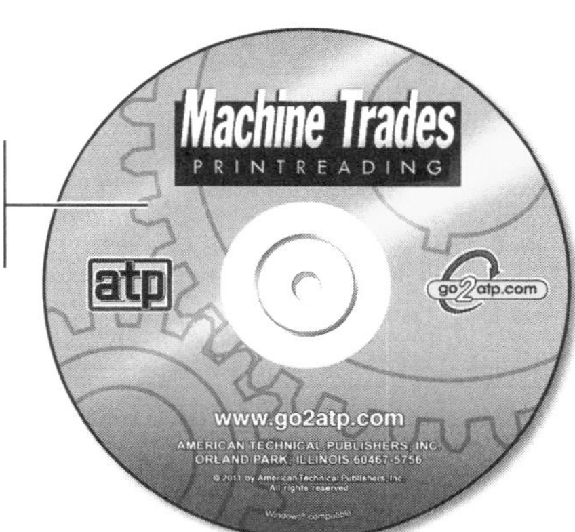

Trade Competency Tests develop skills by applying printreading concepts to actual manufacturing prints.

Eight large-format foldout prints are included for more challenging Trade Competency Tests.

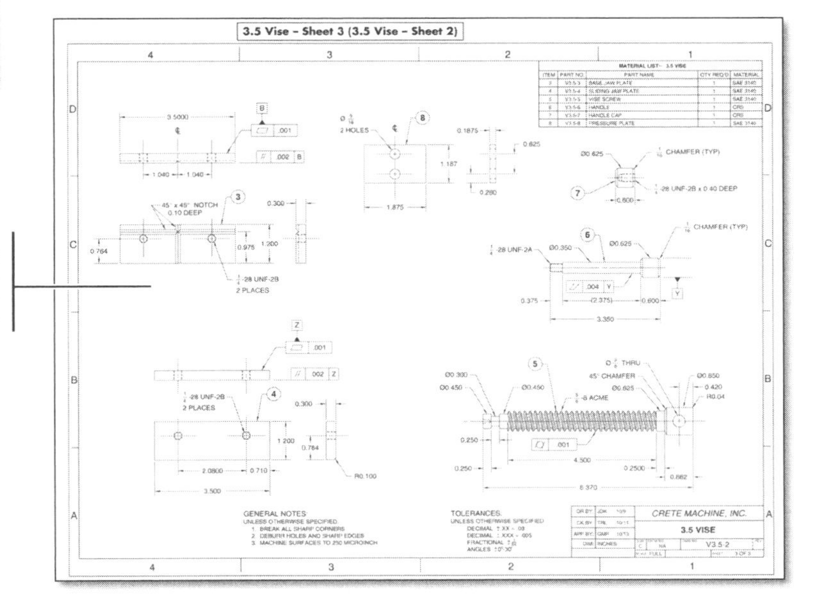

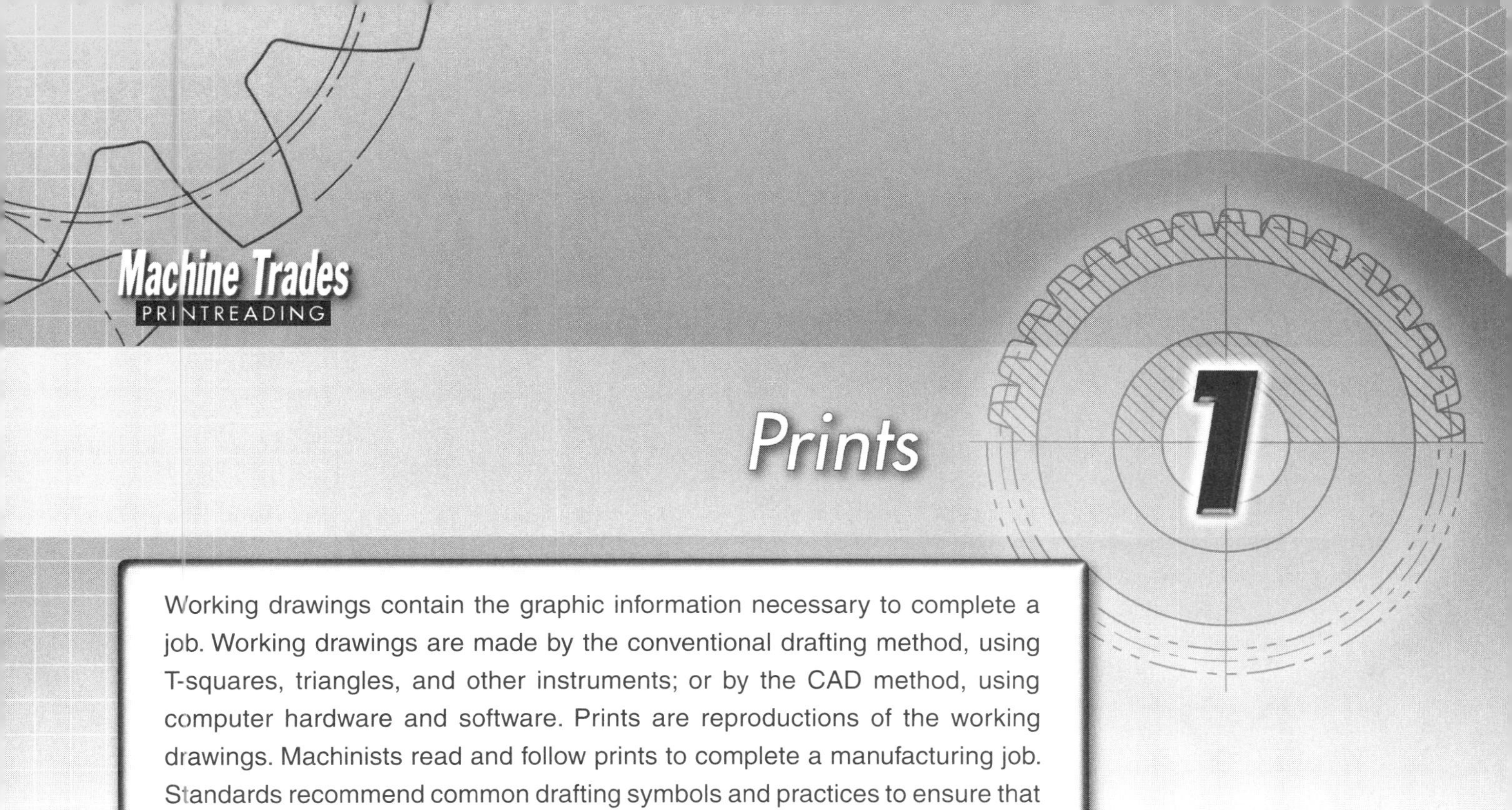

Working drawings contain the graphic information necessary to complete a job. Working drawings are made by the conventional drafting method, using T-squares, triangles, and other instruments; or by the CAD method, using computer hardware and software. Prints are reproductions of the working drawings. Machinists read and follow prints to complete a manufacturing job. Standards recommend common drafting symbols and practices to ensure that manufacturing drawings are universally understandable.

PRINT PRODUCTION

A *print* is a reproduction of a working drawing. A *working drawing* is a set of plans that contains the information necessary to complete a job. Originally, prints were referred to as blueprints because the process used to make them produced white lines on a blue background. Any number of copies could be made from working drawings by using a process similar to the process used for making prints from photographic negatives.

Other types of prints were developed to produce dark lines on a white background, which were generally easier to read and allowed users to add field notes. **See Figure 1-1.** Some processes also made it easier to enlarge or reduce conventional drawings. Currently, conventional drafting and analog reproductions are uncommon. Most working drawings are now produced on computer software and output digitally to large-format plotters or printers.

Blueprints

The use of blueprints began in 1840 when a method was discovered to produce paper sensitized with iron salts that would undergo a chemical change when exposed to light. Drawings made on translucent paper were placed over the sensitized paper and a glass frame held the papers firmly. *Translucent paper* is paper that allows some light to pass through. The frame was then exposed to sunlight. A chemical action occurred wherever the light passed through the translucent paper to the sensitized paper. When the blueprint paper was washed in water, the parts protected by the ink lines on the tracing would show as white lines and the areas exposed to light would show as a blue background. A fixing bath of potassium dichromate, a second rinse with water, and print drying completed the process.

This wet process caused the paper to swell slightly or stretch unevenly. Even after drying, this usually distorted the drawing slightly, which reduced accuracy.

Blueprints have been used in situations with lots of outdoor exposure because they fade less rapidly in sunlight than prints produced by the diazo process. Today, prints made with the blueprint process are rare. However, the term "blueprint" is often used as a general term for any of several types of prints. In most cases, a different process is used to produce a modern "blueprint."

Diazo Prints

Prints made by the diazo process have blue or black lines on a white background. This process also uses photochemical reactions to form images on special paper.

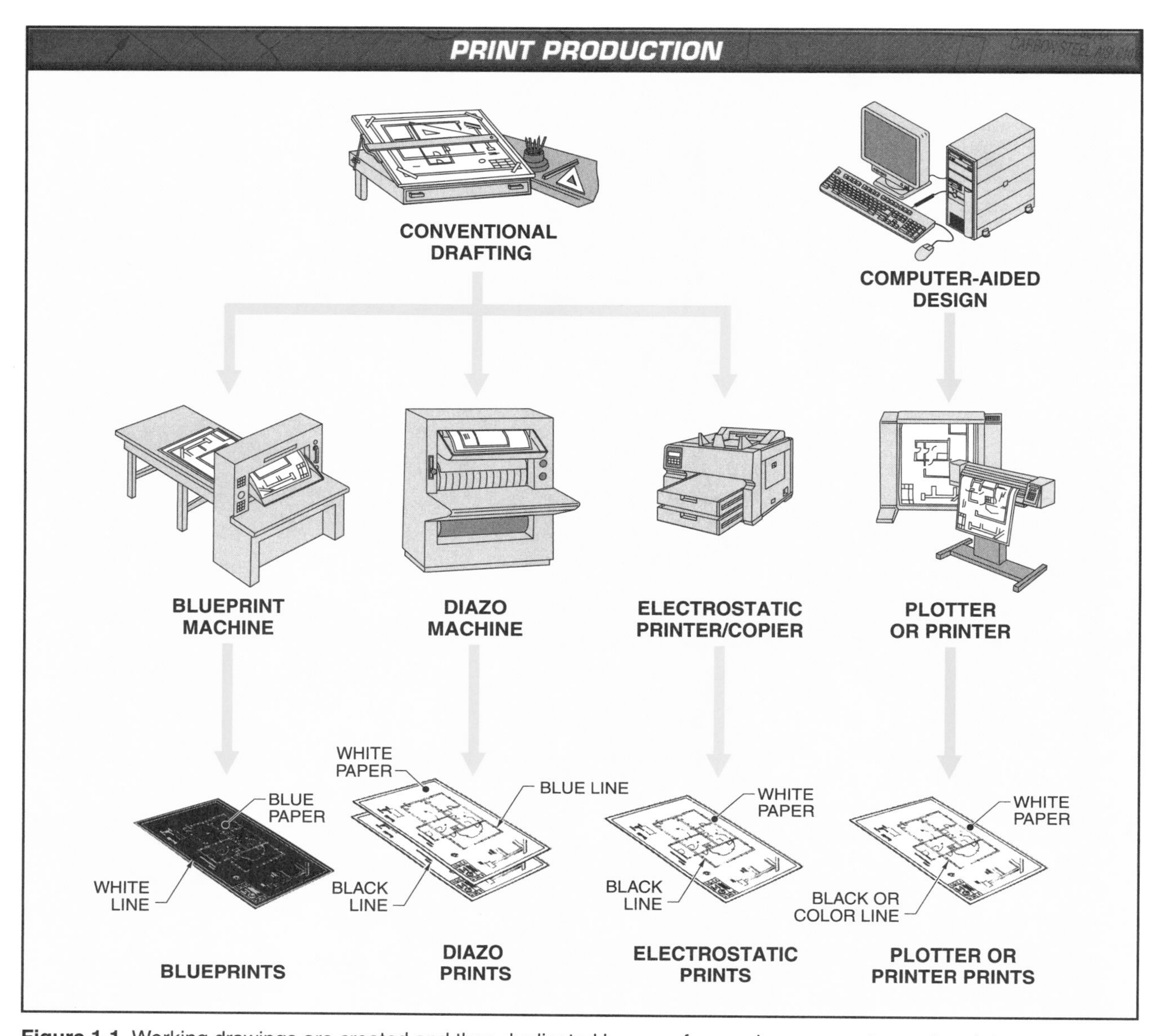

Figure 1-1. Working drawings are created and then duplicated by one of several processes to create prints.

Two development methods are used in the diazo process, each with a special type of sensitized paper. These papers are coated with a chemical that reacts to ultraviolet light. A drawing on translucent material is placed over a sheet of sensitized paper and fed into the print machine. The two sheets are exposed to ultraviolet light as they revolve around a glass cylinder containing an ultraviolet lamp. The sensitized paper is exposed in the areas where the translucent original is clear but not where lines or images block the light. The sheets are separated and the sensitized paper is then developed by either a wet diazo or dry diazo process.

In the wet development method, the sensitized paper passes under a roller that moistens the exposed top surface. This completes the chemical reaction to bring out the image. In the dry development method, the sensitized paper is passed through a heated chamber in which its surface is exposed to ammonia vapor. The ammonia vapor precipitates the dye to bring out the image.

The diazo method can produce prints with either blue or black lines on a white background.

Electrostatic Prints

Electrostatic prints are produced by the same process used in office copiers. Working drawings are exposed to light and the image is projected through a lens onto a negatively charged drum. **See Figure 1-2.** The working drawings can be projected from either full-size paper originals or reduced-size photographic negatives of the original.

Light projected from the drawing discharges areas of the drum. The drum retains the negative charge in the unexposed areas. The drum then turns past a roller where black toner particles are attracted to the negatively charged image areas on the drum surface. As the drum continues to turn in synchronization with positively charged copy paper, the toner particles are attracted to the paper. A fuser fixes the toner to the paper with heat and pressure.

Prints made by this method have black lines on a white background. The advantages of the electrostatic process include easy enlargement and reduction of drawings, small storage size, and quick retrieval and duplication. The major disadvantage is the potential for distortion by projection through a lens.

Plotter Prints

A *plotter* is a computer output device that generates finished drawings with pens. Plotters may use multiple pens in various colors and print on very large sheets of paper. **See Figure 1-3.** Plotters are available in two major styles: rotary-drum and flatbed plotters.

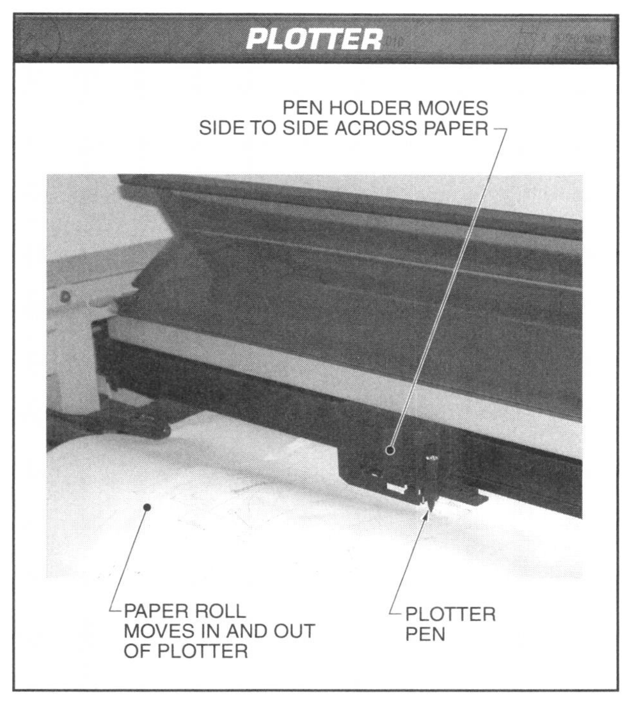

Figure 1-3. A plotter uses pens to produce high-quality drawings.

Since plotters and printers can use many colors, the prints they produce can be color-coded. Different colors can represent different types of information or separate subsystems within a larger system.

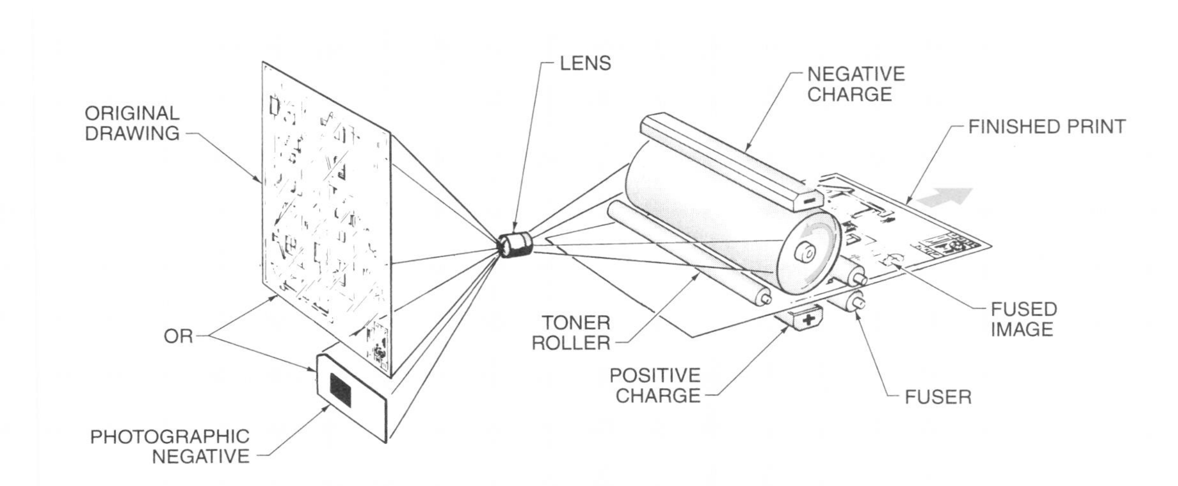

Figure 1-2. Electrostatic printers use an electric charge to transfer a projected image onto paper.

On a rotary-drum plotter, the paper is mounted on a drum and moves with the drum's rotation. The pen moves parallel to the length of the drum. Vertical lines are drawn with the drum remaining in a fixed position while the pen moves along the sheet of paper. Horizontal lines are drawn with the pen remaining stationary and the drum rotating. Curved and diagonal lines are drawn when both the pen and the drum are moving simultaneously.

A flatbed plotter allows the piece of paper to lie flat on a bed. The pen moves along the width and height of the paper while the paper remains stationary.

Inkjet and Laser Printer Prints

A *printer* is a computer output device that generates finished drawings with liquid ink or powder toner. The two types of printers used to produce prints are inkjet and laser printers. Printers using the inkjet process spray superfine droplets of liquid ink onto paper. Inkjet printers are usually capable of multiple-color printing and are relatively inexpensive, but are not suitable for some printreading situations, such as outdoors, because the inks are susceptible to bleeding when exposed to moisture.

Laser printers use a process similar to the electrostatic process, except that a laser transfers the digital drawing information to the negatively charged drum. The drum then attracts the toner powder and transfers the image to paper. Laser printers are typically expensive, but produce crisp, dark black lines on prints that can be used for any application.

DRAFTING

Working drawings for prints may be made using conventional drafting practices or computer-aided design (CAD). *Conventional drafting* is the manual creation of a technical drawing directly onto paper using pens, pencils, and special instruments. Conventional drafting is rarely used in production work, though it is still often taught to students for drafting practice. *Computer-aided design (CAD)* is the creation of a technical drawing in computer software, which can then be output to printers or plotters. Both types should follow general drafting practices and use standard drawing elements.

Conventional Drafting

Conventional drafting begins with taping the paper to a drafting board. All line work is done with penciled construction lines, which are darkened with pen ink to produce the final drawing. Lines are drawn with the aid of several tools, including T-squares, triangles, compasses, dividers, and scales. **See Figure 1-4.** Each of these is available in a variety of sizes and types. Drafting machines (combination T-square, scale, and triangles) and parallel straightedges (combination drafting board and modified T-square) are commonly used in production situations.

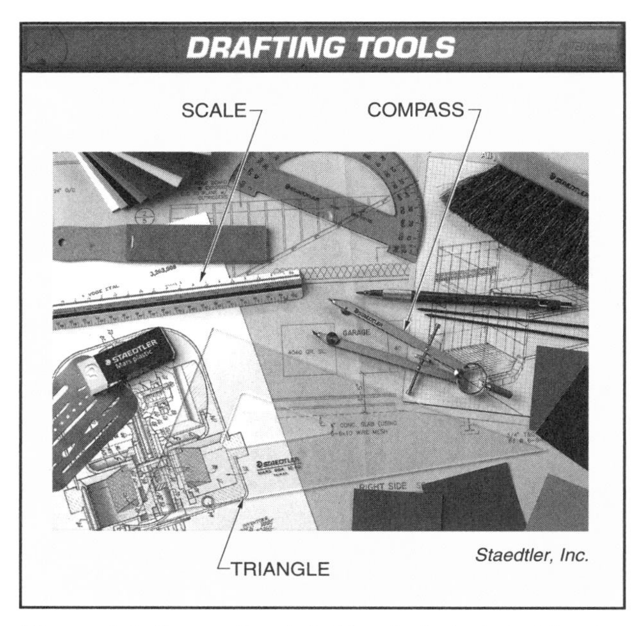

Figure 1-4. Conventional drafting tools include triangles, compasses, and scales.

T-Squares. A *T-square* is a drafting instrument used to draw horizontal lines and as a reference base for positioning triangles. The head of the T-square is held firmly against one edge of the board to ensure accuracy. T-squares are made of wood, plastic, or aluminum and are available in various lengths. The most popular T-squares are 24″ to 36″ in length.

Triangles. A *triangle* is a drafting instrument used to draw vertical and inclined lines. The base of the triangle is held firmly against the blade of the T-square. Two standard triangles, 30°-60° and 45°, are available in a variety of sizes. Triangles are commonly made of clear plastic.

The 30°-60° triangle is used to produce vertical lines and inclined lines of 30° and 60° sloping to the left or right. **See Figure 1-5.** The 45° triangle is used to produce vertical lines and inclined lines of 45° sloping to the left or right. The triangles may be used together to produce inclined lines every 15°.

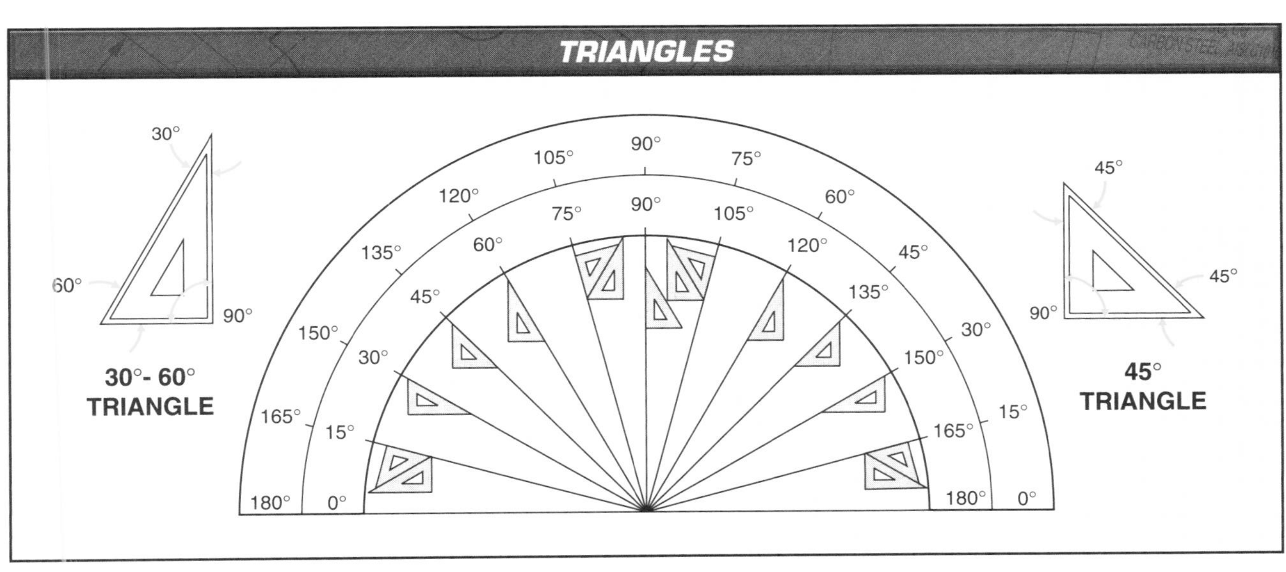

Figure 1-5. The 30°-60° and 45° triangles are used to draw lines 15° apart.

> A protractor is a semicircular instrument with marks at every degree. Drafters can use a protractor to draw angles not produced with triangles.

Compasses. A *compass* is a drafting instrument used to draw arcs and circles. **See Figure 1-6.** One leg of the compass contains a needlepoint that is positioned on the centerpoint of the arc or circle to be drawn. The other leg contains the pencil lead used to draw the line. The two types of compasses are center-wheel and friction. Center-wheel compasses use an adjusting wheel between the legs to change the arc radius. This configuration also reduces the chance of accidentally changing the arc radius while drawing an arc. Friction compasses use resistance in the joint to hold the desired arc radius after it is set by opening or closing the legs.

Dividers. A *divider* is a drafting instrument used to transfer and compare dimensions. It looks like a compass, except that both legs contain a needlepoint. Each needlepoint is positioned at the end of a feature. A divider retains the size of the feature when moved to a scale or other feature for comparison. Also like a compass, a divider can be either the center-wheel or friction type. Because they can be adjusted quickly, friction dividers are usually more useful than center-wheel.

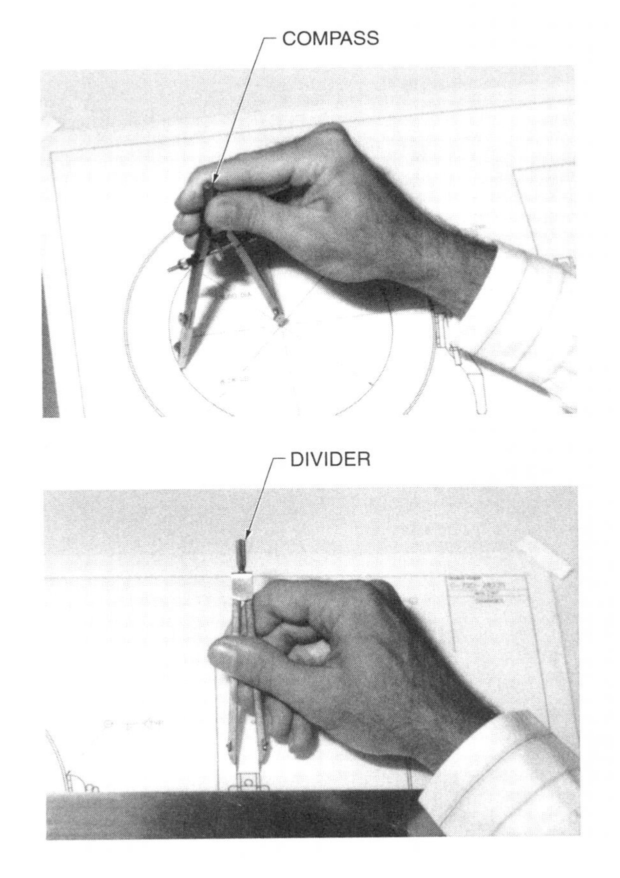

Figure 1-6. The compass is used to draw arcs and circles. The divider is used to transfer dimensions.

Scales. A *scale* is a drafting instrument used to measure lines and reduce or enlarge them proportionally. The three types of scales are the architect's scale, civil engineer's scale, and mechanical engineer's scale. **See Figure 1-7.** Each is available in a variety of sizes.

An architect's scale is used when making drawings of buildings and other structural parts. These scales are usually triangular in shape. One edge of the scale is a standard ruler divided into inches and sixteenths of an inch. The other edges contain 10 scales that are labeled 3, 1½, 1, ¾, ½, ⅜, ¼, 3⁄16, ⅛, and 3⁄32. The ¼ scale means that ¼″ = 1′-0″, and so forth. For larger scale drawings, the 1½″ = 1′-0″, or 3″ = 1′-0″ scales are used.

The civil engineer's scale is used when making maps and survey drawings. Plot plans also may be drawn using this scale. The civil engineer's scale is triangular and graduated in decimal units. One-inch units on the scale are divided into 10, 20, 30, 40, 50, or 60 parts. These units are used to represent the desired measuring unit such as inches, feet, or miles. For example, a building lot line that is 100′-0″ long drawn with the 20 scale (1′ = 20′-0″) measures 5″ on the drawing.

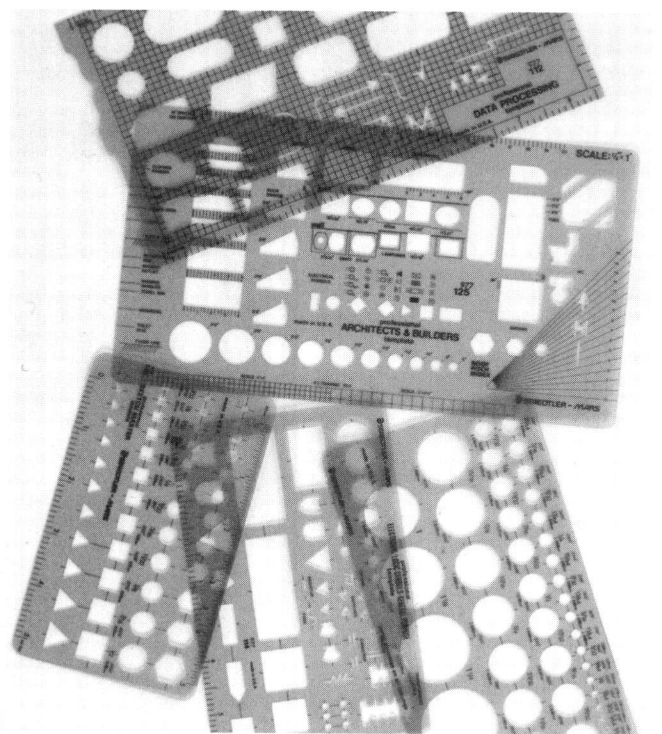

Staedtler, Inc.

Plastic templates for common shapes can be used to quickly draw standard symbols and shapes.

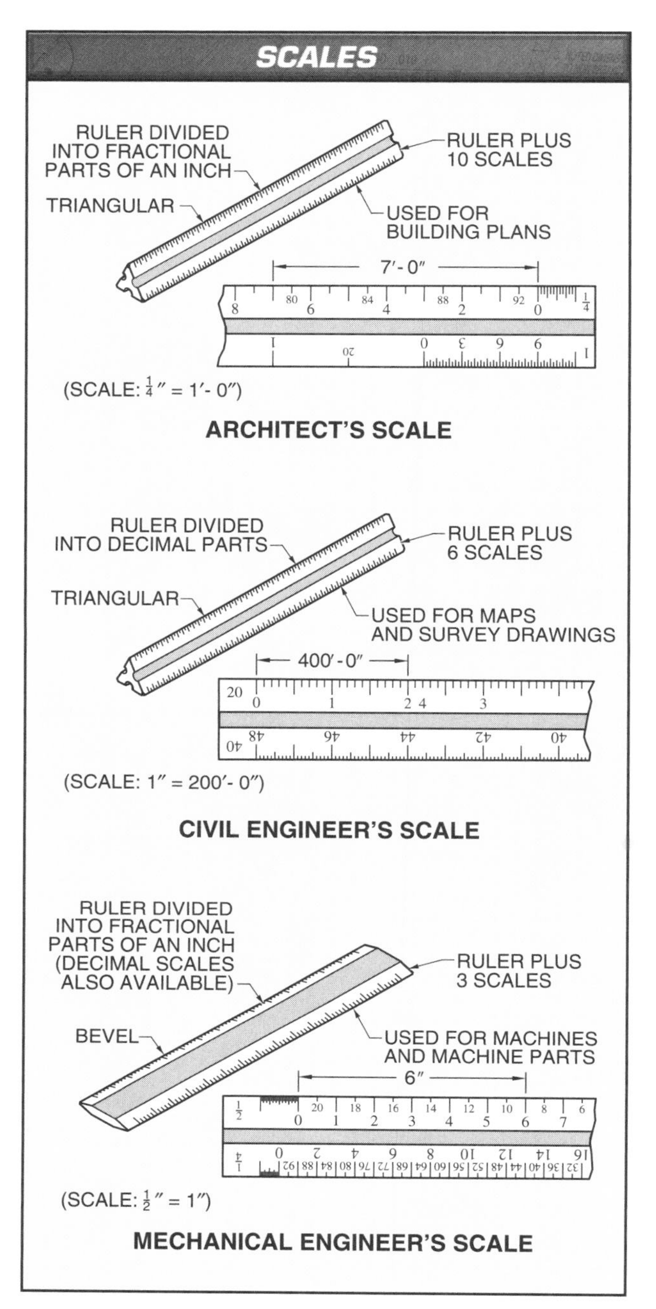

Figure 1-7. Three types of scales are used to produce scaled drawings.

The mechanical engineer's scale is used when drawing machines and machine parts. This scale is similar to the architect's scale, but is flat. The mechanical engineer's scale includes a full size scale and several others, such as ½, ¼, and ⅛ scales. The fractional scales are direct size relationships. For example, ½″ = 1″ in the ½ scale, ¼″ = 1″ in the ¼ scale, and so on. Drawings may be drawn by using the appropriate scale. Fractional inch measurements are made between the zero and the end of the scale.

Pencils. Pencils used in technical drawings are either wood or mechanical. Mechanical pencils produce a more consistent line width than wood pencils, and most styles maintain their point without sharpening. The lead in mechanical pencils can be changed or replaced when needed. **See Figure 1-8.** Lead grade and size should be chosen based on the specific task required.

Lead grades range from hard to soft. Because hard leads do not smudge easily, they are used to draw fine, precise lines and are necessary when extreme accuracy is required. Medium leads are used to draw object lines. Soft leads are used primarily for freehand sketching or to add tone and texture to a sketch. The most common lead grades used for technical drawings are HB, F, H, and 2H. Lead grades 3H or 4H may be used for more precise work or to create light construction lines that will not need to be erased.

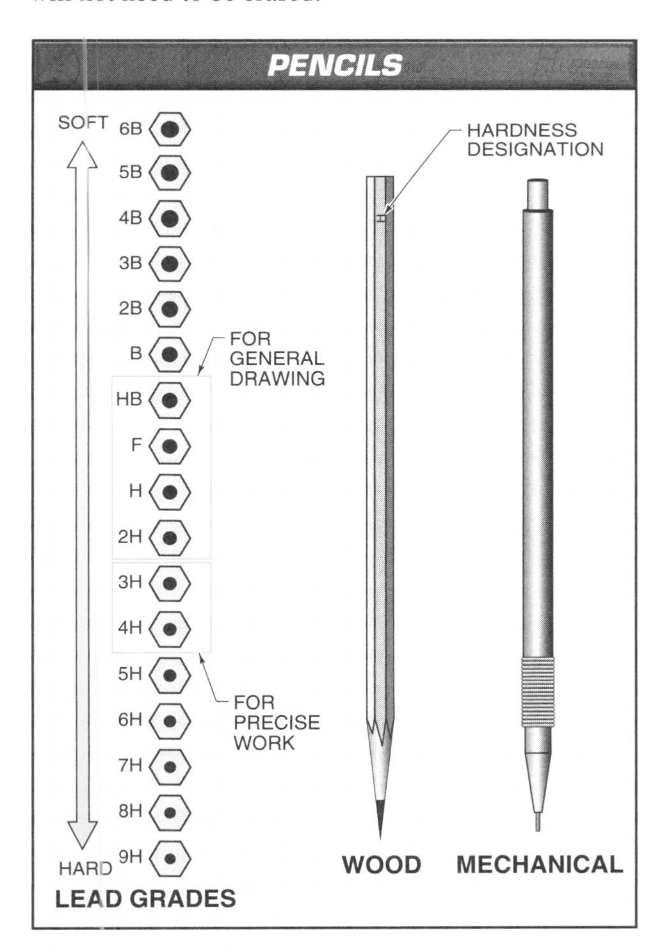

Figure 1-8. Pencil leads range from extremely soft to exceptionally hard.

Pens. A *technical pen* is a drawing instrument that makes ink lines of a consistent width. Technical pens use black, permanent ink and can be used to darken and sharpen lines after a sketch has been drawn in pencil.

Technical pens can have tapered or straight tips. Tapered tips are generally used for artwork or object drawing, while straight tips are generally used for drafting and lettering. Pen tips can vary in width from 0.005″ (0.13 mm) to 0.079″ (2.00 mm). Because technical drawings contain both thick and thin lines, a pair of pens with an approximate 2:1 size ratio is used.

Computer-Aided Design (CAD)

Computer-aided design (CAD) allows for the generation and reproduction of technical drawings and prints with computers. It is also known as computer-aided drafting or computer-aided drafting and design (CADD). This system is popular in architectural and engineering offices.

Machinists reading CAD-generated prints benefit from the quality and consistency of line work, symbol representation, and lettering. **See Figure 1-9.** Designers and engineers benefit from increased drafting productivity achieved in the planning, design, drafting, and reproduction of prints.

Six primary factors contributing to increased productivity are consistency, changeability, layering, modeling, storage, and repeatability.

- Consistency: Constant sameness is achieved in line width, symbol depiction, and representation of drawing components.
- Changeability: Revisions, additions, and deletions are made easily.
- Layering: Similar to using overlays on conventional drawings, this technique allows base work to be used to generate additional related drawings.
- Modeling: A feature that allows viewing the complete part in pictorial form and analysis of its reaction to stress.
- Storage: Drawings stored electronically require minimal physical space.
- Repeatability: An unlimited number of high-quality prints may be reproduced.

CAD systems use hardware and software to generate drawings. *Hardware* is the physical components of a computer system. Hardware includes input devices, the central processing unit (CPU), and output devices. **See Figure 1-10.**

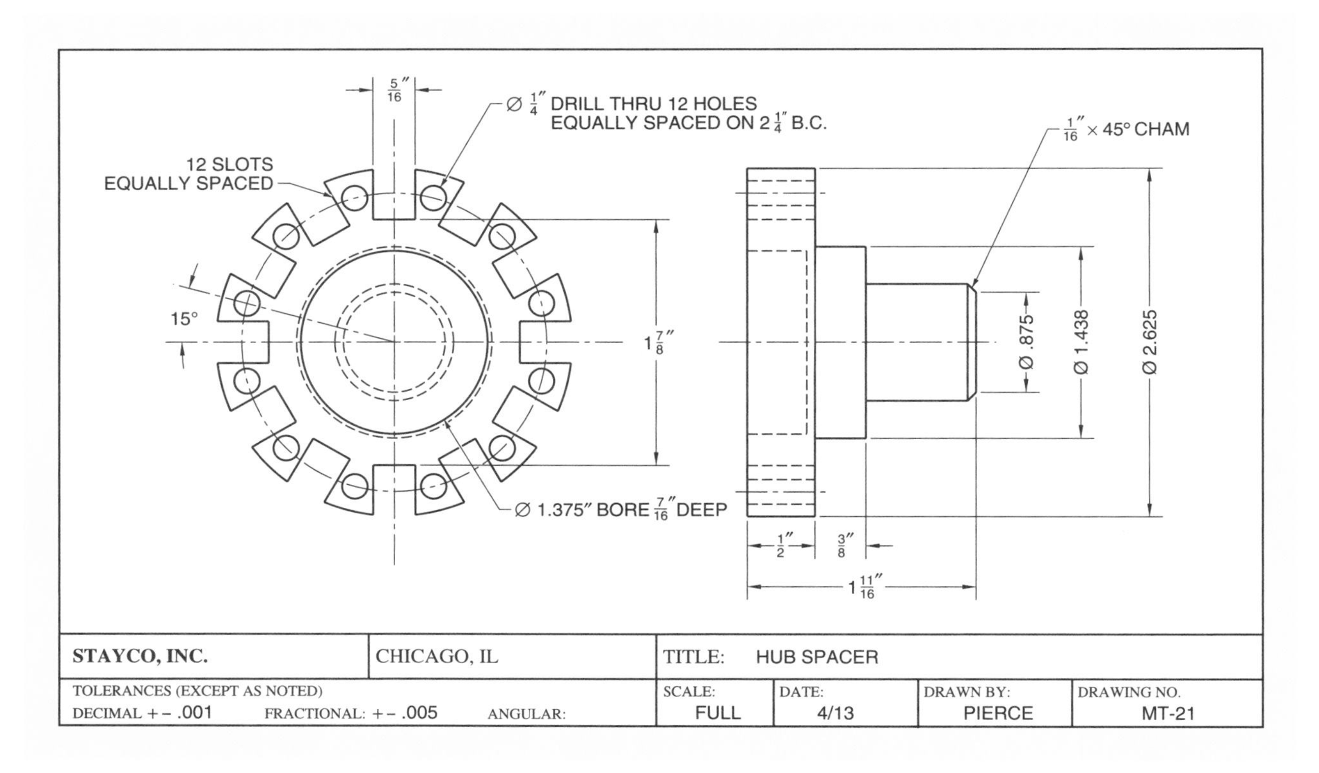

Figure 1-9. Computer-aided design (CAD) systems consistently produce excellent quality drawings.

Figure 1-10. Hardware components of a computer-aided design (CAD) system are the input devices, central processing unit, and output devices.

Software is a collection of programmed instructions in a computer. Operating system software generates the computer interface, allows system hardware components to communicate, and provides system-level access to application software. Application software includes instructions specific to a certain task. In CAD systems, application software is used to generate, modify, and output the manufacturing drawings. **See Figure 1-11.**

Input Devices. An *input device* is hardware used to enter information into a computer system. The input device is interfaced (connected) with the central processing unit, which controls the CAD software.

Specialized CAD software has been developed for different types of technical drawing, each including the standard symbols, linetypes, and templates used in that area. Some software is designed specifically for architectural drawings, mechanical engineering drawings, electrical system diagrams, landscape plans, or other types of drawings.

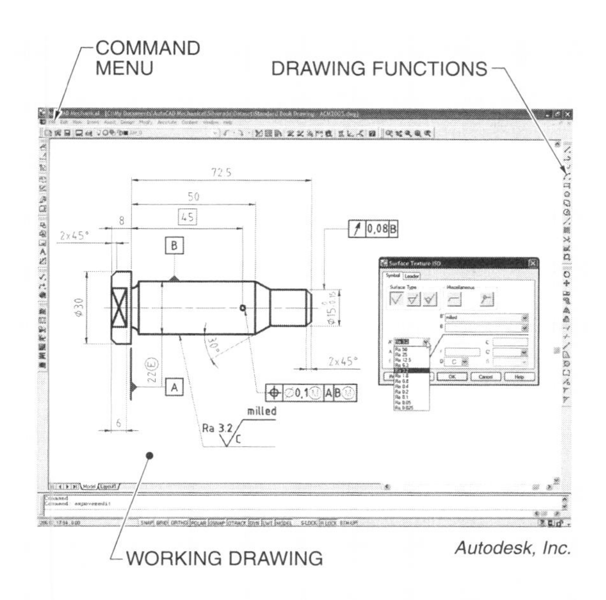

Figure 1-11. Computer-aided design (CAD) software includes commands and drawing functions that allow users to quickly produce high-quality drawings.

Most CAD systems require only a keyboard and mouse to produce drawings and access all the software functions. The keyboard has the capability of inputting notes and dimensions as well as positioning the cursor on the display monitor. The mouse controls the cursor on the display screen through movement on a hard surface. Menu selections are made by moving the cursor to the edge of the display screen and selecting the desired command.

Specialized or older CAD systems may use a graphics tablet or other device designed specifically for use with CAD software. A *graphics tablet* is an electronic input device that consists of a drawing area and a menu. A stylus or mouse is used with the graphics tablet to choose a function from the menu and "draw" the object in the space provided.

In addition to standard commands, the graphics tablet may display a library of symbols. Common symbols include those for welding, surface finish, geometric tolerance, ANSI dimension, and fasteners.

Central Processing Units. The *central processing unit (CPU)* is the control center of a computer. It receives information through the input devices and determines an output. A CPU is classified by its memory capacity and the speed at which it carries out commands. A larger memory capacity generally means a greater capability of producing quality drawings. The central processing unit stores information on magnetic disks and in internal memory. CPUs may be dedicated for CAD systems only or may also run other software, such as word or data processing programs.

Output Devices. An *output device* is hardware that either displays or generates drawings. The basic types of output devices are monitors and printers.

A *monitor* is a video display terminal. The monitor displays the drawing that the operator is developing. Monitors are available in many sizes and are chosen based upon the application. On some types of CAD systems, two monitors are used. One monitor displays the drawing and the other displays the menus.

A printer produces drawings on paper. Printers provide a fast and convenient method of checking the placement of drawing features. Large-format printers are designed to output large CAD drawings and may accept paper up to 44″ wide. **See Figure 1-12.**

Figure 1-12. Large-format printers output high-quality prints.

PRINTS

Conventionally drafted working drawings are commonly produced on film or paper and then copied to produce prints. The two materials used for these working drawings are polyester and vellum. CAD software produces working drawings as electronic files. The files are output to a plotter or printer to produce prints on paper.

Prints are produced on various sizes of paper depending upon the scale used and the complexity of the part being drawn. Standard paper sizes and drawing formats, including title blocks, revision blocks, parts list, supplementary blocks, and drawing numbers, are based on the conventions in ANSI/ASME Y14.1, *Decimal Inch Drawing Sheet Size and Format,* and ANSI/ASME Y14.1M, *Metric Drawing Sheet Size and Format.*

Paper Sizes

Paper can be purchased in sheets or continuous rolls. Flat sheets of paper in the American system are designated by the letters A through F. **See Figure 1-13.** Rolls are designated by the letters G, H, J, and K. Letters earlier in the alphabet indicate smaller paper sizes.

AMERICAN PAPER SIZES

FLAT SHEETS		
SIZE	WIDTH*	LENGTH*
A	8.5	11
B	11	17
C	17	22
D	22	34
E	34	44
F	28	40

ROLLS			
SIZE	WIDTH*	LENGTH*	
		MIN	MAX
G	11	22.5	90
H	28	44	143
J	34	55	176
K	40	55	143

* in in.

Figure 1-13. Standard American paper sizes are designated by letters.

International metric paper sizes are given by letter-number combinations. **See Figure 1-14.** Metric paper sizes are very close to the American sizes, but have slightly different proportions. The length of a metric sheet is $\sqrt{2}$ (approximately 1.414) times the width. For example, the length of an A3 size sheet is 420 mm (297 mm × 1.414 = 420 mm).

INTERNATIONAL PAPER SIZES

SIZE	WIDTH*	LENGTH*	NEAREST USA LENGTH
A4	210	297	A
A3	297	420	B
A2	420	594	C
A1	594	841	D
A0	841	1189	E

* in mm

Figure 1-14. Standard international paper sizes are designated by letter-number combinations.

Because the size systems are very close, a drawing's margins are usually large enough to allow drawings made to one standard size system to be reproduced on the nearest equivalent size on the other standard system.

Basic Formats

Drawing paper is oriented with its base horizontal along the long dimension, with the occasional exception of A size vertical paper. **See Figure 1-15.** The borderline is thick (approximately 0.030″). The vertical margins between the edges of the paper and the borderline vary from 0.25″ to 1.00″, depending on the paper size. The horizontal margins vary from 0.38″ to 1.00″, depending on the paper size.

Lettering. Lettering size and style is based on the conventions in ANSI/ASME Y14.2, *Line Conventions and Lettering*. Uppercase, single-stroke gothic letters are used. They may be inclined or vertical. The preferred slope for inclined letters is 68° to the horizontal. The height of freehand lettering varies from 0.120″ to 0.290″, depending on the paper size.

Zones. Prints may include zones for reference purposes. A *zone* is an area in a print margin that is identified by a letter or number. Commonly, numbers are used in the horizontal margins and letters along the vertical margins. Any area within a print can then be referenced with a letter-number combination. This is particularly useful for large and complex drawings.

The international standard for paper sizes also includes a B series and a C series. The B series is a less common set of sizes based on the geometric mean of A sizes. The C series is used only for envelopes.

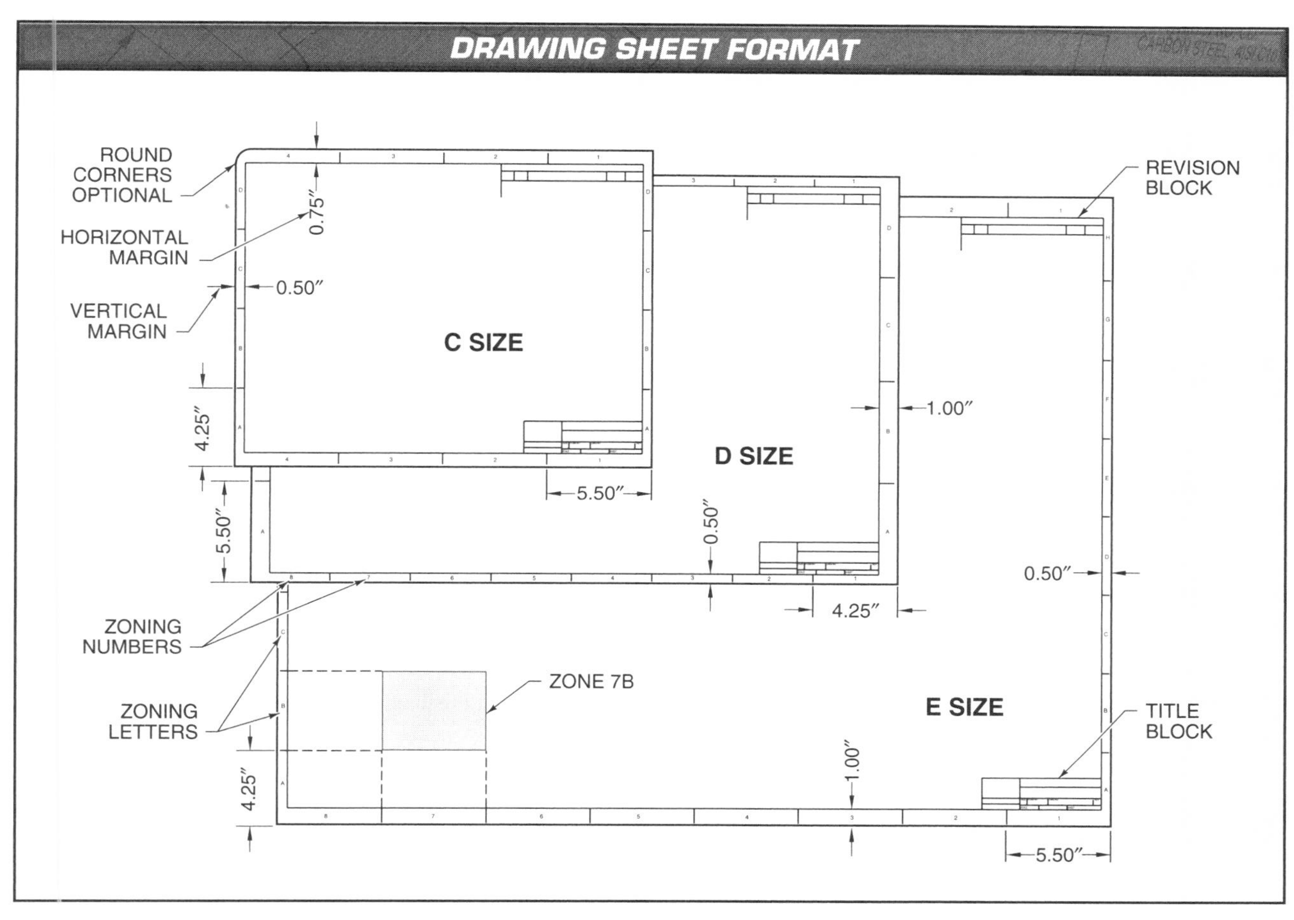

Figure 1-15. The basic format of drawing sheets includes the title block, revision block, margins, and zoning.

Title Blocks. A *title block* is an area located in the lower right-hand corner of a print that contains identifying information about the drawing. Title blocks can be customized to the individual company, but usually include some of the same pieces of information. This information includes the company name and address, drawing title, drawing number, paper size, scale, and sheet number. Initials of the drafter, checker, and person issuing approval are also included, along with the completion date of each task. A Commercial and Government Entity (CAGE) number may also be included in a designated part of the title block. This five-digit code applies to all organizations that either have or are producing items for the United States government.

There are two common sizes for title blocks, one for sheet sizes A, B, C, and G and the other for sheet sizes D, E, F, H, J, and K. The basic difference between the two sizes is the dimensions of the various blocks. See **Figure 1-16.**

Staedtler, Inc.

Paper is available in roll or sheet form.

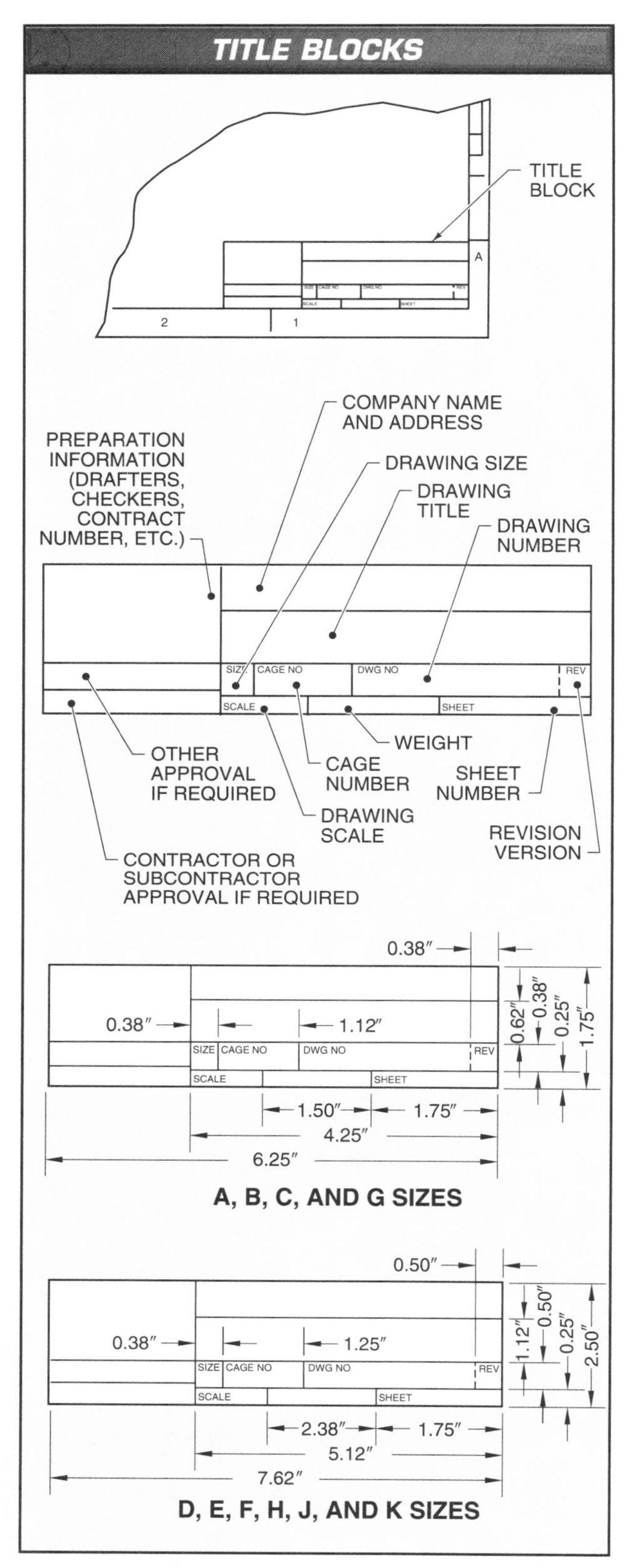

Figure 1-16. Title block size varies for smaller and larger sheet sizes.

Revision Blocks. A *revision block* is an area, located in the upper-right corner of a print, that contains information about changes made to the original drawing. A revision block extends downward as required to accommodate additional rows of information. The revision symbol, description of changes, date, and approvals are included. **See Figure 1-17.** A zone column is added if required. The width of revision blocks may be changed to provide for other columns as necessary. The revision version of a drawing may also be included in the title block next to the drawing number.

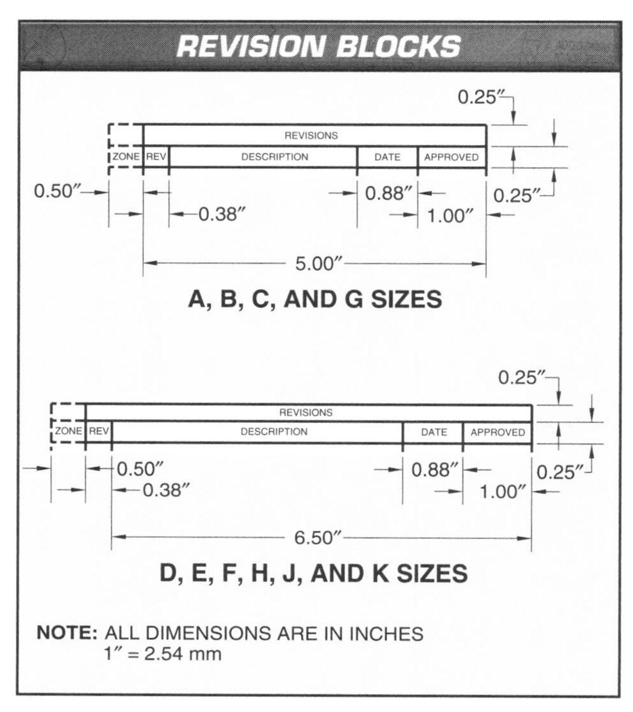

Figure 1-17. Revision blocks are located in the upper-right corner of the sheet.

Parts Lists. A *parts list* is an area located above the title block that contains specifications for any off-the-shelf components used in the assembly shown on the print. **See Figure 1-18.** For example, fasteners are commonly included in parts lists. The parts list contains the name of the part, identifying information, and quantity required. The parts list is also known as a bill of materials.

Supplementary Blocks. A *supplementary block* is an area located to the left of a title block that contains any additional information necessary for reading a print or manufacturing the part. For example, this information may include dimensioning and tolerancing notes, material, and required finish. Multiple supplementary blocks may be included as required.

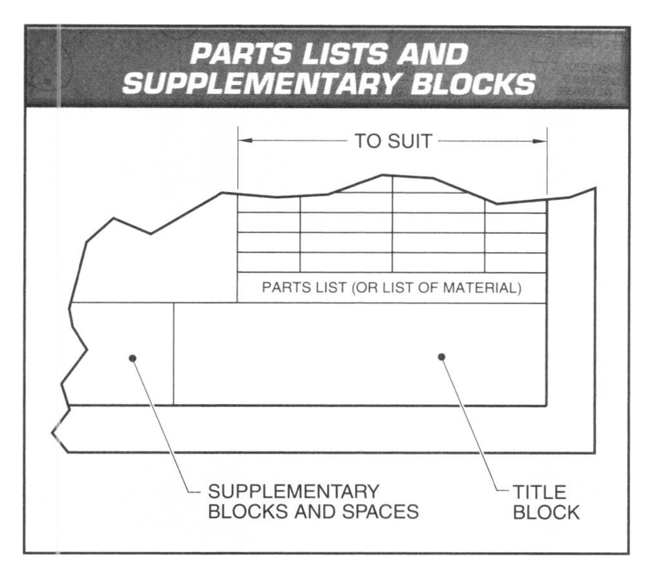

Figure 1-18. Parts lists and supplementary blocks are located next to the title block.

MANUFACTURING STANDARDS

Before large-scale manufacturing developed around the turn on the 20th century, products were usually handmade, often in special sizes. Because of these inconsistencies, assembly parts were usually not interchangeable with parts from other manufacturers, or sometimes even from the same manufacturer. Gradually, some industries founded organizations, such as the American Institute of Electrical Engineers (now IEEE), to help standardize products, processes, and systems among its member companies.

A *standard* is a document, established by consensus, that provides rules, guidelines, or characteristics for activities or their results. Standards establish uniform sizes, shapes, and capacities for products and systems and define terms so there is no misunderstanding among those using the standard. Uniformity and minimum performance requirements also help make homes, workplaces, and public areas safer.

Conformity to standards is voluntary, but is usually advantageous to manufacturers. For instance, if a light bulb manufacturer does not follow the national standard defining the size, shape, and thread pitch of a light fixture socket, their light bulbs will not fit any standard sockets, and their sales will suffer. **See Figure 1-19.**

American National Standards Institute (ANSI)

American manufacturers eventually realized that standards needed to be adopted nationally, and across many industries, in order to avoid duplication, waste, and conflict. In 1918, several industry organizations formed what would later become the American National Standards Institute (ANSI).

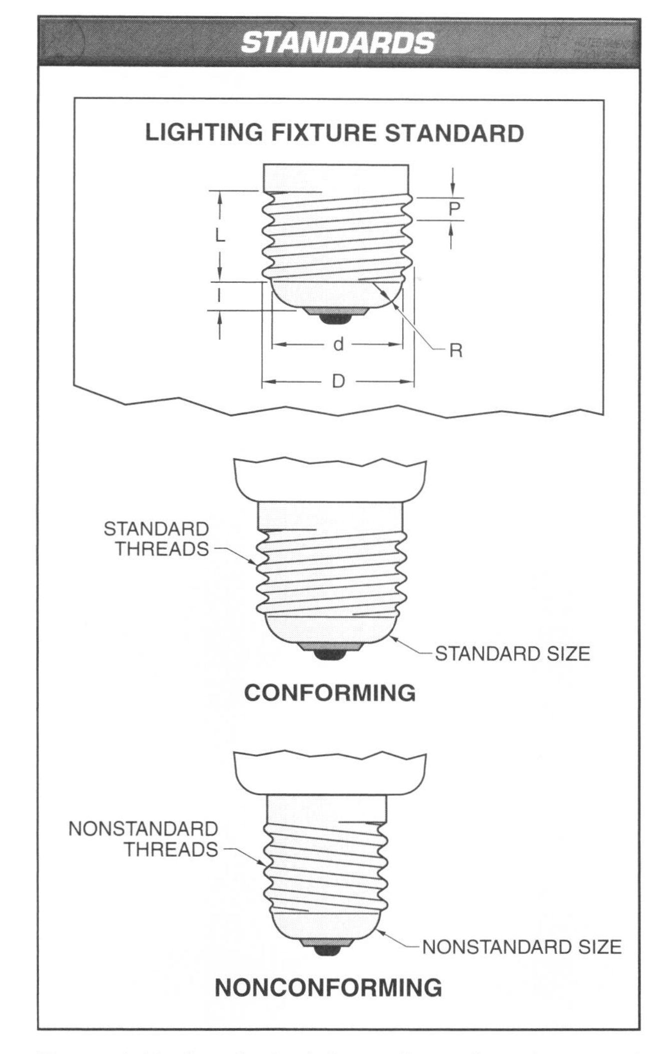

Figure 1-19. Standards define uniform size, shape, and characteristics for products or systems.

The American National Standards Institute (ANSI) does not develop standards, but validates the work of other organizations to compile American National Standards. National standards are developed by industry or support organizations for their respective fields. Following evaluation, ANSI may choose to adopt the industry standard as a national standard and will continue to work with the support organization as the standard is periodically reevaluated and reaffirmed.

Support organizations consist of professional/technical societies, trade associations, and consumer and labor groups. **See Figure 1-20.** Typical support organizations that are directly concerned with machines trades printreading include, but are not limited to, the following:

AGMA	American Gear Manufacturers Association
ASME	ASME International, formerly American Society of Mechanical Engineers
ASTM	ASTM International, formerly American Society for Testing and Materials
AWS	American Welding Society
CAM-I	Consortium for Advanced Manufacturing International
SAE	Society of Automotive Engineers

ANSI standards are designated by their document number and name, such as "ANSI Y14.36M, *Surface Texture Symbols*." The document number may also include the acronym of the organization responsible for the standard and the year the standard was revised, such as "ANSI/ASME Y14.36M-1996." Most standards related to machine trades printreading are developed by ASME International and begin with the designation "Y14."

International Organization for Standardization (ISO)

The International Organization for Standardization, (ISO), is the international counterpart to ANSI. International standards become increasingly important as manufacturers work with suppliers and customers in other countries. The ISO brings together industry and national standards from many countries, including the United States, into a uniform set of standards for the international community to follow. An indirect result of this compilation is that ANSI and ISO standards are moving closer to each other as they change over time. American National Standards are also becoming more inclusive of metric system standards for the international marketplace.

Conformity

Standards are intended as guides to aid manufacturers, consumers, and the general public. However, the existence of a standard does not preclude any company from following procedures that do not conform to the standard. Therefore, while the drawings for many prints are developed based on ANSI standards, other drawings may not always reflect all current standards.

Experienced printreaders recognize that the application of standards varies, so they study prints very carefully to determine the manufacturer's intent. For example, while one manufacturer may use "CENT" as an abbreviation, ANSI/ASME Y1.1, *Abbreviations,* does not include "CENT" as an abbreviation. Careful study of the manufacturer's print shows that "CENT" is intended to be the abbreviation for "center."

INDUSTRY AND STANDARDS ORGANIZATIONS

ANSI
American National Standards Institute
1899 C Street NW, 11th Floor
Washington, DC 20036
www.ansi.org

CSA
Canadian Standards Association
5060 Spectrum Way
Mississauga, ON L4W 5N6
www.csa.ca

ISO
International Organization for Standardization
Case Postale 56 CH - 1211
Geneve 20 Switzerland
www.iso.org

SAE
Society of Automotive Engineers
400 Commonwealth Dr.
Warrendale, PA 15096
www.sae.org

CAM-I
Consortium for Advanced Manufacturing-International
6836 Bee Cave, Suite 256
Austin, TX 78746
www.cam-i.org

ASTM International
Formerly American Society for Testing and Materials
100 Barr Harbor Drive
West Conshohocken, PA 19428
www.astm.org

ASME International
Formerly American Society of Mechanical Engineers
3 Park Ave.
New York, NY 10016
www.asme.org

AGMA
American Gear Manufacturers Association
500 Montgomery St., Suite 350
Alexandria, VA 22314
www.agma.org

AWS
American Welding Society
550 NW LeJenne Rd.
Miami, FL 33126
www.aws.org

Figure 1-20. Many industry and standards organizations are involved in developing standards for printreading.

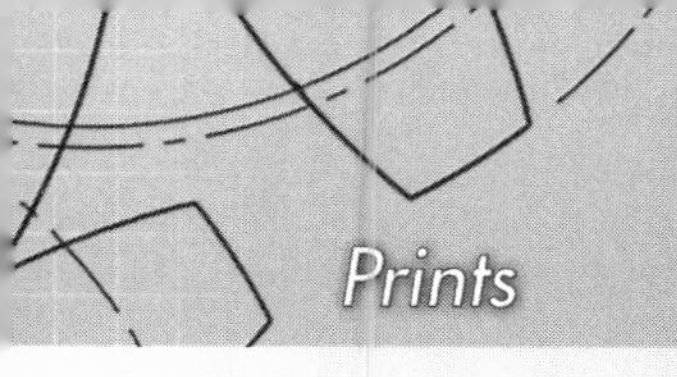

REVIEW QUESTIONS 1

Name ______________________________ Date ______________

Identification — Angles

______ 1. 135°

______ 2. 120°

______ 3. 165°

______ 4. 75°

______ 5. 30°

______ 6. 45°

______ 7. 90°

______ 8. 150°

______ 9. 60°

______ 10. 15°

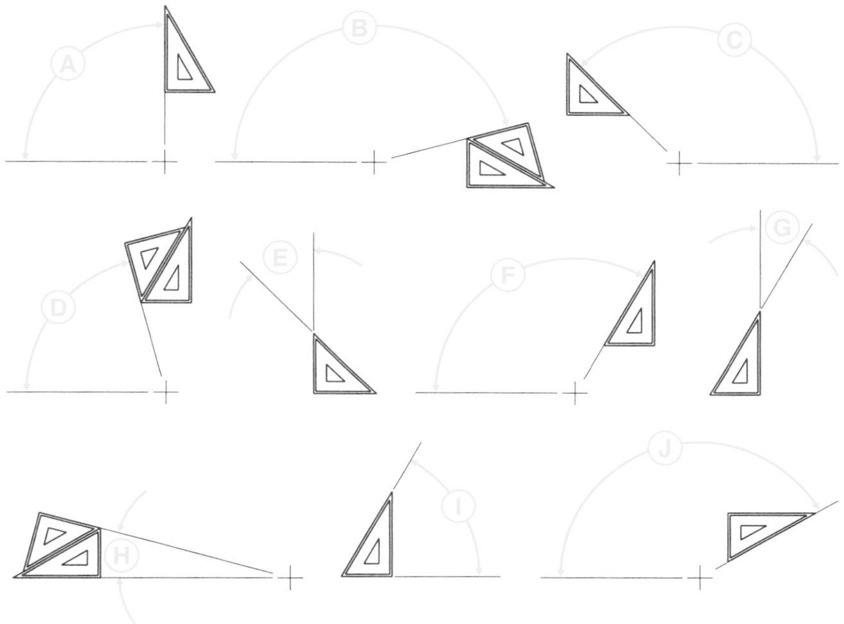

Completion

______ 1. Electrostatic prints have ___ lines on a white background.

______ 2. ___ drafting is the manual creation of a technical drawing directly onto paper using pens, pencils, and special instruments.

______ 3. Triangles may be used together to produce inclined lines every ___°.

______ 4. The ___ is a drafting instrument used to draw arcs and circles.

______ 5. The ___ engineer's scale is used when drawing machines and machine parts.

______ 6. ___ is a collection of programmed instructions in a computer.

______ 7. Machinists reading ___-generated plans benefit from the quality and consistency of line work, symbol representation, and lettering.

______ 8. CAD software produces working drawings as ___ files.

______ 9. Paper can be purchased in sheets or continuous ___.

______ **10.** The preferred slope for inclined letters is ___° to the horizontal.

______ **11.** A(n) ___ is a drafting instrument used to draw vertical and inclined lines.

______ **12.** A(n) ___ is an area in a print margin that is identified by a letter or number.

______ **13.** A(n) ___ block is an area located in the lower right-hand corner of a print that contains identifying information about the drawing.

______ **14.** ANSI validates the work of other organizations to compile American National ___.

______ **15.** A plotter is a computer output device that generates finished drawings with ___.

Multiple Choice

______ **1.** The ___ includes the revision symbol, description of changes, date, and approvals.
- A. revision block
- B. title block
- C. parts list
- D. conformity abbreviation

______ **2.** The ___ does not develop standards.
- A. ASME International
- B. American National Standards Institute
- C. American Welding Society
- D. American Gear Manufacturers Association

______ **3.** The most popular T-squares are ___″ to ___″ in length.
- A. 12; 24
- B. 18; 24
- C. 18; 30
- D. 24; 36

______ **4.** The ___ scale is used when drawing machines and machine parts.
- A. mechanical engineer's
- B. civil engineer's
- C. drafter's
- D. assembler's

______ **5.** Vertical margins between the edges of drawing paper and the borderline vary from ___″ to ___″, depending on the paper size.
- A. .10; .20
- B. .15; .25
- C. .25; 1.00
- D. .25; 1.25

______ **6.** Print lettering is ___.
- A. uppercase
- B. gothic
- C. single-stroke
- D. all of the above

________________ **7.** Rolls of paper are designated for size by ___.

A. letters A through F
B. letters G, H, J, and K
C. letter-number combinations
D. none of the above

________________ **8.** International metric paper sizes are classified by ___.

A. letters A through F
B. letters G, H, J, and K
C. letter-number combinations
D. none of the above

________________ **9.** Any area within a part can be referenced with a ___.

A. letter
B. number
C. letter-number combination
D. symbol

________________ **10.** The length of a metric sheet of paper is ___ times the width.

A. 1.5
B. $\sqrt{2}$
C. 3
D. none of the above

________________ **11.** Laser printers use a process similar to the ___ process.

A. electrostatic
B. inkjet
C. diazo
D. blueprint

________________ **12.** A divider is a drafting instrument used to ___.

A. draw arcs
B. draw circles
C. transfer and compare dimensions
D. transfer letters

________________ **13.** A ___ looks like a compass, except that both legs contain a needlepoint.

A. T-square
B. triangle
C. scale
D. divider

________________ **14.** ___ is the physical components of a computer system.

A. Software
B. Hardware
C. CAD
D. The operating system

_______________ **15.** The acronym "CAGE" refers to the ___.

A. Committee for Advanced General Engineering
B. Computer-Aided Group Engineering
C. Code Association for Government Entities
D. Commercial and Government Entity

True-False

T F **1.** A print is an original of a working drawing.

T F **2.** Laser prints can produce black lines on a white background.

T F **3.** A T-square is a drafting instrument used to draw horizontal lines.

T F **4.** Translucent paper stops all light from passing through.

T F **5.** Hard pencil leads are always used for freehand sketching.

T F **6.** Information is entered into a computer with an input device.

T F **7.** A monitor is an input device.

T F **8.** ANSI standards are designated by document number and name.

T F **9.** The parts list is also known as a bill of materials.

T F **10.** Plotters generate finished drawings by a photographic process.

T F **11.** Most working drawings are now produced on computer software.

T F **12.** Triangles may be used together to produce inclined lines every 15°.

T F **13.** A pencil lead labeled 6B is exceptionally hard.

T F **14.** A graphics tablet is an input device.

T F **15.** Manufacturers are required to follow standards.

Identification — Paper Sizes

_______________ **1.** A sheet

_______________ **2.** A3 sheet

_______________ **3.** A4 sheet

_______________ **4.** G roll

_______________ **5.** H roll

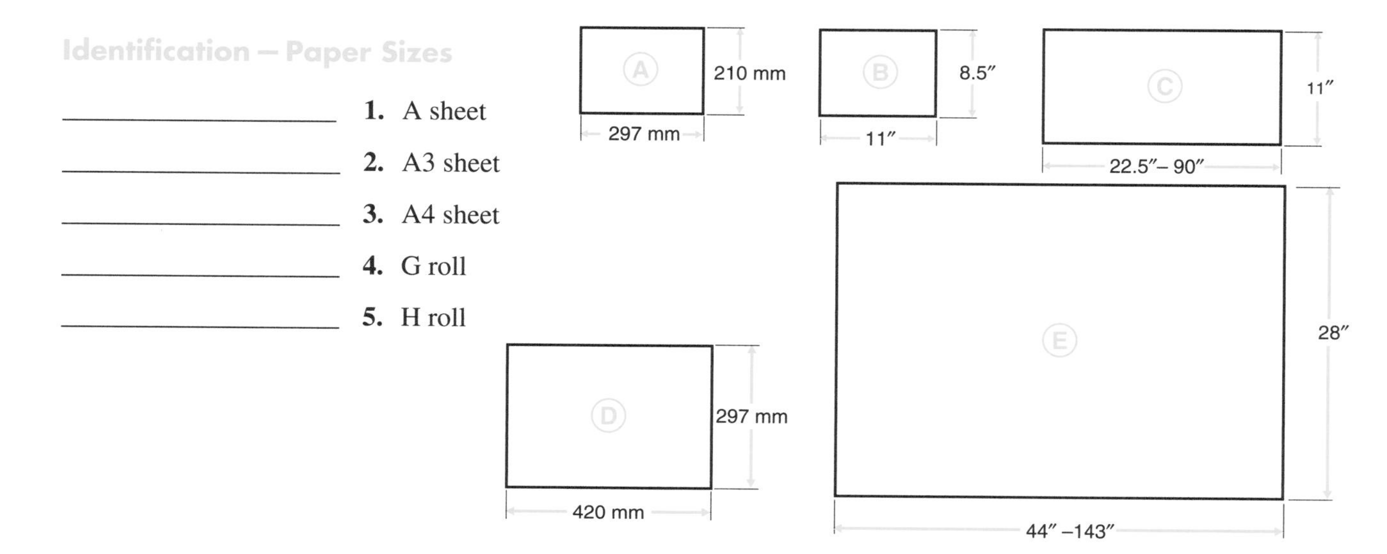

TRADE COMPETENCY TEST 1

Name ______________________ **Date** ______________

Centering Fixture

Refer to title block below.

______	**1.**	The scale of the drawing is ___″ = 1′-0″.
______	**2.**	There are a total of ___ sheets in this set of prints.
______	**3.**	The drawing number is ___.
T F	**4.**	The drawing was drawn and checked by the same person.
______	**5.**	Prints for this drawing are reproduced on size ___ sheets.
T F	**6.**	A CAGE number is not applicable for this drawing.
T F	**7.**	Approval of the drawing was made by TAJ.
______	**8.**	Approval of the drawing was completed on ___.
______	**9.**	The title of the part to be fabricated is ___.
T F	**10.**	Pierce Fixture Co. is located in Detroit, Michigan.
______	**11.**	This sheet is Sheet ___.
T F	**12.**	This particular drawing has been revised.
______	**13.**	The drafter for the drawing was ___.
______	**14.**	The drawing was checked on ___.
T F	**15.**	Checking and approval of the drawing were completed on consecutive days.

DRAFTER	DATE	PIERCE FIXTURE COMPANY			
RJH	5-2-05	CHICAGO, ILLINOIS			
CHECKER	DATE	CENTERING FIXTURE			
TAJ	5-9-05				
APPROVAL	DATE				
CT	5-17-05				
		SIZE C	CAGE NO NA	DWG NO 316915	REV
		SCALE $\frac{1}{4}$″ = 1′- 0″			SHEET 1 OF 3

CENTERING FIXTURE

LH Turret Ass'y

Refer to the print below.

_______________ **1.** The drawing was drawn by ___.

_______________ **2.** All drawing approvals were completed by ___.

_______________ **3.** The tolerance for three-place decimal dimensions is ±___″.

_______________ **4.** The drawing is drawn to ___ scale.

_______________ **5.** Unless otherwise specified, all sharp edges are to be broken a maximum of ___″.

_______________ **6.** Sheet ___ of 6 is shown.

_______________ **7.** ___ gave approval for the Engineering Department.

_______________ **8.** Randall Machine Co. is located in ___, OH.

_______________ **9.** The part shown is called a(n) ___.

_______________ **10.** The print is on a size ___ sheet.

_______________ **11.** The drawing number of the print is ___.

_______________ **12.** The contract number for the job is ___.

_______________ **13.** Angular tolerances are given as ±___.

_______________ **14.** The drawing was completed by the drafter on ___.

_______________ **15.** The number of parts required is ___.

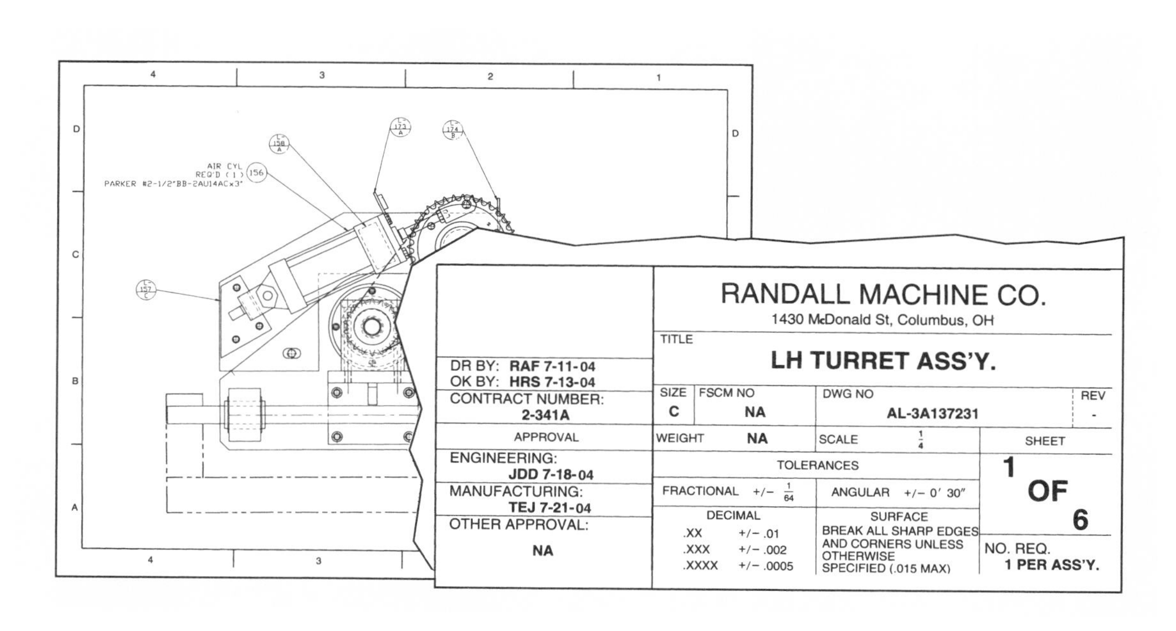

LH TURRET ASS'Y

Machine Trades
PRINTREADING

2 Object Representation

Sketches are used in industry to quickly convey ideas. Sketches may be drawn as pictorial drawings or multiview drawings. Object representation for prints is based on the principles of orthographic projection. The types of lines used and the representation of surface features on multiview drawings are based on standards developed by ASME International and adopted by ANSI. Drawings are dimensioned to show size and location.

SKETCHING TECHNIQUES

Sketching is drawing without instruments. Sketches are made by the freehand method. The only tools required are a pencil, paper, and an eraser. **See Figure 2-1.**

Figure 2-1. Sketching is drawing without instruments.

Sketching pencils are either wooden or mechanical. Wooden pencils must be sharpened, and the lead must be pointed. One type of mechanical pencil contains a thick lead that is sharpened with a file, sandpaper, or lead pointer. Another type contains a thin lead, which does not require pointing. Softer leads such as HB, F, and H are commonly used for sketching.

The paper selected for sketching depends upon the end use of the sketch. Plain paper is commonly used. Tracing vellum is used if the sketch is to be duplicated on a diazo printer. Sketching papers and tracing vellums are available in pads or sheets in standard sizes.

Plain paper is either blank or preprinted with grids to facilitate sketching. Preprinted paper is available in a variety of grid sizes for orthographic and pictorial sketches. A grid size of ¼″, called "quad-rule," is popular. Grids are commonly printed in light blue, nonreproducing inks. This means that if the sheet is duplicated, the grid will not appear on the copy.

Erasers are used to remove pencil marks from paper. Erasers for technical sketching are typically separate tools, not attached to the pencil. As it is rubbed onto the paper, an eraser picks up particles of lead in its soft material and carries it away. Erasers made from different materials have different characteristics. Very soft erasers may damage certain papers. Some also crumble into many tiny pieces, which must be carefully swept away. Erasers made from harder material, or soft material that has hardened with age, may smear some pencil marks. Rubber and vinyl are the most common eraser materials.

The pencil point should be pulled across the paper while sketching. Pushing the pencil point can tear the paper. While pulling the pencil, slowly rotate it to produce lines of consistent width. Horizontal, inclined, vertical, and curved lines are drawn to produce orthographic and pictorial drawings. **See Figure 2-2.**

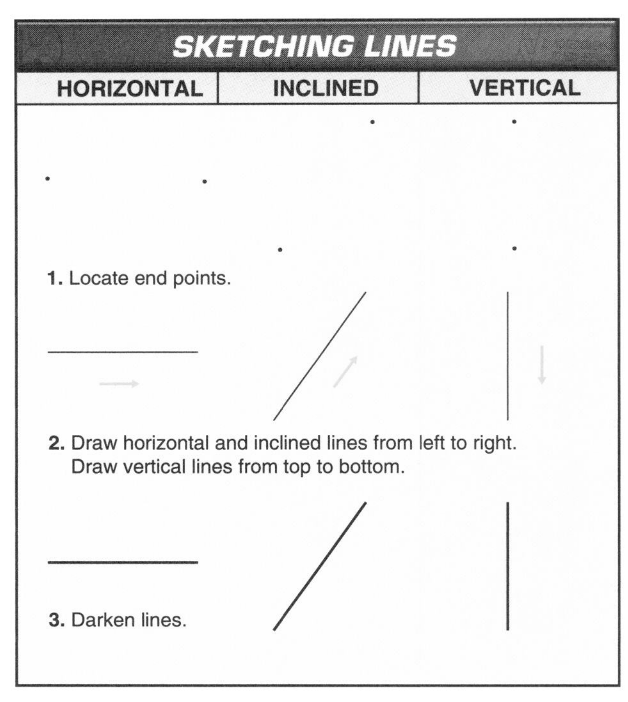

Figure 2-2. Lines are sketched by connecting their end points.

Sketching Horizontal Lines

To sketch horizontal lines, locate the end points with dots to indicate the position and length of the line. For short lines, the end dots are connected with a smooth wrist movement from left to right (for a right-handed person). Long lines may require intermediate dots. If grid paper is used, intermediate dots may not be required. For long lines, a full arm movement may be required to avoid making an arc.

The top or bottom edges of the paper or pad may be used as a guide while sketching horizontal lines. Light, trial lines are drawn first to establish the straightness of the line. The line is then darkened. With sketching experience, the trial lines may be omitted.

Sketching Vertical and Inclined Lines

To sketch vertical lines, locate the end dots and draw from the top to the bottom. The side edges of the paper or pad may be used as a guide while sketching vertical lines.

Staedtler, Inc.

Eraser shields allow drafters to accurately erase only unwanted lines.

Inclined lines are neither horizontal nor vertical. To draw inclined lines, locate end dots and draw from left to right (for a right-handed person). The paper may be rotated so that the inclined lines are in either a horizontal or vertical position to facilitate sketching.

Sketching Plane Figures

Drawings consist of lines, arcs and irregular curves, and plane figures in varying combinations. Common plane figures include circles, triangles, quadrilaterals, and polygons. **See Figure 2-3.**

To sketch circles, locate the centerpoint and draw several intersecting diameter lines. Mark off the radius on these lines, and connect with a series of arcs. The diameter is then commonly dimensioned.

To sketch triangles, draw the base, determine the angles of the sides, and draw straight lines to complete. Generally, one or more of the sides is dimensioned and the angle is noted.

To sketch quadrilaterals, draw the base line and determine corner points. Connect the corner points with straight lines to complete. Dimensions of at least two sides and two angles, depending upon the type of quadrilateral, are required.

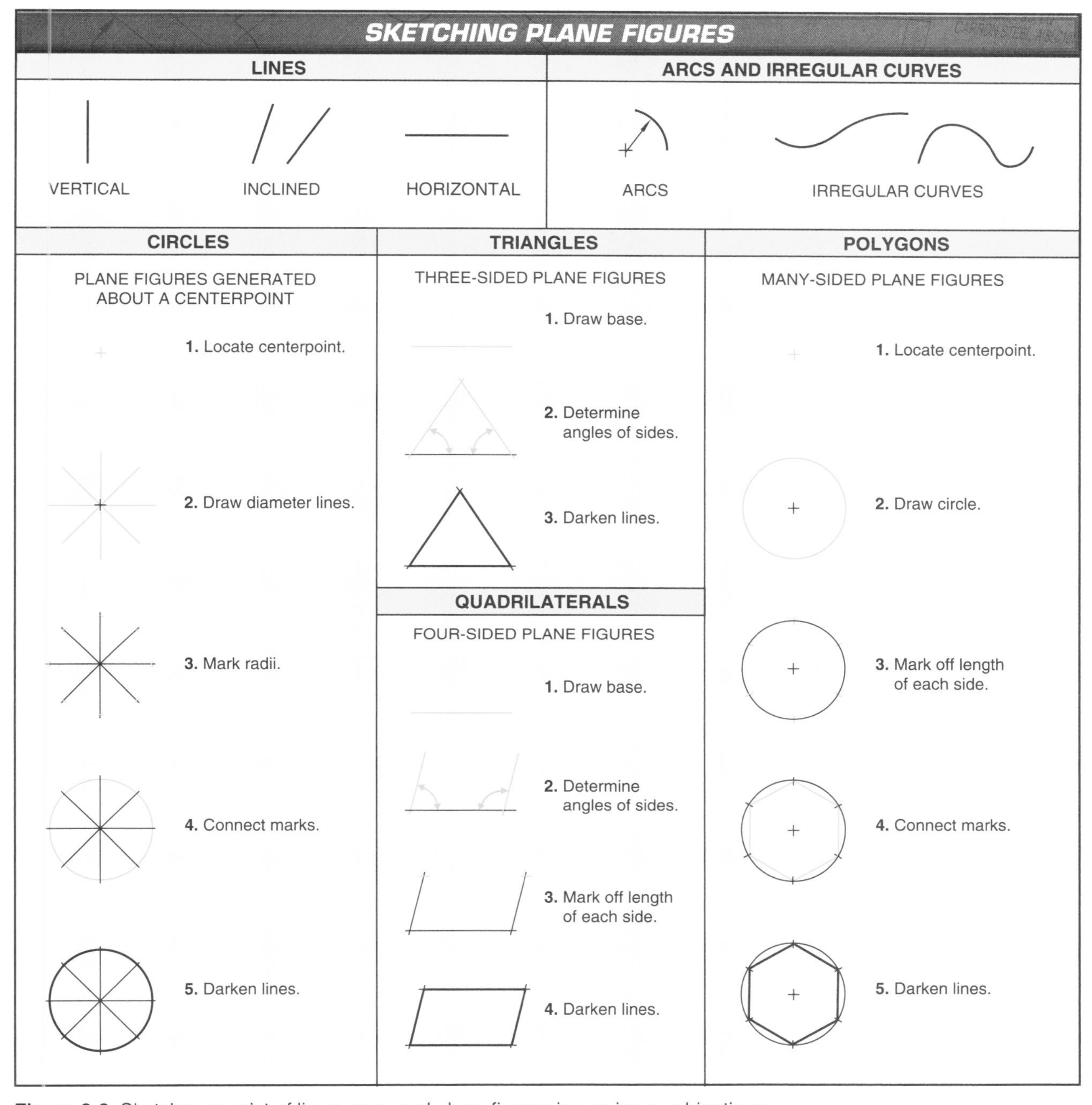

Figure 2-3. Sketches consist of lines, arcs, and plane figures in varying combinations.

To sketch regular polygons, locate the centerpoint, and draw a circle of the appropriate size. Mark off the length of each side on the circumference of the circle, and connect the marks with a series of straight lines. Darken the lines to complete the polygon.

> Grid paper can be used to quickly sketch objects in correct proportions. An easy-to-use scale can be improvised simply by counting squares.

PICTORIAL DRAWINGS

Pictorial drawings look like a "picture" because they convey a sense of perspective and realism of the object being viewed. Height, length, and depth of the object are easily shown with various types of pictorial drawings.

While pictorial drawings are only occasionally used when preparing prints, an understanding of their basic concepts aids in the interpretation of prints. Additionally, the ability to quickly sketch a pictorial drawing of an object or detail helps convey technical information to others. Pictorial drawings are based on the conventions in ANSI/ASME Y14.4M, *Pictorial Drawing*. Three basic types of pictorial drawings are axonometric, oblique, and perspective. **See Figure 2-4.** Perspective drawings are seldom used on machine trades prints.

> Oblique pictorials are best for plate or sheet metal parts because they emphasize one view.

Axonometric Drawings

An *axonometric drawing* is a pictorial drawing that shows three sides of an object at the same scale, but contains no true view of any side. A *true view* is a view in which the line of sight is perpendicular to the surface. The three basic types of axonometric drawings are isometric, dimetric, and trimetric. **See Figure 2-5.** An *isometric drawing* is an axonometric drawing with the axes drawn equally spaced at 120° apart. A *dimetric drawing* is an axonometric drawing with two axes drawn on equal angles to the third, but the angle between them is larger or smaller. A *trimetric drawing* is an axonometric drawing with all axes drawn at different angles. Of these, the isometric is the most commonly used.

Isometric Drawings. Isometric drawings contain three equal axes that are drawn 120° apart. Because of this 120° angle, no surface appears as a true view; however, the object has a natural appearance because three sides are seen.

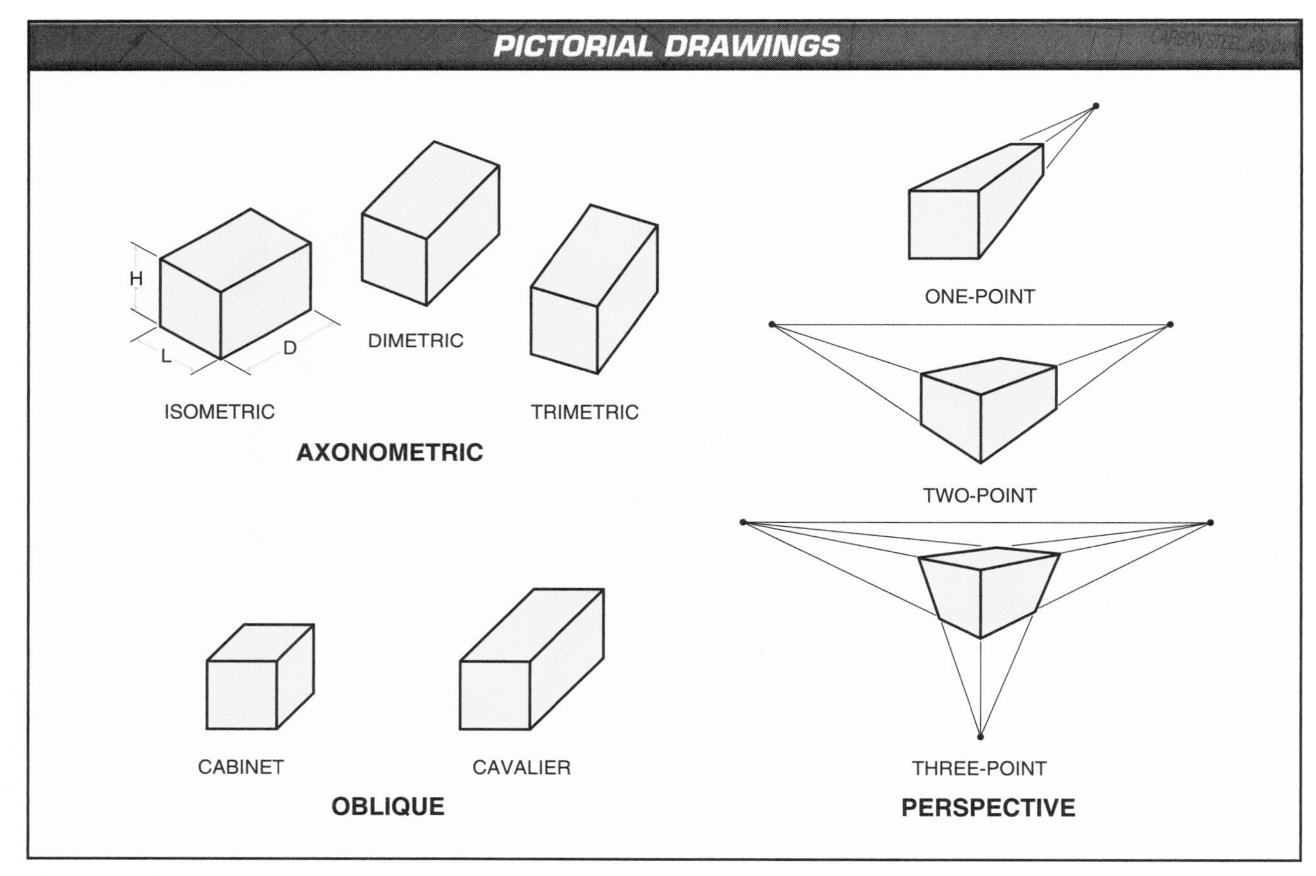

Figure 2-4. Three basic types of pictorial drawings are axonometric, oblique, and perspective.

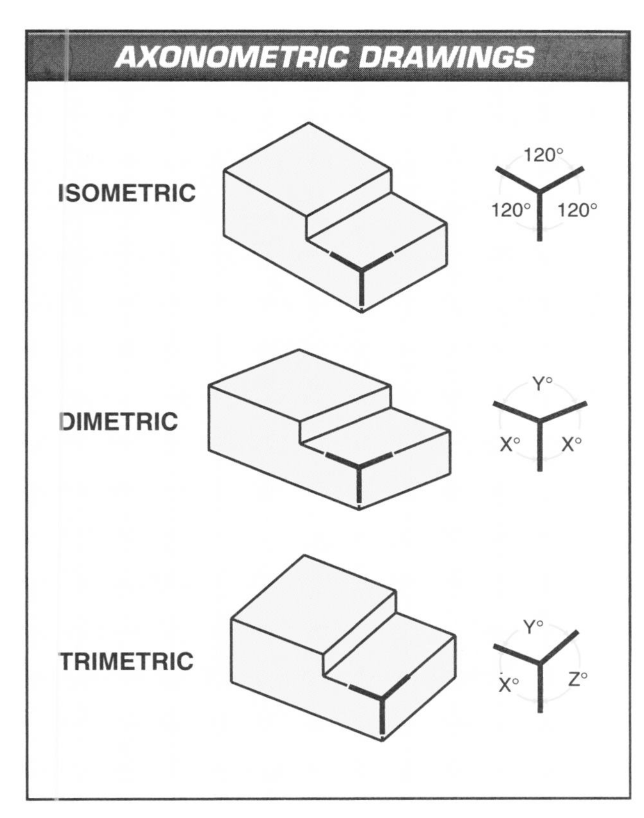

Figure 2-5. Isometric drawings are the most commonly used type of axonometric drawing.

Because of the skewed sides, circles (drilled holes, counterbores, or other round features) appear as ellipses on isometric surfaces. **See Figure 2-6.** Additionally, arcs appear as portions of ellipses. All surfaces not in one of the three principal isometric planes must be drawn by locating end points of the skewed surface. The end points are connected to complete the skewed surface.

Sketching Isometric Drawings. Isometric sketches are made with the following procedure. **See Figure 2-7.**

1. Locate the isometric axes and block in the front view using length and height measurements from the multiview drawing. There are two front surfaces on this particular object. These two surfaces are parallel to one another. The depth dimension used to establish the location of the second front surface is taken from either the top or right side view of the multiview.
2. Sketch the outline shape of the front surfaces. Measurements are taken from the multiview. Notice that the arc on the second front surface is drawn as a portion of an ellipse.
3. Locate the centerpoint of the drilled hole on the second front surface. Construct an ellipse using measurements from the front view of the multiview.
4. Draw receding lines long enough to mark the depth of the object. Receding lines are parallel to the isometric axis.
5. Draw lines to establish the back surface. These lines are parallel to the isometric axis. The skewed line on the back surface representing the V portion is drawn to its corresponding line in the front surface.
6. Find the depth through the drilled hole to determine if the back portion of the drilled hole will show through on the front view. Draw a portion of the ellipse as required.
7. Darken all lines to complete the isometric sketch.

Circles on Isometric Drawings. Circles on isometric surfaces are drawn as ellipses. An *ellipse* is a plane curve with two focal points. The sum of the distances from these two focal points to any point on the ellipse determines the shape of the ellipse. As the distance between the focal points decreases, the ellipse becomes more circular in shape.

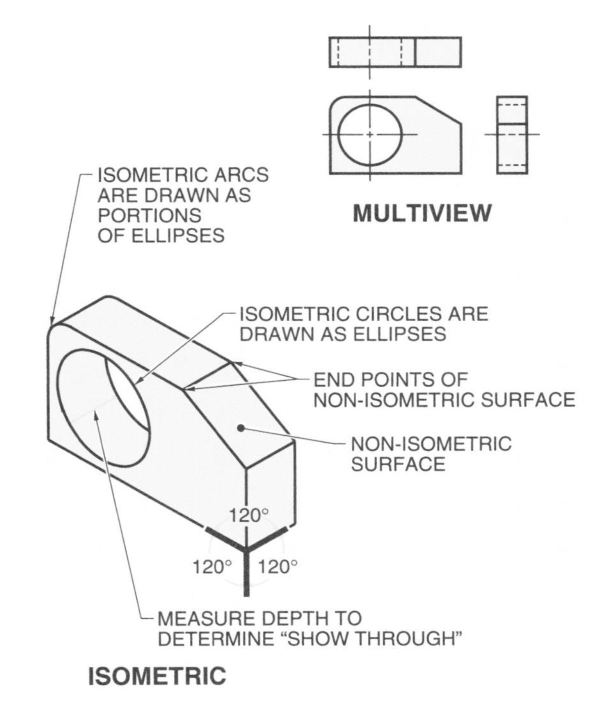

Figure 2-6. Isometric drawings contain three equal axes drawn 120° apart.

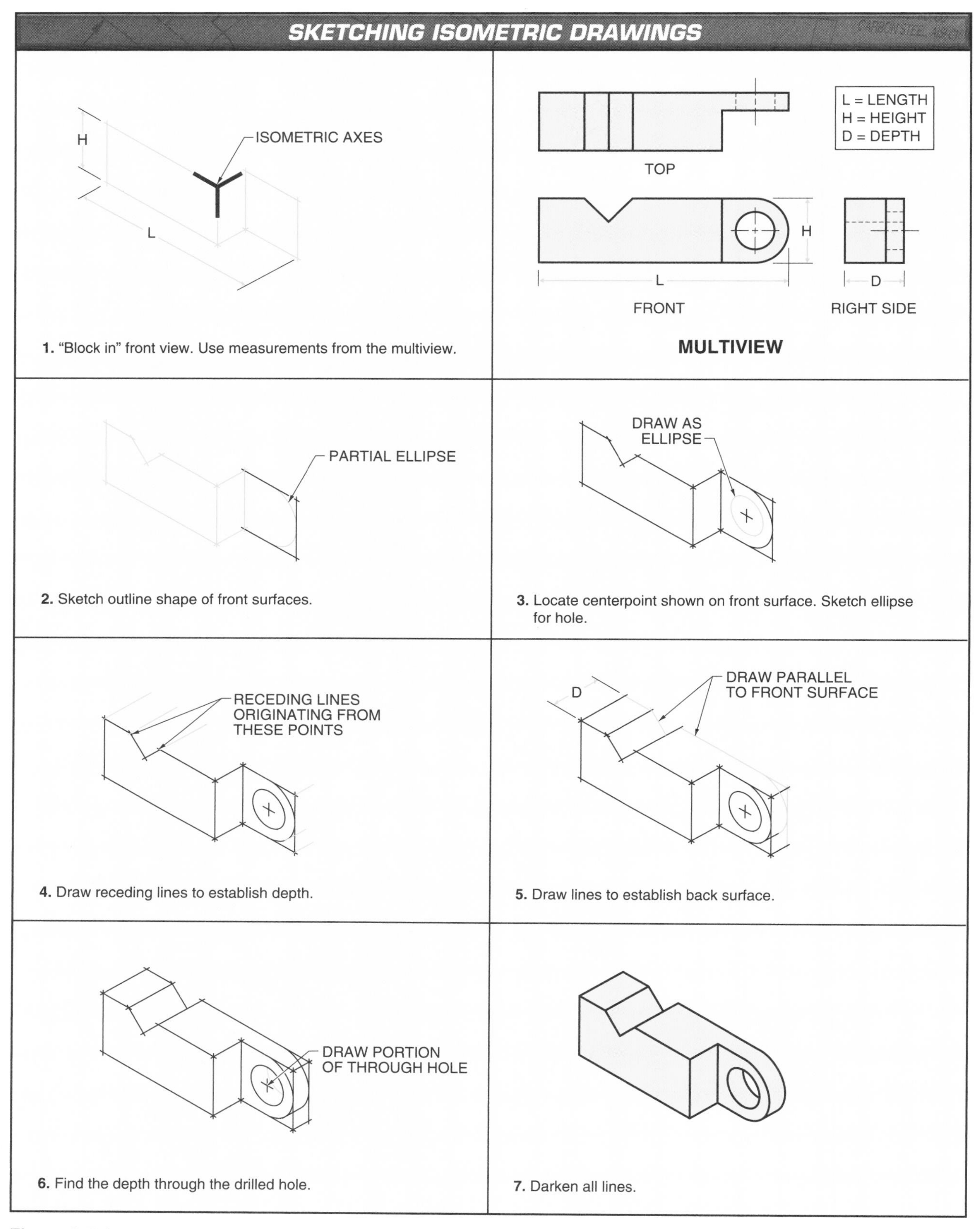

Figure 2-7. Isometric drawings can be quickly sketched using measurements from the multiview.

For sketching, the parallelogram method of constructing ellipses is often used. **See Figure 2-8.** In this method, dimensions from the multiview are used to determine the size of the parallelogram. Arcs are then drawn or sketched using the intersecting points as centerpoints.

Oblique Drawings

An *oblique drawing* is a pictorial drawing that shows one surface of an object as a true view. All other surfaces of the object are distorted by the angle of the receding oblique lines. All features shown on the face containing the true view are drawn as they appear. Additionally, right angles are shown at 90° on surfaces having a true view. For example, a drilled hole shown on the true view face of an oblique drawing is shown as a circle. Drilled holes on any other surface of the oblique drawing appear as ellipses. The angle at which an ellipse is drawn on these surfaces is determined by the angle of the receding lines. Normally, these lines are drawn on a 30° or 45° angle.

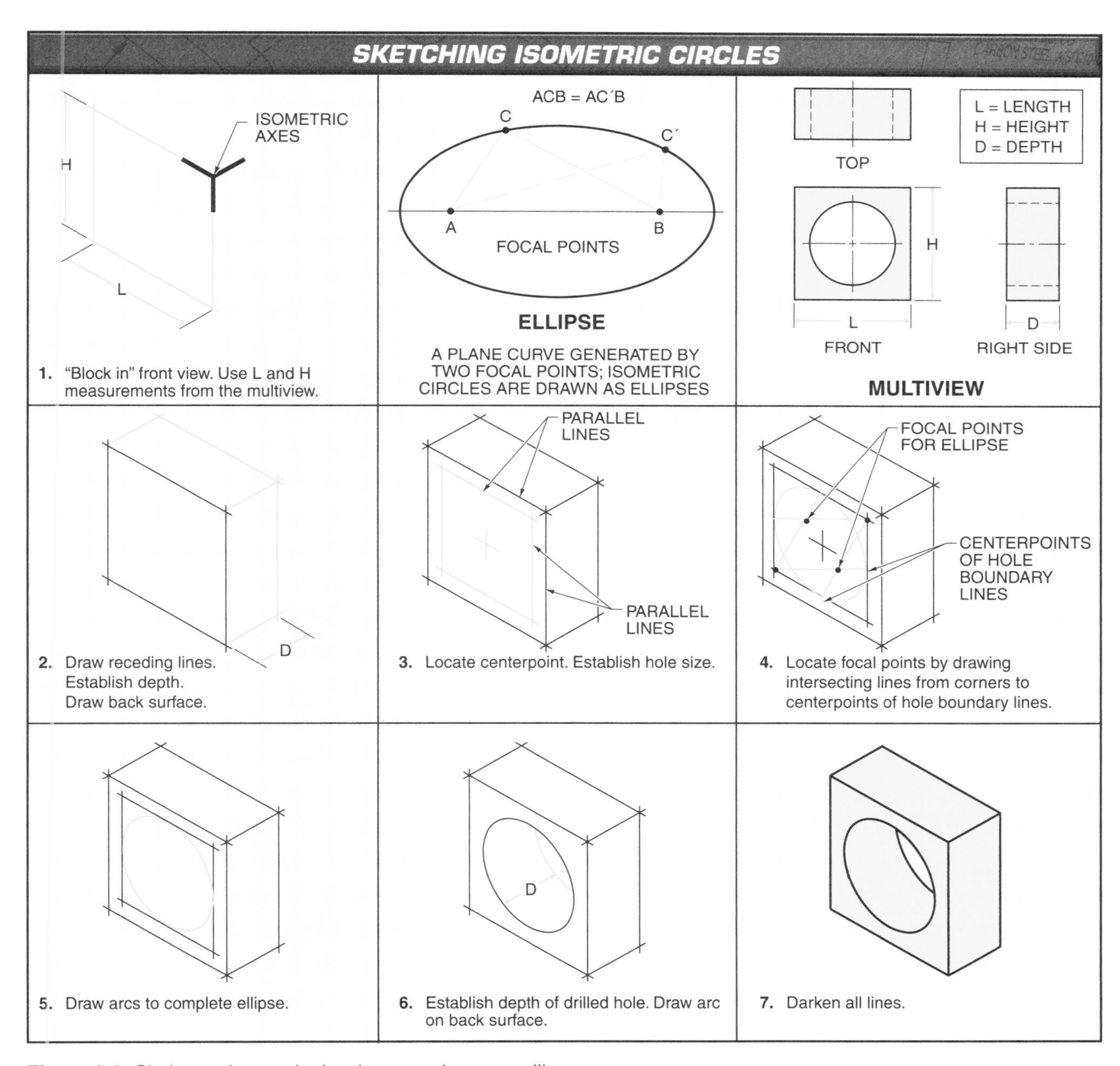

Figure 2-8. Circles on isometric drawings are drawn as ellipses.

The front view of an object is generally the view that shows the most shape or detail, or is commonly thought of as the front view. The two types of oblique drawings are cabinet and cavalier.

A *cabinet drawing* is an oblique drawing with receding lines drawn to one-half the scale of the true view. This is the most commonly used type of oblique drawing. A *cavalier drawing* is an oblique drawing with receding lines drawn to the same scale as the true view. Using the same scale to draw all parts of an oblique drawing results in an image that appears distorted, so the cavalier type of drawing is seldom used.

Sketching Oblique Drawings. Oblique drawings are made with the following procedure. **See Figure 2-9.**

1. Determine which oblique drawing type will be sketched. (An oblique cabinet is shown.)
2. Block in the front view (usually the view with the most detail). Use length and height measurements of the multiview.
3. Complete the outline shape.
4. Locate all centerpoints of circles and arcs. Sketch circles and arcs.
5. Draw receding lines long enough to mark the depth of the object. Receding lines shown are drawn at a 45° angle. They could also be drawn at a 30° angle.
6. Establish depth dimension from the right side or top view of the multiview. Note that only one-half the dimension shown is drawn for the depth of the oblique cabinet.
7. Draw lines to establish the object's back surface. Draw receding lines. Find the depth through the drilled hole to determine if the back portion of the drilled hole will show through on the front view. Draw portion of hole as required.
8. Darken all lines to complete the oblique sketch.

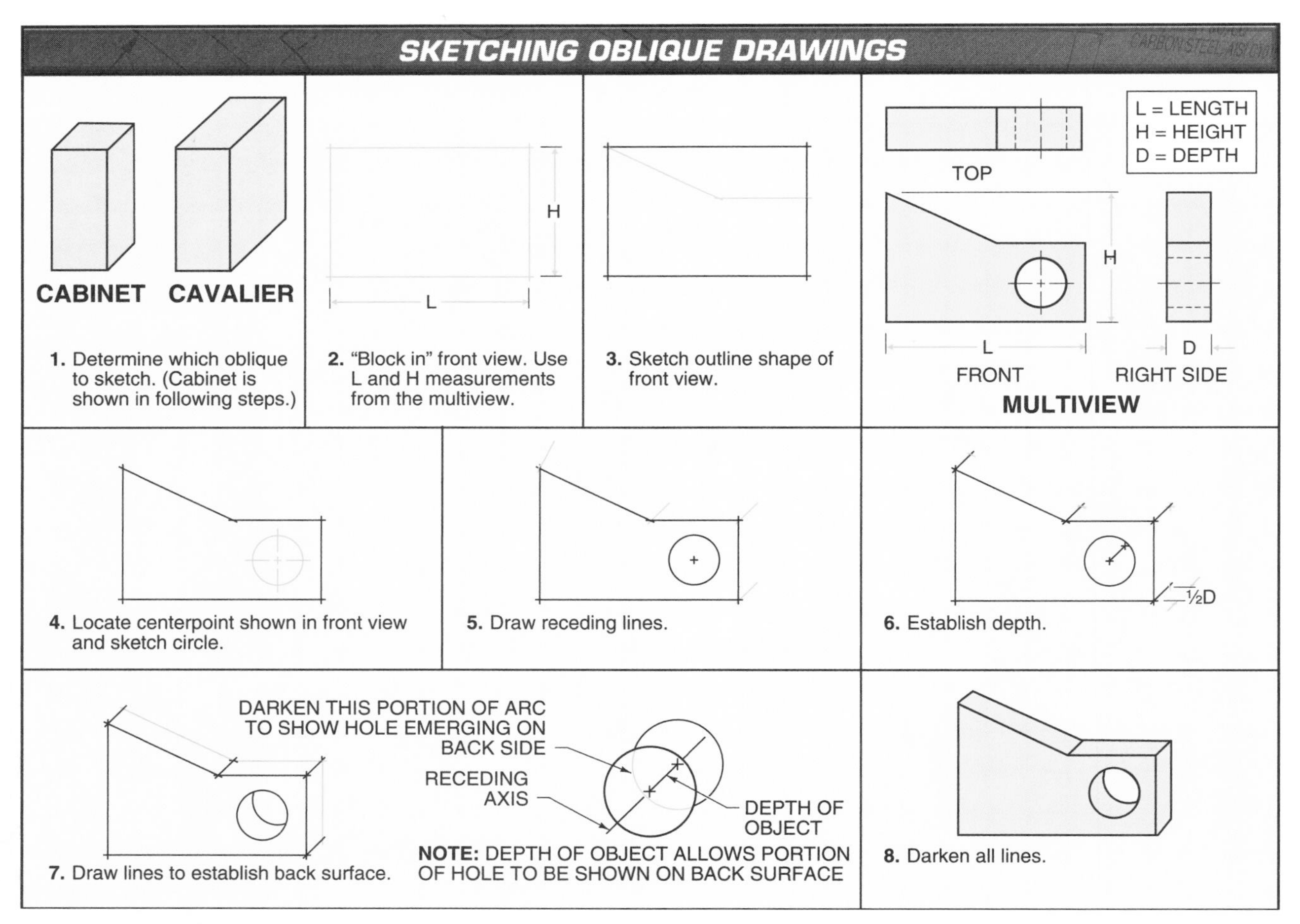

Figure 2-9. Oblique drawings show one surface on an object as a true view.

MULTIVIEW DRAWINGS

Manufacturing prints usually use multiview drawings to show every feature of the object in true view. A *multiview drawing* is a collection of two-dimensional drawings of a three-dimensional object that are shown in true view. **See Figure 2-10.** Multiview drawings are necessary to give a complete picture of an object by showing multiple views in detail and without distortion. These drawings use orthographic projection to form the views. Each view has a relationship with each of the other views and should be placed on a drawing using specific guidelines. The construction and arrangement of multiview drawings is based on ANSI/ASME Y14.3, *Multiview and Sectional View Drawings*.

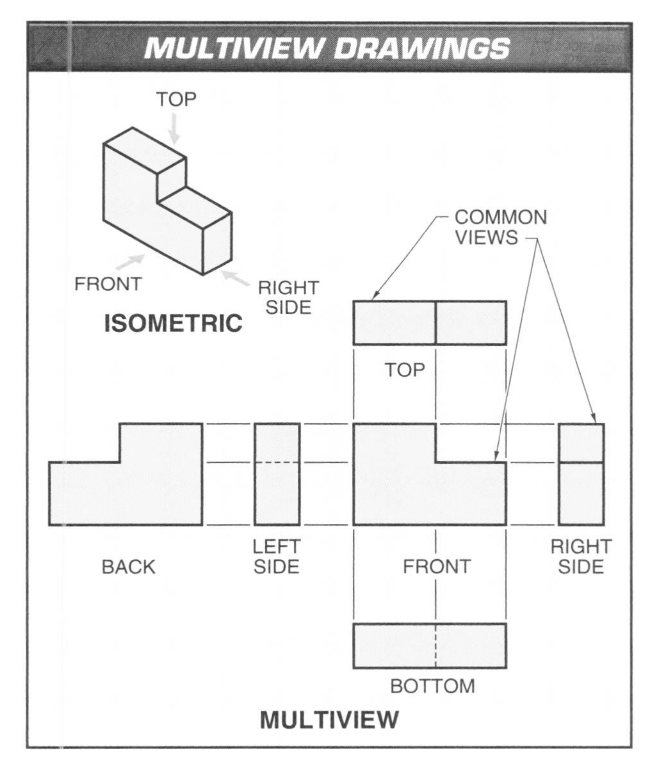

Figure 2-10. Multiview drawings are a set of true views of an object.

Orthographic Projection

Orthographic projection is a method of representing the true shape of one view of an object onto a single plane. A *plane of projection* is an imaginary surface on which the shape of an object from that view is drawn. In orthographic projection, a plane of projection is parallel to the surface of an object, which shows that surface in true view. **See Figure 2-11.** All edges and contours of an object are projected onto the plane of projection. Edges that are not visible from that point of view are drawn as hidden lines.

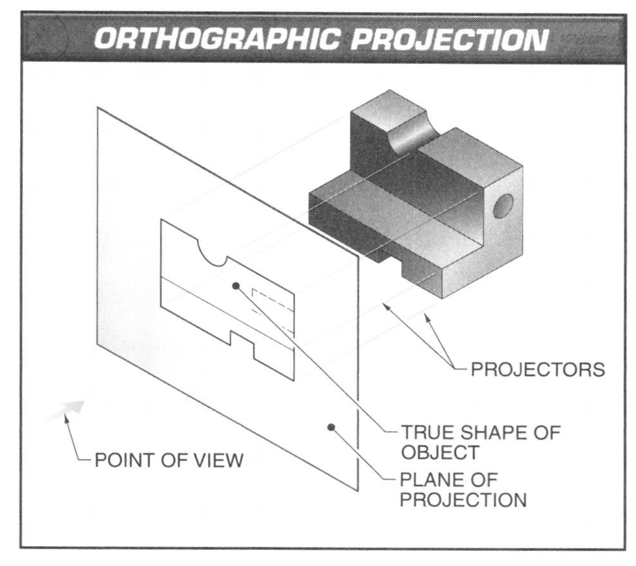

Figure 2-11. Orthographic projection forms a true view of an object by projecting an image of the object onto an imaginary parallel plane.

Principal Views

There are six principal planes of projection, which form an imaginary "glass box" around an object. The principal views on the sides of the glass box are the front, top, right side, back, left side, and bottom views. **See Figure 2-12.** If the edges of the box are considered "hinged," the box can be unfolded to lay the views side by side.

The choice of which side is considered the "front" may be arbitrary. Usually it is the side that most characteristically shows the shape of the object. The orientation of the object should be chosen to show its principal features in all selected views with a minimum number of hidden lines.

Certain views are related based on the height, width, and depth dimensions of the object. **See Figure 2-13.** The back, left side, front, and right side views of an object share the same height. The top, front, and bottom views of an object share the same width. The top, right side, bottom, and left side views of an object share the same depth. The features and surfaces of each view should be aligned to the same features and surfaces in adjacent views. This minimizes the number of dimensions that need to be included. For example, lining up the back, left side, front, and right side views, which all have the same height, allows that dimension to be listed only once.

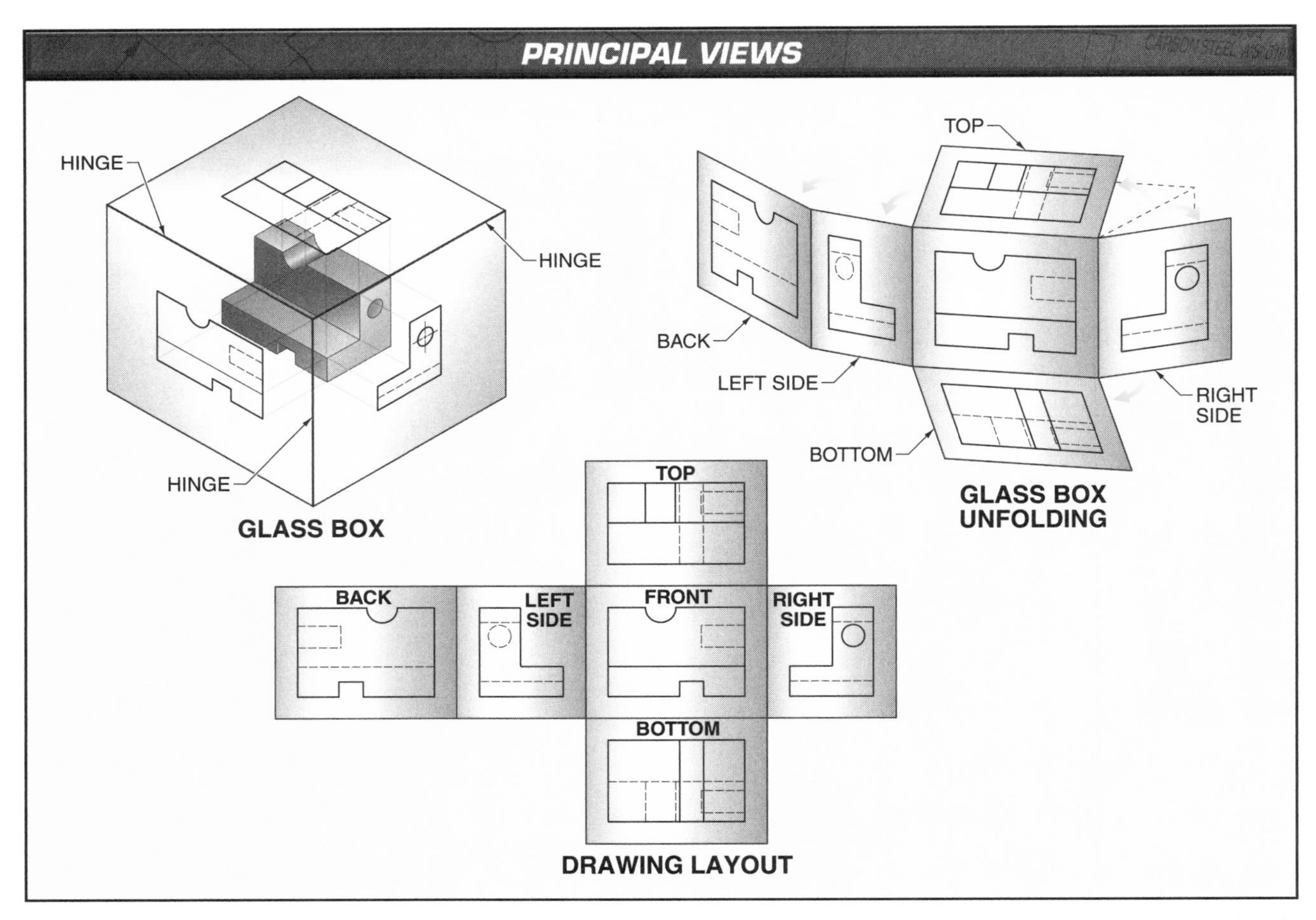

Figure 2-12. The six planes of projection form an imaginary box around an object. The box is unfolded to arrange the six principal views into a multiview drawing.

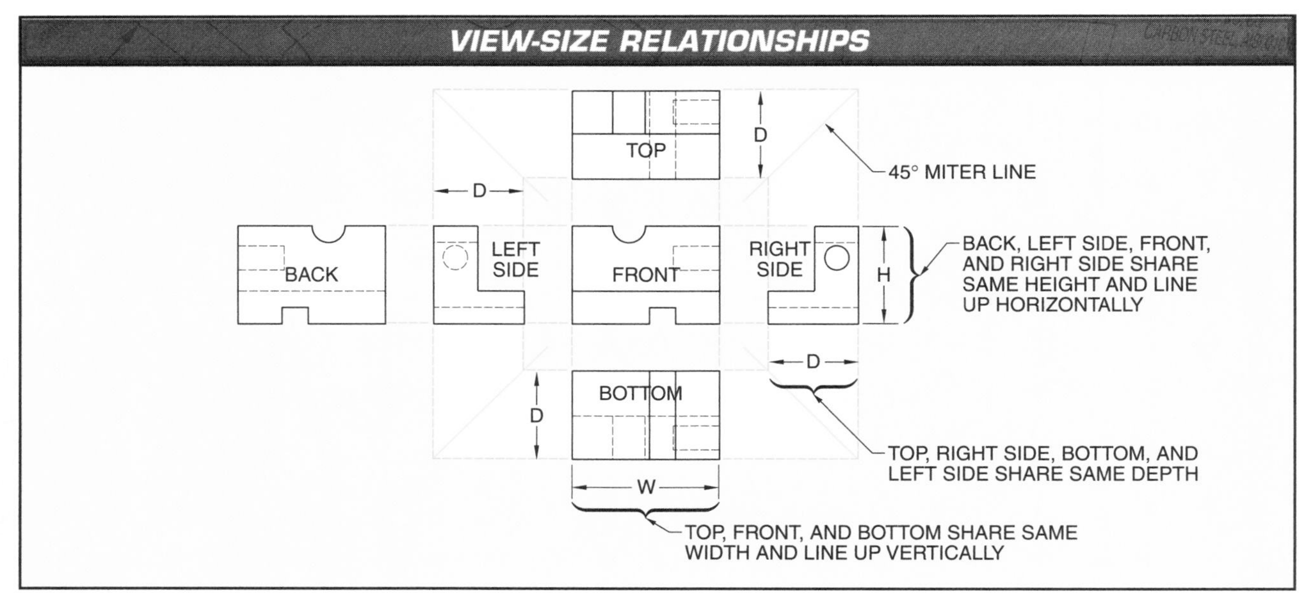

Figure 2-13. The six principal views of an object are directly related to each other based on height, width, and depth.

It is often not necessary to include all six principal views in order for an object's shape and features to be fully understandable. Usually, only the front, top, and right side are included. Depending on the object, even fewer may be needed. For example, a cylindrical shape can be fully described with only two views. On the other hand, complex objects may require more than three views.

Projection Systems

Views should be placed on a drawing in a certain arrangement to avoid confusion as to which is which. There are two common schemes for view arrangements that are based on different concepts for how the views are formed on the planes of projection. The results of each arrangement show the same information in the same views, but the views are arranged differently. The most common arrangement in the U.S., U.K., Canada, and Australia is called third-angle projection. In most other parts of the world multiviews are arranged according to first-angle projection.

When sharing prints between organizations, especially internationally, it is important to be clear on the projection system. Some prints will clearly name each view as FRONT VIEW, TOP VIEW, or SIDE VIEW. However, views are often left unlabelled in order to avoid interference with dimensioning, so view arrangement may then be used for identification. A standard symbol is often included on the print to indicate the projection system used. **See Figure 2-14.**

Third-Angle Projection. *Third-angle projection* is a projection system that places the plane of projection between the object and the observer. **See Figure 2-15.** The view is imagined to be projected from the object onto a transparent plane in a direction opposite to the point of view. When only the three common views are included, third-angle projection can be identified by the front view being in the lower left-hand corner of the layout.

Projection system names are based on a reference system formed by two intersecting planes that divide space into four quadrants. The first quadrant is the space in front of and above the planes. Therefore, a projection of a shape in this space onto the reference planes is called first-angle projection. Similarly, a projection of a shape in the third quadrant, which is behind and below the planes, is a third-angle projection.

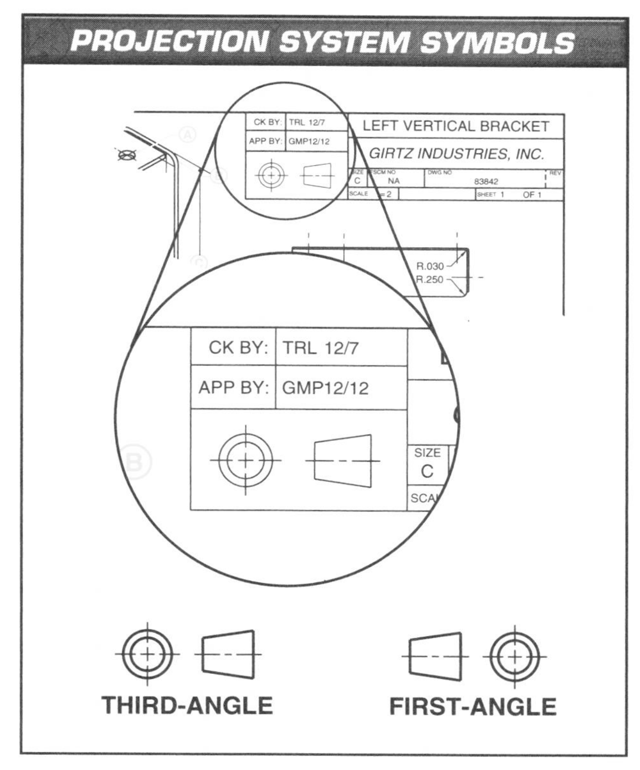

Figure 2-14. The projection system used for a multiview drawing is often represented in or near the title block by a symbol.

Understanding views and projection systems is critical to visualizing how to machine features into a part.

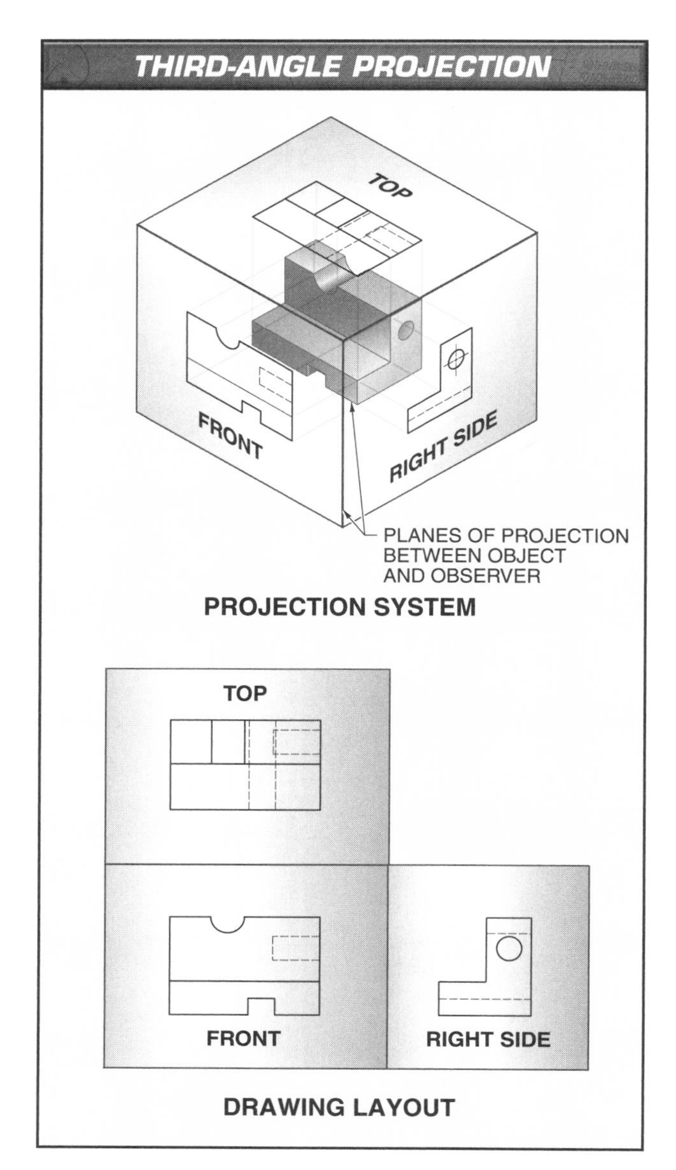

Figure 2-15. In third-angle projection, the plane of projection is between the object and the observer.

First-Angle Projection. *First-angle projection* is a projection system that places the object between the plane of projection and the observer. **See Figure 2-16.** The view is imagined to be projected through the object and onto the plane behind it. First-angle projection can be identified by the front view being in the upper right-hand corner of the layout.

Auxiliary Views

If the six principal views are not sufficient to easily depict certain object features, then one or more auxiliary views may be added. **See Figure 2-17.** An *auxiliary view* is a view that shows the true shape of an object surface that is not parallel to one of the six principal planes of projection. Auxiliary views show the inclined or oblique surface in true view as if the viewing plane is parallel to the auxiliary surface. On the print, an auxiliary view should be arranged so that one side is parallel and aligned with the corresponding edge on another view. If the multiview layout does not allow this, the auxiliary view is removed to another area of the print and labeled with reference arrows.

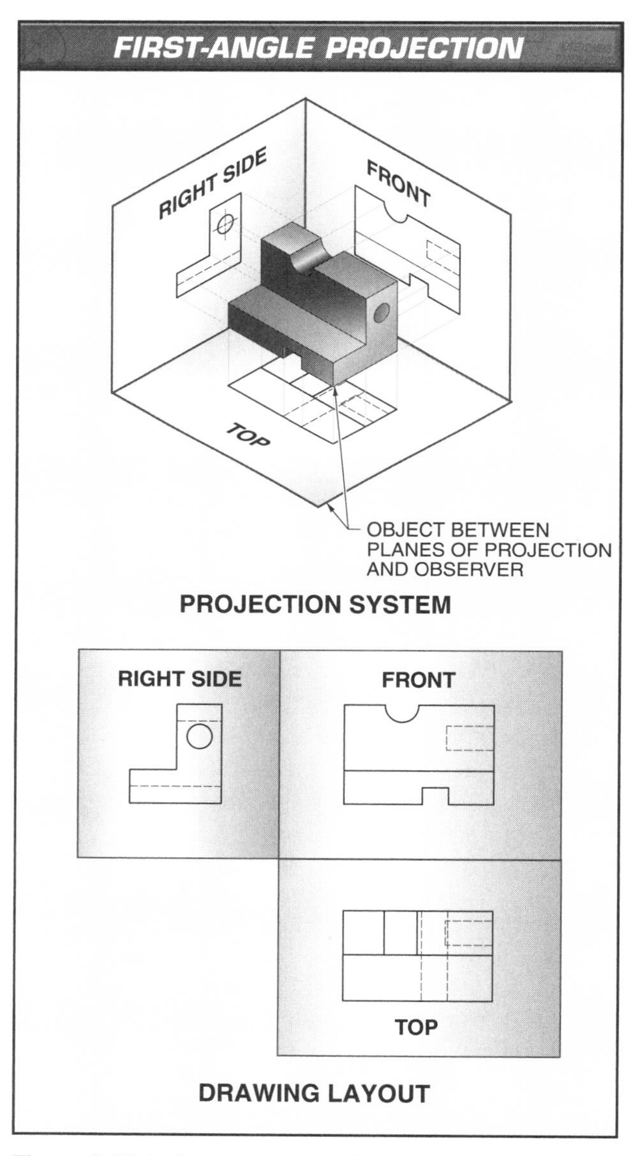

Figure 2-16. In first-angle projection, the object is between the plane of projection and the observer.

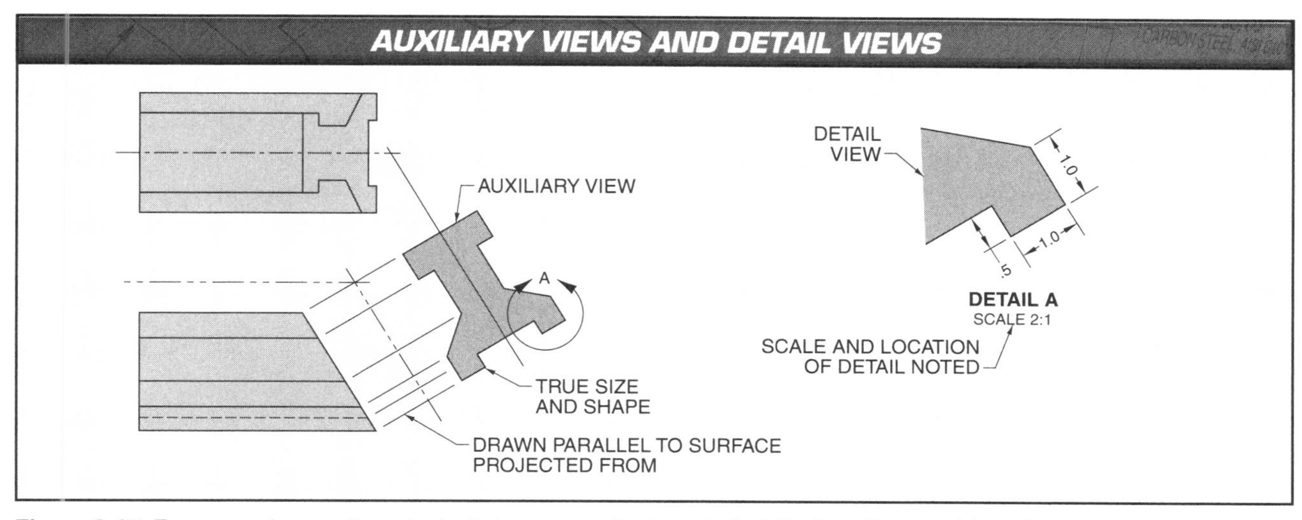

Figure 2-17. For some shapes, the principal views are not adequate to fully describe the object shape. Therefore, auxiliary views and detail views are added as needed.

Detail Views

A *detail view* is a portion of a drawing that is enlarged to show intricate features more clearly. An area of one of the normal-scale views is marked with the detail symbol, which is a circular arrow. An uppercase letter identifies the detail. The detail view is arranged at some convenient place on the print and labeled with the same letter and the scale. A print may include multiple detail views. In this case, each is identified with its own letter and scale.

Reference Arrows

An alternative method of view identification is described in ISO 128, *Technical Drawings: Basic Conventions for Views*. This convention uses reference arrows and uppercase letters with one or more views, often axonometric. Each individual view is labeled with the uppercase letter, which is associated with a reference arrow indicating viewing direction. **See Figure 2-18.** This method is useful when space constraints on a print do not allow conventional view arrangement.

Sketching Multiviews

Multiviews are often drawn from pictorial drawings. **See Figure 2-19.** To create a multiview drawing using third-angle projection, apply the following procedure:

1. Choose the orientation that best shows the principal features of the object. Block in the width and height dimensions of the front view. Place the front view on the drawing paper in a way that allows space for the additional views.
2. Project the width and height dimensions of the front view to the top and right-side views.
3. Establish depth dimensions on the top and right side views. The depth dimension is shown with a 45° miter line on the right side of the top view to establish the same depth dimension in the right side view.
4. Lightly project lines from the remaining features, such as holes or inclined planes, to their appropriate views. Complete the visible lines of the remaining features.
5. Draw hidden lines as required. Add auxiliary views or details as needed.
6. Darken all visible lines to complete the drawing.

Steel rules measuring dimensions in fractional inches include marks for 32nds or 64ths of an inch.

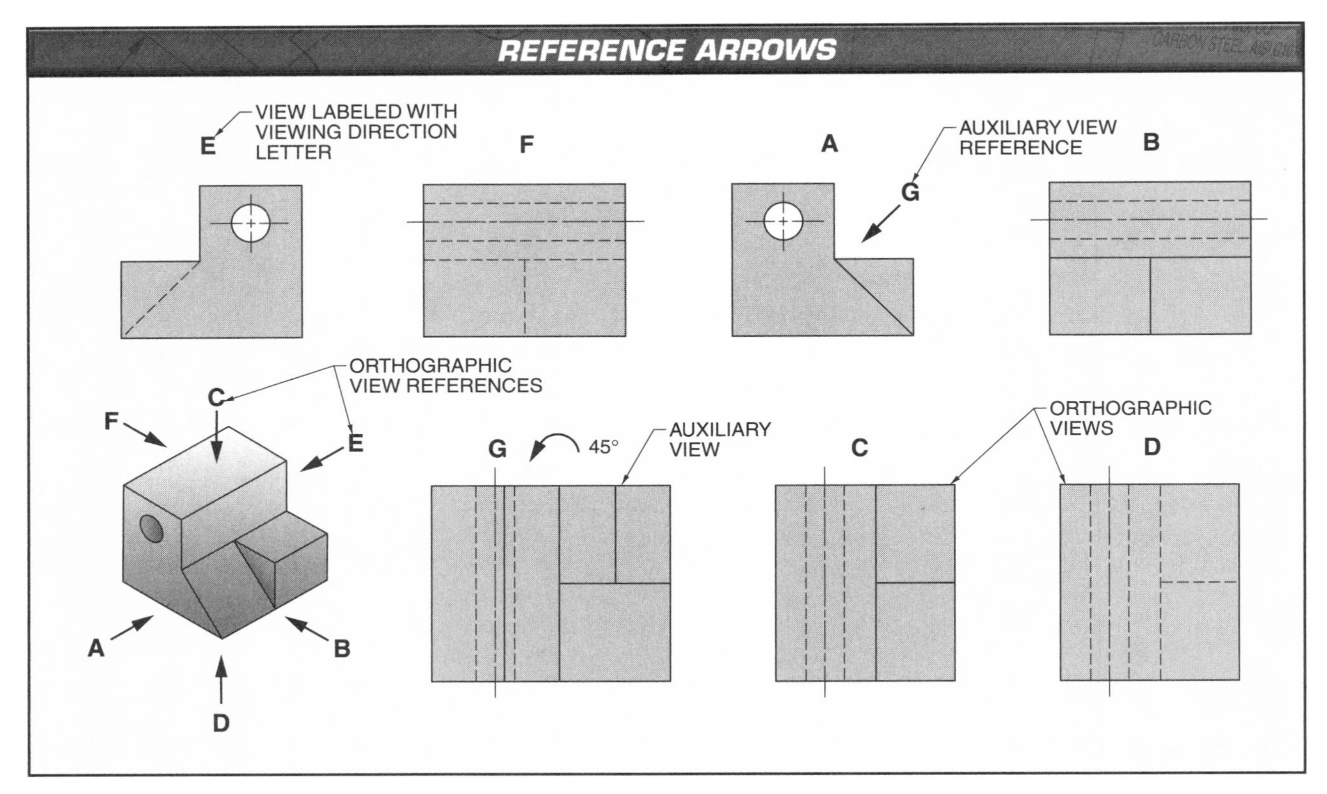

Figure 2-18. Reference arrows represent points of view. Letters are used to identify each view.

LINE CONVENTIONS

Drawings are composed of lines to show the shape of the drawn object. In printreading, a variety of line styles have certain meanings, which are standardized in ANSI/ASME Y14.2, *Line Conventions and Lettering*. **See Figure 2-20.** The same types of lines are used, regardless of whether they are drawn by conventional drafting methods or by CAD software.

First, all lines should be one of two widths. Thick lines are a minimum of 0.024″ (0.6 mm) wide. Thin lines are a minimum of 0.012″ (0.3 mm) wide. Patterns of dots and dashes, waviness, cross marks, arrowheads, and zigzags are used to further differentiate line types. The unique style for each line allows it to convey additional meaning in a concise way.

> Standardized line conventions are important for conveying information in a very compact but precise way. Lines are given extra meaning by way of their thickness and pattern instead of requiring extra notes and descriptions that can clutter a drawing.

The lengths of dashes and spaces are no longer defined in the standard. Instead, these line specifications should be selected so that they are easily identifiable, consistent, and appropriate for the size and scale of the drawing.

Visible Lines

A *visible line* is a thick line that represents an edge or contour that can be seen from the view of an object. Visible lines are thick and dark and are the most common lines on a print.

Hidden Lines

A *hidden line* is a thin dashed line that represents an edge or contour that cannot be seen from the view of an object. Hidden lines are drawn with a series of thin, short, evenly spaced dashes. Hidden lines should always begin and end with a dash that contacts the object or other line from which they start or end.

Views to be drawn should be selected so that the use of hidden lines is minimized. A view with a large number of hidden lines is difficult to read. Instead, a different view that shows the same features with visible lines is a better choice.

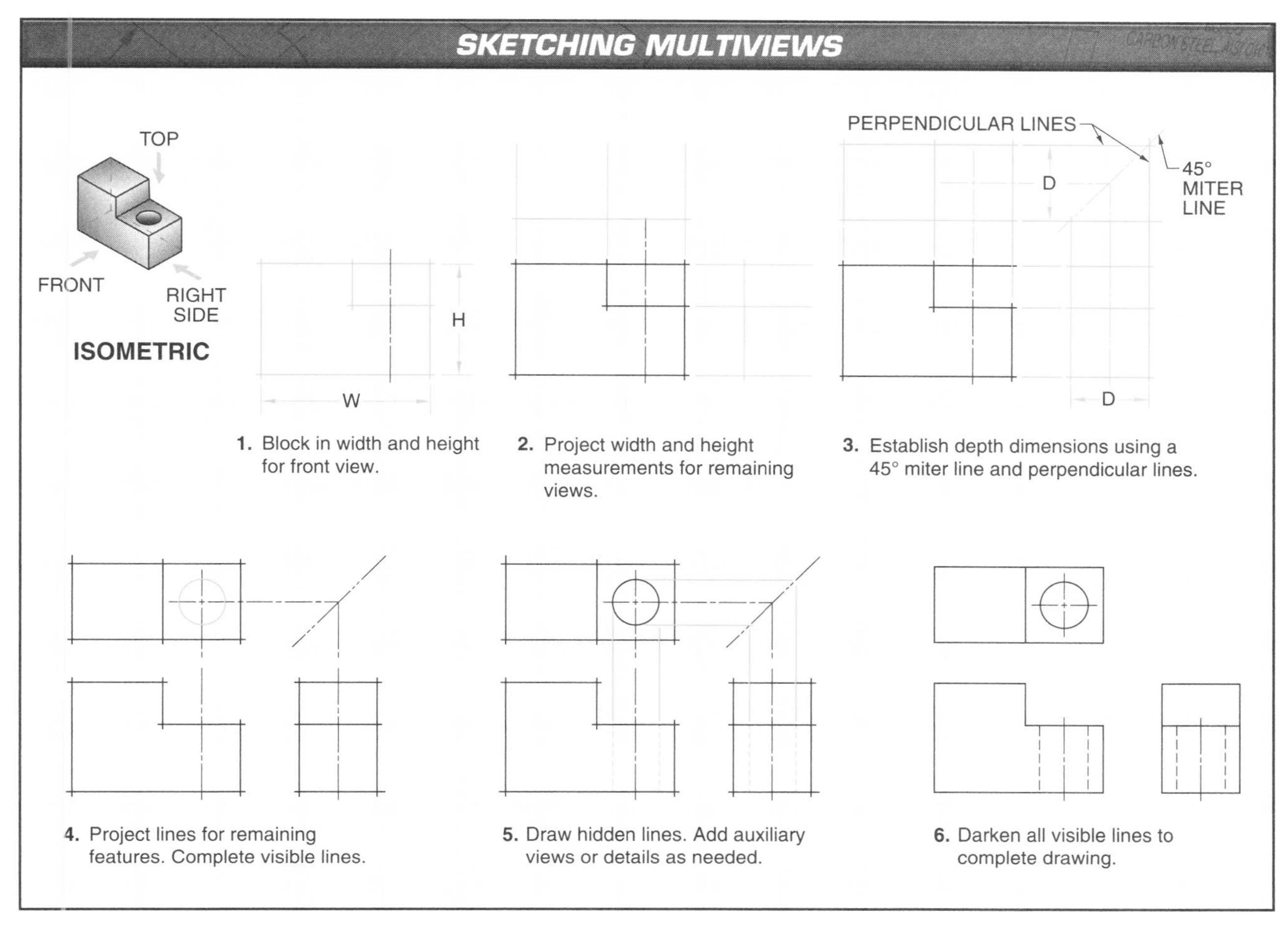

Figure 2-19. Multiview drawings are produced from an object's dimensions.

Centerlines

A *centerline* is a thin dashed line that locates axes or centerpoints of arcs and circles. Centerlines are composed of a series of long and short dashes separated by small spaces. Centerlines should extend at least a short distance beyond the edge of the object or feature. An older convention also included the symbol ℄ (a combination of the letters C and L) at one end of a line to identify it as a centerline, though this is no longer recommended.

Centerpoints are drawn in the centers of all circles and arcs. They are formed by the intersection of the short dashes of perpendicular centerlines.

Symmetry Lines

A *symmetry line* is a centerline that defines a plane of symmetry for a partial view. To save space, prints may show only one-half of symmetrical objects. The other, identical half is indicated by a centerline with two short parallel lines drawn at right angles through the centerline near each end.

Autodesk, Inc.

The use of different types of lines helps machinists understand the size and shape of a part.

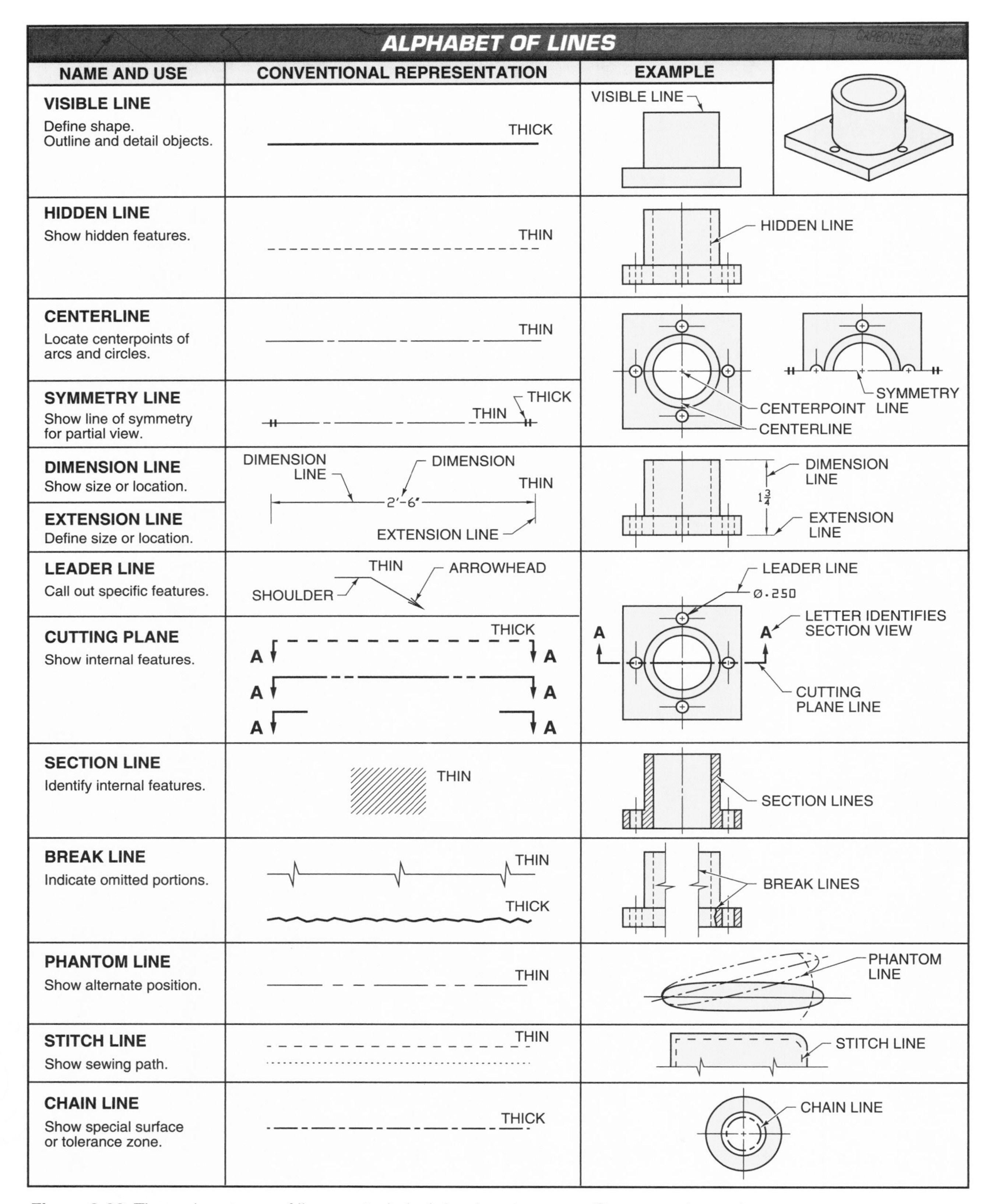

Figure 2-20. The various types of lines on technical drawings have specific uses and meanings.

Dimension Lines

A *dimension line* is a thin line that indicates the extent and direction of dimensions. **See Figure 2-21.** Each dimension line is terminated with arrowheads. Dimension lines are commonly broken near the middle for the placement of measurement numerals. If a horizontal dimension line is not broken, the numeral is placed above the dimension line. When there is not adequate space, a pair of very short dimension lines, each terminated in a single arrowhead, may be placed outside the indicated dimension.

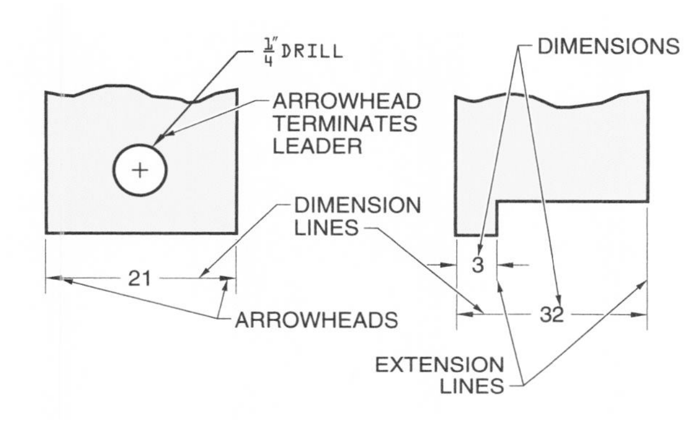

Figure 2-21. Dimensioning uses dimension lines, arrowheads, and extension lines to show the size of object features.

An *arrowhead* is a symbol that indicates direction or identifies the endpoint of a line. Arrowheads should be three times as long as they are wide and proportional to the thickness of the dimension line. **See Figure 2-22.** Arrowheads may be drawn in one of four different styles, though the same style should be used throughout a drawing.

Extension Lines

An *extension line* is a thin line that extends from a feature in order to facilitate dimensioning. Extension lines should not touch the features from which they extend. A small gap should separate the extension line from the part.

Generally, extension lines should not cross other extension lines or dimension lines. If they must cross, they are not broken. If extension lines must cross arrowheads, they are broken. Extension lines extend a short distance beyond the dimension line.

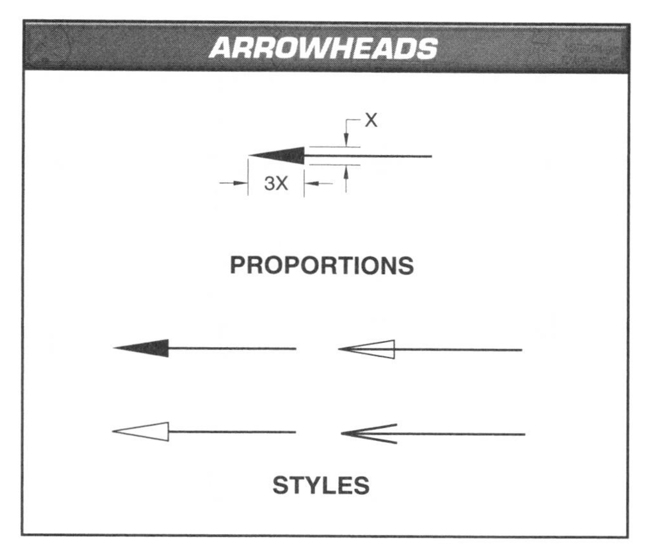

Figure 2-22. Arrowheads should be of a certain 3:1 proportion and standard style.

Leader Lines

A *leader line* is a thin line that connects a dimension, note, or specification with a particular feature. **See Figure 2-23.** A leader line is composed of a short horizontal line connected to a longer inclined line. The horizontal shoulder is centered on the text at the beginning or ending of the note. The other end of the leader line terminates at the feature. When ending on a dimension line, the two lines simply connect. Otherwise, a leader line may terminate with either an arrowhead or a dot. Arrowheads are used to terminate on a line. Dots are used to indicate an area.

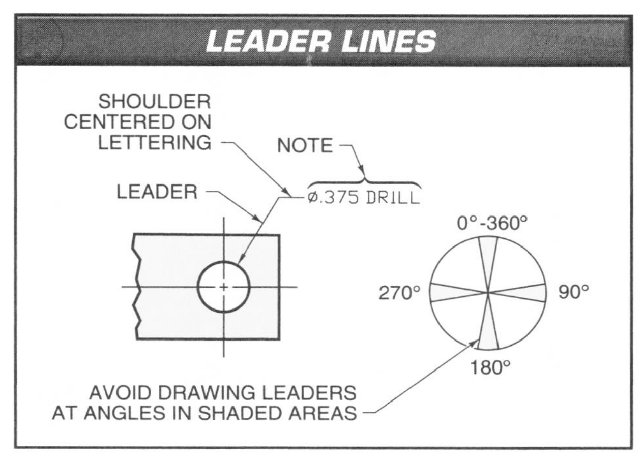

Figure 2-23. Leader lines connect notes with a particular object feature. The inclined portion of the leader line should stand out from the rest of the drawing.

The inclined portion of a leader line should not be excessively long, cross other leader lines, or be drawn in a way that makes them blend in with other lines. This means leader lines should not be horizontal, vertical, or parallel with adjacent non-leader lines. Leader lines are drawn at an angle so they will show up easily.

Cutting Plane and Viewing Plane Lines

Both cutting plane and viewing plane lines use the same line styles. A *cutting plane line* is a thick, dashed or broken line, terminating in right angle arrowheads, that indicates where an object is imagined to be cut in order to view internal features. A *viewing plane line* is a thick, dashed or broken line, terminating in right angle arrowheads, that indicates the direction of an alternate external view of an object. The difference between the two applications is whether the view is internal or external.

These lines are drawn in one of three styles: with evenly spaced short dashes, with alternating long dashes and pairs of short dashes, or with the middle portion omitted. Arrowheads on the ends of the cutting plane line indicate the direction of the view. Uppercase letters near the arrowheads identify the view.

Section Lines

A *section line* is a thin line used in a hatch pattern fill that identifies an area as being a cutting plane surface. Section lines are always drawn at an angle. The preferred angle for general section lines is 45°, but other angles are also used when multiple adjacent parts are sectioned. This helps each part stand out separately. Specific patterns of section lines may also be used to identify the particular type of material. When sectioned areas are large, the section line pattern may be included at only the inside edge of the boundary.

Break Lines

A *break line* is a line that indicates the omission of unnecessary portions of a drawing. There are two standard styles for break lines: a thick, irregularly wavy line and a thin line with occasional zigzags. The wavy line can be used to outline any shape, such as a particularly complex corner of an otherwise featureless part. The zigzag line, however, is always straight.

Most breaks are used to condense drawings with long, uniform sections. This allows a drawing to show end details of long parts or assemblies without taking up room by showing the simple portions. The drawing is still dimensioned as if the complete part is shown. **See Figure 2-24.** A common variation for round objects is to draw the edge with large semicircular breaks.

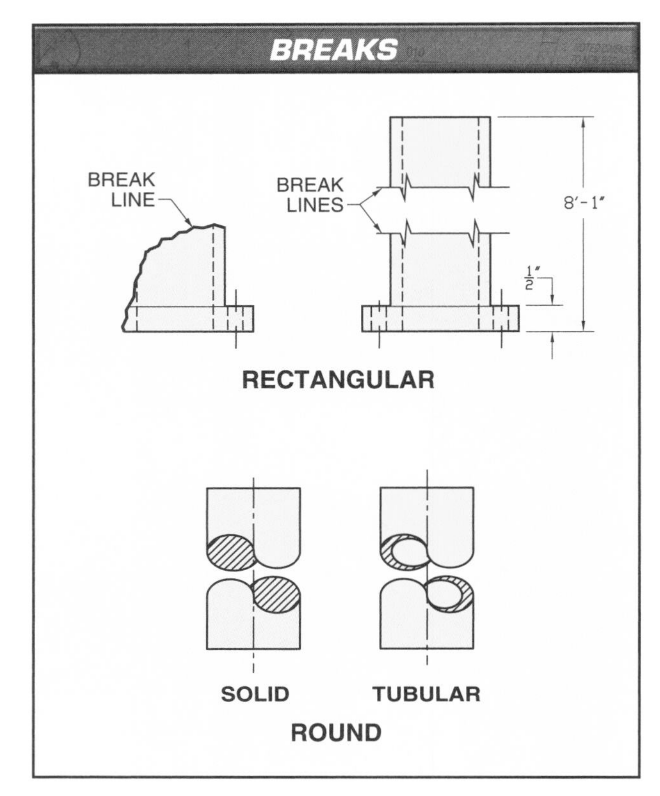

Figure 2-24. Breaks are used to avoid including a large portion of a drawing that has few features.

Phantom Lines

A *phantom line* is a thin, dashed line that indicates additional reference information. Phantom lines are also known as reference lines. These lines are used for several purposes: showing a moving part's alternate position, outlining adjacent parts in assemblies, indicating the omission of repeating detail, and indicating filleted and rounded corners. Phantom lines are alternating long dashes with pairs of short dashes.

Stitch Lines

A *stitch line* is a thin, dashed or dotted line that indicates a sewing path. Stitch lines are used for specifying the assembly of fabric parts. Stitch lines are composed of either short dashes and spaces of equal length or a series of dots.

Chain Lines

A *chain line* is a thick, dashed line that identifies a surface requiring special treatment or outlines a tolerance zone. Chain lines consist of alternating long and short dashes.

SURFACES

The surface of a formed metal part may be rough. Smooth machined surfaces are required where parts mate or fit together. A *normal surface* is a plane surface parallel to a principal plane of projection. **See Figure 2-25.** It appears as a true view in the orthographic view to which it is parallel. An *oblique surface* is a plane surface not parallel to any principal plane of projection. It does not appear as a true view in any orthographic view. An oblique surface appears foreshortened in all orthographic views.

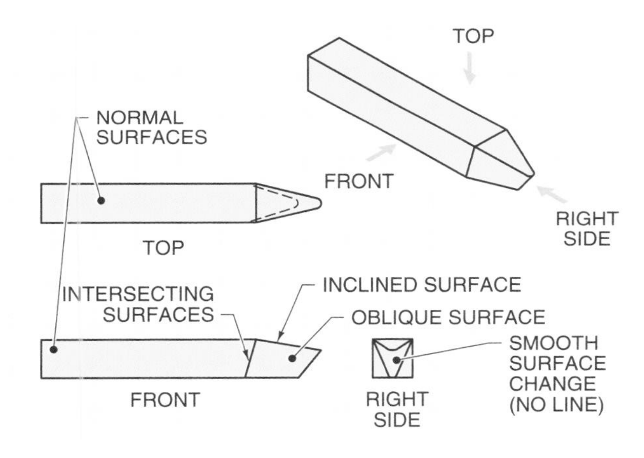

Figure 2-25. Surfaces are defined by lines.

An *inclined surface* is a plane surface perpendicular to one principal plane of projection and inclined to the others. *Intersecting surfaces* are surfaces that meet at an edge. Distinct changes in surfaces are represented by visible lines. Smooth changes in surfaces, such as when a curved surface becomes tangent to a flat surface, are not represented by visible lines.

SURFACE FEATURES

Much of a drawing's information is used to technically describe surface features of an object. A *surface feature* is an intentional deviation in an otherwise flat surface. Common surface features include various types of holes and edge treatments.

Abbreviations and symbols are used for surface features in drawings to conserve space, to promote consistency, and because they are easily recognizable. An *abbreviation* is a shortened version of the letters forming a word. For example, an abbreviation for diameter is DIA. Because abbreviations are representative of words in a specific language, they may vary in different languages.

A *symbol* is a simplified graphic representation of an object or idea. Symbols are not based on any specific language, so they can be easily recognized. For example, the symbol for diameter is ∅. Many symbols have been incorporated into the national and international standards for machine trades drawings. **See Figure 2-26.** The use of symbols is preferred over the use of abbreviations whenever possible.

ABBREVIATIONS AND SYMBOLS

ABBREVIATION	SYMBOL	MEANING
CL	℄	CENTERLINE
DIA	∅	DIAMETER
	×	REPEATING FEATURE
DEEP OR DP	↧	DEPTH
CBORE OR SFACE	⌴	COUNTERBORE OR SPOTFACE
CSK OR CSINK	⌵	COUNTERSINK
R		RADIUS
SR		SPHERICAL RADIUS
SDIA	S∅	SPHERICAL DIAMETER

Figure 2-26. Drawings may include abbreviations and symbols as shorthand notations. Symbols are preferred.

Holes

Holes appear as circles in one view and in profile in adjacent views. Hole sizes are specified by their diameters, never by their radii. **See Figure 2-27.** The symbol for diameter is ∅ and is always placed before the number. Older drawings may use the abbreviation DIA and place it after the number. Notes give the size and type of hole. For example, a typical note that reads ∅.750 DRILL indicates that the diameter of the drill used to drill the hole is ¾″. The diameter of the hole (0.750) is given first, followed by the operation (DRILL). However, most hole notes without the name of an operation may also be assumed to indicate drilling.

Several holes of the same size may be dimensioned only once by adding the repetitive feature symbol. The repetitive feature symbol indicates the number of times the given instructions or dimensions should be applied to a part with the multiplication symbol ×. For example, if four identical holes 1/4″ in diameter are to be drilled, the note may include 4× ∅.250. This symbol should only be used when there are no other features of different sizes that the repeating feature may be confused with.

A *through hole* is a drilled hole passing completely through the material. A *blind hole* is a drilled hole that does not completely pass through the material. The depth of blind holes is specified in the notes. The symbol for depth is ⌶, though older drawings may use the word DEEP instead. The depth is measured only along the cylindrical portion of the drilled hole. The point of the drill leaves a cone shape in the bottom of the drilled hole. This cone shape is generally drawn as a 90° angle, although it may be from approximately 80° to approximately 118°. Common operations to produce holes include drilling, counterboring, countersinking, counterdrilling, and spotfacing.

Holes are often counterbored so that fastener heads sit below the part surface.

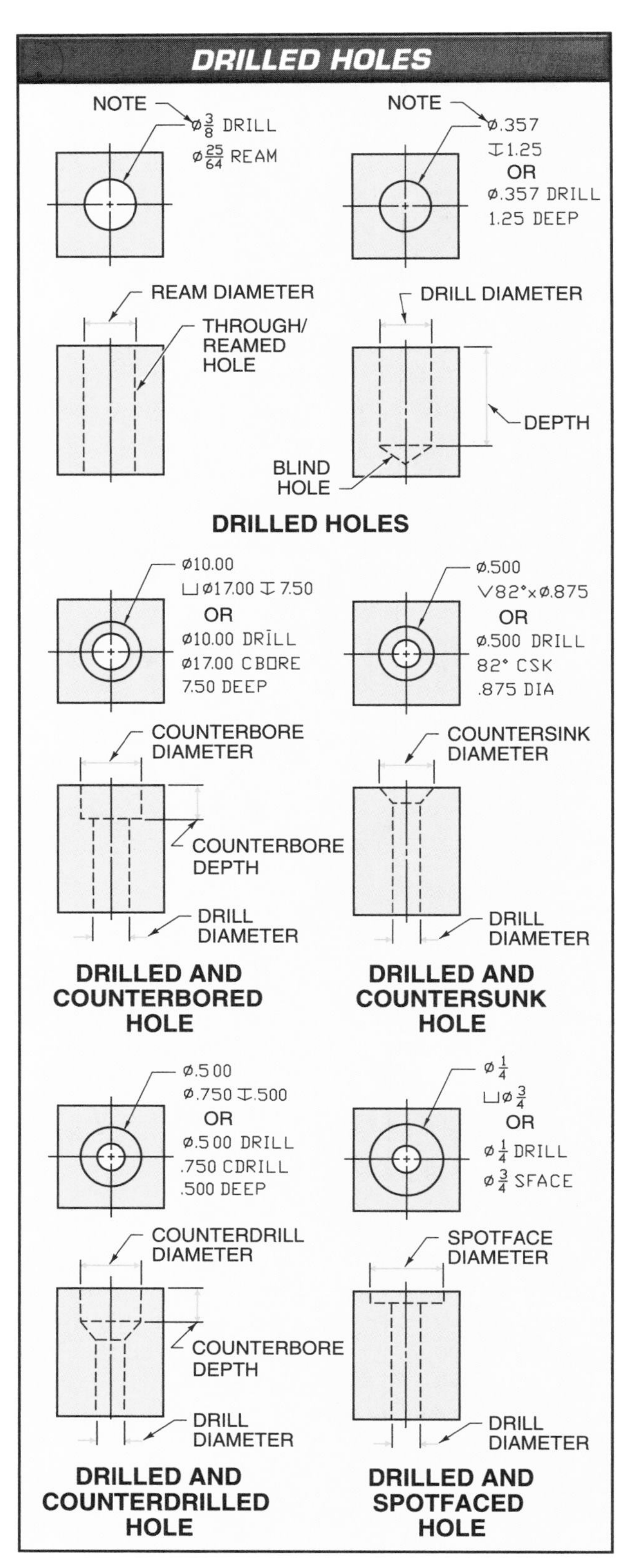

Figure 2-27. Hole sizes are specified by diameters, depths, and operations.

Drilled Holes. A *drilled hole* is a round hole in a material produced by a twist drill. A hole appears as a circle in one view and as parallel hidden lines in the adjacent view. The diameter of the hole is dimensioned in the circular view by one of several methods, such as a note containing the dimension, extension lines from the diameter that contain the dimension, or a leader from a note that contains the dimension.

The centerpoint is used on all circular views of holes. The centerline is used on all side views of holes. The centerpoint and centerline represent the axis of the concentric surfaces.

The depth of a hole is measured from the outer surface to the bottom of the cylindrical hole. The cone shape at the bottom that is formed by the head of the drill is not included in the depth measurement. The depth dimension may be included in the drill note, if not already otherwise dimensioned. If it is not clear that a hole passes completely through the material, the dimension for the hole may be followed by the abbreviation THRU. Drilled holes may also be counterbored, countersunk, counterdrilled, or spotfaced.

Counterbored Holes. A *counterbored hole* is an enlarged and recessed hole with square shoulders. Counterbored holes permit screw heads or other mating parts to be recessed below the surface of the part.

The diameter of the smaller hole is given first. The diameter of the larger (counterbored) hole is given next, followed by the depth of the counterbore, if given. The symbol for a counterbore is ⌴ and is placed before the counterbore diameter. Older drawings may use the abbreviation CBORE on notes instead.

Countersunk Holes. A *countersunk hole* is a hole with a cone-shaped recess at the outer surface. Countersunk holes permit the flush seating of parts with angled or tapered heads. For example, a flat head screw can be positioned in a countersunk hole so that the head is flush with the surface. A *countersink* is the tool that produces a countersunk hole. Countersinks are commercially available with standard included angles of 60° or 82°.

Countersunk holes are dimensioned by giving their drill diameter first. The diameter of the countersunk portion is given next, followed by the angle of the countersunk portion. The counterbore symbol ⌴ may be used to indicate the countersink diameter. The symbol for a countersink angle is ⌵ and is placed before the countersink angle. Older drawings may use the abbreviation CSK instead.

Counterdrilled Holes. A *counterdrilled hole* is a hole with a cone-shaped opening below the outer surface. Counterdrilled holes permit the recessed seating of parts with angled or tapered heads. For example, a flat head screw can be positioned in a counterdrilled hole so that the head is recessed below the surface.

Counterdrilled holes are dimensioned by giving the drill diameter first. The diameter of the counterdrilled portion is given next, followed by the depth of the counterdrilled portion. The included angle of the counterdrill is optional. These dimensions can all be specified with the drilling symbols, or with the older usage of the abbreviation CDRILL.

Spotfaces. A *spotface* is a flat surface machined at a right angle to a drilled hole. Spotfaces may be recessed below the surrounding surfaces or may be slightly above the surrounding surfaces. They provide an area for tight fits of square-shouldered parts.

Spotfaces are dimensioned by giving their drill diameter first. The diameter of the spotface is given next. The depth of the spotface or the remaining thickness of the material is optional. The counterbore ⌴ and depth ↧ symbols are also used when dimensioning spotfaces. The abbreviations SF or SFACE are commonly used on older drawings.

Edges and Corners

An *edge* is the intersection of two surfaces. A *corner* is the angular space at the intersection of surfaces. Dimensions are required to note the sizes of the surfaces and geometric characteristics of the edges and corners. Fillets, rounds, and runouts are examples of corners. Bevels and chamfers are examples of edges.

Some surface features may be very small or intricate in proportion to the whole part, and therefore may be difficult to show on a regular orthographic view. In these cases, special close-up detail views may be used to show complex features. Detail views are associated with orthographic views using sequential capital letters, beginning with A. The area on a view to be detailed is circled and identified with a letter. The detail view of the circular area is drawn elsewhere on the print, and its title identifies the detail area and scale, such as "DETAIL A, SCALE 2:1."

Fillets and Rounds. A *fillet* is a rounded interior corner. A *round* is a rounded exterior corner. Both fillets and rounds are used to avoid sharp corners on objects. **See Figure 2-28.** Fillets and rounds are both defined by the radius of their curve. Notes with leader lines pointing to rounded features include the letter R and radius value. (Three-dimensional rounded features, such as sections of a sphere shape, may use the symbols SR or S∅, for spherical radius and spherical diameter respectively.) When a drawing contains several fillets and/or rounds of the same radius, a general note on the print may indicate that all fillets and rounds, unless otherwise specified, are of a particular radius.

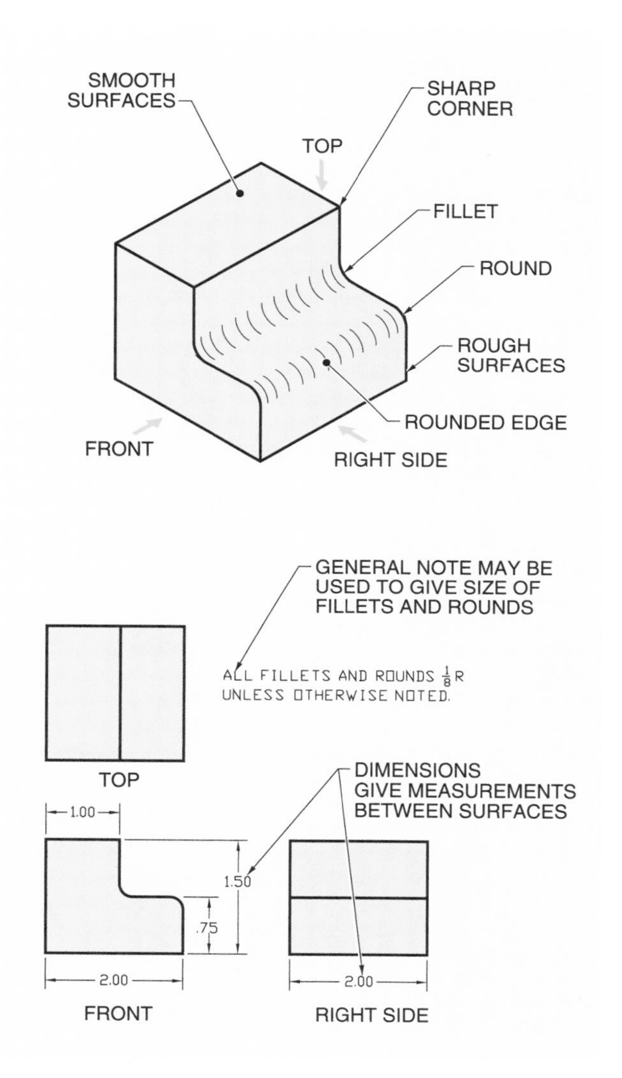

Figure 2-28. Fillets and rounds are used to avoid sharp corners on objects.

Runouts. A *runout* is the curve produced by a plane surface tangent to a cylindrical surface. The radius of a runout is commonly equal to that of a fillet and is dimensioned in the same way. The direction of the runout is determined by the intersecting surfaces that form the runout. **See Figure 2-29.**

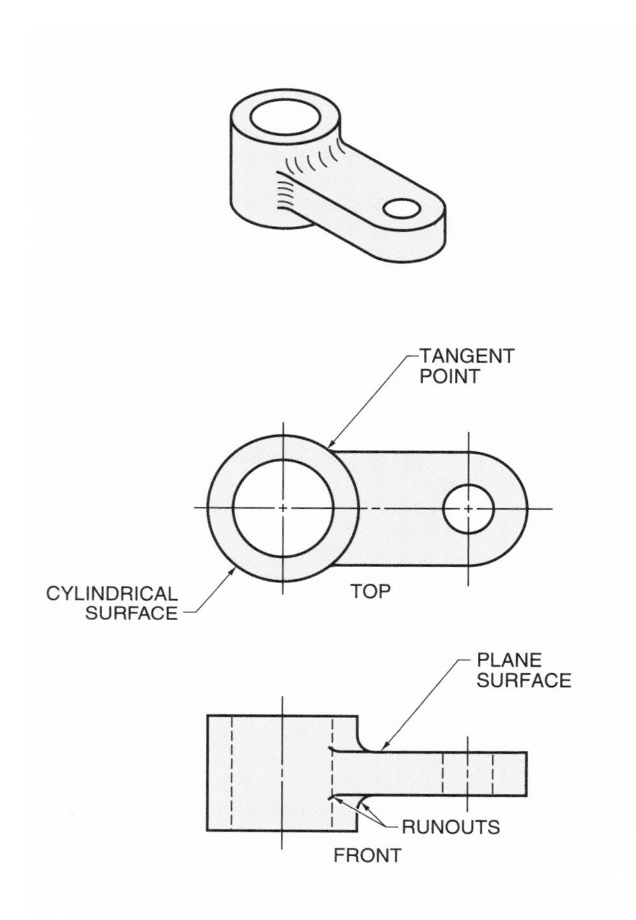

Figure 2-29. Runouts are curves produced by a plane surface tangent to a cylindrical surface.

Bevels. A *bevel* is a sloped surface. Neither intersection with the surfaces of the object is 90°. A bevel is commonly dimensioned by an angle and a linear dimension or by two linear dimensions. **See Figure 2-30.**

Chamfers. A *chamfer* is a beveled edge. Chamfers are used to reduce sharp corners and are often used to aid assembly of cylindrical parts to be placed into holes. They are commonly dimensioned by an angle and a linear dimension or by two linear dimensions.

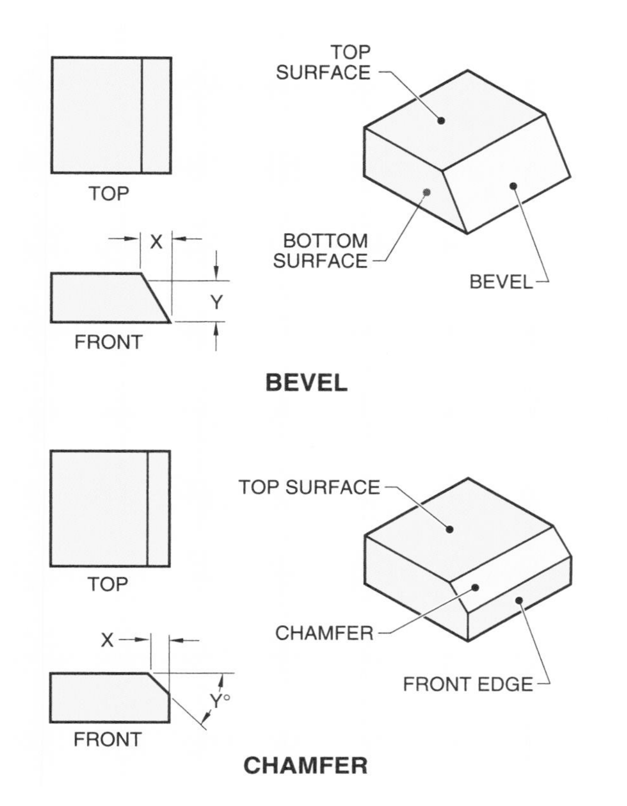

Figure 2-30. A bevel extends from surface to surface. A chamfer extends from surface to edge.

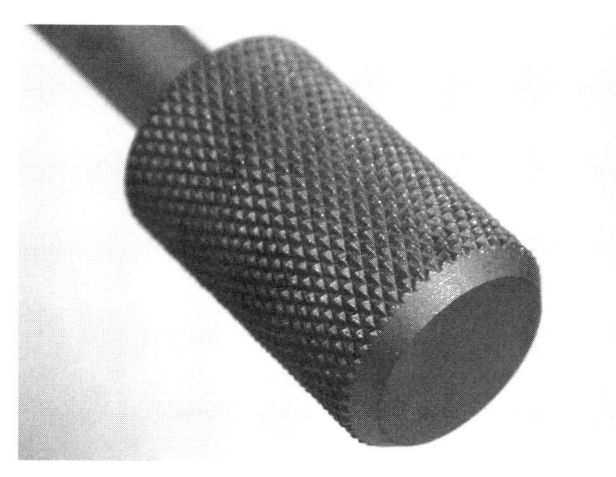

Since machined edges can be sharp, chamfers are often cut onto the edges of handles and knobs.

Sharp machined edges are prone to chipping, which can damage the adjacent surfaces on a part. If an edge is not required to be sharp, this kind of damage can be avoided by purposely chamfering the edge. Chamfers may also be added for parts assembly or aesthetic reasons.

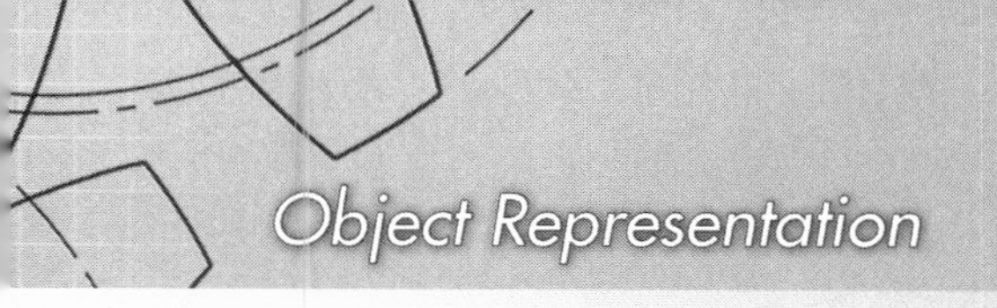

Name ______________________________ **Date** ____________________

Sketching — Isometrics

Sketch an isometric view of each part from its multiview.

①

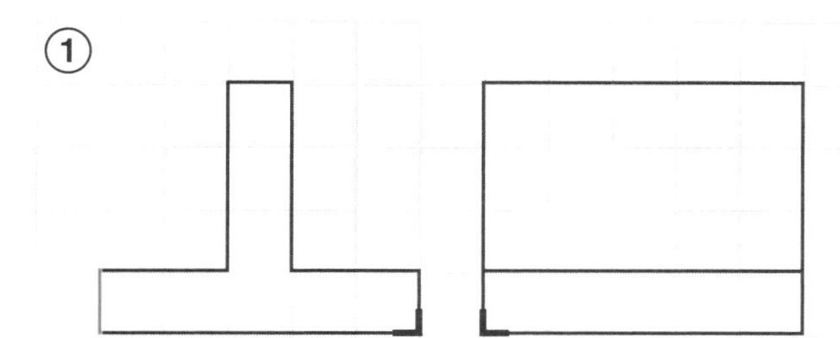

②

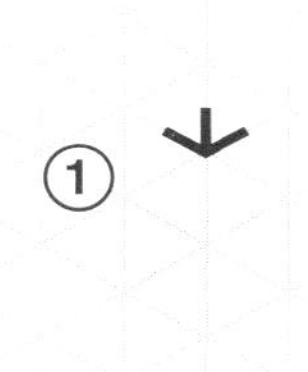

①

③

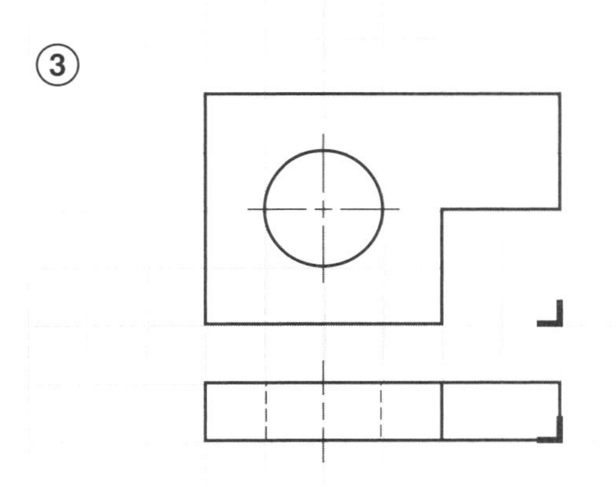

②

③

④

④

Sketching — Obliques

Sketch an oblique cabinet view of each part from its multiview.

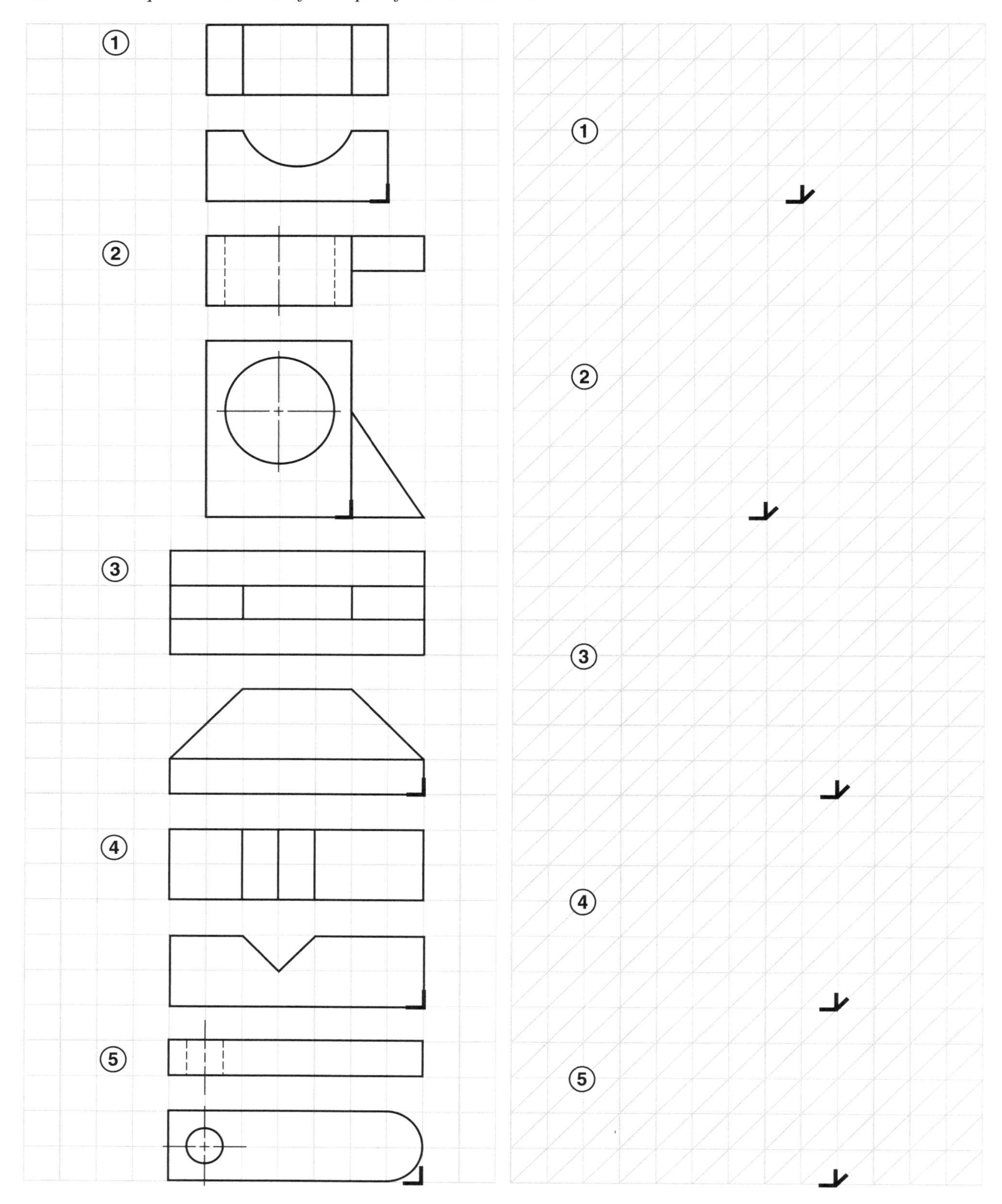

Sketching — Front and Right Side Views

Sketch the front and right side views of each part from its isometric view.

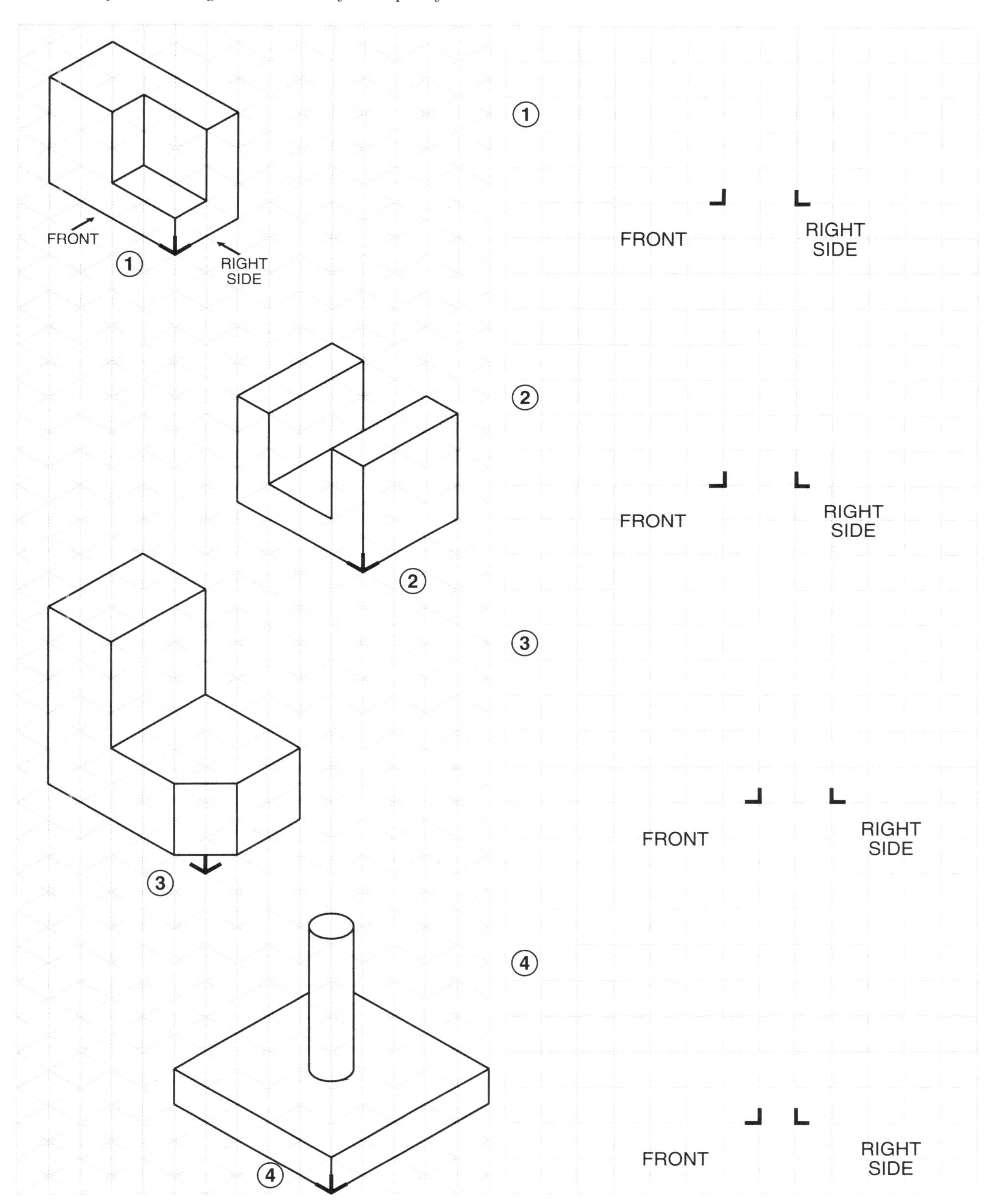

Sketching — Multiviews (Missing Line)

For each multiview, sketch the missing line(s) on the indicated view.

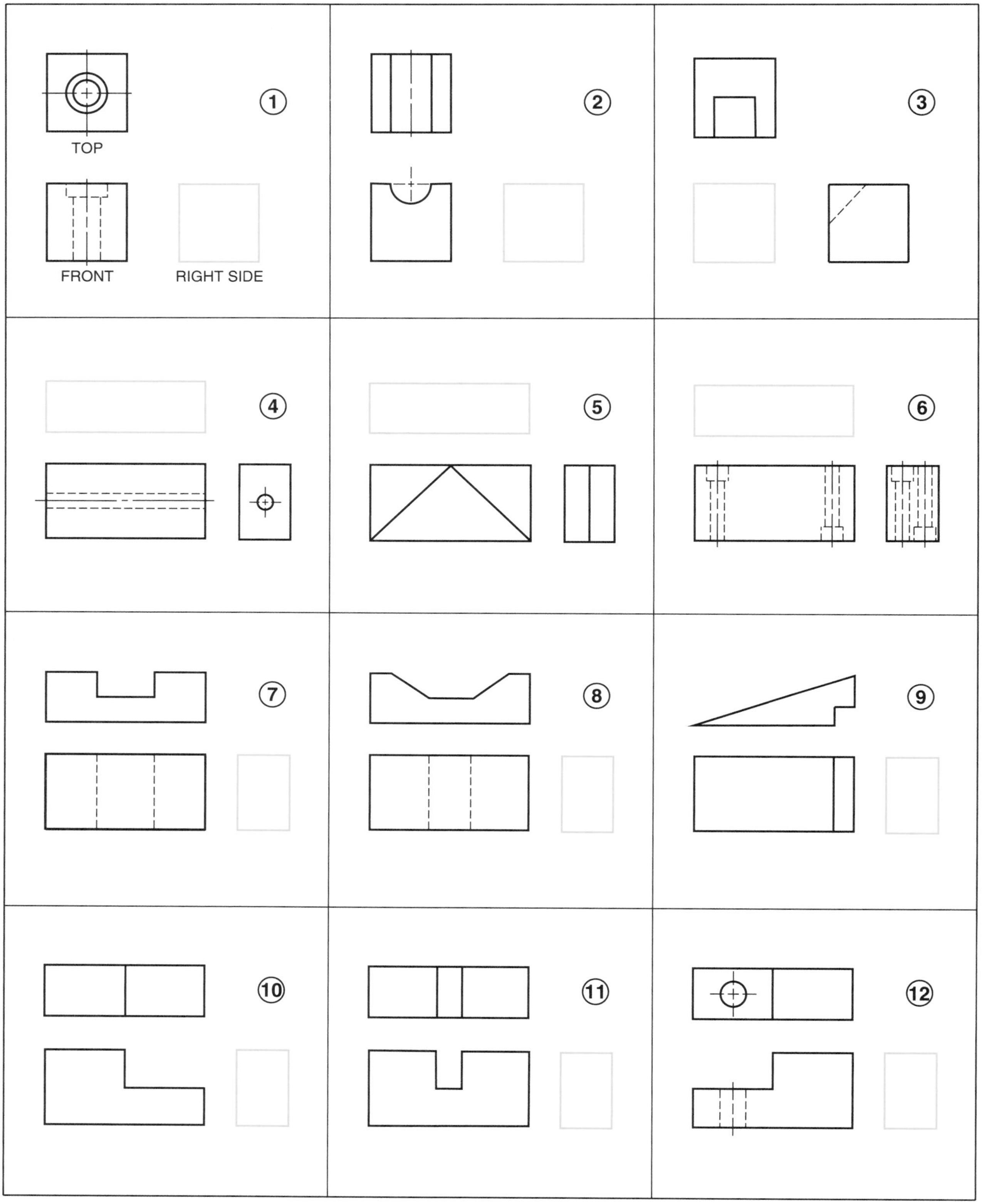

1 TOP FRONT RIGHT SIDE

2

3

4

5

6

7

8

9

10

11

12

Sketching — Multiview Drawings

Sketch the front, top, and right side views of each part from its isometric view. Each square equals ¼″. Allow ½″ between views.

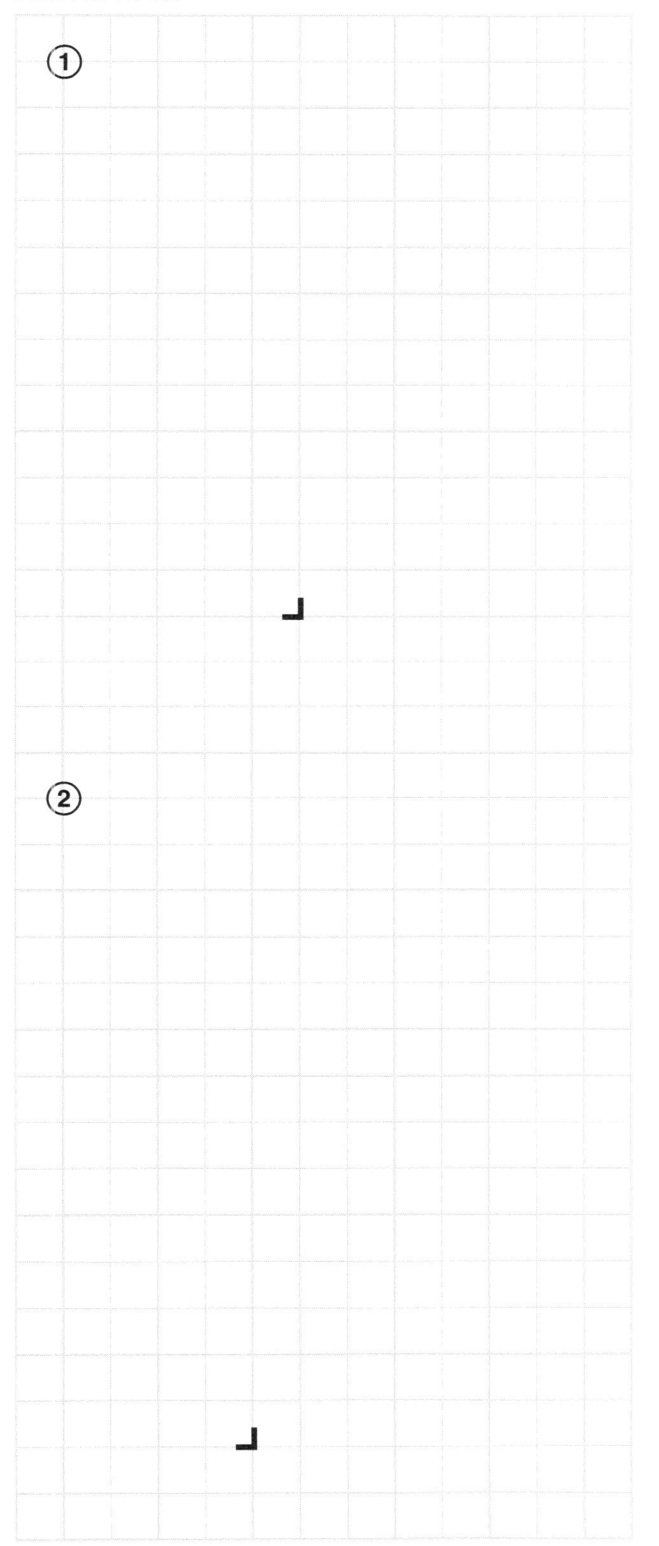

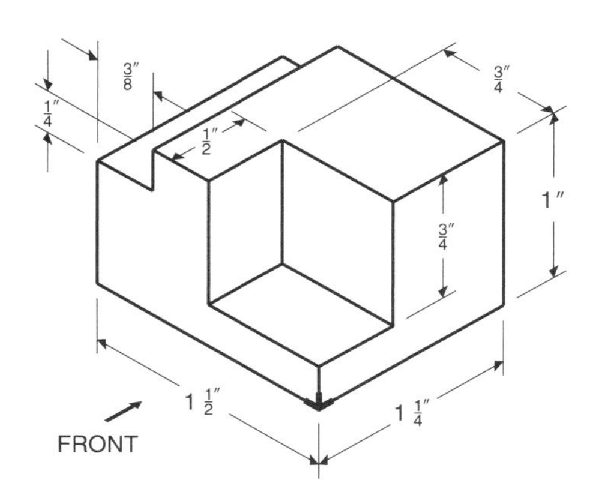

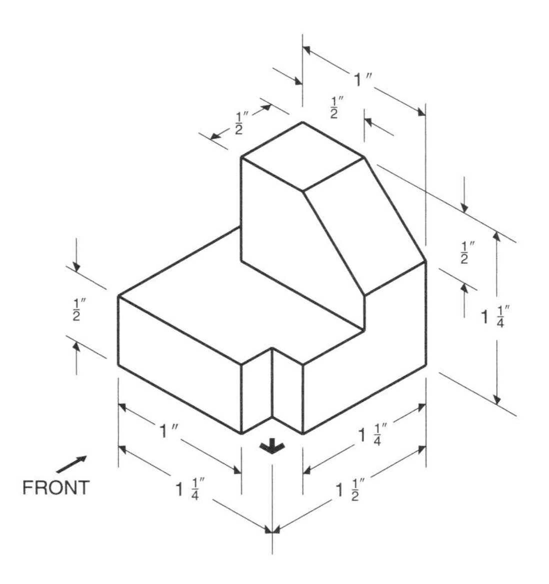

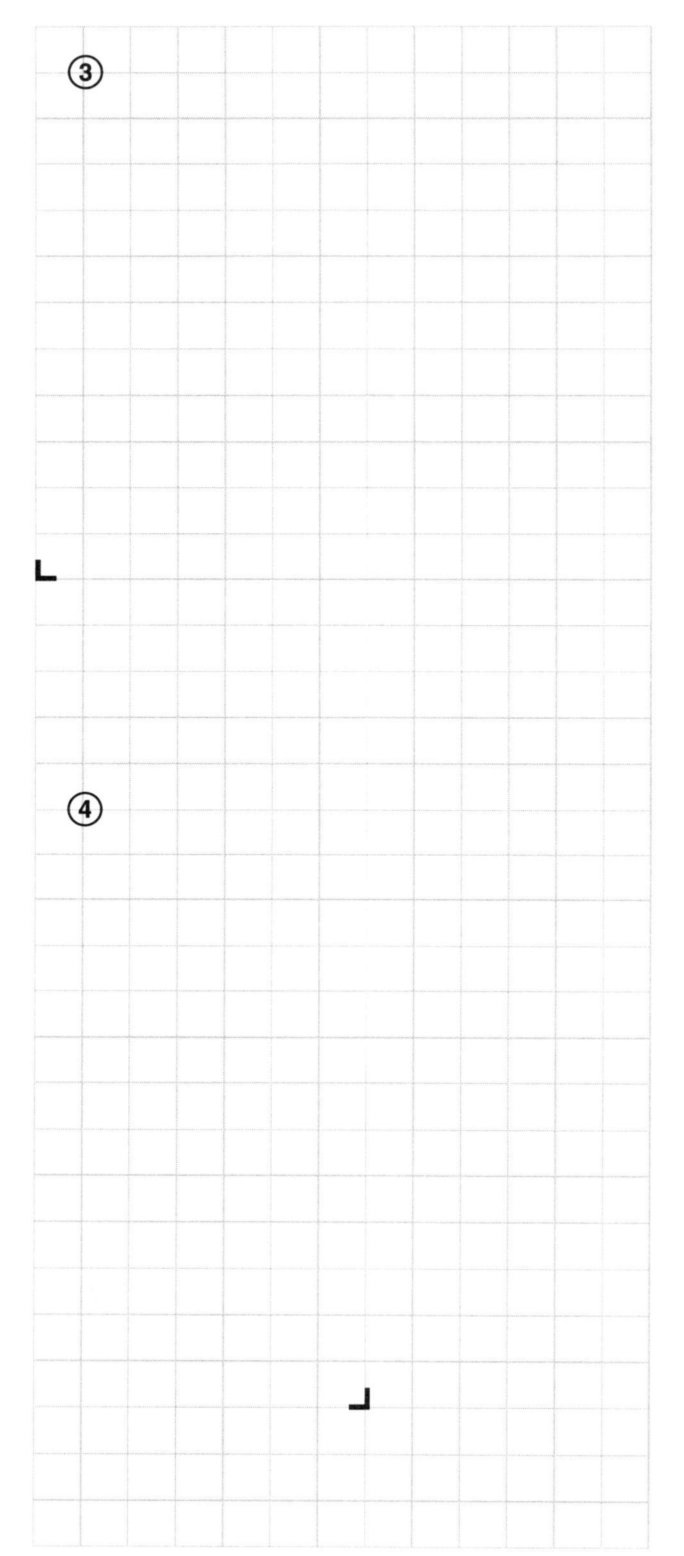
3
4

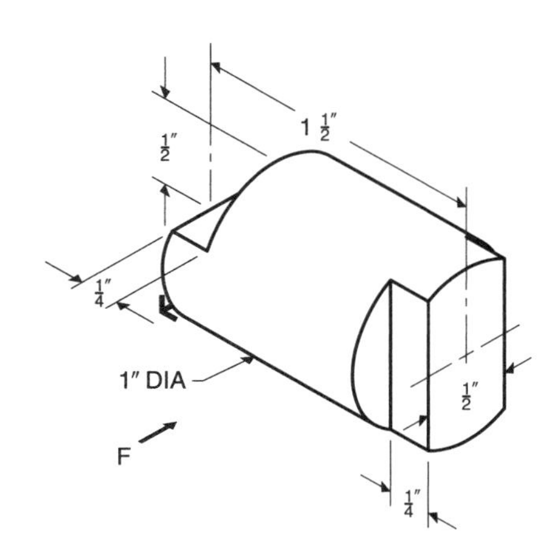
1 1/2"
1/2"
1/4"
1" DIA
F
1/2"
1/4"

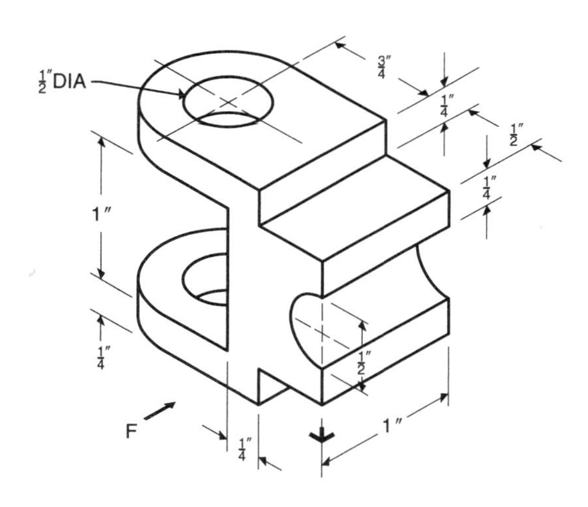
1/2" DIA
3/4"
1/4"
1/2"
1/4"
1"
1/4"
1/2"
F
1/4"
1"

REVIEW QUESTIONS 2

Name ______________________________ Date ____________________

True-False

T F **1.** Sketching pencils are either wooden or mechanical.

T F **2.** Paper may be preprinted with grids to facilitate sketching.

T F **3.** The pencil point should be pulled across the paper while sketching.

T F **4.** Shading techniques are used with orthographic drawings.

T F **5.** Perspective drawings are seldom used on machine trades prints.

T F **6.** An oblique drawing shows one surface of an object as a true view.

T F **7.** Visible lines are thin and dark.

T F **8.** Break lines are used to condense drawings.

T F **9.** Hole sizes are specified by their diameters.

T F **10.** Spotfaces are always ¼″ or deeper.

T F **11.** A circle on an isometric drawing appears as an ellipse.

T F **12.** Multiview drawings show every feature of an object in true view.

T F **13.** A cavalier drawing is a type of isometric drawing.

T F **14.** Multiview drawings use orthographic projection to form the views.

T F **15.** There are eight principal planes of projection.

Completion

______________ **1.** An axonometric drawing shows ___ sides of an object.

______________ **2.** Isometric drawings contain three equal axes that are drawn ___° apart.

______________ **3.** A(n) ___ is a plane curve with two focal points.

______________ **4.** A(n) ___ drawing has receding lines drawn to one-half the scale of lines in the true view.

______________ **5.** ___ projection shows the true shape of one view of an object onto a single plane.

______________ **6.** Edges that are not visible from a point of view are drawn as ___ lines.

______________ **7.** ___ lines represent an edge or contour that can be seen from the view of an object.

_____ **8.** ___ lines are terminated by arrowheads on both ends.

_____ **9.** A(n) ___ surface is a plane surface parallel to a plane of projection.

_____ **10.** A(n) ___ hole is a drilled hole that does not completely pass through the material.

_____ **11.** A(n) ___ hole is an enlarged and recessed hole with square shoulders.

_____ **12.** A(n) ___ is a rounded interior corner.

_____ **13.** A(n) ___ is a rounded exterior corner.

_____ **14.** A(n) ___ is a thin dashed line that locates axes or centerpoints of arcs and circles.

_____ **15.** ___ are used to condense drawings with long, uniform sections.

Identification — Alphabet of Lines

_____ **1.** Visible line

_____ **2.** Cutting plane line

_____ **3.** Break line

_____ **4.** Phantom line

_____ **5.** Hidden line

_____ **6.** Centerline

_____ **7.** Dimension line

_____ **8.** Section line

_____ **9.** Extension line

_____ **10.** Arrowhead

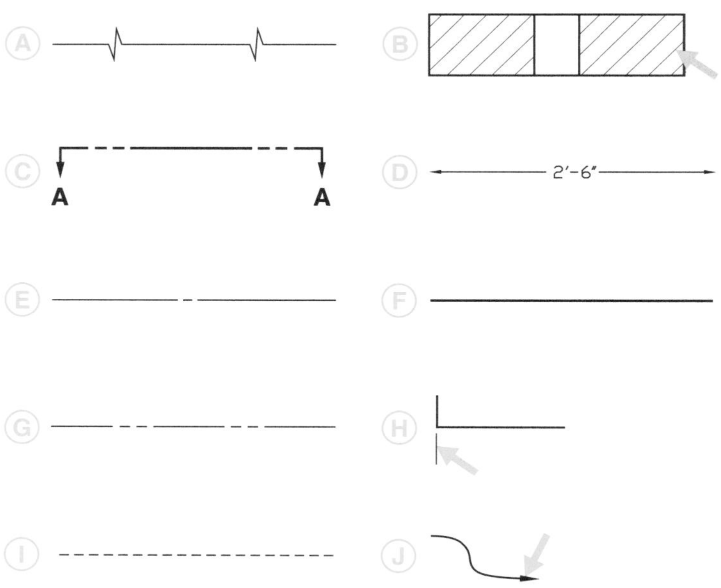

A

B

C

A A

D

2′-6″

E

F

G

H

I

J

Identification — Orthographic Projection

_____ **1.** Front view: height

_____ **2.** Front view: width

_____ **3.** Top view: depth

_____ **4.** Right side view: depth

_____ **5.** Top view: width

_____ **6.** Right side view: height

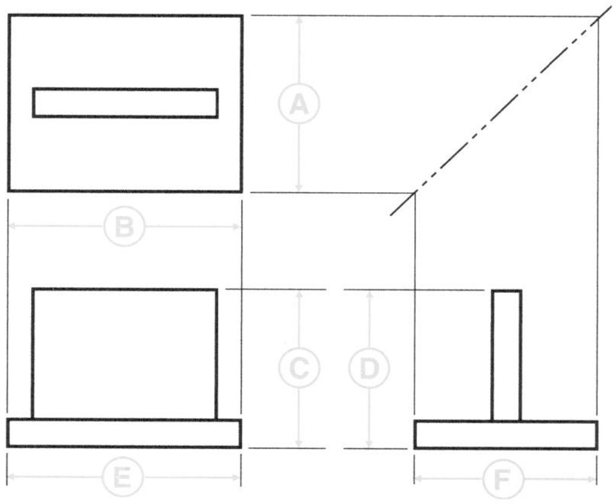

A

B

C

D

E

F

Multiple Choice

_______________ **1.** A true view is a view in which the ___.
A. line of sight is parallel to the surface
B. surface can be completely seen
C. line of sight is perpendicular to the surface
D. none of the above

_______________ **2.** Arrowheads on the ends of the cutting plane line indicate the ___.
A. size of the section view
B. direction from the multiview to the section view
C. direction of view
D. all of the above

_______________ **3.** A spotface is ___.
A. a flat surface
B. either slightly above or below the surrounding surface
C. sometimes abbreviated by the letters SF
D. all of the above

_______________ **4.** A ___ is the curve produced by a plane surface tangent to a cylindrical surface.
A. fillet
B. bevel
C. chamfer
D. runout

_______________ **5.** ___ surfaces meet at an edge.
A. Inclined
B. Intersecting
C. Oblique
D. Normal

_______________ **6.** Break lines may be drawn as ___.
A. thin lines with zigzags
B. thick lines with arrowheads
C. thin, dashed lines
D. thin, dotted lines

_______________ **7.** First-angle projection can be identified by the ___ view being in the upper right-hand corner of the layout.
A. right-side
B. left-side
C. top
D. front

_______________ **8.** Isometric drawings contain three equal axes that are drawn ___° apart.
A. 45
B. 90
C. 120
D. none of the above

_______________ **9.** ___ holes have a cone-shaped recess at the outer surface.

A. Countersunk
B. Counterbored
C. Drilled
D. Counterdrilled

_______________ **10.** Uppercase letters near the ends of a cutting plane line ___.

A. indicate the viewing direction
B. indicate the order of operation
C. identify the view
D. none of the above

Identification – Holes

_______________ **1.** Drilled hole

_______________ **2.** Drilled and counterbored hole

_______________ **3.** Drilled and countersunk hole

_______________ **4.** Drilled and counterdrilled hole

_______________ **5.** Drilled and spotfaced hole

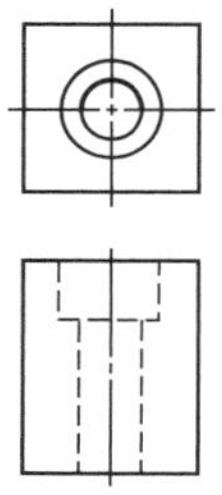

A

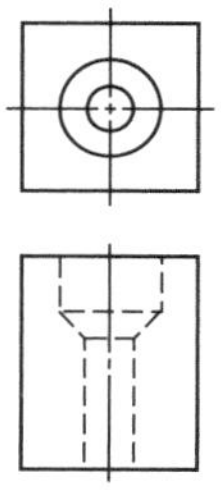

B

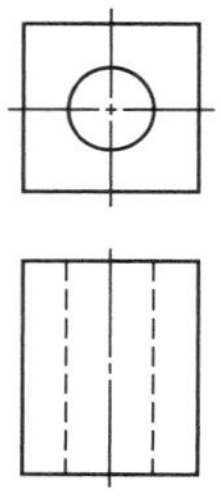

C

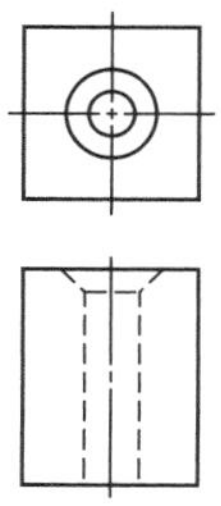

D

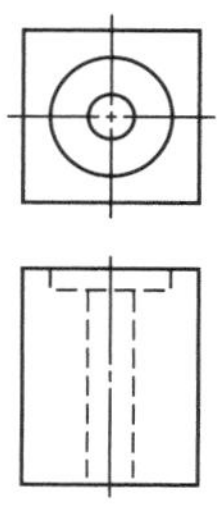

E

TRADE COMPETENCY TEST 2

Name ______________________ Date ______________

Flat Wrench

Refer to print on page 56.

________ **1.** The Flat Wrench is designed for use with a(n) ___ head bolt.

________ **2.** The overall length of the Flat Wrench is ___″.

T F **3.** Both ends of the Flat Wrench are squared off.

T F **4.** Two orthographic views of the Flat Wrench are shown.

________ **5.** The drawing was drawn by ___.

T F **6.** Sharp corners are to be ground.

T F **7.** The Flat Wrench will not have a coating.

T F **8.** The offset in the Flat Wrench is specified with an angular dimension.

________ **9.** Line A is a(n) ___ line.

________ **10.** Line B is a(n) ___ line.

________ **11.** Line C is a(n) ___ line.

________ **12.** Line D is a(n) ___ line.

T F **13.** Decimal and fractional dimensions are used to dimension the Flat Wrench.

________ **14.** The overall width of the Flat Wrench is ___″.

________ **15.** The company logo is to be stamped onto the Flat Wrench per spec ___.

________ **16.** The hexagon is ___″ across the flats.

________ **17.** A total of ___ gusset(s) are specified.

________ **18.** The maximum thickness of the Flat Wrench is ___″.

T F **19.** The Flat Wrench is heat-treated.

T F **20.** The maximum allowable offset angle is 67°.

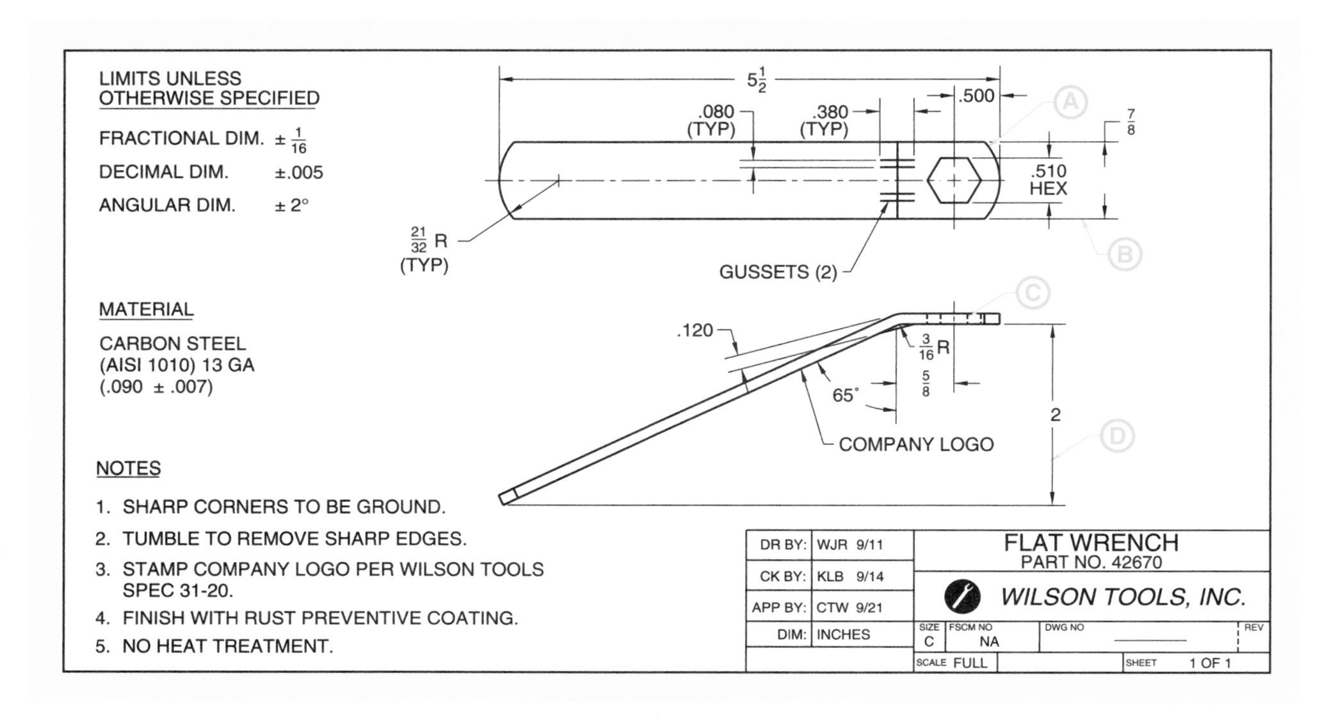
LIMITS UNLESS OTHERWISE SPECIFIED
FRACTIONAL DIM. ± 1/16
DECIMAL DIM. ±.005
ANGULAR DIM. ± 2°
MATERIAL
CARBON STEEL
(AISI 1010) 13 GA
(.090 ± .007)
NOTES
1. SHARP CORNERS TO BE GROUND.
2. TUMBLE TO REMOVE SHARP EDGES.
3. STAMP COMPANY LOGO PER WILSON TOOLS SPEC 31-20.
4. FINISH WITH RUST PREVENTIVE COATING.
5. NO HEAT TREATMENT.
5 1/2
.080 (TYP)
.380 (TYP)
.500
7/8
.510 HEX
21/32 R (TYP)
GUSSETS (2)
A
B
C
D
.120
3/16 R
5/8
65°
2
COMPANY LOGO
DR BY: WJR 9/11
CK BY: KLB 9/14
APP BY: CTW 9/21
DIM: INCHES
FLAT WRENCH
PART NO. 42670
WILSON TOOLS, INC.
SIZE C
FSCM NO NA
DWG NO
REV
SCALE FULL
SHEET 1 OF 1

FLAT WRENCH

Left Vertical Bracket

Refer to print below.

_______________ **1.** The diameter of the hole at A is ___″.

_______________ **2.** The radius dimension at B is ___″.

_______________ **3.** The dimension at C is ___″.

_______________ **4.** The edge-to-edge dimension at D is ___″.

_______________ **5.** The center-to-center dimension at E is ___″.

_______________ **6.** The center-to-center dimension at F is ___″.

_______________ **7.** The edge-to-center dimension at G is ___″.

_______________ **8.** The radius dimension at H is ___″.

_______________ **9.** The edge-to-center dimension at I is ___″.

_______________ **10.** The center-to-center dimension at J is ___″.

_______________ **11.** The general linear tolerance is ±___″.

T F **12.** The orthographic views shown include the top, front, and right side.

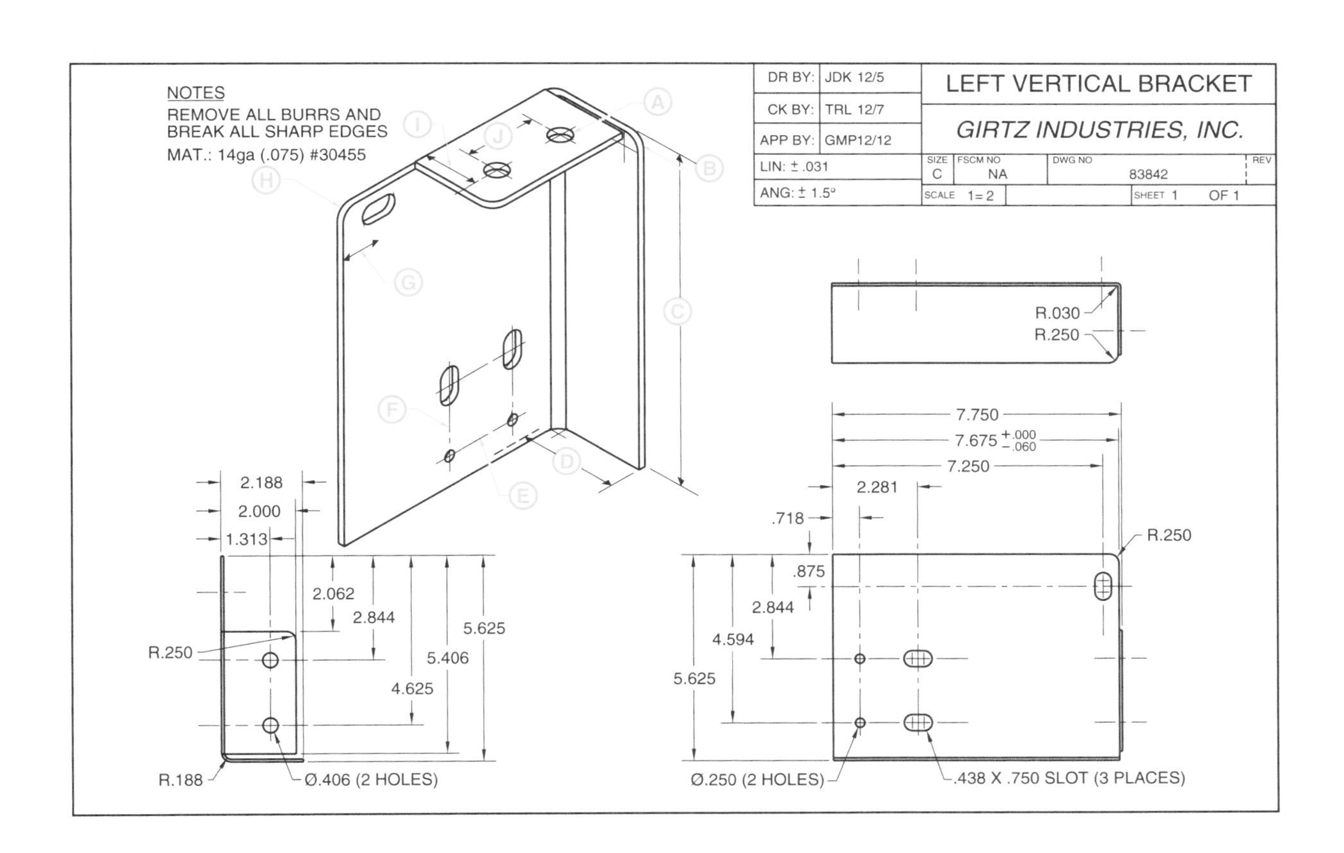

Sectional and Auxiliary Views

3

Sectional views show the internal shapes of objects. The types of sectional views include full section, half section, offset section, broken-out section, revolved section, removed section, auxiliary section, and thin section. Auxiliary views show the shape of surfaces that are not parallel to one of the three principal planes. The two types of auxiliary views are primary and secondary auxiliary views.

SECTIONAL VIEWS

For objects that are comparatively simple in design, the need to show complete construction details can be met by orthographic representation. Many objects, however, have internal shapes that are so complicated that it is virtually impossible to show their true shape without using numerous hidden lines. A sectional view reveals the actual internal shape of an object and retains the significant outline of the external contour. **See Figure 3-1.**

A *sectional view* is the view of a cross section of an object. It is obtained by passing an imaginary cutting plane through the object, and the cut part is removed. **See Figure 3-2.** The types of sectional views include full section, half section, offset section, broken-out section, revolved section, removed section, auxiliary section, and thin section. Sectional view practices are based on the conventions in ANSI/ASME Y14.3, *Multiview and Sectional View Drawings*.

Cutting Plane Lines

The cutting plane is shown on the orthographic view as a cutting plane line. There are three standard styles for cutting plane lines, all using thick lines. **See Figure 3-3.** The arrowheads point in the direction of sight in which the object is viewed when the sectional view is made. Cutting plane lines are based on the conventions in ANSI/ASME Y14.2, *Line Conventions and Lettering*.

Carrier Corporation

Figure 3-1. Pictorial sectional views help the viewer understand the internal parts of an assembly.

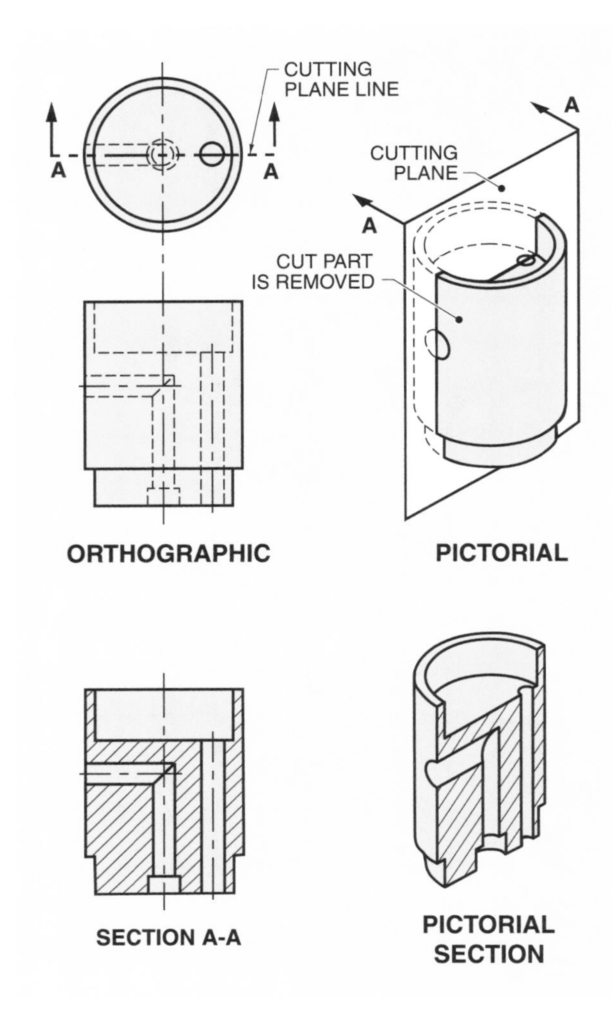

Figure 3-2. A sectional view is obtained by passing a cutting plane through the object.

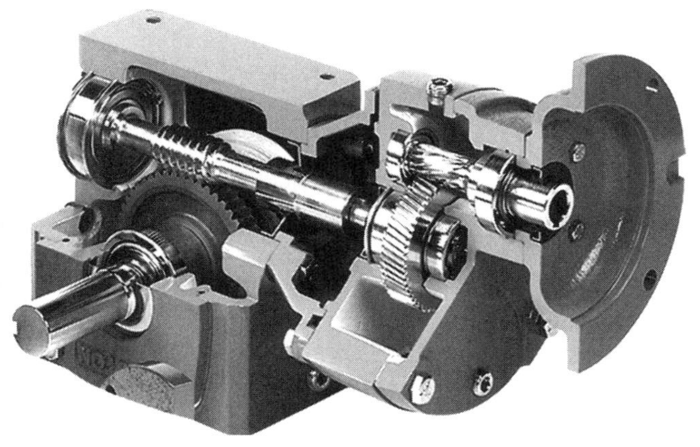

Boston Gear

Cutaway assemblies show how parts fit and work together.

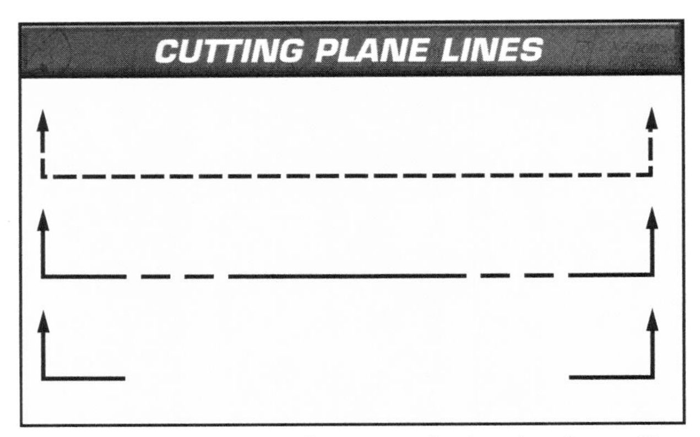

Figure 3-3. There are three standard styles for cutting plane lines.

Uppercase letters such as A-A, B-B, C-C, and so on are used to identify the section. **See Figure 3-4.** The letters should be located adjacent to the arrowheads and read horizontally, and should not be underlined. A notation is also placed under the view, such as SECTION A-A.

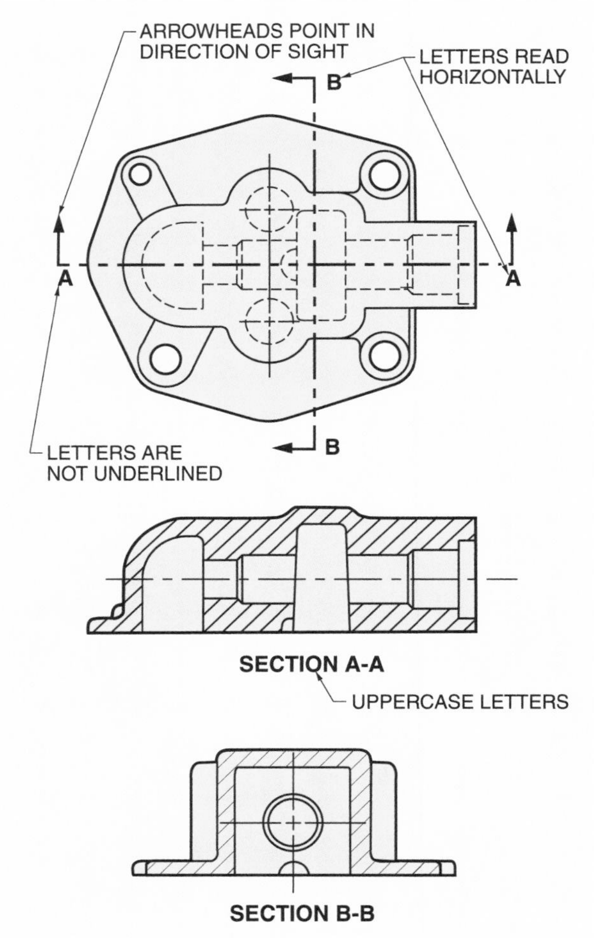

Figure 3-4. Uppercase letters are used to identify sections.

If two or more sections appear on the same sheet, they should be arranged in alphabetical order from left to right, if possible. Section letters should be used in alphabetical order, omitting the letters I, O, and Q. This avoids confusing I with the number 1, and O or Q with zero. If more than 23 sections are used, the additional sections should be indicated by double letters in alphabetical order: AA-AA, BB-BB, and so on.

On objects with major centerlines, the cutting plane can be assumed to pass through the axis of symmetry. **See Figure 3-5.** The cutting plane line may then be omitted since it is already clear that the section is taken along that centerline.

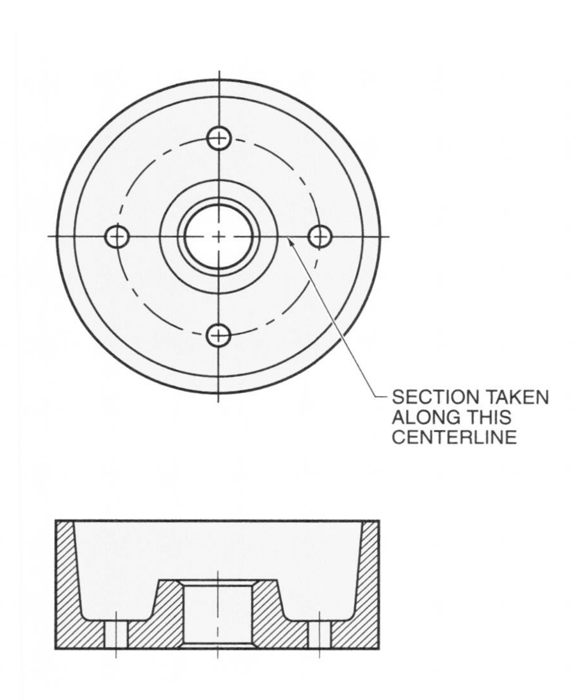

Figure 3-5. The cutting plane may be omitted if a section is taken along a major centerline.

Full Sections

A *full section* is a sectional view in which the cutting plane passes entirely across the object. **See Figure 3-6.** The portion of the object between the observer and the cutting plane is considered removed and the remaining portion is exposed to view. Full sections are commonly used.

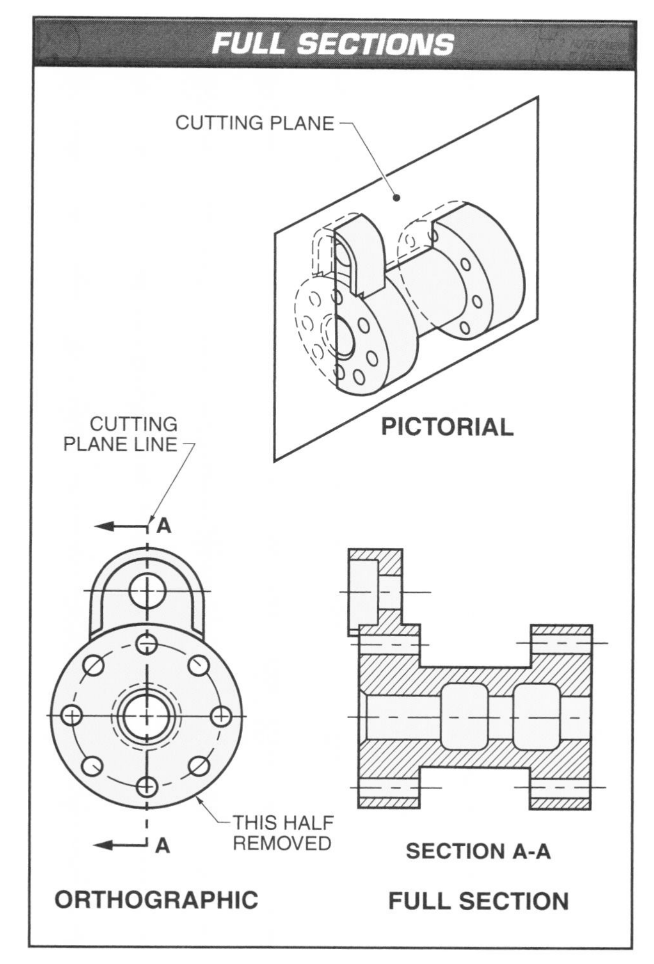

Figure 3-6. A full section is a sectional view in which the cutting plane passes entirely across the object.

Half Sections

A *half section* is a sectional view in which two cutting planes are passed at right angles to each other along centerlines or symmetrical axes. **See Figure 3-7.** Passage of the cutting planes permits the removal of one-quarter of the object, and a half section of the interior is exposed to view.

A half section has the advantage of showing the interior of the object and at the same time maintaining the shape of the exterior. It is used with symmetrical objects only. A *symmetrical object* is an object in which one half is the mirror image of the other half. An *asymmetrical object* is an object that cannot be divided in such a way that one half is the mirror image of the other half. Half sections are not used with asymmetrical objects.

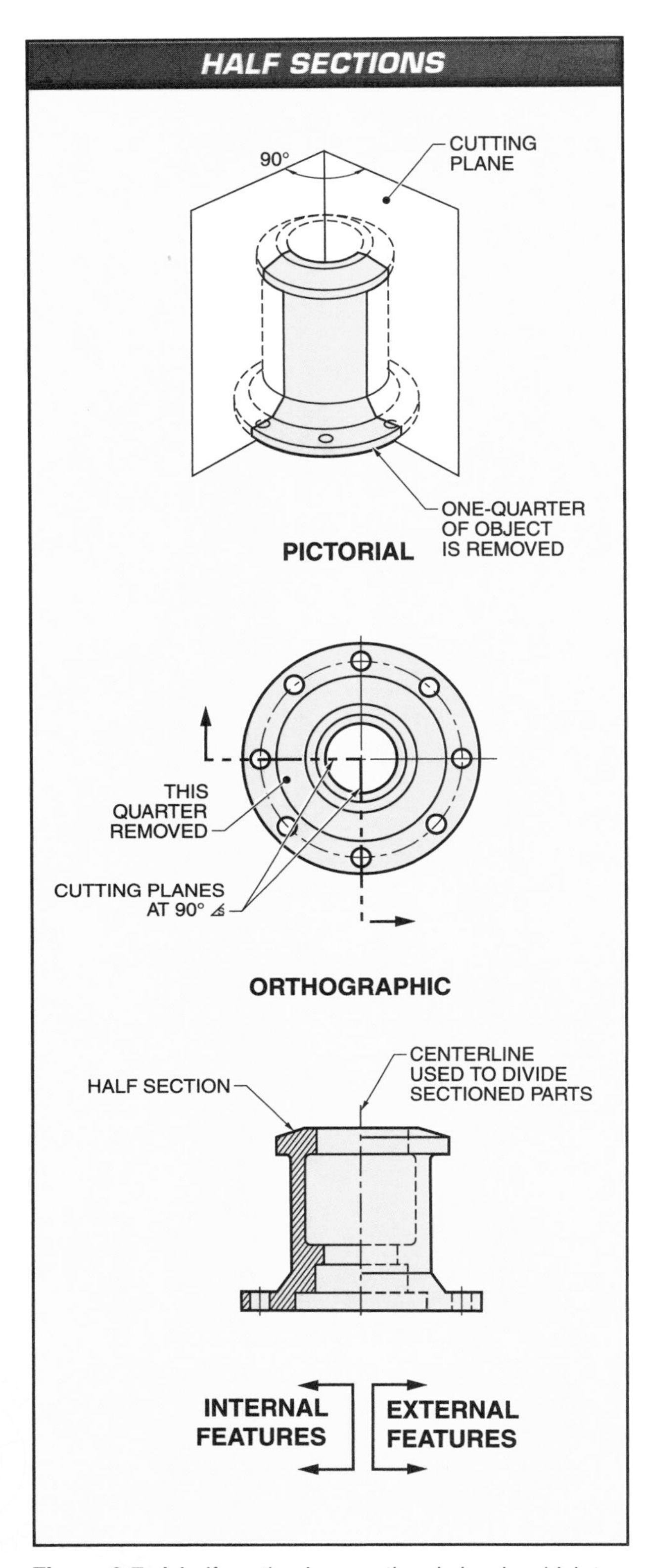

Figure 3-7. A half section is a sectional view in which two cutting planes are passed at right angles to each other along centerlines.

Because it is often difficult to completely dimension the internal shape of a half section, this type of section view is not widely used in detail drawings. It is used in assembly drawings where it is necessary to show both internal and external construction on the same view.

Offset Sections

An *offset section* is a sectional view in which the cutting plane line changes direction in order to include features that are not located in a straight line. The cutting plane is offset to pass through these features. **See Figure 3-8.**

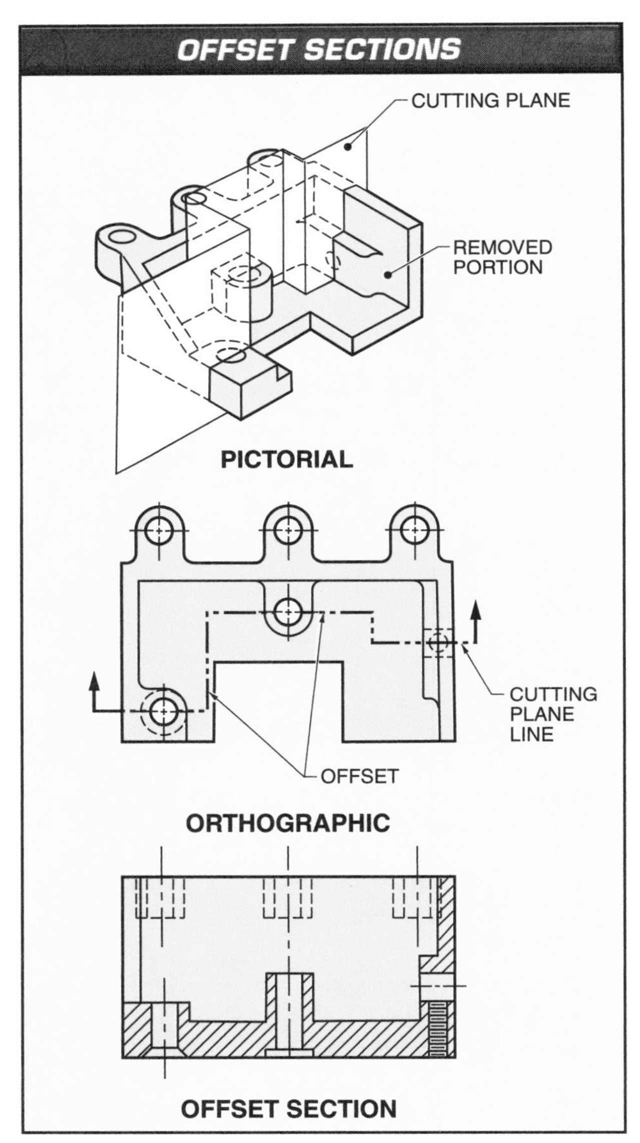

Figure 3-8. An offset section is a sectional view in which the cutting plane line changes direction to pass through features not located in a straight line.

By offsetting the cutting plane in several places, the shape of the openings and recesses that normally would not be seen with a regular full section can be viewed. In making an offset section, the offsets are not shown in the sectional view but only in the view indicating the cutting plane line.

Broken-Out Sections

A *broken-out section* is a sectional view in which a small portion designated by a freehand break line is removed. **See Figure 3-9.** By removing only a small portion, it is possible to preserve detail of the object that otherwise would be eliminated in a full or half section. No cutting plane line is necessary to show a broken-out section.

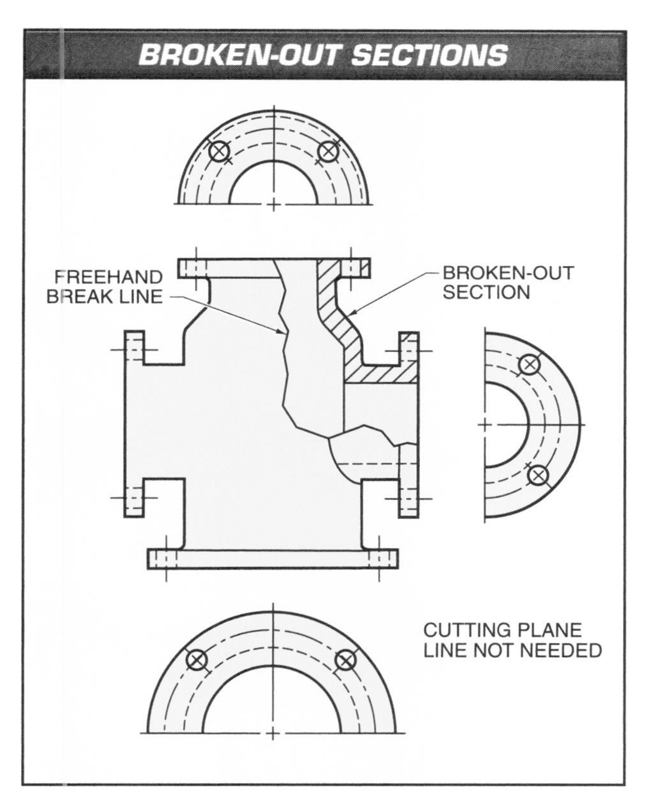

Figure 3-9. A broken-out section is a sectional view in which a small portion designated by a freehand break line is removed.

Revolved Sections

A *revolved section* is a sectional view in which a cross-sectional shape is shown at the cutting plane location. Typical parts commonly shown with revolved sections are elongated parts, such as bars, spokes, arms, ribs, and other long parts.

The cutting plane is passed perpendicular to the axis of the piece and then revolved in place 90° into the plane of the sheet. **See Figure 3-10.** The visible lines on each side of the adjacent view may be removed and broken lines used to leave the revolved section area clear.

The true shape of the exposed revolved section should always be retained regardless of the direction of the contour lines of the object. No cutting plane line is necessary to show a revolved section.

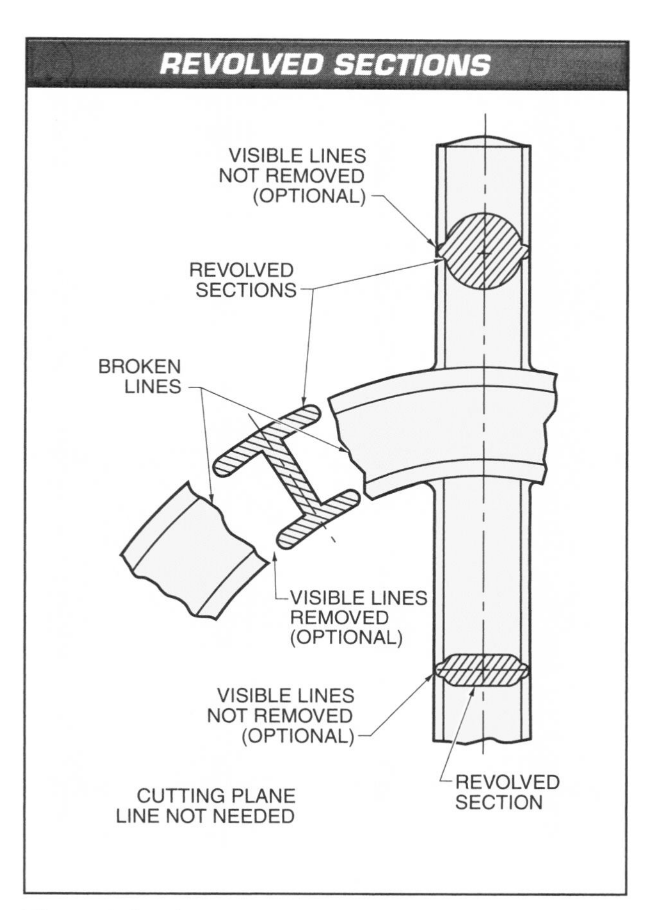

Figure 3-10. A revolved section is a sectional view that shows the cross-sectional shape of elongated parts.

Removed Sections

A *removed section* is a sectional view that is detached from the projected view and located elsewhere on the sheet. **See Figure 3-11.** By removing the section, the regular view can be left intact and the removed section drawn to a larger scale to facilitate more complete dimensioning.

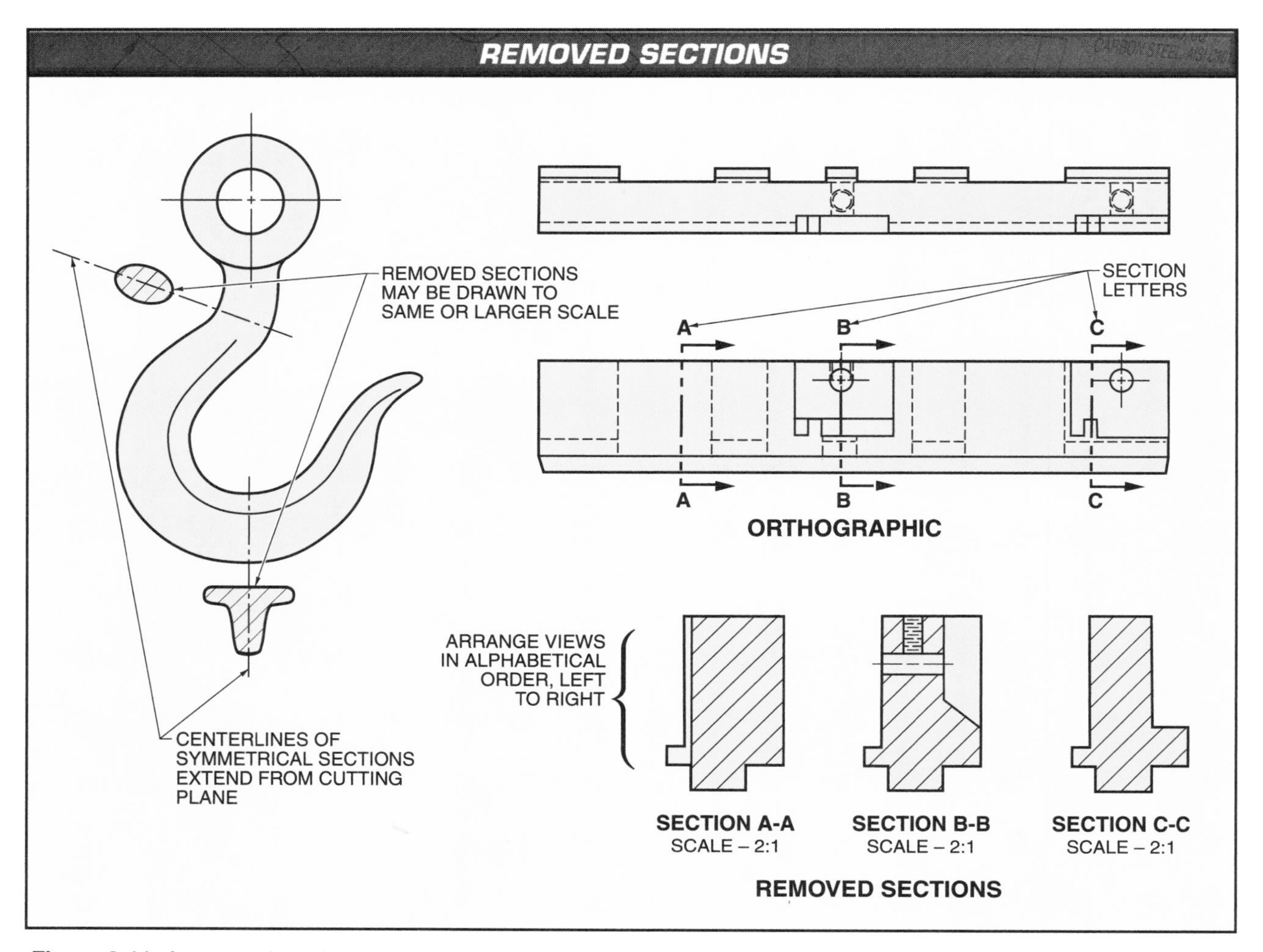

Figure 3-11. A removed section is a sectional view that is detached from the projected view.

A removed section should be labeled in uppercase lettering, such as SECTION B-B, to identify it, with the cutting plane line labeled with corresponding letters at its ends. The exception is when there is a small number of sections and they are clearly identifiable with a cutting plane by their location on the drawing.

A removed section should be placed in a convenient location, preferably on the same sheet with the regular view. On multiple-sheet drawings where it is not practical to place a removed section on the same sheet with the regular views, identification and zoning references should be indicated for related sheets. Below the section title, the sheet and zone numbers where the cutting plane line is located should be given as follows:

SECTION B-B
ON SHEET 4, ZONE A3

A similar note should be placed on the drawing where the cutting plane is shown, with a leader pointing to the cutting plane, referring to the sheet and zone where the section is located.

A removed section may be drawn to a larger scale, if necessary. In this case, the scale should be shown under the section title. Removed sections may be placed on centerlines extended from the section cuts.

Auxiliary Sections

An *auxiliary section* is a sectional view that is not one of the principal planes. **See Figure 3-12.** An auxiliary section may be full, half, broken-out, removed, or revolved. The section should be shown in its normal auxiliary position and clearly identified with a cutting plane and appropriate letters.

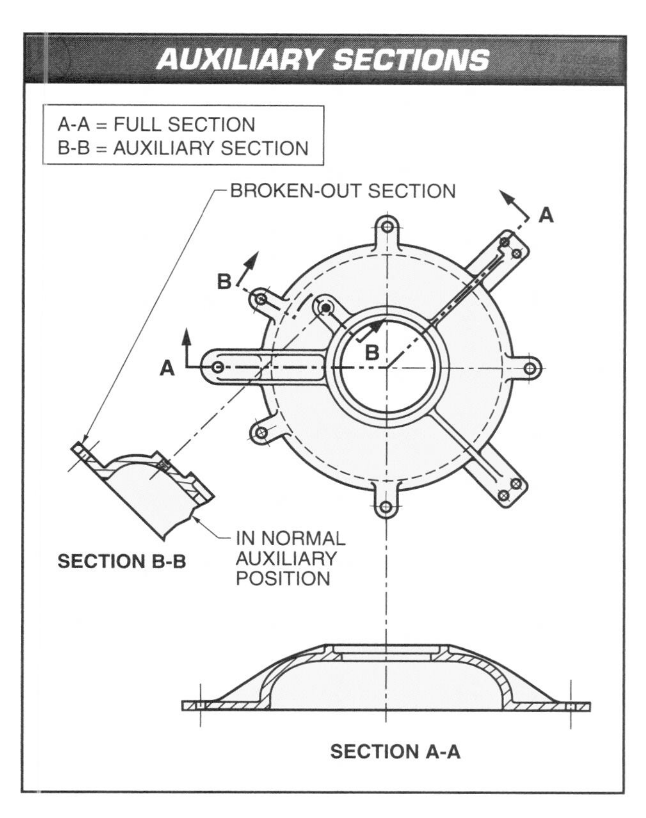

Figure 3-12. An auxiliary section is a sectional view that is not one of the principal planes.

Thin Sections

A *thin section* is a sectional view of a part that is too thin to be shown by the ordinary cross-sectioning convention. **See Figure 3-13.** Materials commonly drawn as thin sections include sheet metal, gaskets, packing, and other thin parts.

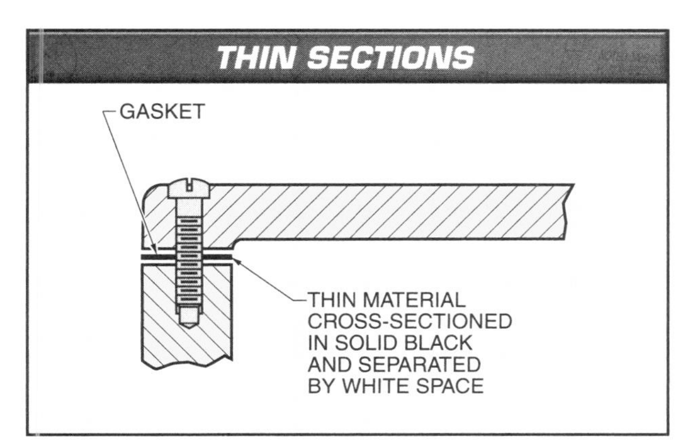

Figure 3-13. A thin section is a sectional view of a part that is too thin to be shown by the ordinary cross-sectioning convention.

Cross sections of thin material are drawn in solid black. If two or more pieces are adjacent to each other, a white space is left to separate the parts.

Some industries specify that where two or more adjacent thin sections are used, they shall be shown solid, but an additional exploded view of that portion shall be included to properly define the arrangement of parts.

Section Lining

Section lining is the pattern of section lines that fills the area of the surfaces formed by a cutting plane. **See Figure 3-14.** Section lining is a method used to easily distinguish between full and sectional portions of a part or assembly. The section lining forms a pattern used to fill cross-sectional shapes.

The recommended angle for drawing section lines is 45°. However, there are situations when other section lining angles may be appropriate. Alternative section lining must be chosen only for the purpose of improving clarity.

If the shape of a sectional area is such that conventional section lines would be nearly parallel or perpendicular to the dominant visible lines of the section, the section lines must be drawn to some other angle. This avoids confusion between the two types of lines.

Boston Gear

Cutaway assemblies are the physical representations of sectional views, usually half sections.

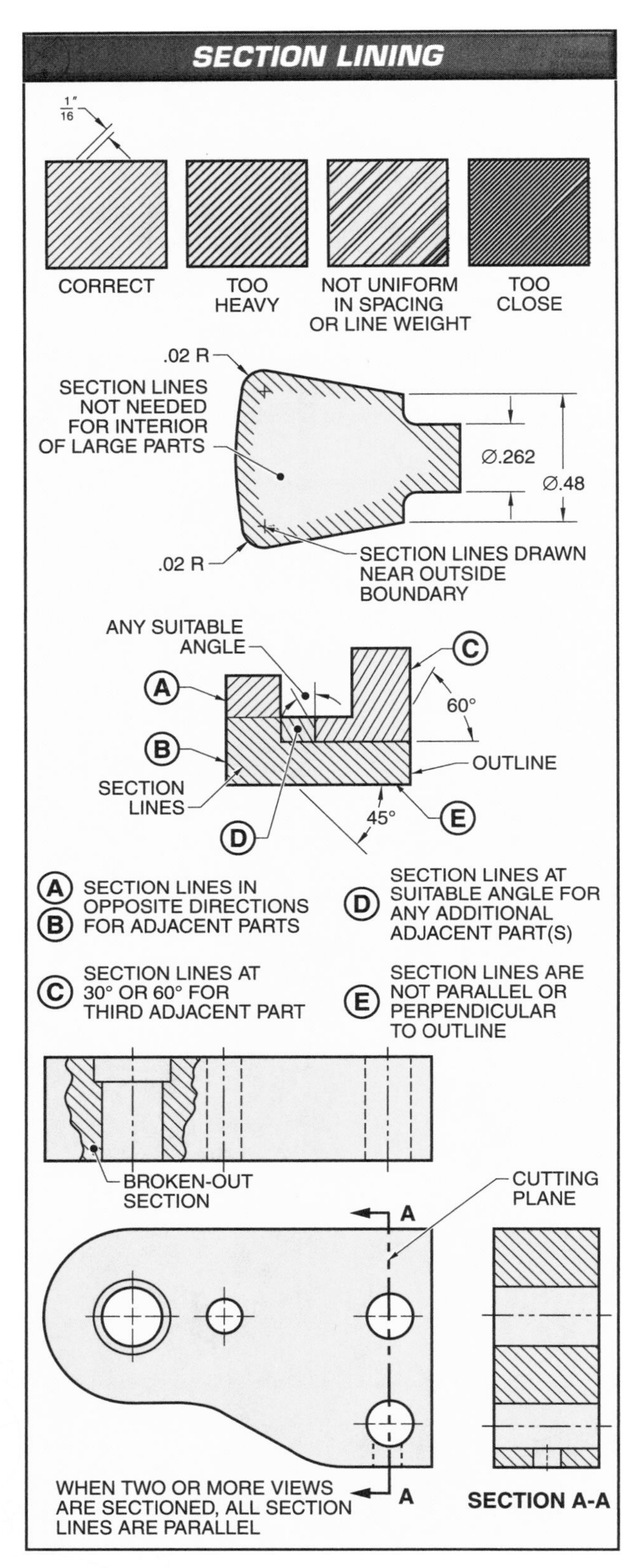

Figure 3-14. Section lining is a series of thin, evenly spaced diagonal lines that show the surfaces through which the cutting plane has passed.

If the section consists of multiple parts, such as in an assembly drawing, the section lines in adjacent parts should run in different directions in order to provide contrast. Angles of 45° are typically chosen first, followed by 30° or 60°. The section lines may be drawn at any suitable angle so that each part stands out separately and clearly. Also, the lines in adjacent parts should not meet at the part boundaries.

A particular section lining pattern is associated with a particular part. When a part appears in multiple sectional views, the same sectional lining angle is used.

It is permissible to use section lines only near the boundary of the sectioned area, especially for large areas. The interior portion is left clear.

Section lines should be uniformly spaced at least 1/16″ apart. The actual spacing may be greater, depending upon the size of the drawing.

A print may use different patterns of section lining to differentiate parts in assemblies using multiple materials. Formerly, ANSI/ASME Y14.2 illustrated and defined symbols for this purpose. These symbols have since been removed from the standard, and their use is no longer recommended. **See Figure 3-15.** The elaborate patterns may not reproduce well, causing readability problems. Also, the patterns are of limited usefulness, since they define only broad categories and notes are still needed to specify material details. However, material-specific section lining symbols may still be found on prints that are older or prints produced by engineers who continue to use the symbols. Instead, the standard recommends the use of only a general-purpose pattern, with variation only in the angle of the lines.

Hidden Lines in Sectional Views

As a rule, all hidden lines should be omitted from a sectional view. The only exception is when hidden lines are indispensable for clarification or for dimensioning. **See Figure 3-16.** In half sections, hidden lines should be used only on the unsectioned side, providing they are necessary for dimensioning or clarity.

ANSI removed material-specific section lining symbols from Y14.2, *Line Conventions and Lettering,* in 1992, although the symbols are still often used when they aid printreading comprehension. Instead, the current standard encourages the use of general purpose section lining with adjacent or general notes explaining any material requirements.

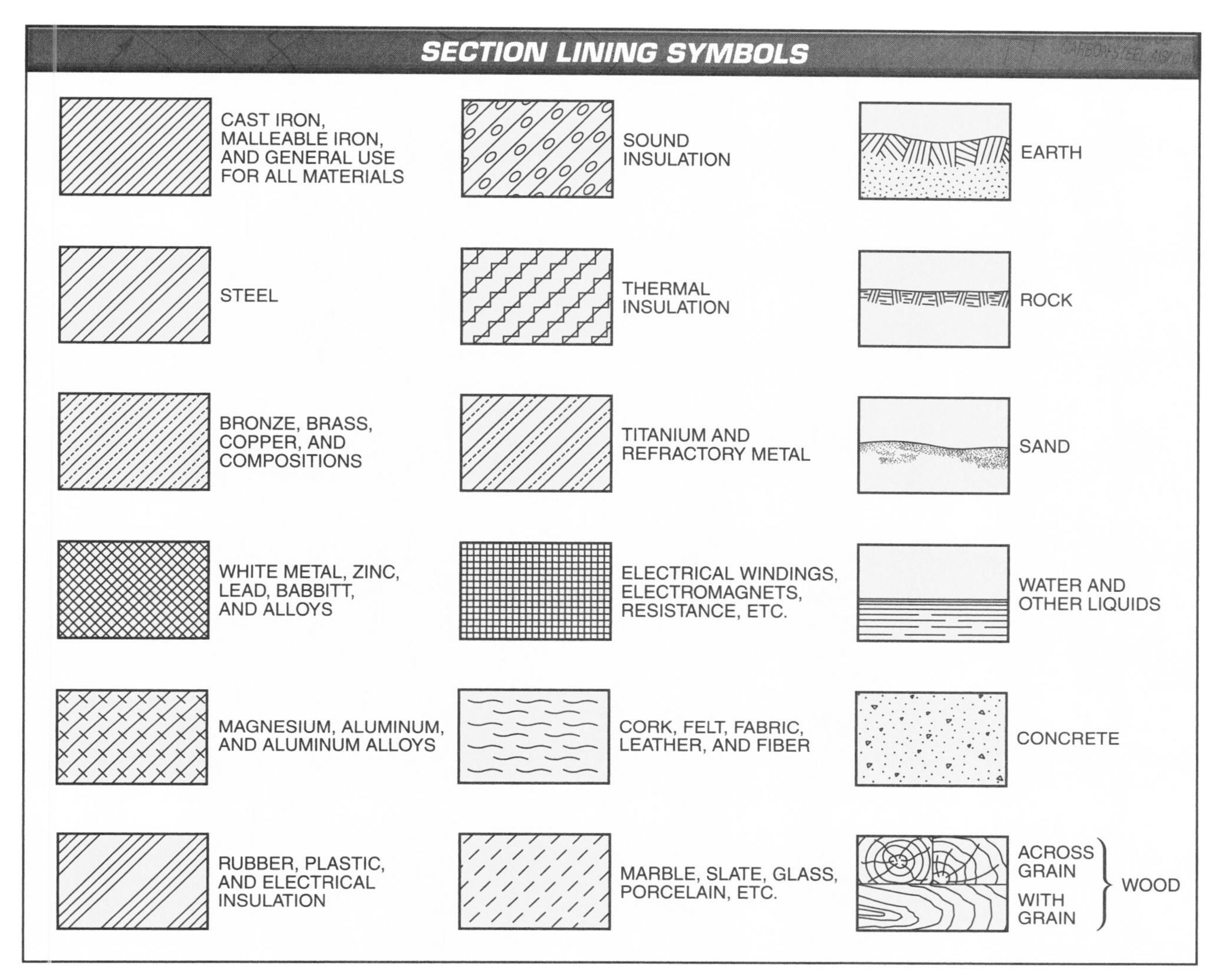

Figure 3-15. Section lining symbols show material types.

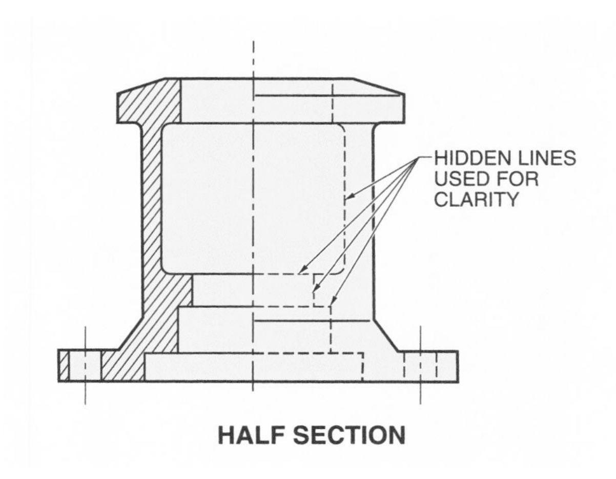

Figure 3-16. Hidden lines are placed on a section view only for clarity or dimensioning.

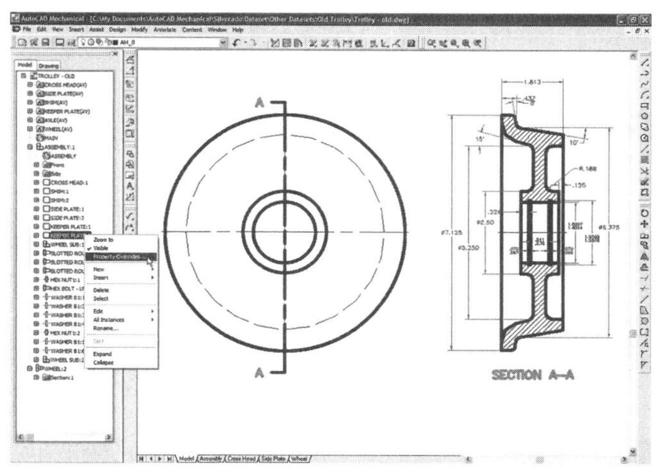

Autodesk, Inc.

CAD software can be used to create sectional views from other views.

Sections through Webs or Ribs

When the cutting plane passes through the long dimension of a web, rib, gear tooth, or other similarly thin and flat element, the element should not be sectioned to avoid presenting a false impression of thickness or size. **See Figure 3-17.**

Alternate section lining may be used in cases where the actual presence of a flat element is not sufficiently clear without section lining, or where clear description of the feature may be improved. For example, if the presence of the ribs is not immediately apparent in the sectional view, alternate section lining may be used to show the ribs. Alternate section line spacing is twice as wide as in normal sections.

Sections through Shafts, Bolts, or Pins

When the cutting plane contains the centerlines of such elements as shafts, bolts, nuts, rods, rivets, keys, pins, spokes, screws, ball or roller bearings, or similar shapes, no sectioning is needed for those elements. **See Figure 3-18.** If the cutting plane cuts across the axes of these elongated parts, they should be sectioned in the usual manner.

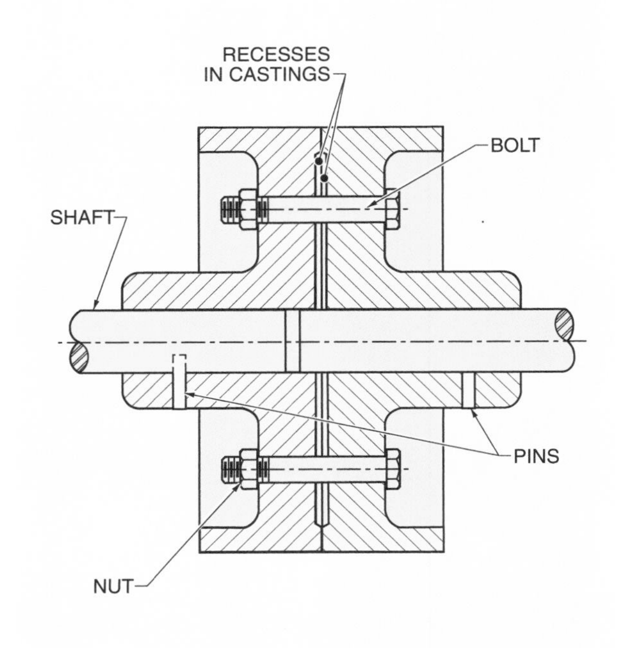

Figure 3-18. Elements such as bolts, screws, and pins should not be sectioned.

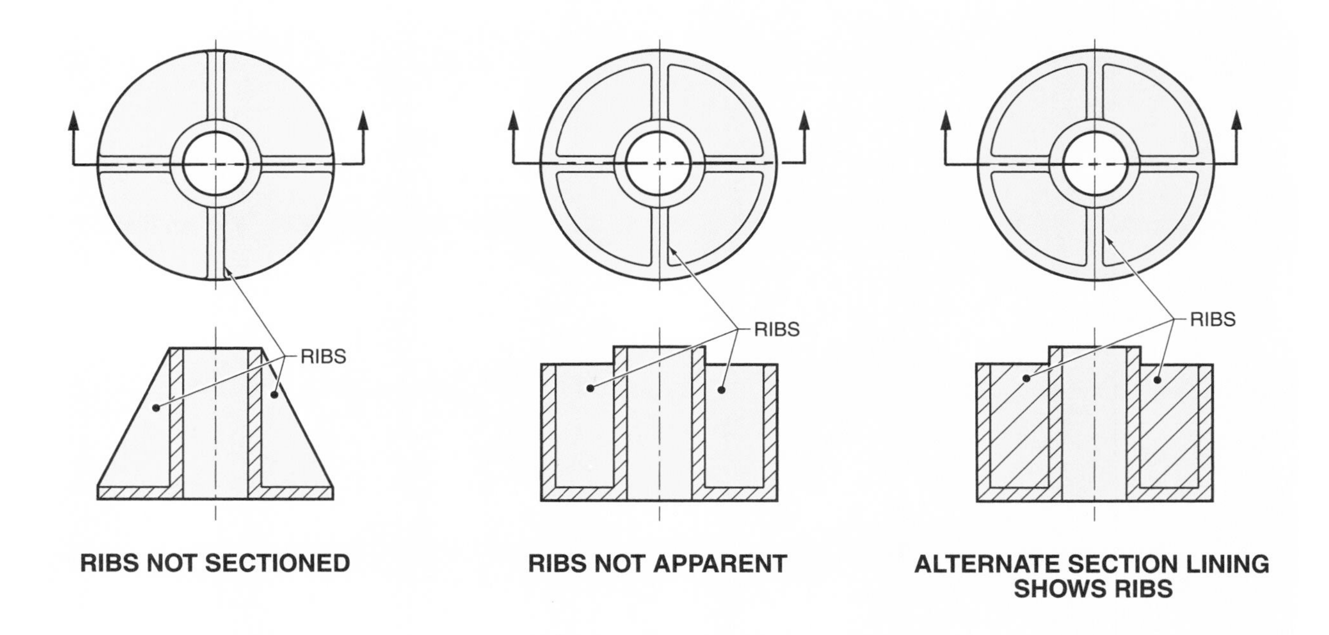

Figure 3-17. Webs, ribs, gear teeth, or other thin features in which the cutting plane passes flatwise are not sectioned.

Foreshortened Projections and Related Features

If the true projection of inclined elements results in foreshortening, the elements should be rotated into the plane of the paper. **See Figure 3-19.** *Foreshortening* is the apparent shortening of inclined parts when shown in an orthographic view.

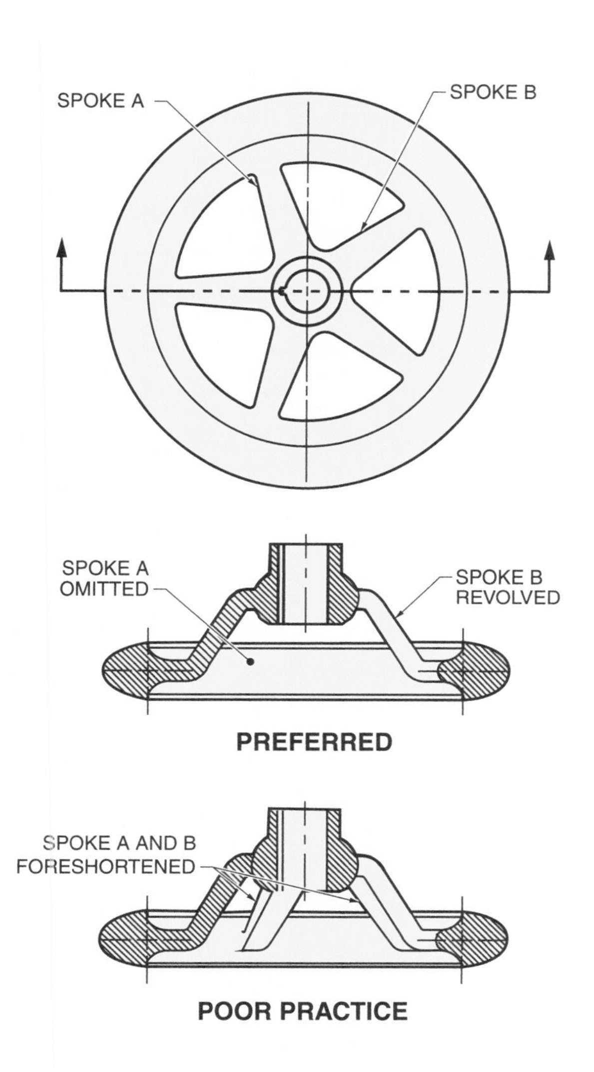

Figure 3-19. Foreshortening is the apparent shortening of particular parts.

In drawings of drilled flanges, the holes may be rotated for clarity to show their true distance from the center rather than in true projection. **See Figure 3-20.** To include features not along a straight line, the cutting plane may be bent or changed in direction to pass through these features and the sections drawn as if they were rotated into a plane. Such sections are known as aligned sections, whether features are rotated into the cutting plane or the cutting plane is bent to pass through them.

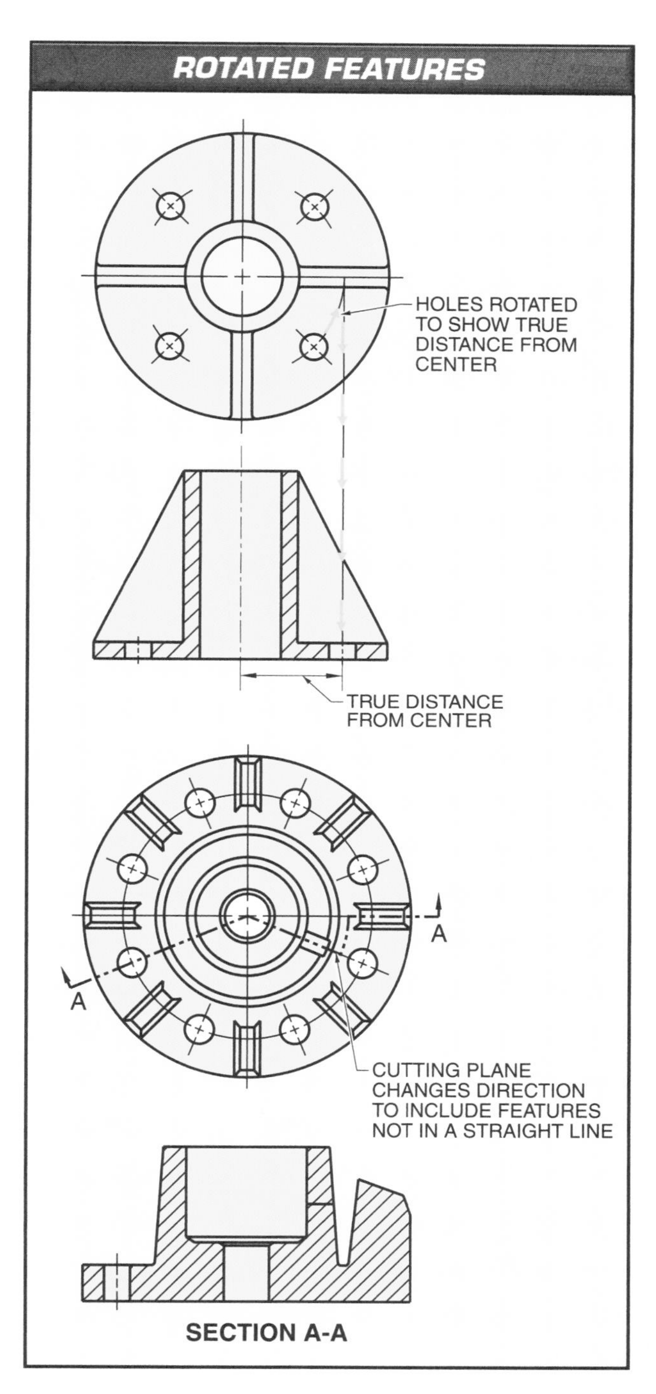

Figure 3-20. To show true distances and shapes, features are rotated or cutting planes change direction.

AUXILIARY VIEWS

Objects with surfaces not parallel to one of the six principal planes may require auxiliary views. The complete details of the true shape of the inclined surfaces are shown on the auxiliary views.

Drawings with auxiliary views still require one or more principal views, which provide the information needed to understand the location and orientation of the auxiliary views. For most objects, two principal views are sufficient. The general rule for multiview drawings still applies: only as many views are included as are needed in order to provide a complete description of the object. **See Figure 3-21.**

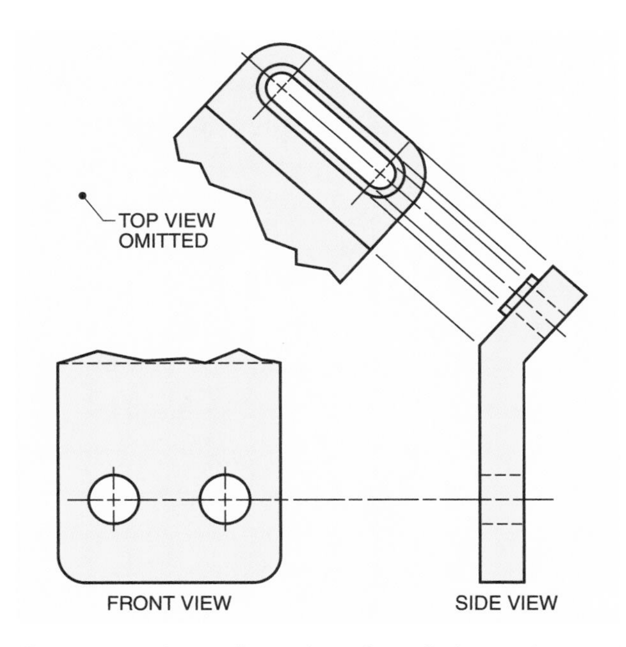

Figure 3-21. An auxiliary view often eliminates the need for one of the principal views.

The two types of auxiliary views are primary auxiliary views and secondary auxiliary views. The difference between the two is their relationship to the principal views.

Primary Auxiliary Views

A *primary auxiliary view* is an auxiliary view that is projected to a plane that is perpendicular to one of the three principal planes and inclined to the other two. **See Figure 3-22.** The true shape of an inclined surface is obtained only by projecting onto a plane parallel to the surface. This auxiliary plane is considered hinged to the plane to which it is perpendicular and then revolved into the frontal plane in much the same way the other views are rotated to their principal planes of projection.

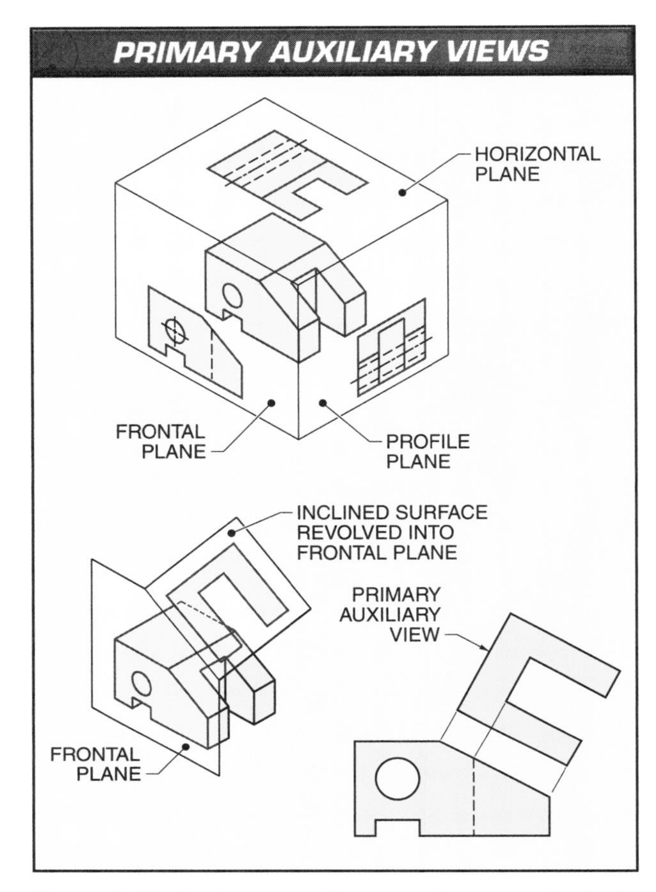

Figure 3-22. A primary auxiliary view is perpendicular to one plane and inclined to the other two.

Partial Auxiliary Views. In making a partial auxiliary view, the practice is to show the shape of only the inclined surface, which is a true view. **See Figure 3-23.** The projection of the entire view usually adds very little to the shape description. The additional lines needed to present a complete view often detract from the true intent of the partial auxiliary view.

Primary Auxiliary View Groups. Generally, primary auxiliary views may be classified into three groups: front auxiliary, top auxiliary, and side auxiliary groups. **See Figure 3-24.** The views are determined according to the plane to which the auxiliary surface is perpendicular, or hinged.

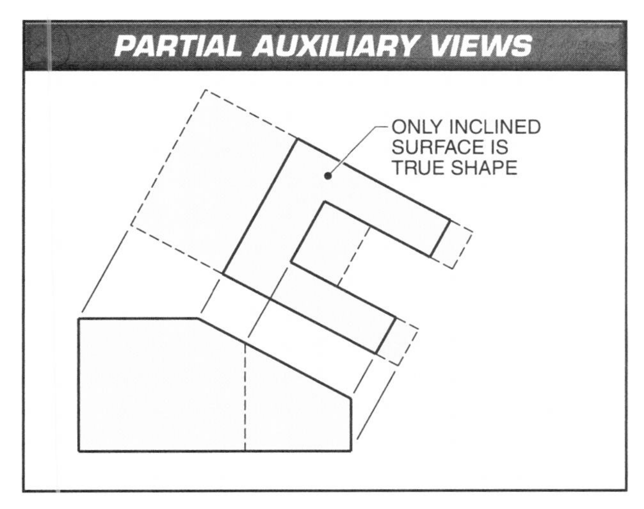

Figure 3-23. Partial auxiliary views show the shape of the inclined surface only.

A *front auxiliary view* is a primary auxiliary view with the inclined surface perpendicular to the frontal plane. A *top auxiliary view* is a primary auxiliary view with the inclined surface perpendicular to the top plane. A *side auxiliary view* is a primary auxiliary view with the inclined surface perpendicular to the side plane.

Auxiliary View Lines. Extra lines are sometimes included to clarify the relationships between principal and auxiliary views. A *reference line* is a line representing the imaginary hinge on which an auxiliary view was rotated into the plane of the print. A *projection line* is a line that connects the edges of an object in a principal view and an auxiliary view. Projection lines always cross reference lines at right angles. Centerlines of circular features may also be extended from a principal view into an auxiliary view.

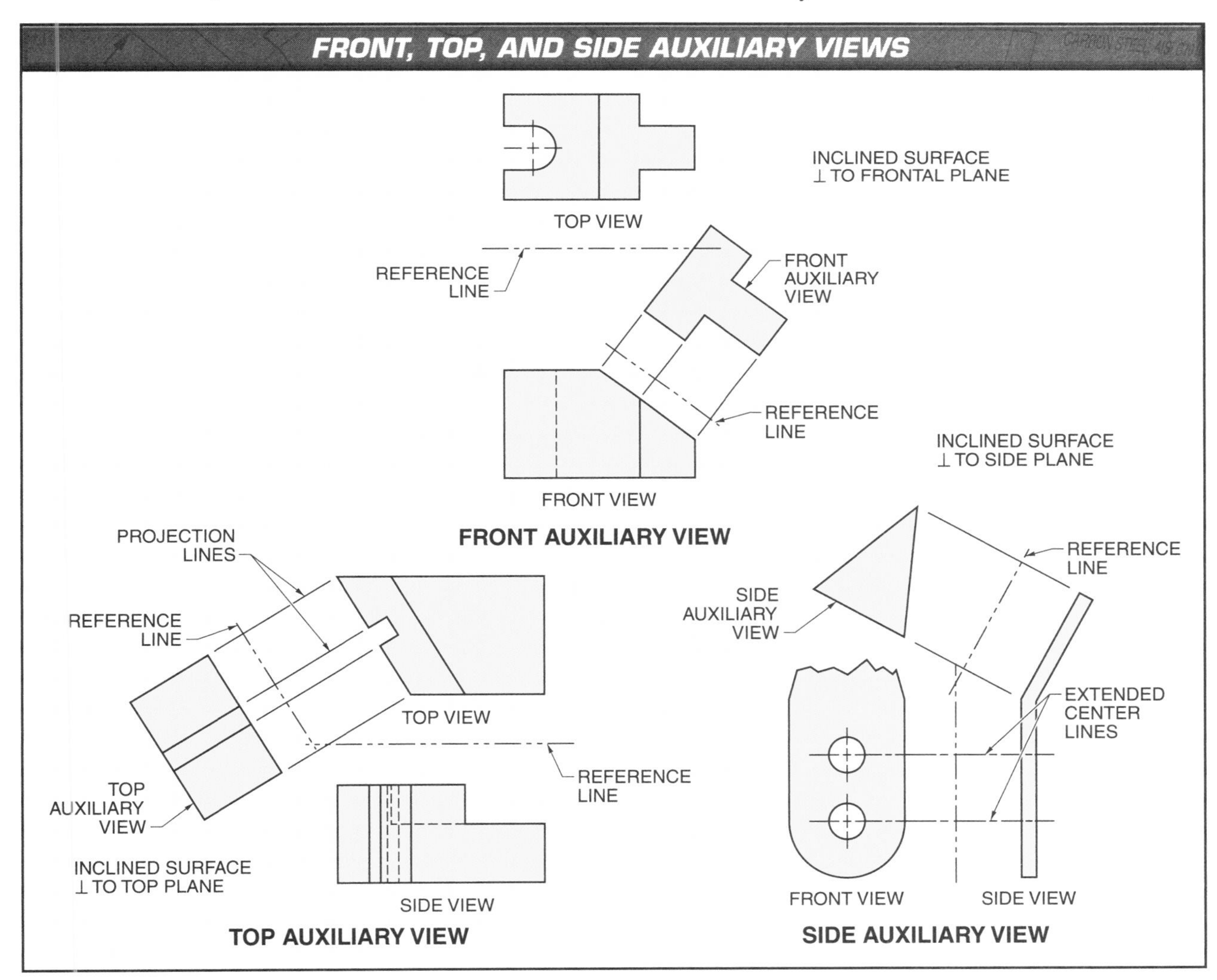

Figure 3-24. Primary auxiliary views are classified according to the plane to which the auxiliary view is hinged.

Symmetrical Auxiliary Views. If an auxiliary view is symmetrical on either side of a line parallel to the reference line, the line can serve as a centerline for a half section view. **See Figure 3-25.** In that case, only half of the auxiliary view is drawn since the other portion is simply a mirror image. The auxiliary view is developed from the right or left of the centerline.

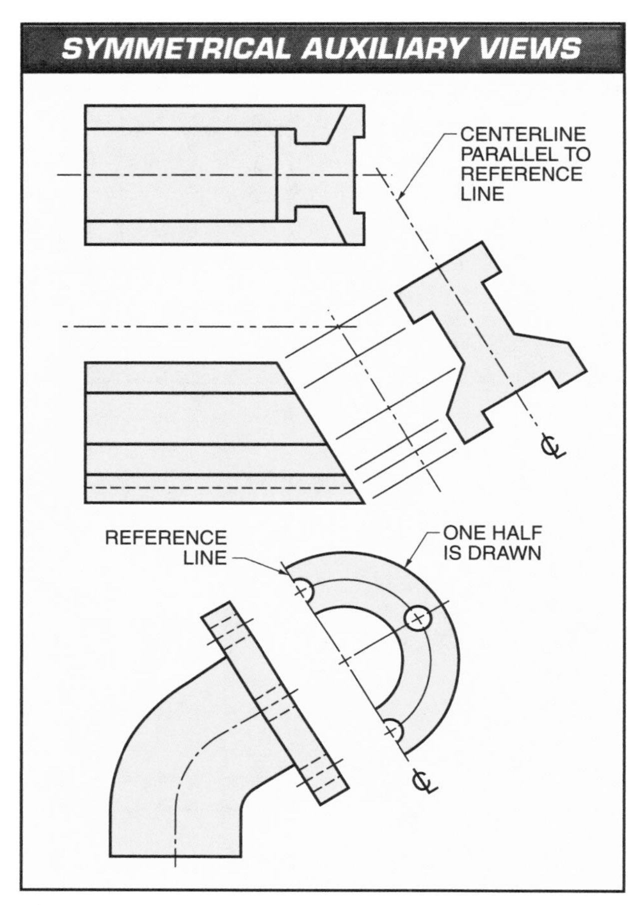

Figure 3-25. In a symmetrical auxiliary, the view is developed from the right and left of the centerline.

An auxiliary view should be arranged adjacent to the related principal view. If it must be in a different area, or on a different sheet, its location can be referenced with zoning and/or sheet numbers.

Secondary Auxiliary Views

A *secondary auxiliary view* is an auxiliary view that is projected to a plane that is oblique to all of the principal planes. Whereas a primary auxiliary view is projected from a principal view, a secondary auxiliary view is projected from a primary auxiliary view. **See Figure 3-26.**

When the true shape of a surface cannot be shown in a front, side, or primary auxiliary view, a secondary auxiliary view is necessary. The actual shape of the object is produced by first drawing a primary auxiliary view and then projecting a secondary auxiliary view from the primary auxiliary view. A secondary auxiliary view may be projected from a front, top, or side auxiliary view.

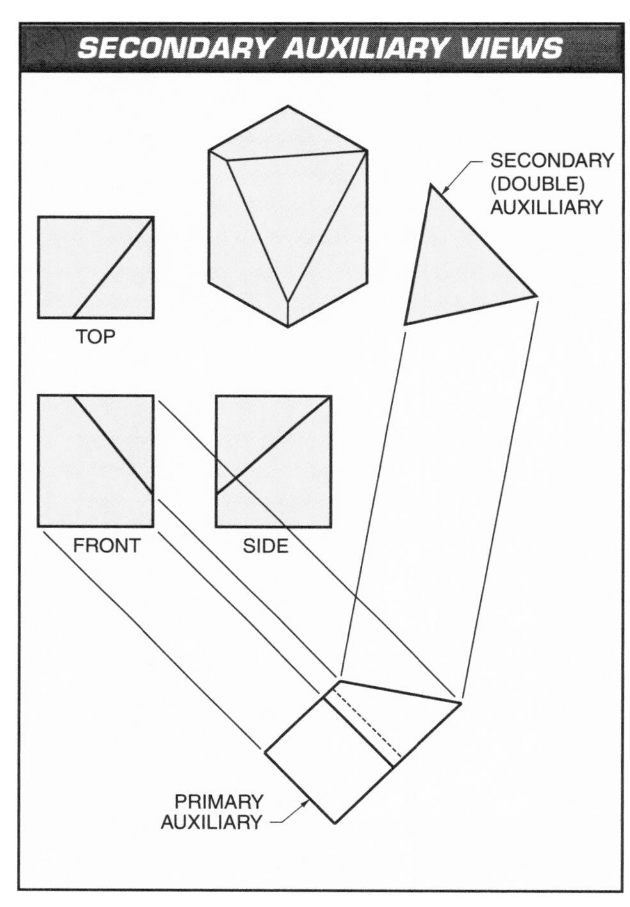

Figure 3-26. A secondary auxiliary view is oblique to all principal planes.

SKETCHING 3

Name ______________________________ Date ______________

Sketching — Sectional Views

Sketch views as indicated.

1 FULL SECTION

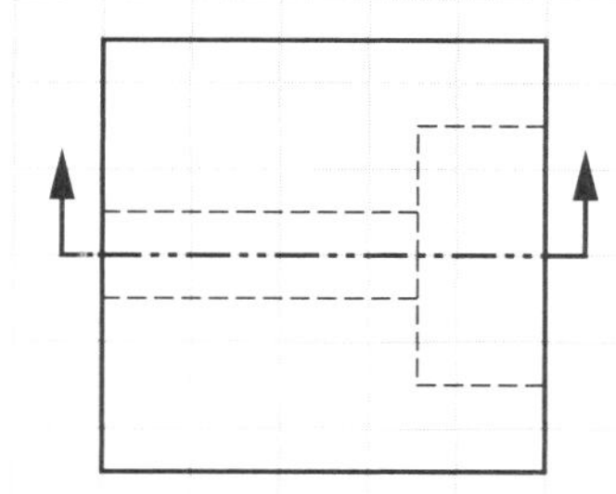

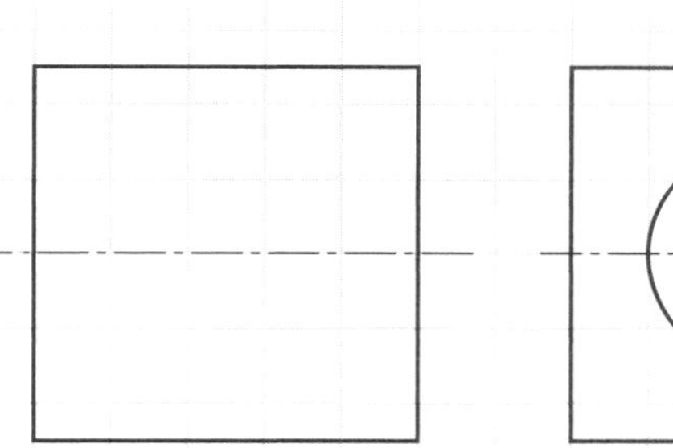

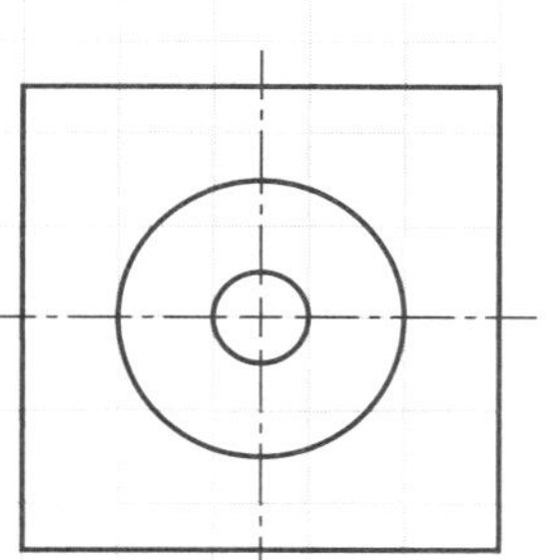

2 OFFSET SECTION

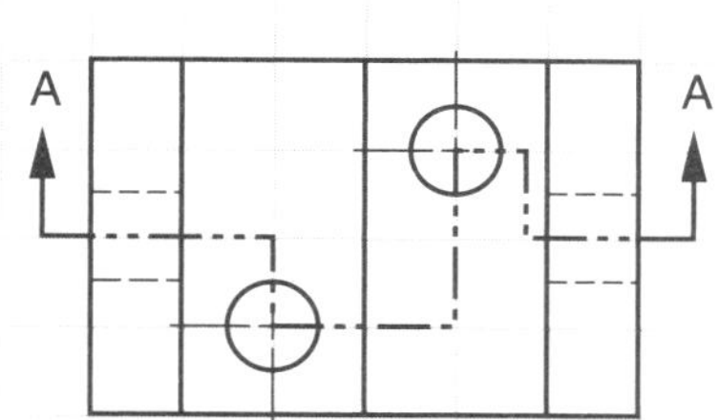

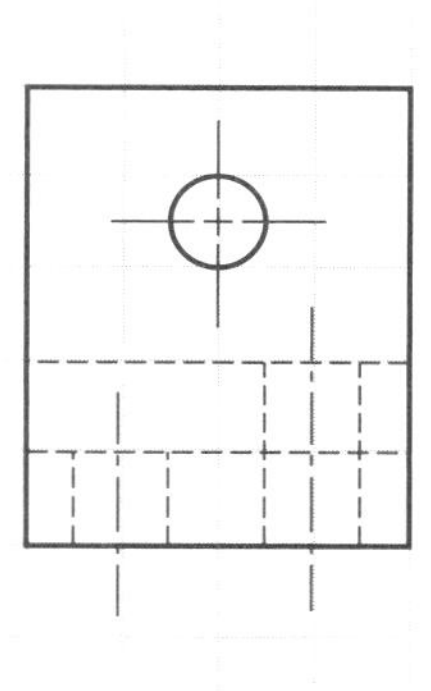

SECTION A-A

3 FULL SECTION – Material: Brass

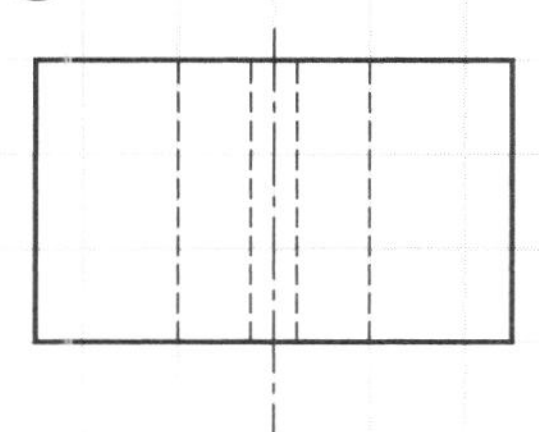

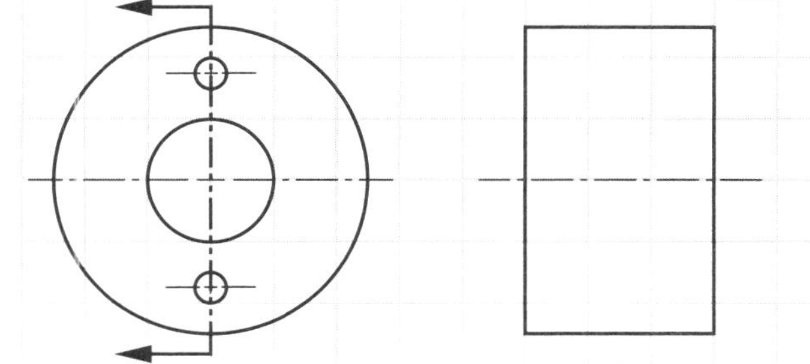

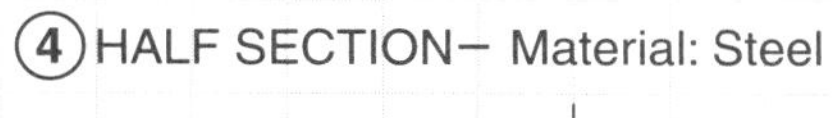

4 HALF SECTION – Material: Steel

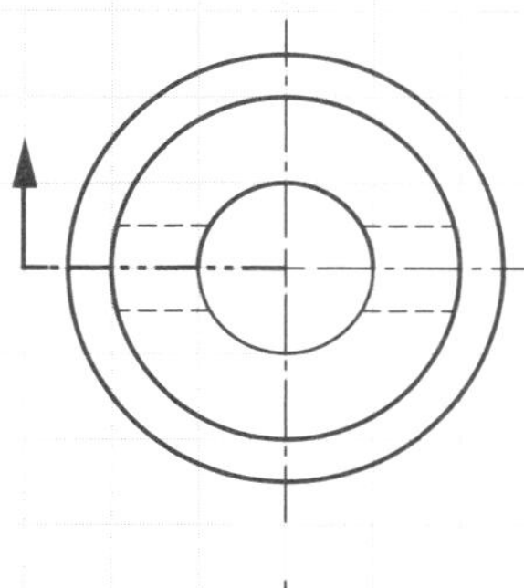

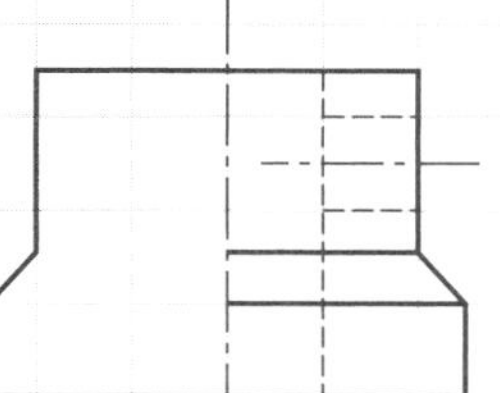

Sketching — Dimensioned Sectional Views

Sketch views as indicated.

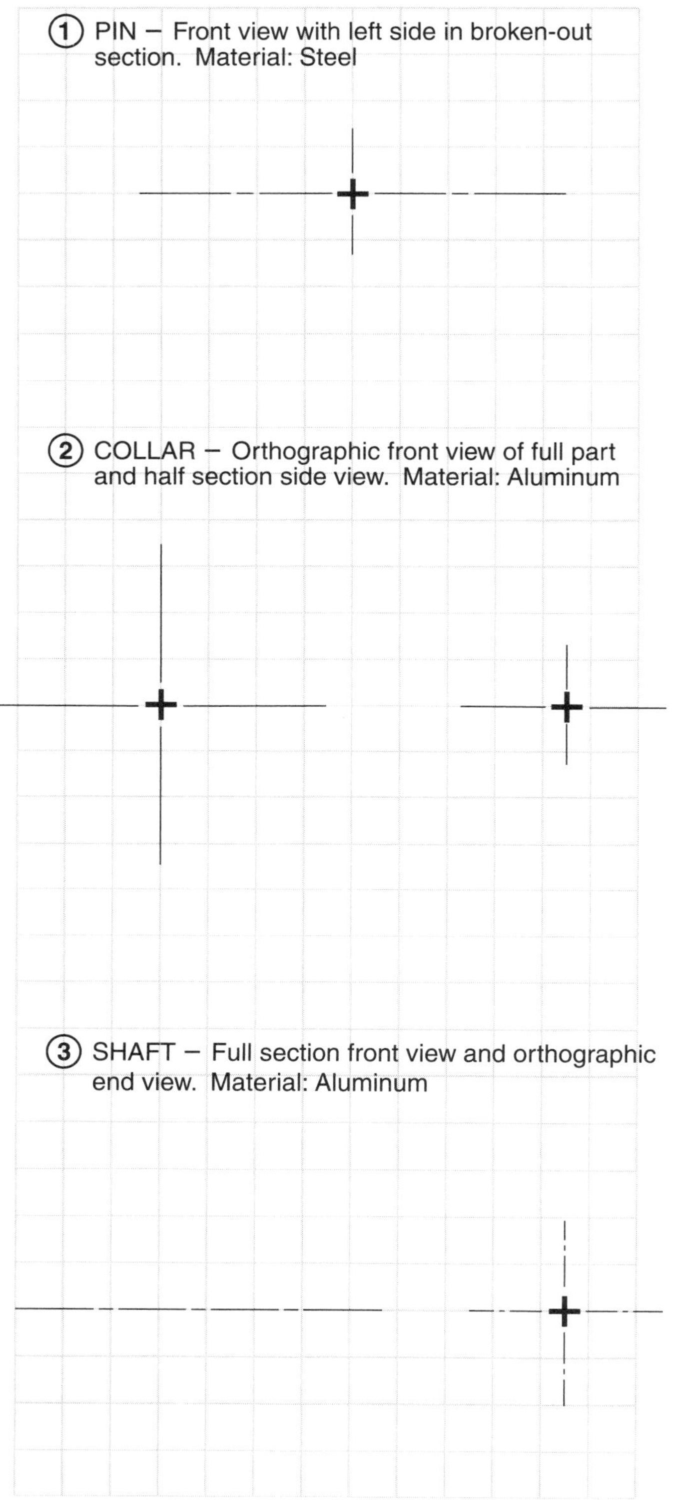

1. PIN – Front view with left side in broken-out section. Material: Steel

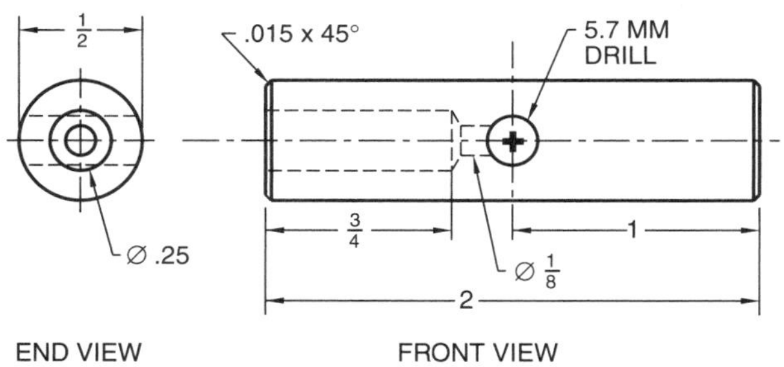

2. COLLAR – Orthographic front view of full part and half section side view. Material: Aluminum

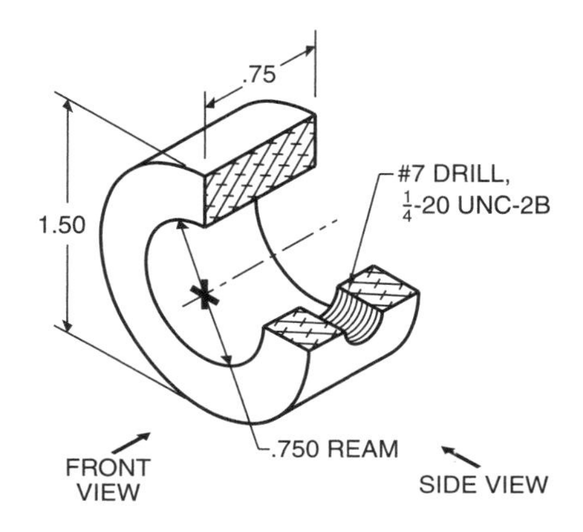

3. SHAFT – Full section front view and orthographic end view. Material: Aluminum

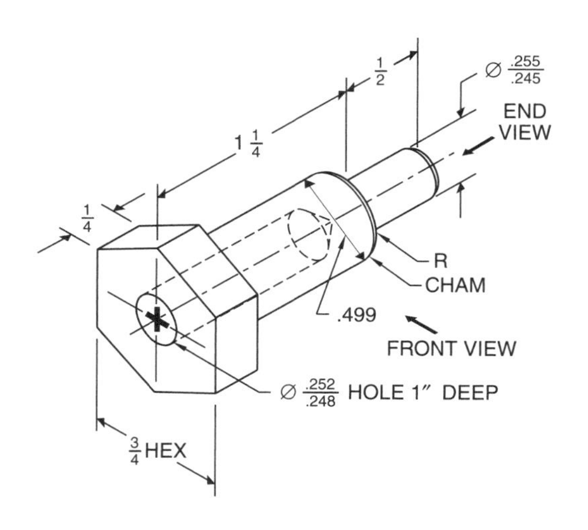

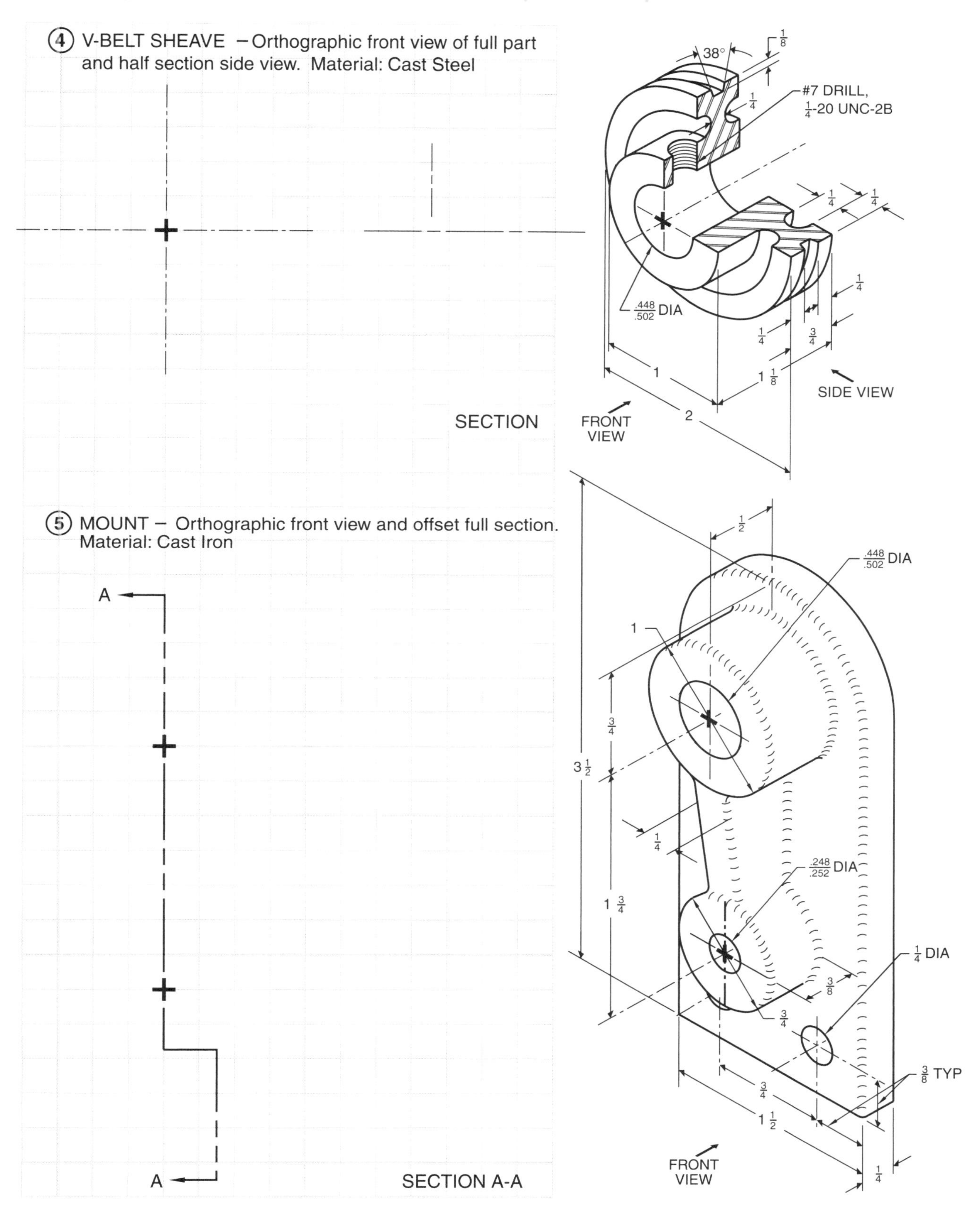

4 V-BELT SHEAVE – Orthographic front view of full part and half section side view. Material: Cast Steel
SECTION
38°
1/8
1/4
#7 DRILL,
1/4-20 UNC-2B
1/4
1/4
.448/.502 DIA
1/4
3/4
1/4
1
1 1/8
2
FRONT VIEW
SIDE VIEW
5 MOUNT – Orthographic front view and offset full section. Material: Cast Iron
A
A
SECTION A-A
1/2
.448/.502 DIA
1
3/4
3 1/2
1/4
.248/.252 DIA
1 3/4
1/4 DIA
3/8
3/4
3/8 TYP
3/4
1 1/2
FRONT VIEW
1/4

Sketching — Primary Auxiliary Views

Sketch the frontal auxiliary view.

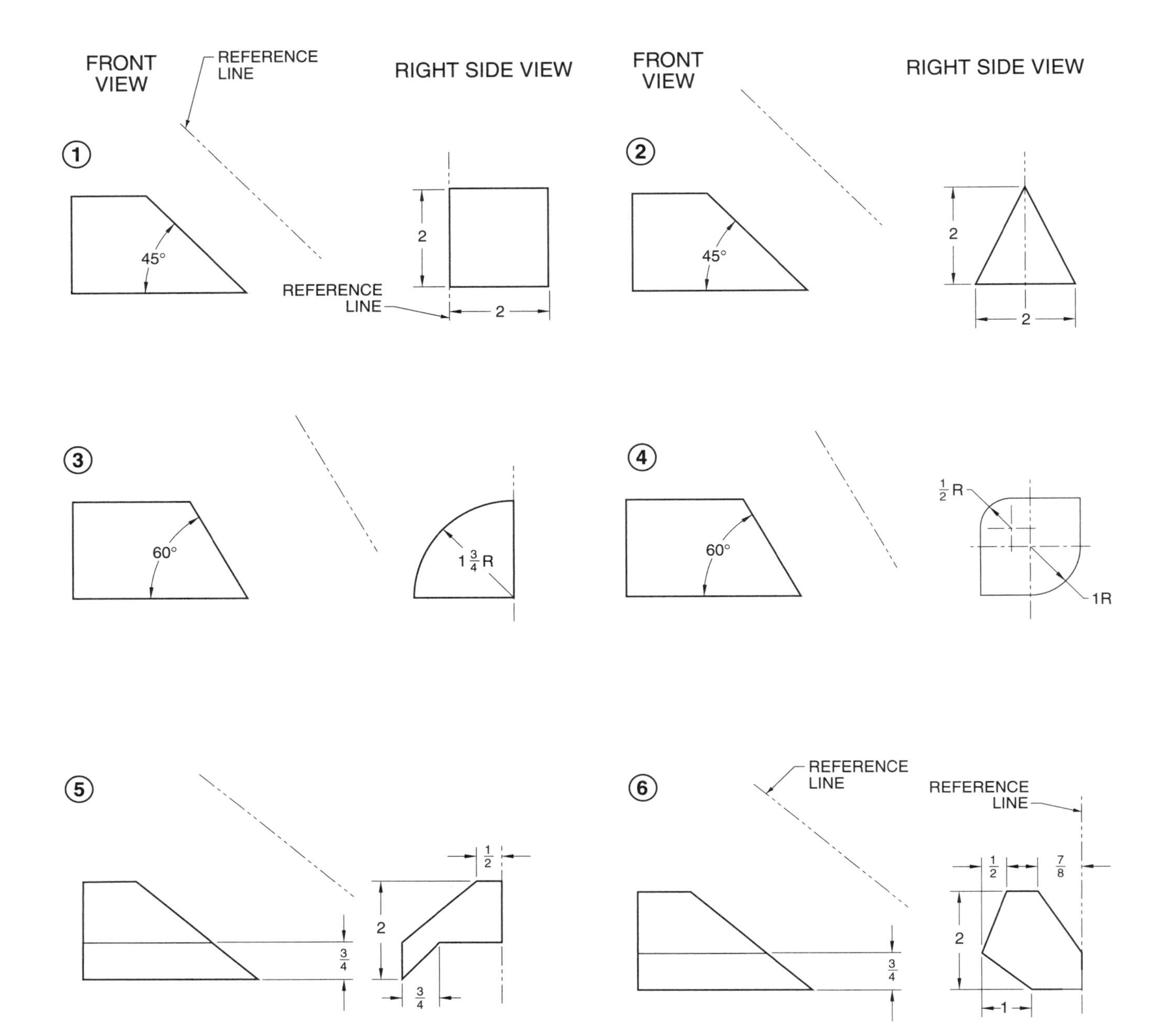

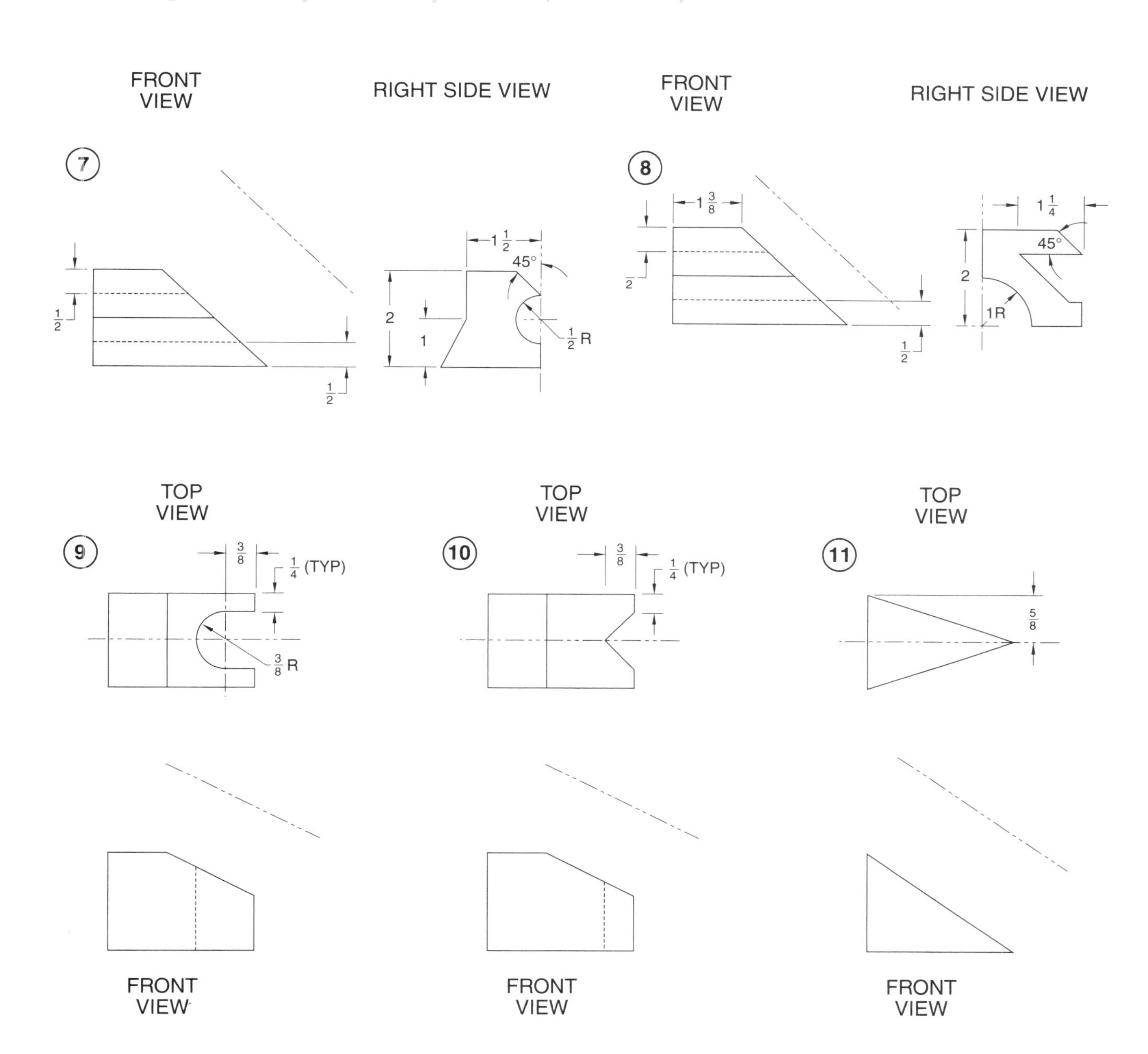

FRONT VIEW
RIGHT SIDE VIEW
FRONT VIEW
RIGHT SIDE VIEW
7
1 1/2
45°
2
1
1/2 R
1/2
1/2
8
1 3/8
1 1/4
45°
2
1R
1/2
1/2
TOP VIEW
TOP VIEW
TOP VIEW
9
3/8
1/4 (TYP)
3/8 R
10
3/8
1/4 (TYP)
11
5/8
FRONT VIEW
FRONT VIEW
FRONT VIEW

REVIEW QUESTIONS 3

Name ______________________________ Date ____________________

Completion

______________ 1. A(n) ___ section is a sectional view in which the cutting plane passes entirely across the object.

______________ 2. A(n) ___ section is a sectional view in which two cutting planes are passed at right angles to each other along the centerlines or symmetrical axes.

______________ 3. A(n) ___ object is an object in which one half is the mirror image of the other half.

______________ 4. A(n) ___ section is a sectional view in which a cross-sectional shape is shown at the cutting plane location.

______________ 5. Section ___ is a series of thin, evenly spaced diagonal lines that show the surfaces formed by a cutting plane.

______________ 6. As a rule, all ___ lines should be omitted from a sectional view.

______________ 7. Cross sections of ___ material are drawn in solid black.

______________ 8. A(n) ___ section shows the interior of an object while maintaining the shape of the exterior.

______________ 9. A(n) ___ section is a sectional view in which the cutting plane line changes direction in order to include features that are not located in a straight line.

______________ 10. A(n) ___ section is detached from the projected view and located elsewhere on the print.

______________ 11. Objects with surfaces that are not ___ to one of the six principal planes may require auxiliary views.

______________ 12. A(n) ___ auxiliary view is a view projected to a plane that is perpendicular to one of the three principal planes and is inclined to the other two.

______________ 13. In making an auxiliary view, the practice is to show the shape of only the ___ surface.

______________ 14. A(n) ___ auxiliary view is projected to a plane that is oblique to all of the principal planes.

______________ 15. A(n) ___ section is a sectional view that is not one of the principal planes.

______________ 16. A(n) ___ auxiliary view has the inclined surface perpendicular to the frontal plane.

______________ 17. ___ is the apparent shortening of inclined parts when shown in an orthographic view.

True-False

T	F	**1.** A sectional view is the view of a cross section of an object.
T	F	**2.** Letters identifying sections should read vertically.
T	F	**3.** A cutting plane line may be omitted if it is clear that the section is taken along the centerline.
T	F	**4.** Half sections are not widely used in detail drawings.
T	F	**5.** A cutting plane line is required to show a broken-out section.
T	F	**6.** In half sections, hidden lines should be used only on the sectioned side.
T	F	**7.** There are three standard styles for cutting plane lines, all using thick lines.
T	F	**8.** A secondary auxiliary view may be projected from a front auxiliary, top auxiliary, or side auxiliary view.
T	F	**9.** If an auxiliary view is symmetrical on either side of a line parallel to the reference line, the line can serve as a centerline for a half section view.
T	F	**10.** The difference between primary and secondary auxiliary views is their relationship to the principal views.

Identification — Section Lining Symbols

_______________ **1.** Earth

_______________ **2.** White metal, zinc, lead, babbitt, and alloys

_______________ **3.** Concrete

_______________ **4.** Bronze, brass, copper, and compositions

_______________ **5.** Rubber, plastic, and electrical insulation

_______________ **6.** Steel

_______________ **7.** Cast iron, malleable iron, and general use for all materials

_______________ **8.** Magnesium, aluminum, and aluminum alloys

_______________ **9.** Electric windings, electromagnets, resistance, etc.

_______________ **10.** Cork, felt, fabric, leather, and fiber

A B C D E

F G H I J

Sectional and Auxiliary Views

TRADE COMPETENCY TEST 3

Name ______________________ Date ______________

Holder–Punch

Refer to print on page 82.

______ **1.** The maximum overall diameter of the Holder–Punch is ___″.

______ **2.** The minimum overall diameter of the Holder–Punch is ___″.

T F **3.** A half section is shown.

T F **4.** The cutting plane line passes through the centerpoint.

______ **5.** Surface E must be ___ to surface datum B.

______ **6.** Surface F is concentric with datum A within a ___″ tolerance zone.

______ **7.** The centers of the counterbored holes are ___″ from the centerpoint.

T F **8.** All external corners are chamfered.

______ **9.** Datum B is ___ to within a 0.0002″ tolerance zone.

______ **10.** Datum A is ___ to datum B.

______ **11.** Surface F is perpendicular to datum B within a ___″ tolerance zone.

______ **12.** The minimum diameter at B is ___″.

______ **13.** The most critical diameter on the print is shown at ___.

T F **14.** Hole G is centered a maximum of 1.4375″ from datum A.

T F **15.** The ∅.750 hole has a maximum allowable depth of 2.237″.

T F **16.** All chamfers are the same size.

T F **17.** Both ∅3⁄16 holes are drilled through.

T F **18.** The maximum distance the center of hole G can vary in any direction is 0.00025″.

T F **19.** Surface F and surface datum A are parallel with each other within 0.0010″.

______ **20.** The diameter at C is ___″.

______ **21.** A total of ___ holes measuring 11⁄32″ in diameter are drilled through the Holder–Punch.

______ **22.** The maximum diameter at D is ___″.

______ **23.** The minimum diameter at D is ___″.

T F **24.** Hole G is located with reference to datum A at maximum material condition.

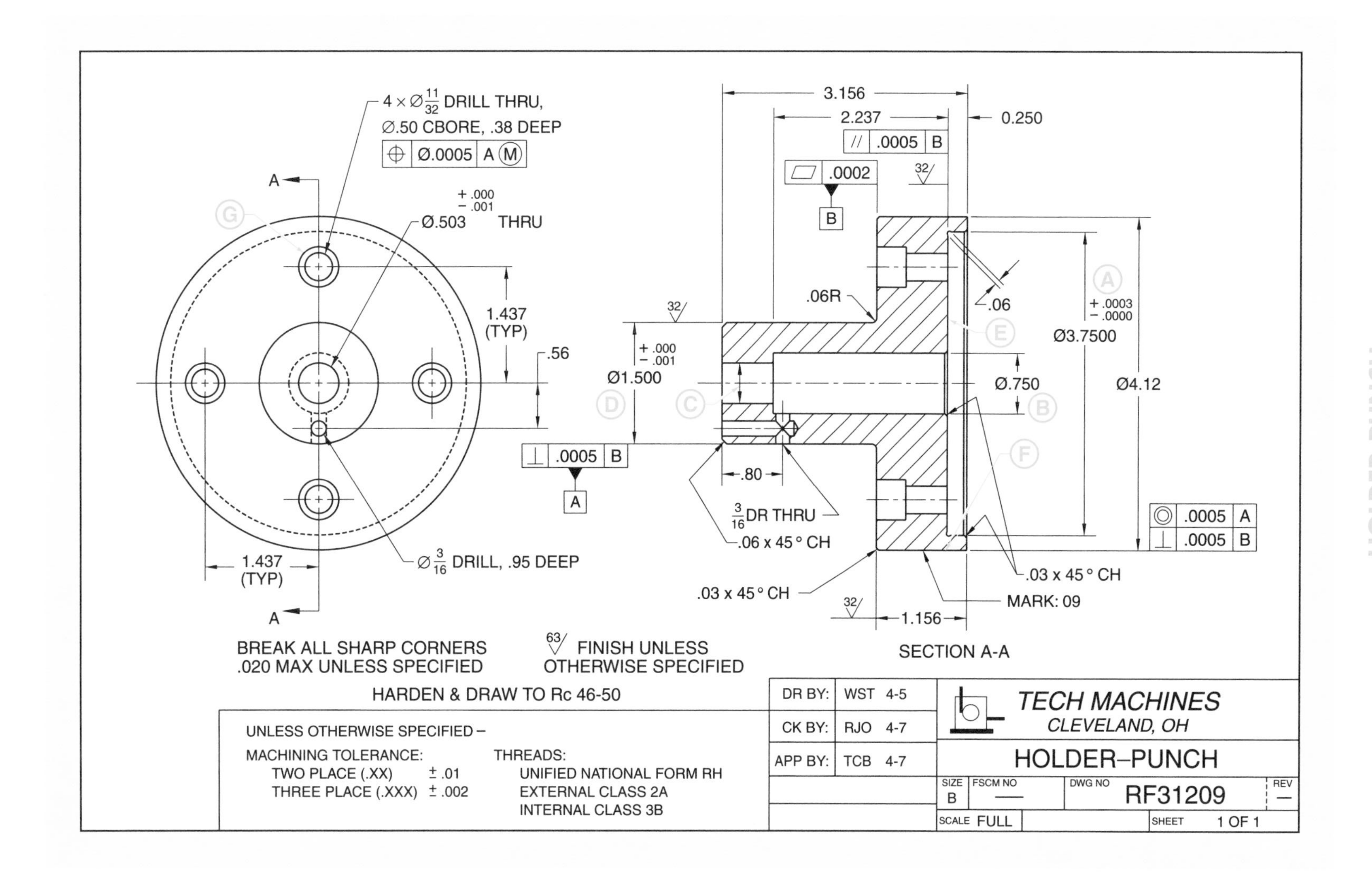

HOLDER-PUNCH

Pressure Clevis

Refer to print on page 84.

________________ **1.** A(n) ___ auxiliary view is shown of the Pressure Clevis.

________________ **2.** Front and top ___ views of the Pressure Clevis are shown.

________________ **3.** The threaded hole is on a surface that is ___° to the top surface.

________________ **4.** ___ lines are used to indicate that a full auxiliary is not shown.

T F **5.** The threaded hole is counterbored.

________________ **6.** The distance from the centerpoint of the drilled hole to the base is ___″.

________________ **7.** The maximum diameter of the drilled hole is ___″.

________________ **8.** The overall length of the Pressure Clevis is ___″.

________________ **9.** The overall height of the Pressure Clevis is ___″.

________________ **10.** All holes are square to the machined surface to ±___″.

________________ **11.** The only reference dimension on the print is ___″.

________________ **12.** The scale of the drawing is ___.

T F **13.** The threaded hole is ¾″ deep.

________________ **14.** The radius around the drilled-through hole is ___″.

T F **15.** The pictorial view is not to scale (NTS).

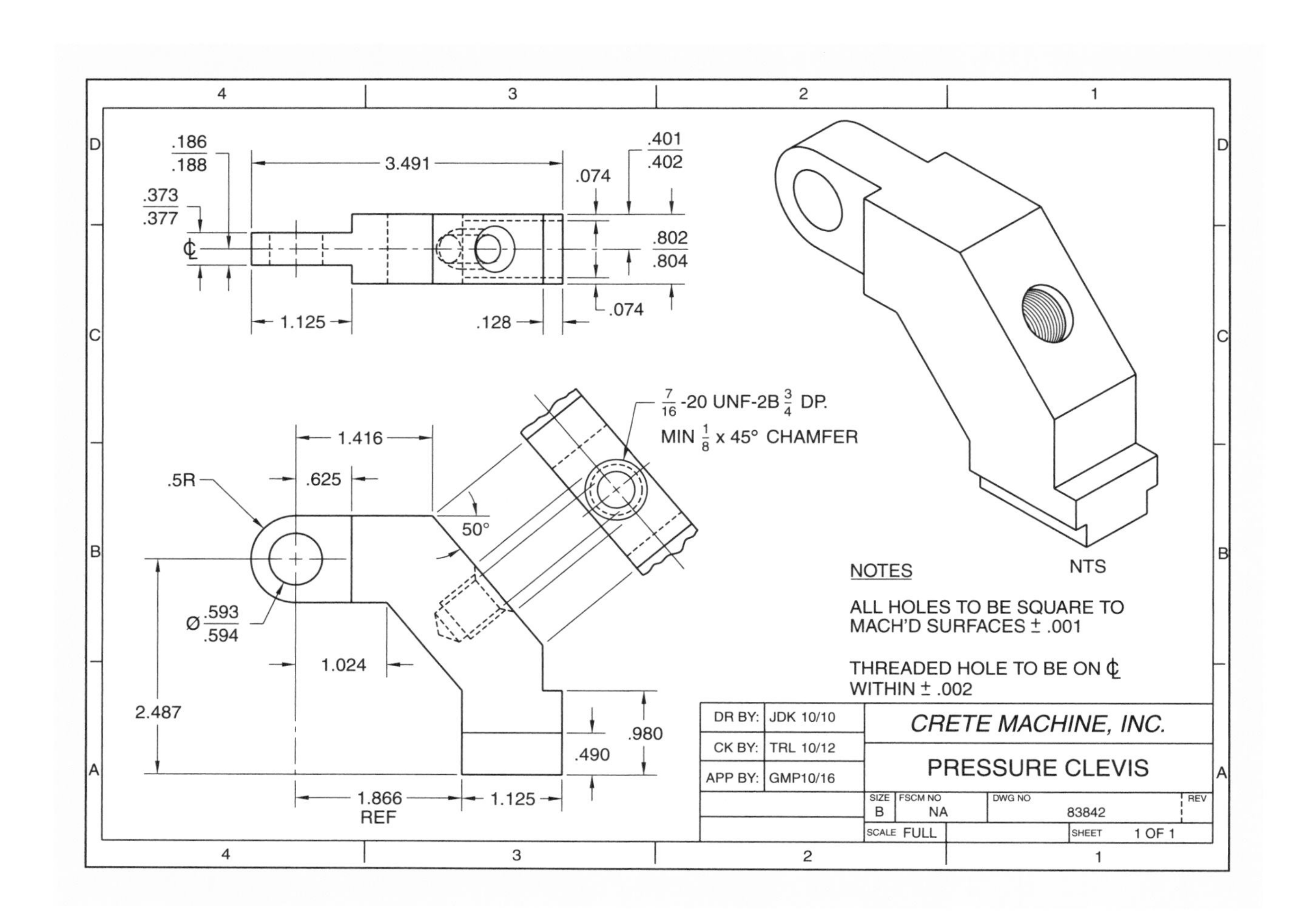

PRESSURE CLEVIS

Shop Math 4

Machinists use basic math concepts to estimate material and labor costs and to perform layout and machining operations. Whole numbers and common or decimal fractions are added, subtracted, multiplied, or divided to find solutions to problems. Sums, remainders, percentages, measurements, areas of plane figures, and volumes of solid figures are found by applying the proper math functions.

WHOLE NUMBERS

A *whole number* is a number with no fractional or decimal parts. **See Figure 4-1.** For example, numbers such as 1, 2, 19, 46, 67, and 328 are whole numbers. Whole numbers are also called integers. An *odd number* is a number that cannot be divided by 2 an exact number of times. For example, numbers such as 1, 3, 5, 57, and 109 are odd numbers. An *even number* is a number that can be divided by 2 an exact number of times. For example, numbers such as 2, 4, 6, 48, and 432 are even numbers.

A *prime number* is a number greater than 1 that can be divided an exact number of times only by itself and the number 1. For example, numbers such as 2, 3, 5, 7, 11, 13, 17, 19, 23, and so on, are prime numbers.

Addition

Addition is the process of uniting two or more numbers to make one number. It is the most common operation in mathematics. The sign + (plus) indicates addition and is used when numbers are added horizontally or vertically. **See Figure 4-2.** When more than two numbers are added vertically, the addition operation is implied and no sign is required. A *sum* is the result obtained from adding two or more numbers.

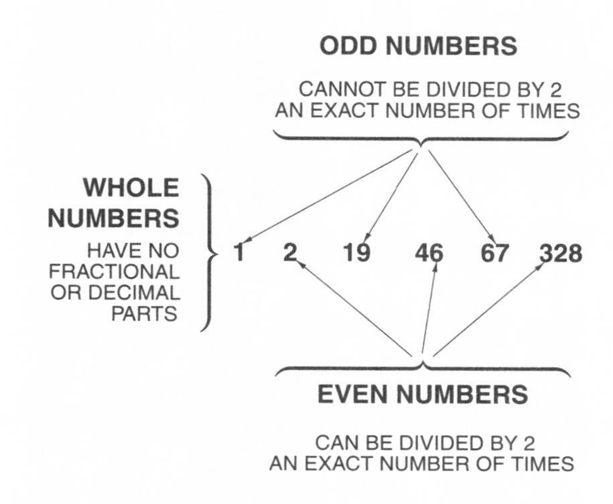

Figure 4-1. Whole numbers have no fractional or decimal parts.

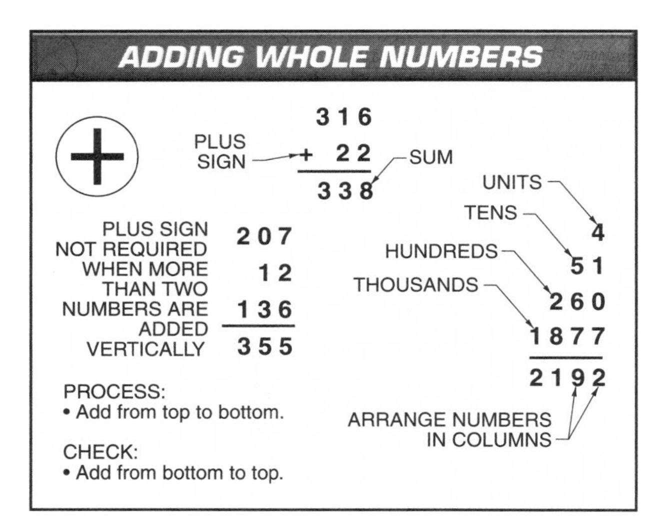

Figure 4-2. Addition is the process of uniting two or more numbers to make one number.

To add whole numbers vertically, place all numbers in aligned columns. The units must be in the ones (units) column, tens in the tens column, hundreds in the hundreds column, and so on. Add the columns from top to bottom, beginning with the ones column. When the sum of the numbers in one column is 0 through 9, record the sum and move to the next column. When the sum of the numbers in a column is 10 or more, record the last digit and carry the remaining digit(s) to the next column. Follow this same procedure for remaining columns.

Adding whole numbers horizontally is more difficult than adding them vertically. For example, 25 + 120 + 37 + 3 = 185 shows whole numbers added horizontally. This method is not as commonly used as the vertical alignment method because mistakes can occur more easily.

Check vertically aligned addition problems by adding the numbers from bottom to top. Check horizontally aligned addition problems by adding the numbers from right to left. The same sum will result both times if addition has been correctly performed in both directions.

Subtraction

Subtraction is the process of removing a quantity from another quantity. It is the opposite of addition. The sign – (minus) indicates subtraction. A *minuend* is a number from which a subtraction is made. A *subtrahend* is a number that is subtracted. A *remainder* is the difference between a minuend and subtrahend. Place the minuend above the subtrahend when vertically aligning numbers. **See Figure 4-3.**

As in addition, the first column of numbers represents ones, the second column represents tens, the third column represents hundreds, and so on. Whenever a subtrahend digit is larger than the corresponding minuend digit, borrow one unit from the column immediately to the left and continue the operation. For example, when subtracting 8 from 24, borrow a 1 from the tens column, subtract 8 from 14, record the 6 in the units column, and record the remaining 1 in the tens column for a remainder of 16.

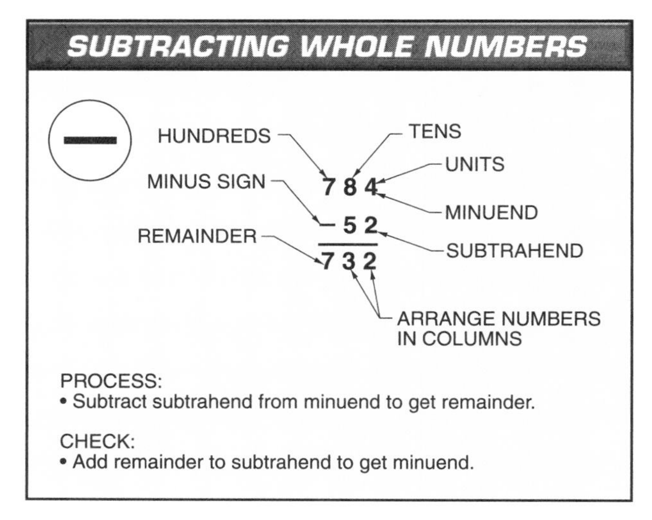

Figure 4-3. Subtraction is the process of removing a quantity from another quantity.

> The numerals of 0 through 9 that are used in modern Western cultures are known as Arabic numerals, though they may have actually originated in India. The number system spread to the Middle East by the seventh century. It was not until the fifteenth century that Arabic numerals were regularly used in Europe.

Multiplication

Multiplication is the process of adding one number as many times as there are units in the other number. For example, 3 × 4 = 12 produces the same result as adding 4 + 4 + 4 = 12. The sign × (times or multiplied by) indicates multiplication. The symbol ·, a small dot centered vertically, also indicates multiplication. A *multiplicand* is a number that is multiplied. A *multiplier* is a number by which multiplication is done. A *product* is the result of multiplication. **See Figure 4-4.**

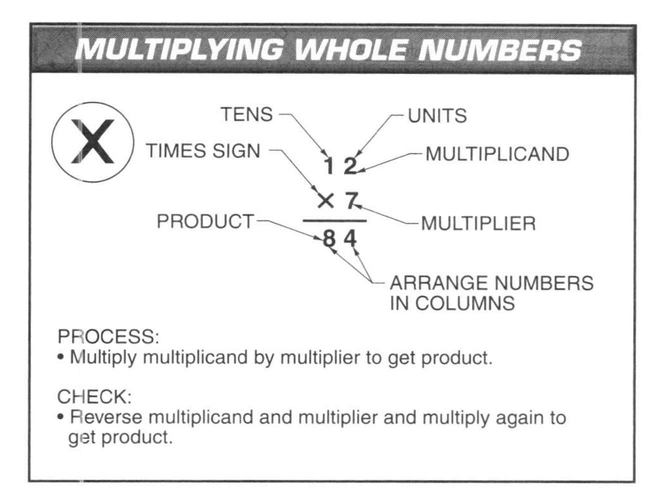

Figure 4-4. Multiplication is the process of adding one number as many times as there are units in another number.

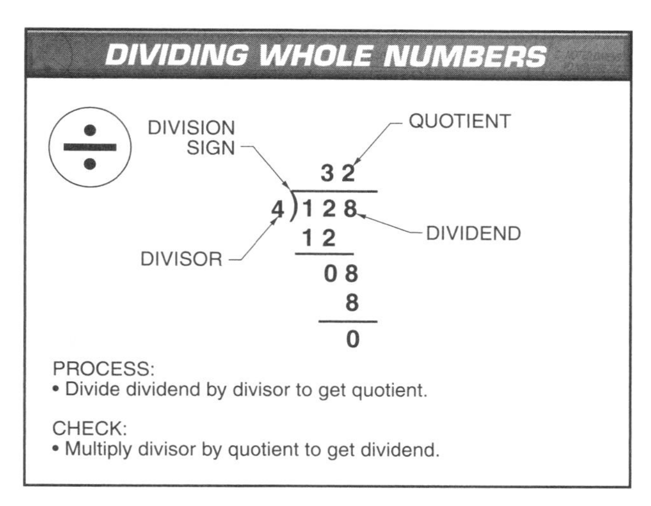

Figure 4-5. Division is the process of determining how many times one number contains another number.

The larger number is commonly used as the multiplicand when the units being multiplied are the same. For example, 8′ × 4′ = 32 sq ft. Numbers may be arranged vertically (preferred) or horizontally when multiplying. An effective method of checking the product is to reverse the multiplicand and the multiplier and perform the operation again. The same product will result if both operations have been multiplied correctly.

Zeros have no value; therefore, any number multiplied by zero equals zero. For example, 21 × 0 = 0. To multiply a multiplicand by 10, add one zero. For example, to multiply 74 by 10, add one zero to the 74 to get 740 (74 × 10 = 740). Add two zeros to multiply by 100, and so on.

Division

Division is the process of determining how many times one number contains the other number. It is the reverse of multiplication. The sign ÷ (divided by) indicates division. The / and) symbols also indicate division. A *dividend* is a number to be divided. A *divisor* is a number by which division is done. A *quotient* is the result of division. A *remainder* is the part of a quotient left over when the quotient is not a whole number. **See Figure 4-5.**

To divide a number by 10, 100, 1000, or another multiple of 10, remove as many places from the right of the dividend as there are zeros in the divisor. For example, 500 ÷ 10 = 50. Notice that one zero was removed from the dividend (500) to yield the quotient of 50.

Any remainder is placed over the divisor and expressed as a fraction. For example, 27 ÷ 4 = 6¾. Notice that 4 goes into 27 six times with a remainder of 3. The 3 is then placed over the 4 (divisor) as the remainder.

To check division, multiply the divisor by the quotient. For example, 48 ÷ 4 = 12. To check this problem, multiply 4 (divisor) by 12 (quotient). For example, 4 × 12 = 48.

Fractions may represent three different concepts: a division operation, a portion of a whole, and a ratio. They can all be evaluated in the same way.

COMMON FRACTIONS

A *fraction* is a portion of a whole number. The number 1 is the smallest whole number. Anything smaller than 1 is a fraction and can be divided into any number of fractional parts. Fractions are written above and below or on both sides of a fraction bar. Fraction bars may be horizontal or inclined.

A *denominator* is the number that shows the size of the parts in a fraction. The denominator is the lower (or right-hand) number of a fraction. The *numerator* is the number of parts in a fraction. The numerator is the upper (or left-hand) number. **See Figure 4-6.** For example, the fraction ¾ shows that a whole number is divided into four equal parts (denominator), and three of these parts (numerator) are present.

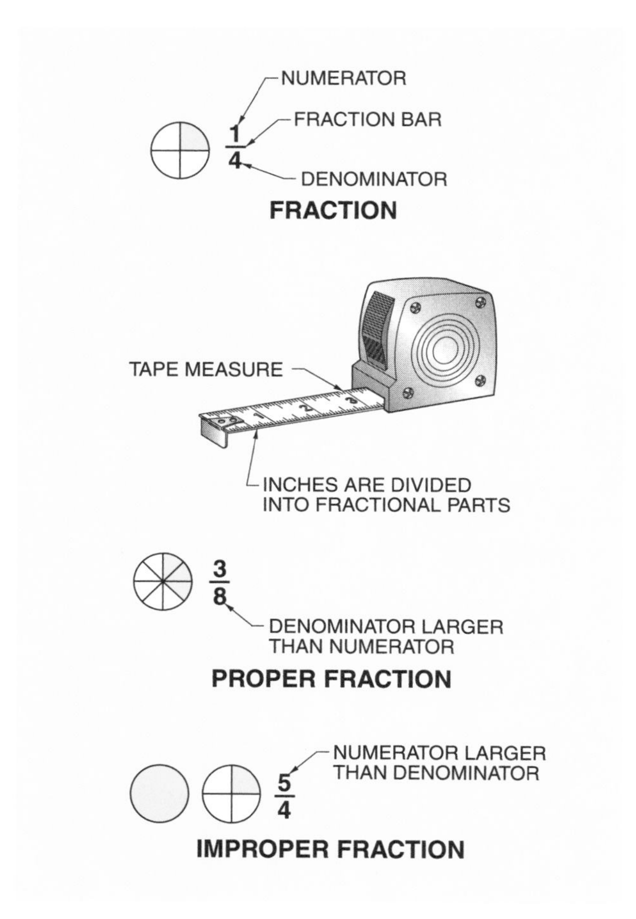

Figure 4-6. A fraction is a portion of a whole number represented by numbers on either side of a fraction bar.

A *proper fraction* is a fraction with a denominator larger than its numerator. An *improper fraction* is a fraction with a numerator larger than its denominator. A *mixed number* is a combination of a whole number and a fraction. For example, ¾ is a proper fraction, 5⁄4 is an improper fraction, and 1¼ is a mixed number.

Any type of unit can be divided into fractional parts. For example, inches are commonly divided into fractional parts of an inch based on halves, fourths, eighths, sixteenths, thirty-seconds, and sixty-fourths. Fractional parts of an inch are always expressed in their lowest terms.

The lowest term is found by dividing the fraction by the highest number that will divide equally into the denominator and numerator. For example, the lowest term of the fraction 12⁄16 is ¾. This is obtained by dividing 4 into 12 and 4 into 16. Always reduce fraction results to their lowest terms.

Adding Fractions

Fractions may be added horizontally or vertically. Horizontal placement is the most common, as identification of numerators and denominators is easier. Fractions that may be added include fractions with like denominators, fractions with unlike denominators, and mixed numbers. There is a different rule for each of these three combinations. **See Figure 4-7.**

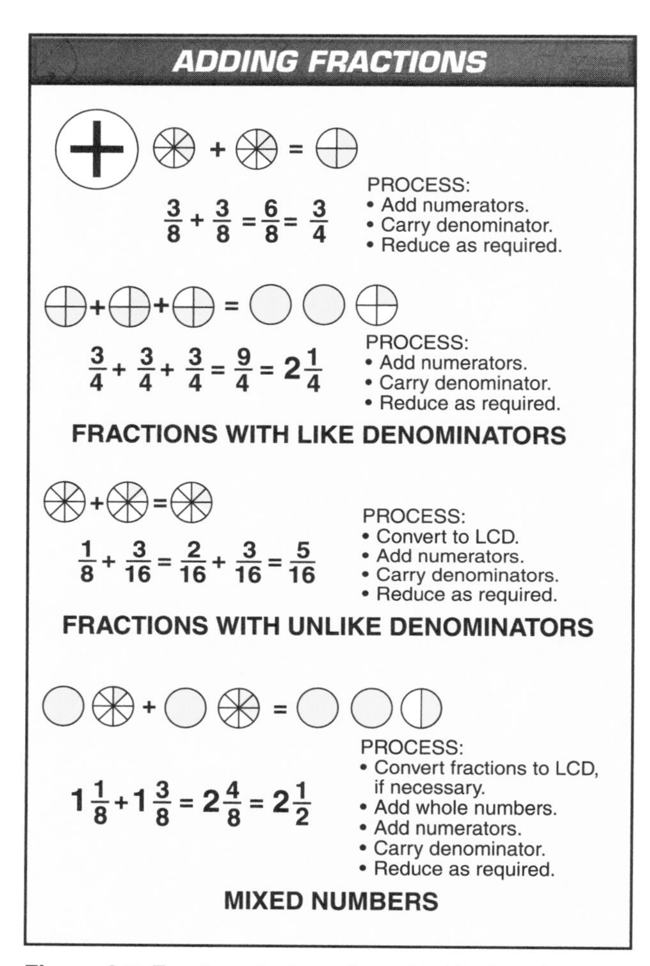

Figure 4-7. Fractions that may be added include fractions with like denominators, fractions with unlike denominators, and mixed numbers.

Adding Fractions with Like Denominators. Fractions having the same denominator are added by adding the numerators and placing them over the denominator. For example, in the problem ⅓ + ⅓ = ⅔, the numerators (1 + 1) are added to produce 2. The denominator (3) remains constant.

Fractions that produce a sum in which the numerator is larger than the denominator (improper fractions) are changed to a mixed number by dividing the numerator by the denominator, recording the quotient obtained, and treating the remainder as a numerator placed over the original denominator. For example, in the problem $3/8 + 3/8 + 3/8 = 9/8$, the improper fraction ($9/8$) is changed to $1\,1/8$ by dividing 9 by 8. The addition is then written $3/8 + 3/8 + 3/8 = 9/8 = 1\,1/8$.

Adding Fractions with Unlike Denominators. To add fractions in which the denominators are not the same, change the denominators to the lowest common denominator, add the numerators, and carry the denominator. A *lowest common denominator (LCD)* is the smallest number that can be used as a common denominator for a group of fractions. For example, to add $3/8 + 1/2 + 3/4$, the denominators should be changed to 8 because each original denominator can be multiplied by a whole number to equal 8. Multiply each numerator by the same whole number used to change its denominator to 8 to get $3/8 + 4/8 + 6/8$. Add the numerators $3 + 4 + 6 = 13$ and place over the common denominator to get $13/8$. Change the improper fraction $13/8$ by dividing 13 by 8. Thirteen can be divided by 8 one time with a remainder of 5, which is placed over the 8 to produce $1\,5/8$ ($3/8 + 1/2 + 3/4 = 3/8 + 4/8 + 6/8 = 13/8 = 1\,5/8$).

Adding Mixed Numbers. To add fractions containing mixed numbers, convert the fractions to their LCD (if necessary), add the whole numbers, add the numerators, and carry the denominator. For example, in the problem $1\,1/4 + 3\,1/4 + 4\,1/4 = 8\,3/4$, the whole numbers $(1 + 3 + 4)$ are added to produce 8. The numerators $(1 + 1 + 1)$ are added to produce 3, which is placed over the denominator 4.

Subtracting Fractions

Subtraction of fractions is similar to addition of fractions. Fractions may be subtracted horizontally or vertically. Horizontal placement is the most common, as identification of numerators and denominators is easier. All fractions must have a common denominator before one can be subtracted from another. Fractions that may be subtracted include fractions with like denominators, fractions with unlike denominators, and mixed numbers. There is a different rule for each of these three combinations. **See Figure 4-8.**

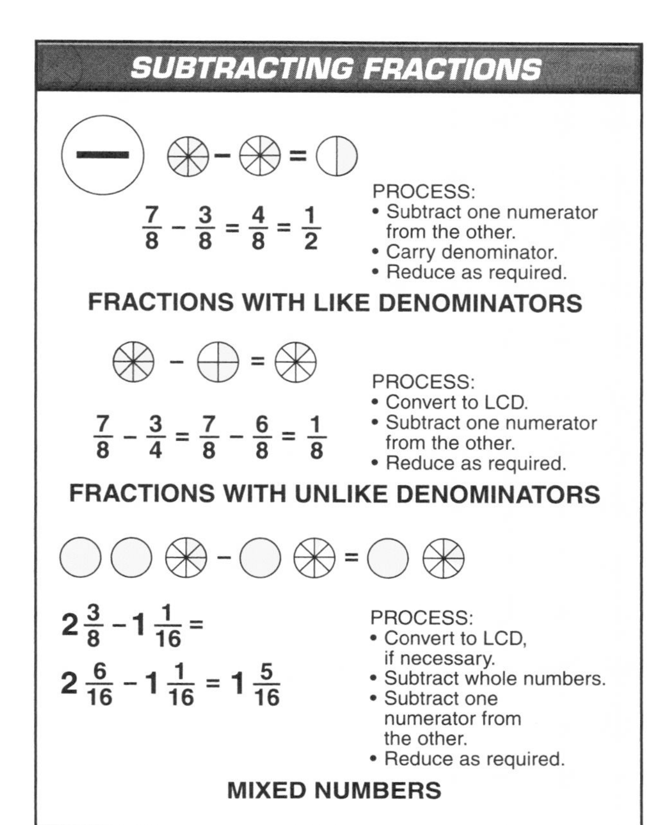

Figure 4-8. Fractions that may be subtracted include fractions with like denominators, fractions with unlike denominators, and mixed numbers.

Subtracting Fractions with Like Denominators. To subtract fractions having the same denominators, subtract one numerator from the other numerator and place over the common denominator. For example, to subtract $7/16$ from $11/16$, subtract the numerator 7 from the numerator 11 to get 4. Place the 4 over the denominator 16. Reduce $4/16$ by dividing by the largest number that goes into the numerator and denominator an even number of times. In this example, divide the numerator and denominator by 4 to get $1/4$ ($11/16 - 7/16 = 4/16 = 1/4$).

The lowest common denominator is also known as the least common denominator. The method of adding or subtracting fractions can be accomplished with any common denominator. However, using the lowest common denominator minimizes the size of the numbers involved, which generally makes the math operations easier.

Subtracting Fractions with Unlike Denominators. To subtract fractions having unlike denominators, first convert the fractions to their LCD and then subtract one numerator from the other. For example, to subtract $\frac{7}{16}$ from $\frac{3}{4}$, convert $\frac{3}{4}$ to $\frac{12}{16}$ and subtract $\frac{7}{16}$ to get $\frac{5}{16}$ ($\frac{3}{4} - \frac{7}{16} = \frac{12}{16} - \frac{7}{16} = \frac{5}{16}$).

Subtracting Mixed Numbers. To subtract fractions having mixed numbers, follow the applicable procedure for denominators, subtract the numerators, subtract the whole numbers, and, if necessary, reduce the result. For example, to subtract $1\frac{1}{4}$ from $3\frac{1}{2}$, convert the fractions to their LCD and subtract one numerator from another to get $\frac{2}{4} - \frac{1}{4} = \frac{1}{4}$. Subtract the whole numbers to get $3 - 1 = 2$. Add the whole number and the fraction to get $2 + \frac{1}{4} = 2\frac{1}{4}$.

Multiplying Fractions

Fractions may be multiplied horizontally or vertically. Horizontal placement of fractions is the most common, as identification of numerators and denominators is easier. Fractions that may be multiplied include two fractions, fractions and a whole number, a mixed number and a whole number, and two mixed numbers. There is a different rule for each of these four combinations. **See Figure 4-9.**

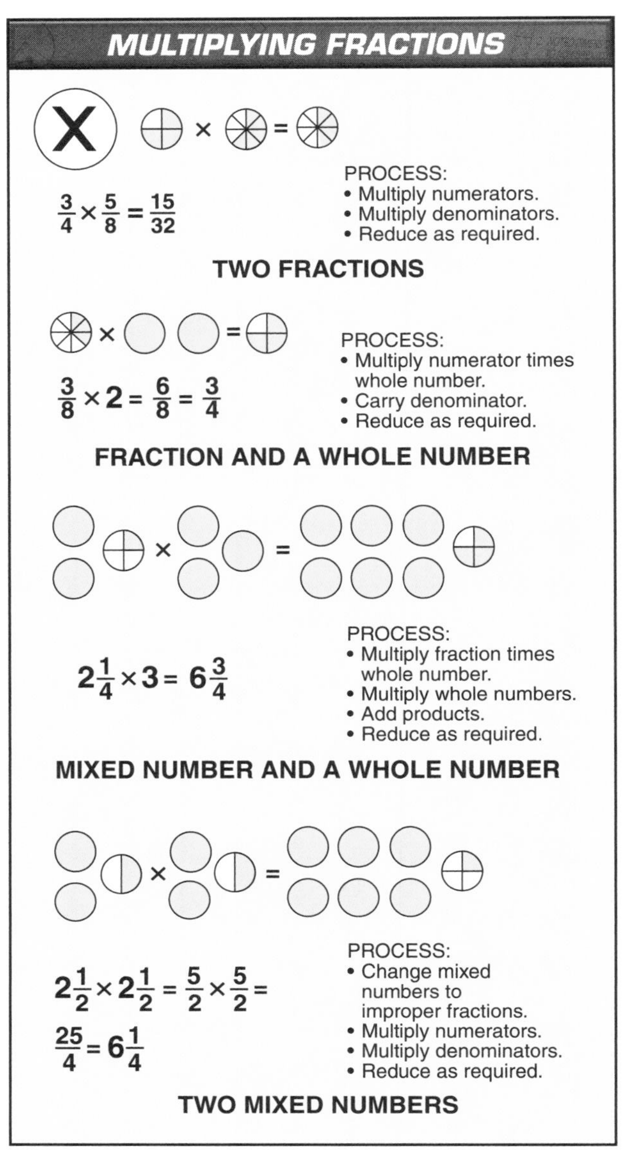

Figure 4-9. Fractions that may be multiplied include two fractions, a fraction and a whole number, a mixed number and a whole number, and two mixed numbers.

Multiplying Two Fractions. To multiply two fractions, multiply the numerator of one fraction by the numerator of the other fraction. Do the same with the denominators. Reduce the answer as required. For example, to multiply $\frac{3}{8}$ by $\frac{1}{8}$, multiply the numerators to get 3 ($3 \times 1 = 3$) and multiply the denominators to get 64 ($8 \times 8 = 64$). Thus, $\frac{3}{8} \times \frac{1}{8} = \frac{3}{64}$.

Multiplying a Fraction and a Whole Number. To multiply a fraction and a whole number, multiply the numerator of the fraction by the whole number and place over the denominator. Reduce the answer as required. For example, to multiply $\frac{1}{8} \times 3$, multiply the numerator 1 by 3 (whole number) to get 3 ($1 \times 3 = 3$) and place the 3 over the denominator 8 to get $\frac{3}{8}$. Thus, $\frac{1}{8} \times 3 = \frac{3}{8}$.

Multiplying a Mixed Number and a Whole Number. To multiply a mixed number and a whole number, multiply the fraction of the mixed number by the whole number, multiply the whole numbers, and add the two products. For example, to multiply $4\frac{7}{8} \times 3$, multiply $\frac{7}{8}$ (fraction of the mixed number) by 3 (whole number) to get $2\frac{5}{8}$ ($\frac{7}{8} \times 3 = \frac{21}{8} = 2\frac{5}{8}$). Multiply the whole numbers to get 12 ($4 \times 3 = 12$) and add the two products to get $14\frac{5}{8}$ ($2\frac{5}{8} + 12 = 14\frac{5}{8}$). Thus, $4\frac{7}{8} \times 3 = 14\frac{5}{8}$.

Multiplying Two Mixed Numbers. To multiply two mixed numbers, change both mixed numbers to improper fractions and multiply. For example, to multiply $3\frac{1}{4}$ by $4\frac{1}{2}$, change the mixed number $3\frac{1}{4}$ to the improper fraction $\frac{13}{4}$ by multiplying the whole number 3 by the denominator 4 and adding the 1 ($3 \times 4 = 12$; $12 + 1 = 13$). Change the mixed number $4\frac{1}{2}$ to the improper fraction $\frac{9}{2}$ by multiplying the whole number 4 by the denominator 2 and adding the 1 ($4 \times 2 = 8$; $8 + 1 = 9$). Multiply the improper fractions to get $12\frac{1}{8}$ ($\frac{13}{4} \times \frac{9}{2} = \frac{117}{8} = 14\frac{5}{8}$). Thus, $3\frac{1}{4} \times 4\frac{1}{2} = 14\frac{5}{8}$.

Dividing Fractions

Fractions are divided horizontally. Fractions that may be divided include a fraction by a whole number, a mixed number by a whole number, two fractions, a whole number by a fraction, and two mixed numbers. There is a different rule for each of these five combinations. **See Figure 4-10.**

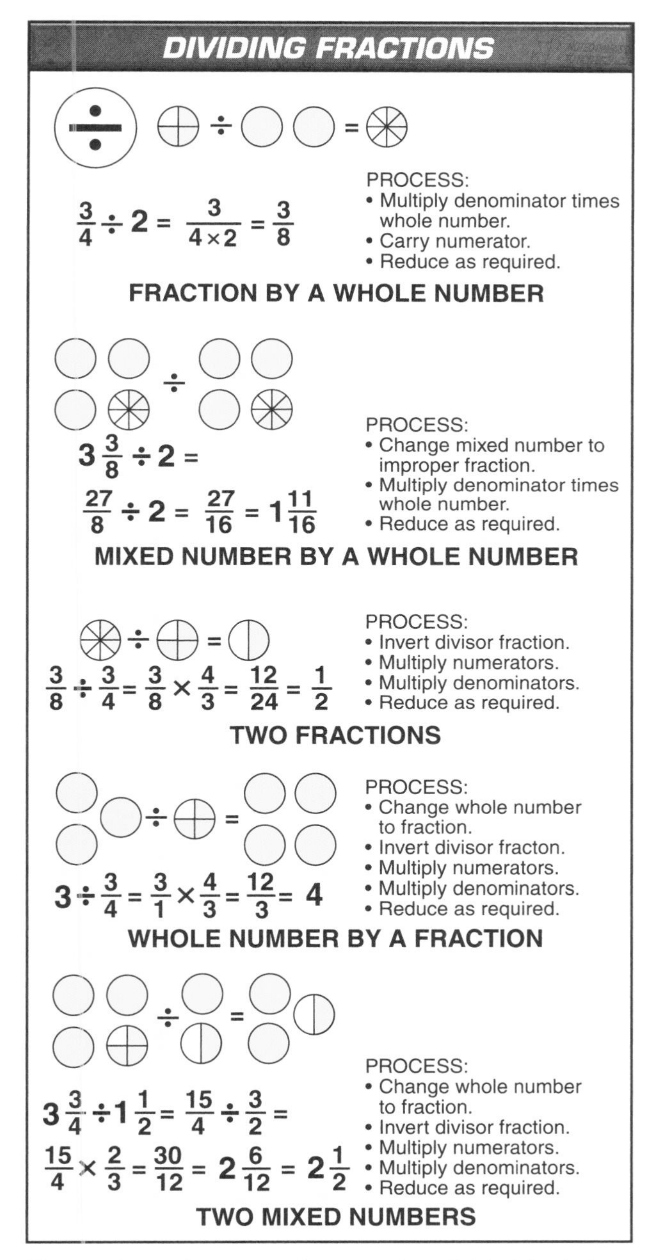

Figure 4-10. Fractions that may be divided include a fraction by a whole number, a mixed number by a whole number, two fractions, a whole number by a fraction, and two mixed numbers.

Dividing a Fraction by a Whole Number. To divide a fraction by a whole number, multiply the denominator of the fraction by the whole number. For example, to divide $3/8$ by 4, multiply the denominator 8 by the whole number 4 to get 32 ($8 \times 4 = 32$). Place the numerator 3 over the 32 to get $3/32$. Thus, $3/8 \div 4 = 3/32$.

Dividing a Mixed Number by a Whole Number. To divide a mixed number by a whole number, change the mixed number to an improper fraction and multiply the denominator of the improper fraction by the whole number. For example, to divide $2\,7/8$ by 3, change the mixed number $2\,7/8$ to $23/8$. Multiply the denominator of the improper fraction $23/8$ by the whole number 3 to get $23/24$. Thus, $2\,7/8 \div 3 = 23/24$.

Dividing Two Fractions. To divide two fractions, invert the divisor fraction and multiply the numerator by the numerator and the denominator by the denominator. For example, to divide $3/8$ by $1/4$, invert the divisor fraction $1/4$ to $4/1$ and multiply by $3/8$ to get $1\,1/2$ ($3/8 \times 4/1 = 12/8 = 1\,4/8 = 1\,1/2$). Thus, $3/8 \div 1/4 = 1\,1/2$.

Dividing a Whole Number by a Fraction. To divide a whole number by a fraction, change the whole number into fraction form, invert the divisor fraction, and multiply the numerator by the numerator and the denominator by the denominator. For example, to divide 12 by $3/4$, change the whole number 12 to fraction form $12/1$. Invert the divisor fraction 3/4 and multiply to get 16 ($12/1 \times 4/3 = 48/3 = 16$). Thus, $12 \div 3/4 = 16$.

Dividing Two Mixed Numbers. To divide two mixed numbers, change both mixed numbers to improper fractions, invert the divisor fraction, and multiply the numerator by the numerator and the denominator by the denominator. For example, to divide $12\,1/2$ by $3\,1/8$, change the mixed number $12\,1/2$ to $25/2$ and the mixed number $3\,1/8$ to $25/8$. Invert the divisor fraction $25/8$ and multiply to get 4 ($25/2 \times 8/25 = 200/50 = 4$). Thus, $12\,1/2 \div 3\,1/8 = 4$.

DECIMALS

A *decimal* is a number expressed in base 10 notation. This means that the placement of a number in relation to a decimal point indicates its relative weight in powers of 10. A *decimal point* is a period that separates a decimal's whole number from its fraction. All numbers to the left of the decimal point are whole numbers. All numbers to the right of the decimal point are a fraction.

For example, the decimal 16.485 contains 1 tens-place unit and 6 ones-place units. The first digit to the right of the decimal point indicates 4 tenths-place units, the second digit indicates 8 hundredths-place units, and the third digit indicates 5 thousandths-place units.

Working with decimals involves four mathematical operations: addition, subtraction, multiplication, and division. Certain procedures are followed to perform these operations with decimals. **See Figure 4-11.**

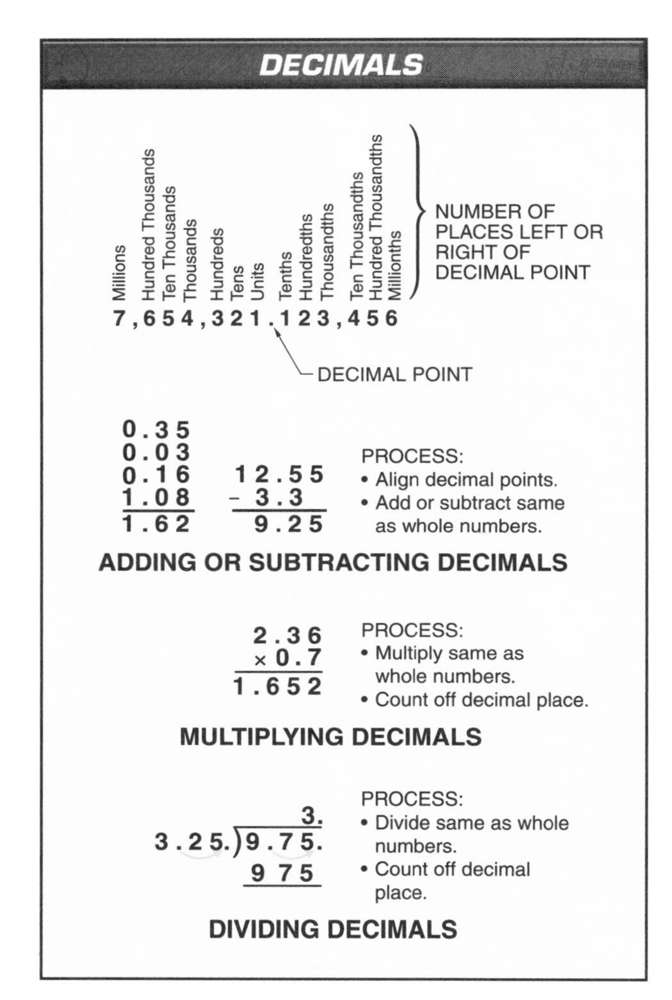

Figure 4-11. Decimal numbers represent portions of a whole number with numbers to the right of a decimal point.

Adding or Subtracting Decimals

To add or subtract decimals, align the numbers vertically on the decimal points. Thus, units will be added to or subtracted from units, tenths to tenths, hundredths to hundredths, and so on. Add or subtract as with whole numbers, and place the decimal point of the sum or remainder directly below the other decimal points. For example, to add 27.08 and 9.127, align the numbers vertically on the decimal points and add to get 36.207.

Multiplying Decimals

To multiply decimals, multiply as with whole numbers. Then begin at the right of the product and count to the left the total number of decimal places in the quantities multiplied. This is the location of the decimal point. Prefix zeros when necessary. For example, to multiply 20.45 by 3.15, align the numbers vertically on the decimal points, multiply as in whole numbers, and count four places from the right of the decimal point to get 64.4175.

Dividing Decimals

To divide decimals, divide as though the dividend and divisor are whole numbers. Then, for the decimal point, count from right to left as many decimal places as the difference between the number of decimal places in the dividend and divisor. If the dividend has fewer decimal places than the divisor, add zeros to the dividend. There must be at least as many decimal places in the dividend as in the divisor. For example, to divide 2.5 into 16.75, divide as in whole numbers and count off one decimal place from right to left to get 6.7.

Decimals are numbers that use 10 as a base. This means decimals use a set of ten symbols (digits 0 through 9) to represent numbers, and the pattern of digits repeats for every multiple of ten. Numbers can also be represented in binary (base 2), heximal (base 6), octal (base 8), hexadecimal (base 16), and other bases.

Converting Between Fractions and Decimals

Converting fractions into decimals or decimals into fractions is essential for printreading and manufacturing tasks. Part dimensions on a print may be in one system, while the necessary measuring tools or instruments may use the other system.

Any whole number components of a fraction or decimal are not involved in the conversion. Only the fractional portion must be converted. **See Figure 4-12.**

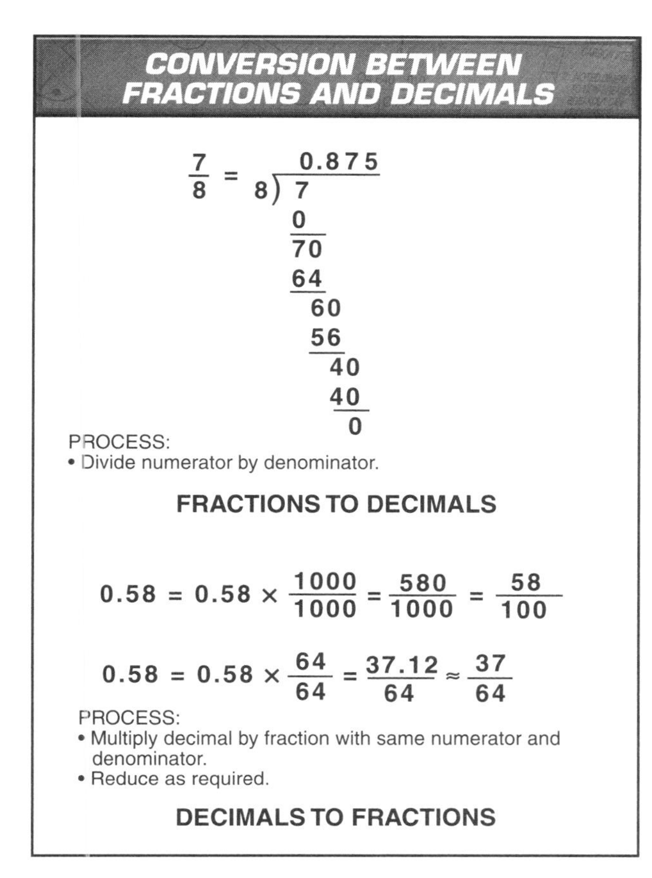

Figure 4-12. Numbers are converted between fraction and decimal form using simple math.

To convert a fraction into a decimal number, the numerator is divided by the denominator. Depending upon the numbers, the division may carry on indefinitely. The process is continued to as many decimal places as appropriate for the application.

To convert a decimal into a fraction, the decimal is multiplied by a fraction that has a denominator with the same numerical value as its numerator. There are multiple possible outcomes, depending upon the chosen denominator. A choice that simplifies the math uses a denominator that is a power of 10, such as 100 or 1000. For example, 0.58 equals $^{58}/_{100}$.

However, it may be more useful to use other denominators. For example, denominators commonly associated with inch measurements are eighths, sixteenths, thirty-seconds, and sixty-fourths. The same process is followed using one of these values as both the numerator and denominator in the fraction. When the decimal measurement does not easily convert to a practical fraction, it may be adequate to round its fraction value to the nearest whole numerator.

Significant Digits

Zeros added to the right side of a decimal point do not alter a number's value or affect calculation results. For example, 0.5, 0.50, 0.500, and 0.5000 are all the exact same value. However, the zeros convey degrees of precision. For example, a measurement of 0.5 indicates that the measurement is precise to the nearest tenth. The actual value could be anywhere between 0.45 and 0.55. However, a measurement of 0.500, while being numerically equivalent, indicates that the measurement is far more precise (to the nearest one-thousandth).

A *significant digit* is a digit that indicates the precision of a measured value. Significant digits include all digits of a number except leading and trailing zeros that are only added to indicate the placement of the decimal point. **See Figure 4-13.**

SIGNIFICANT DIGITS

NUMBER	NUMBER OF SIGNIFICANT DIGITS
345	3
56,000,000	2
5603	4
0.230	3
0.0000891	3
7.280×10^{12}	4

Figure 4-13. Significant digits indicate the precision of a number.

When performing arithmetic, results can often be calculated to many places, especially with electronic calculators. However, this may imply more precision to the value than is appropriate. Therefore, there are rules for how to round results appropriately. **See Figure 4-14.**

- When adding or subtracting numbers, the result is rounded to the same place as the least precise number. The importance is the position of the significant digits, not the number of significant digits.
- When multiplying or dividing numbers, the result is rounded to the same number of significant digits as the number with the fewest significant digits. In this case, the importance is only on the number of significant digits, not their position.

It is not necessary to consider significant digits for all mathematical problems. However, when working with measured values in scientific and technical fields, significant digits are often important.

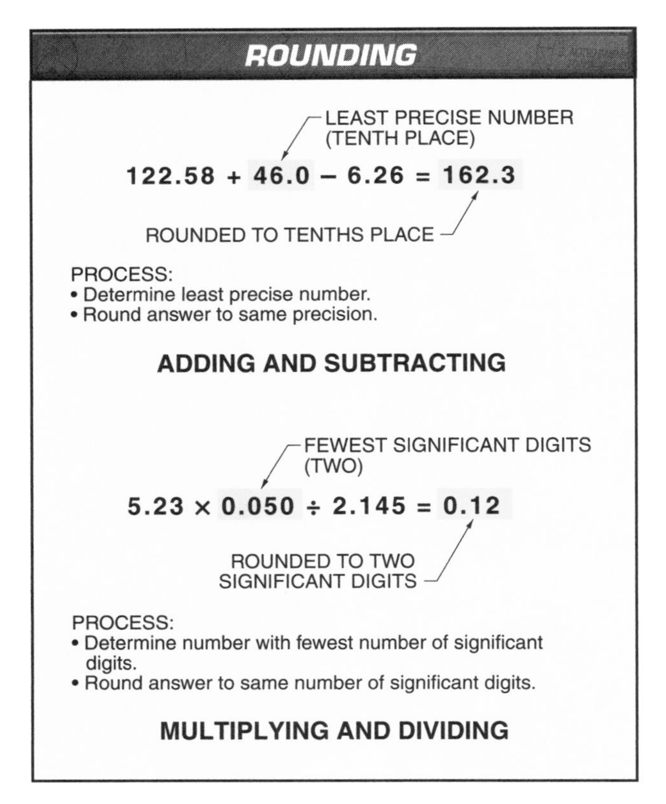

Figure 4-14. Rounding rules may require the consideration of significant digits.

PLANE FIGURES

A *plane figure* is a two-dimensional figure. Plane figures are the basis for sketching and are formed from curves or straight lines. The most common types of plane figures include circles and certain polygons.

Circles

A *circle* is a plane figure in which all points are equal distance from its center. **See Figure 4-15.** All circles contain 360°. A *circumference* is the perimeter of a circle. A *diameter* is the distance across a circle through the centerpoint. A *radius* is the distance from a circle's centerpoint to its circumference. A circle's radius is equal to one-half of its diameter.

A *chord* is a line across a circle, but not through the centerpoint. An *arc* is a portion of a circle's circumference. A *quadrant* is one-fourth of a circle. Quadrants include a right angle. A *semicircle* is one-half of a circle. *Concentric circles* are two or more circles that have different diameters but the same centerpoint. A *tangent* is a straight line touching a circle's circumference at only one point. A *secant* is a straight line touching a circle's circumference at two points.

Portions of circles can also be defined as plane figures. A *sector* is a plane figure enclosed between two radii and an arc of a circle. A *segment* is a plane figure enclosed between the arc of a circle and a chord that connects the two ends of the arc.

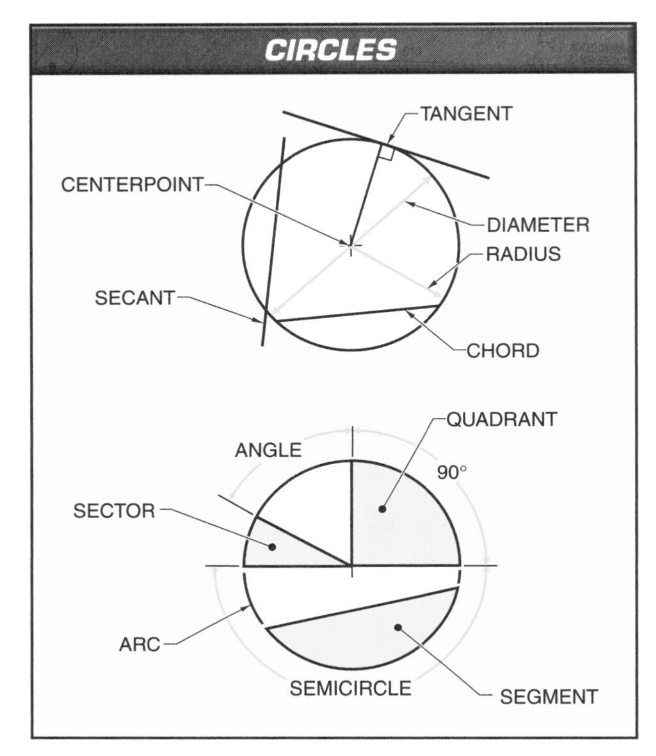

Figure 4-15. A circle is a plane figure generated around a centerpoint.

Polygons

A *polygon* is any plane figure with three or more straight sides. **See Figure 4-16.** Polygons are categorized by their number of sides, the length of their sides, and the parallelism of opposite sides. A *regular polygon* is a polygon with sides that are all of equal length.

Some polygons are defined by their included angles. The type of angle is defined by the distance between its two legs, which is measured in degrees. **See Figure 4-17.** An *acute angle* is any angle smaller than 90°. A *right angle* is an angle that is exactly 90°. Right angles may be identified on drawings with a small square in the corner. An *obtuse angle* is any angle greater than 90°. A *straight angle* is an angle that is exactly 180°.

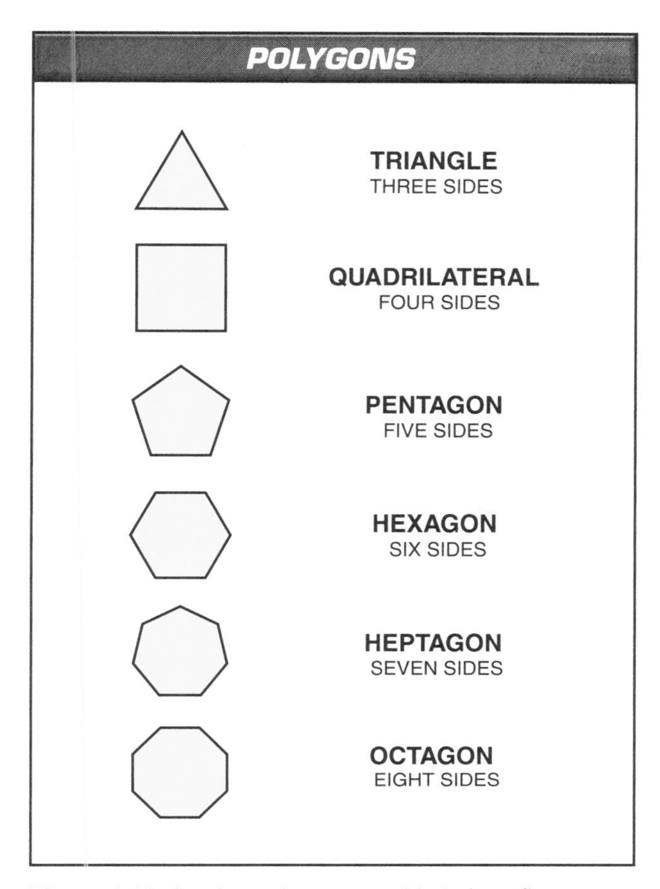

Figure 4-16. A polygon is a many-sided plane figure.

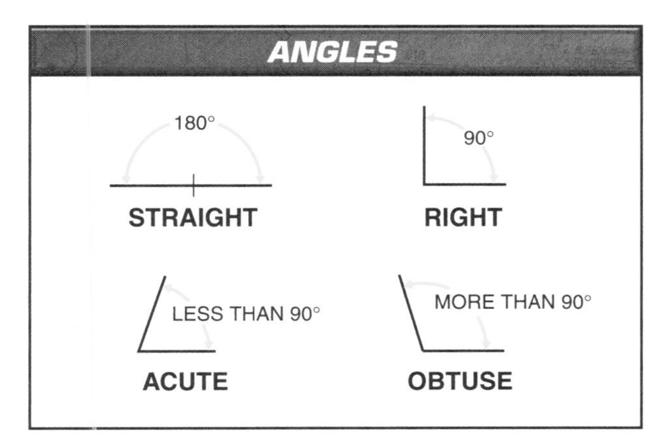

Figure 4-17. Angles are named for the distance between their two legs.

Some triangles may fit more than one description for angle or side length. For example, a triangle may be obtuse isosceles or right scalene.

The most common polygons are the triangle (three sides) and the quadrilateral (four sides). Polygons with five or more sides are named by combining a prefix, which indicates the number of sides, with the ending "gon." For example, a pentagon has five sides, a hexagon has six sides, and an octagon has eight sides.

Triangles. A *triangle* is a polygon with three sides. **See Figure 4-18.** The sum of a triangle's three angles always equals 180°. The symbol Δ indicates a triangle. The points of a triangle are named with uppercase letters. The sides and angles of a triangle are named by lowercase letters. For example, a triangle may be named ΔABC and contain sides x, y, and z and angles a, b, and c.

The *altitude* is the dimension of a triangle that is perpendicular to the base. The *base* is the horizontal side of a triangle. Any side can be potentially considered the base.

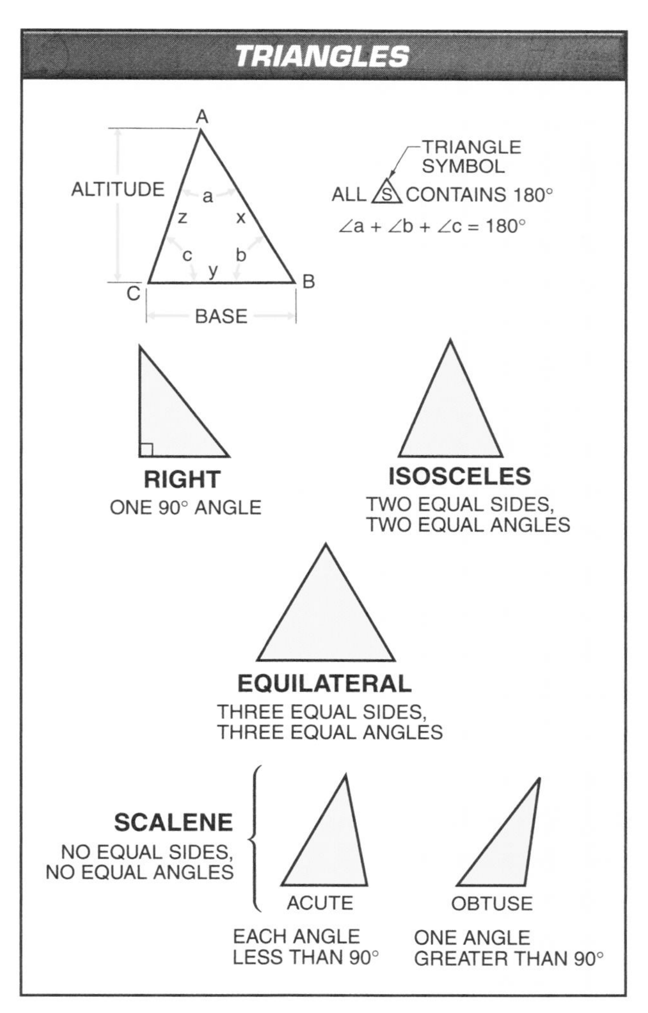

Figure 4-18. A triangle is a three-sided plane figure.

A *right triangle* is a triangle with one right angle. An *oblique triangle* is a triangle that does not contain a right angle. An *isosceles triangle* is a triangle that contains two equal angles and two equal sides. An *equilateral triangle* is a triangle that has three equal angles and three equal sides. Each angle of an equilateral triangle is 60°. A *scalene triangle* is a triangle that has no equal angles or equal sides.

Triangles may be acute or obtuse. An *acute triangle* is a triangle with each angle less than 90°. An *obtuse triangle* is a triangle with one angle greater than 90°.

Quadrilaterals. A *quadrilateral* is a polygon with four sides. The sum of the four angles of a quadrilateral is always 360°. The types of quadrilaterals include rectangles, squares, rhombuses, rhomboids, trapezoids, and trapeziums. **See Figure 4-19.**

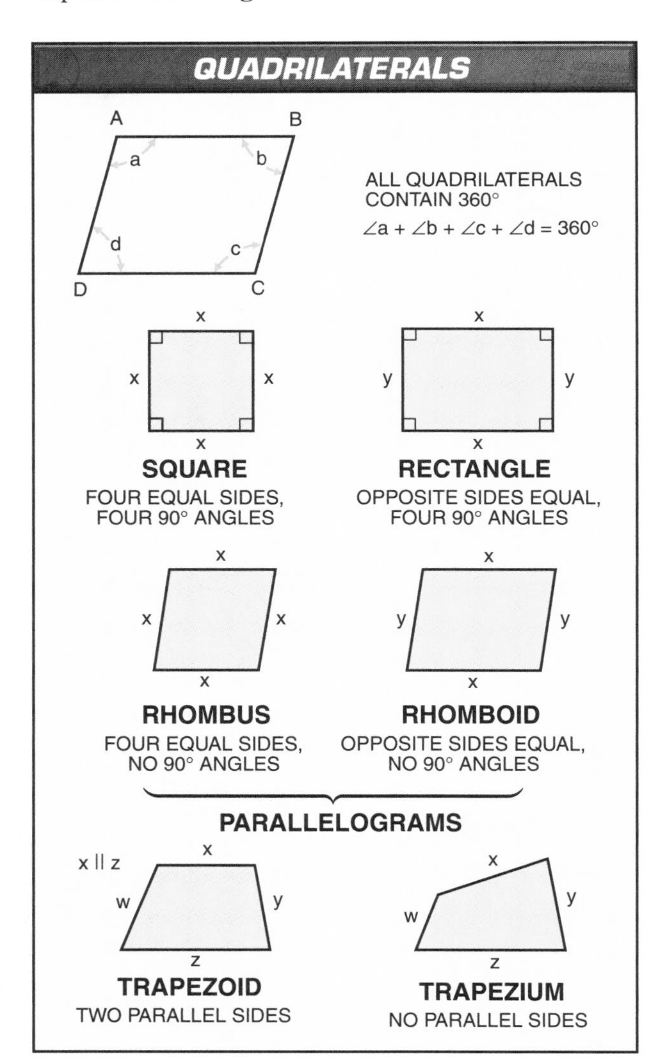

Figure 4-19. A quadrilateral is a four-sided plane figure.

A *rectangle* is a quadrilateral with opposite sides that are equal and four right angles. A *square* is a quadrilateral with sides that are all equal and four right angles. A *rhombus* is a quadrilateral with sides that are all equal and no right angles. A *rhomboid* is a quadrilateral with opposite sides that are equal and no right angles. The rectangle, square, rhombus, and rhomboid are all parallelograms. A *parallelogram* is a four-sided plane figure with opposite sides that are parallel and equal.

A *trapezoid* is a quadrilateral with only two sides that are parallel. A *trapezium* is a quadrilateral with no sides that are parallel. Trapezoids and trapeziums are not parallelograms because not all opposite sides are parallel.

SOLID FIGURES

A *solid figure* is a three-dimensional figure. **See Figure 4-20.** These figures have length, height, and depth. Solid figures are represented in two dimensions using various combinations of plane figures.

A *regular solid figure* is a solid figure with faces that are regular polygons. A *tetrahedron* is a regular solid figure formed from four triangles. This is commonly called a triangular pyramid. A *hexahedron* is a regular solid figure formed from six squares. It is commonly referred to as a cube. Other regular solids include the octahedron, dodecahedron, and icosahedron.

Other common solid figures include prisms, cylinders, pyramids, cones, and spheres. A *prism* is a solid figure with two bases that are identical, parallel polygons. A *base* is a polygon at an end of a solid figure. A prism can be triangular, rectangular, pentagonal, and so on, according to the shape of its bases.

A *cylinder* is a solid figure with circles as its bases. A *pyramid* is a solid figure with a base that is a polygon and sides that are triangles. The *vertex* is the common point of the triangular sides that form a pyramid. A *cone* is a solid figure with a circular base and a surface that tapers from the base to the vertex. A *sphere* is a solid figure generated by a circle revolving about an axis.

A *lateral face* is the side of a solid figure. For prisms, there are as many of these lateral faces as there are sides of one of the bases. The altitude of a solid is the perpendicular distance between the two bases or between the base and vertex. When the bases are perpendicular to the faces, the altitude equals the length of an edge of a lateral face.

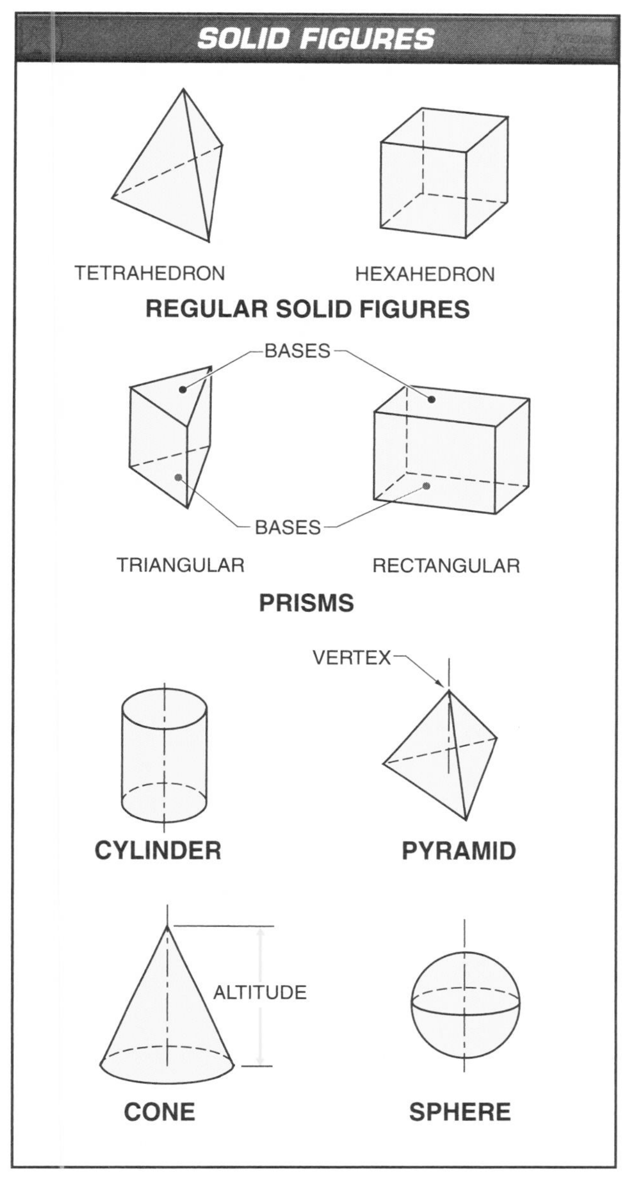

Figure 4-20. Solid figures have length, height, and depth.

COMMON MATH FORMULAS

An *equation* is a statement of equality between two mathematical expressions. In every equation, the numerical value on the left side of the equal sign is equal to the numerical value on the right side of the equal sign. For example, $2 + \frac{1}{2} = \frac{4}{8} \times 5$.

Equations with unknown values (variables) are solved by rearranging and converting the terms into an expression that is equivalent to a single unknown value. A *variable* is a symbol used as a substitute for any real number. Variables are usually represented by Latin or Greek letters. For example, the equation $a + 2 = b - 5$ can be rearranged in a way to define the value of the variable a in relation to the other terms. In this case, $a = b - 7$. Alternatively, the expression can be solved as $b = a + 7$.

A *formula* is an equation involving multiple variables that has been solved for one of the variables. Many formulas are already well established as a method for calculating a certain quantity. For example, the conversion of measurements from one unit to another involves a common type of formula. **See Figure 4-21.** The conversion of millimeters to inches requires the following formula:

$L_{in.} = L_{mm} \times 0.0394$

where

$L_{in.}$ = length (in in.)

L_{mm} = length (in mm)

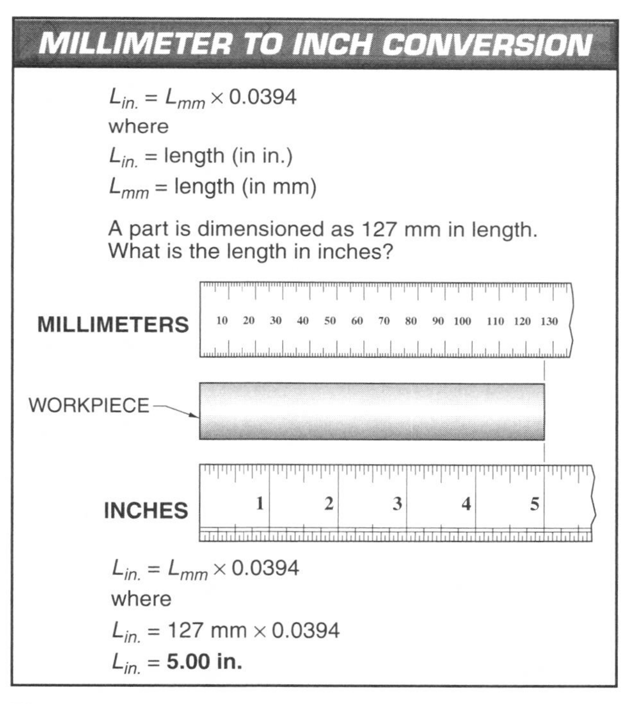

Figure 4-21. Machinists often use simple formulas to convert measurement units.

Formulas may also be rearranged to solve for one of the other represented values. For example, the millimeter to inch conversion formula may be rearranged to convert inches to millimeters instead:

$$L_{mm} = L_{in.} \times \frac{1}{0.0394} = L_{in.} \times 25.4$$

In the metal trades, common formulas related to plane and solid figures are used when laying out jobs. **See Figure 4-22.** For example, a welder may be required to lay out and build a cylindrical tank to hold a specified number of gallons of liquid. By applying the volume formula for cylinders, the welder can determine the size of the cylindrical tank.

The value known as pi (π) is a constant equal to the circumference of a circle divided by its diameter. This ratio is identical for a circle of any size. The value is a never-ending and never-repeating decimal. For most applications using just 3.1416 is sufficient.

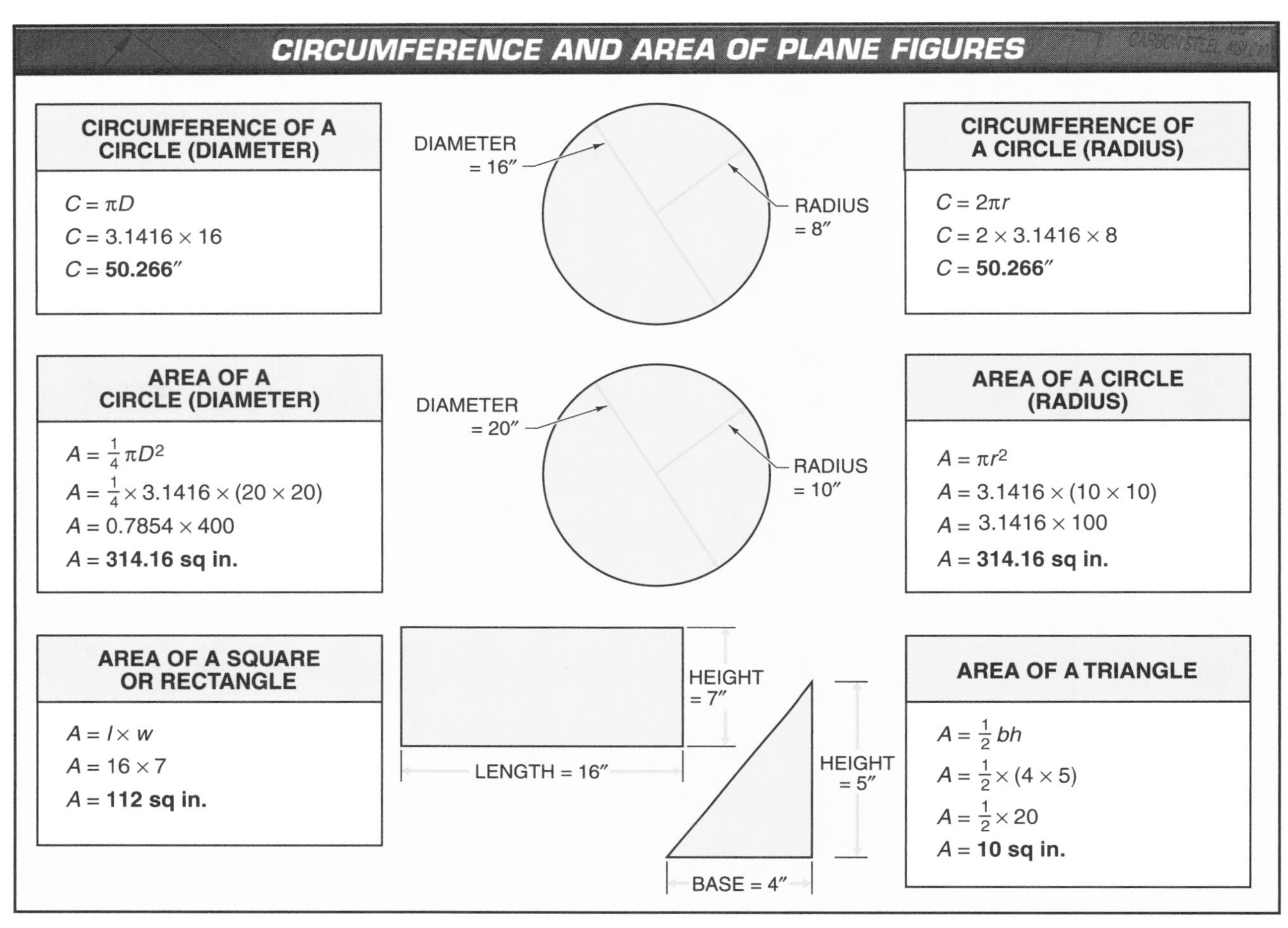

Figure 4-22. Plane figure formulas are used to find circumference and area.

Circumference of a Circle

When the diameter is known, the circumference of a circle is calculated using the following formula:

$C = \pi D$

where

C = circumference

π = 3.1416

D = diameter

For example, what is the circumference of a 20″ diameter circle?

$C = \pi D$

$C = 3.1416 \times 20$

$C = $ **62.83″**

The diameter of a circle equals two times its radius. When the radius is known instead of the diameter, the circumference of a circle is found by substituting $2r$ for D in the above formula, resulting in the following formula:

$C = 2\pi r$

where

C = circumference

π = 3.1416

r = radius

For example, what is the circumference of a 10″ radius circle?

$C = 2\pi r$

$C = 2 \times 3.1416 \times 10''$

$C =$ **62.83″**

If either radius or diameter is known, the other can be found by doubling or halving, respectively.

Area

Area is a quantity that represents the size of a surface or two-dimensional shape. Area is commonly expressed in square units. For example, a standard size piece of plywood contains 32 sq ft (4′ × 8′ = 32 sq ft). A square inch measures 1″ × 1″ or its equivalent. A square foot contains 144 sq in. (12″ × 12″ = 144 sq in.). The area of any plane figure can be determined by applying the proper formula.

Area of a Circle. When the diameter is known, the area of a circle is calculated using the following formula:

$A = \frac{1}{4} \pi D^2$

where

A = area

$\pi = 3.1416$

D = diameter

For example, what is the area of a 28″ diameter circle?

$A = \frac{1}{4} \pi D^2$

$A = \frac{1}{4} \times 3.1416 \times (28 \times 28)$

$A = 0.7854 \times 784$

$A =$ **615.75 sq in.**

When the radius is known, the area of a circle is found by substituting $2r$ for D in the above formula and reducing the terms, resulting in the following formula:

$A = \frac{1}{4} \pi (2r)^2 = \frac{1}{4} \pi 4r^2 = \pi r^2$

where

A = area

$\pi = 3.1416$

r = radius

For example, what is the area of a 14″ radius circle?

$A = \pi r^2$

$A = 3.1416 \times (14 \times 14)$

$A = 3.1416 \times 196$

$A =$ **615.75 sq in.**

Area of a Square or Rectangle. The area of a square or the area of a rectangle is calculated using the following formula:

$A = l \times w$

where

A = area

l = length

w = width

For example, what is the area of a 22′-0″ × 16′-0″ storage room?

$A = l \times w$

$A = 22 \times 16$

$A =$ **352 sq ft**

Area of a Triangle. The area of a triangle is calculated using the following formula:

$A = \frac{1}{2} bh$

where

A = area

b = base

h = height (or altitude)

For example, what is the area of a triangle with a 10″ base and a 12″ height?

$A = \frac{1}{2} bh$

$A = \frac{1}{2} \times (10 \times 12)$

$A = \frac{1}{2} \times 120$

$A =$ **60 sq in.**

Pythagorean Theorem

The *Pythagorean theorem* is a theorem that states that the square of the hypotenuse of a right triangle is equal to the sum of the squares of the other two sides. The *hypotenuse* is the side of a right triangle opposite the right angle. One special case of the Pythagorean theorem shows that the sides of a right triangle can have a perfect 3-4-5 relationship (or multiples of 3-4-5). This simple relationship is often used in laying out right angles and checking corners for squareness. **See Figure 4-23.** The length of the hypotenuse of a right triangle is calculated using the following equation:

$c^2 = a^2 + b^2$

where

c = length of hypotenuse

a = length of one side

b = length of other side

For example, what is the length of the hypotenuse of a triangle having sides of 3′ and 4′?

$$c^2 = a^2 + b^2$$
$$c = \sqrt{a^2 + b^2}$$
$$c = \sqrt{(3 \times 3) + (4 \times 4)}$$
$$c = \sqrt{9 + 16}$$
$$c = \sqrt{25}$$
$$c = \mathbf{5'}$$

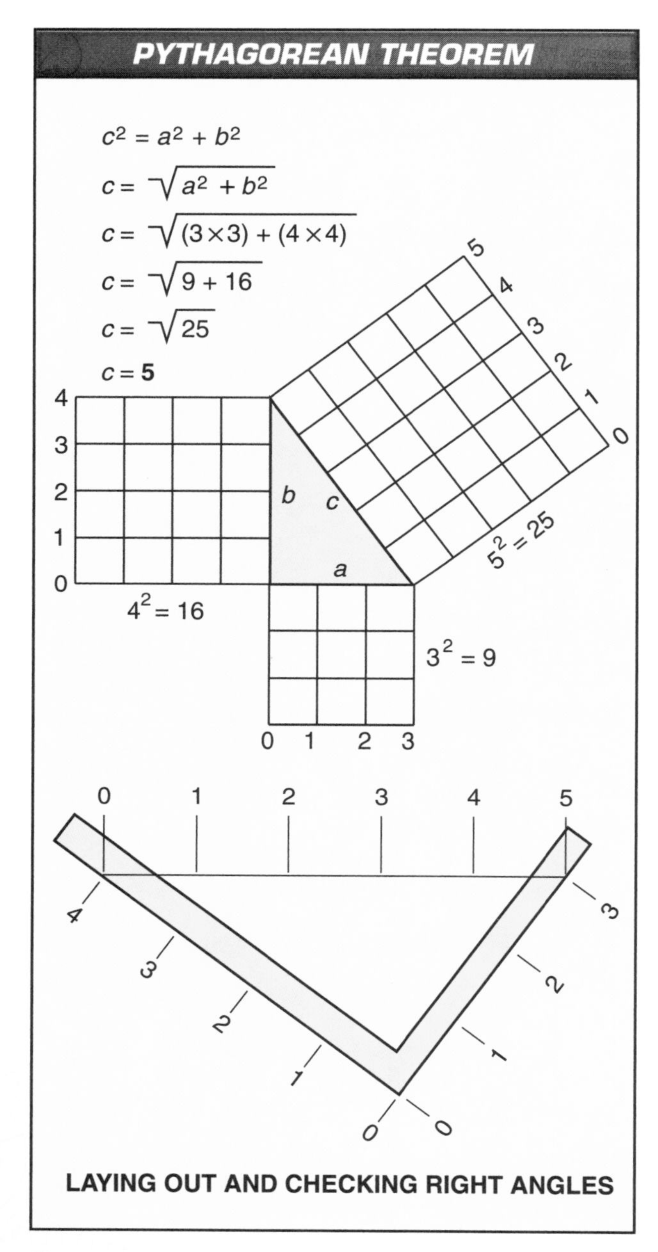

Figure 4-23. The square of the hypotenuse of a right triangle is equal to the sum of the squares of the other two sides of the triangle.

Volume

Volume is a quantity that represents the size of a three-dimensional object. Volume is commonly expressed in cubic units. For example, the volume of a standard size concrete block is 1024 cu in. (8″ × 8″ × 16″ = 1024 cu in.). A cubic inch measures 1″ × 1″ × 1″ or its equivalent. A cubic foot contains 1728 cu in. (12″ × 12″ × 12″ = 1728 cu in.). A cubic yard contains 27 cu ft (3′ × 3′ × 3′ = 27 cu ft). The volume of a solid figure can be determined by applying the proper formula. **See Figure 4-24.**

Volume of a Rectangular Solid. The volume of a rectangular solid is calculated using the following formula:

$V = l \times w \times h$

where

V = volume

l = length

w = width

h = height

Volume of a Cylinder. The formula for the volume of a cylinder includes the formula for the area of the base circle, multiplied by the height of the cylinder. When the diameter of the circle is known, the volume of a cylinder is calculated using the following formula:

$V = \frac{1}{4}\pi D^2 \times l$

where

V = volume

π = 3.1416

D = diameter

l = length (or height)

When the radius of the circle is known, the volume of a cylinder is calculated using the following formula:

$V = \pi r^2 \times l$

where

V = volume

π = 3.1416

r = radius

l = length (or height)

The first proof of the Pythagorean theorem is attributed to the Greek philosopher and mathematician Pythagoras, who lived in the sixth century BCE. However, there is evidence that the idea behind the theorem was already known to the Babylonians and Indians in prior centuries.

VOLUME OF SOLIDS

VOLUME OF A RECTANGULAR SOLID

$V = l \times w \times h$

$V = 19 \times 10 \times 7$

$V = \mathbf{1330\ cu\ in.}$

$V = 1330\text{ cu in.} \times \frac{1\text{ cu ft}}{1728\text{ cu in.}}$

$V = \mathbf{0.77\ cu\ ft}$

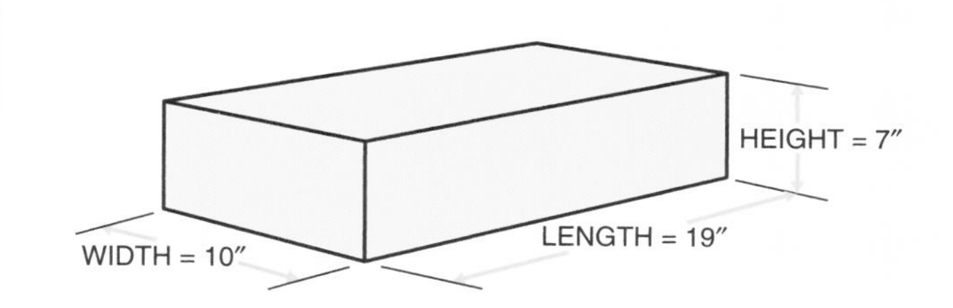

VOLUME OF A CYLINDER (DIAMETER)

$V = \frac{1}{4}\pi D^2 \times l$

$V = \frac{1}{4} \times 3.1416 \times (16 \times 16) \times 60$

$V = 0.7854 \times 256 \times 60$

$V = \mathbf{12{,}064\ cu\ in.}$

$V = 12{,}064\text{ cu in.} \times \frac{1\text{ cu ft}}{1728\text{ cu in.}}$

$V = \mathbf{6.98\ cu\ ft}$

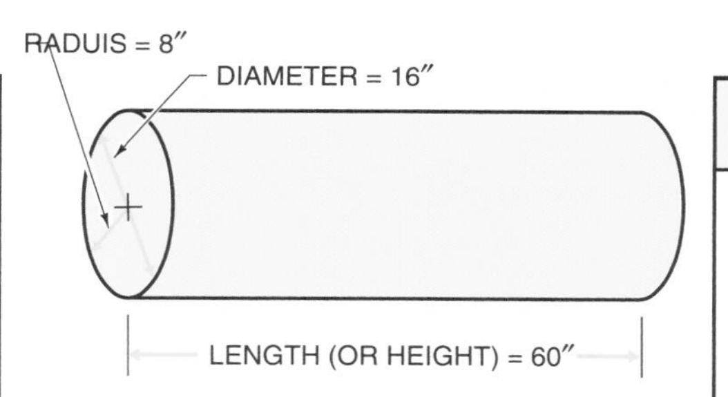

VOLUME OF A CYLINDER (RADIUS)

$V = \pi r^2 \times l$

$V = 3.1416 \times (8 \times 8) \times 60$

$V = 3.1416 \times 64 \times 60$

$V = \mathbf{12{,}064\ cu\ in.}$

$V = 12{,}064\text{ cu in.} \times \frac{1\text{ cu ft}}{1728\text{ cu in.}}$

$V = \mathbf{6.98\ cu\ ft}$

VOLUME OF A SPHERE (DIAMETER)

$V = \frac{1}{6}\pi D^3$

$V = \frac{1}{6} \times 3.1416 \times (7 \times 7 \times 7)$

$V = 0.5236 \times 343$

$V = \mathbf{179.59\ cu\ ft}$

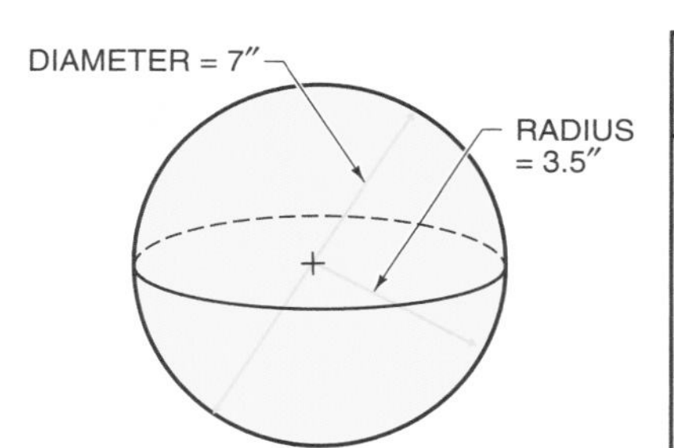

VOLUME OF A SPHERE (RADIUS)

$V = \frac{4}{3}\pi r^3$

$V = \frac{4}{3} \times 3.1416 \times (3.5 \times 3.5 \times 3.5)$

$V = 4.1888 \times 42.875$

$V = \mathbf{179.59\ cu\ ft}$

VOLUME OF A CONE

1. Solve for area of base

$A_b = \frac{1}{4}\pi D^2$

$A_b = \frac{1}{4} \times 3.1416 \times (3.25 \times 3.25)$

$A_b = 0.7854 \times 10.5625$

$A_b = \mathbf{8.30\ sq\ ft}$

2. Solve for volume

$V = \frac{1}{3}A_b a$

$V = \frac{1}{3} \times 8.30 \times 5$

$V = \frac{1}{3} \times 41.5$

$V = \mathbf{13.83\ cu\ ft}$

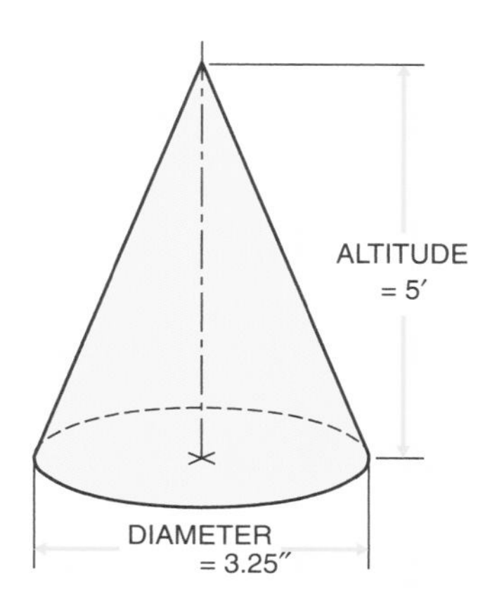

Figure 4-24. Volume is the three-dimensional size of an object.

Volume of a Sphere. When the diameter is known, the volume of a sphere is calculated using the following formula:

$V = \frac{1}{6}\pi D^3$

where

V = volume

π = 3.1416

D = diameter

When the radius is known, the volume of a sphere is calculated using the following formula:

$V = \frac{4}{3}\pi r^3$

where

V = volume

π = 3.1416

r = radius

Volume of a Cone. The volume of a cone is found by first solving for the area of the base and then solving for volume. The area of the base is calculated using the following formula:

$A_b = \frac{1}{4}\pi D^2$

or

$A_b = \pi r^2$

where

A_b = area of base

π = 3.1416

D = diameter

r = radius

The volume of the cone is then calculated using the following formula:

$V = \frac{1}{3}A_b \times a$

where

V = volume

A_b = area of base

a = altitude

REVIEW QUESTIONS 4

Name ______________________________ Date ________________

True-False

T F **1.** Division is the process of adding one number as many times as there are units in the other number.

T F **2.** An odd number cannot be divided by 2 an exact number of times.

T F **3.** A proper fraction has a numerator larger than its denominator.

T F **4.** An improper fraction can be changed to a mixed number.

T F **5.** The denominator is the number that shows the size of the parts in a fraction.

T F **6.** The placement of a number in relation to a decimal point indicates its relative weight in powers of 10.

T F **7.** The diameter of a circle equals one-half its radius.

T F **8.** Any number multiplied by a zero equals that number.

T F **9.** To check division, multiply the sum by the quotient.

T F **10.** A decimal point is a period that separates a decimal's whole number from its fraction.

T F **11.** A circle is a plane figure.

T F **12.** Plane figures are formed from curves or straight lines.

T F **13.** Area is commonly expressed in square units.

T F **14.** A lowest common denominator is the smallest numerator that can be used as a common denominator for a group of fractions.

T F **15.** To convert a fraction into a decimal number, the numerator is divided by the denominator.

T F **16.** A trapezoid is a parallelogram.

T F **17.** A scalene triangle contains two equal sides and two equal angles.

T F **18.** An octagon is a plane figure with six sides.

T F **19.** The hypotenuse is the side of a right triangle opposite the right angle.

T F **20.** A cubic foot contains 1278 cu in.

Completion

_______________ **1.** A(n) ___ number is a number with no fractional or decimal parts.

_______________ **2.** A(n) ___ number is a number that can be divided by 2 an exact number of times.

_______________ **3.** A(n) ___ angle is an angle that is exactly 90°.

_______________ **4.** A(n) ___ is the result of division.

_______________ **5.** ___ is the process of uniting two or more numbers to make one number.

_______________ **6.** ___ is the opposite of addition.

_______________ **7.** A decimal is a number expressed in base ___ notation.

_______________ **8.** A(n) ___ is any plane figure with three or more straight sides.

_______________ **9.** An equation is a statement of ___ between two mathematical expressions.

_______________ **10.** A(n) ___ is a symbol used as a substitute for any real number.

Identification — Math

_______________ **1.** Minuend

_______________ **2.** Remainder

_______________ **3.** Multiplier

_______________ **4.** Divisor

_______________ **5.** Sum

_______________ **6.** Dividend

_______________ **7.** Quotient

_______________ **8.** Subtrahend

_______________ **9.** Product

_______________ **10.** Multiplicand

(A)

217
+ 31
248

(B) (C) (D)

853
−142
711

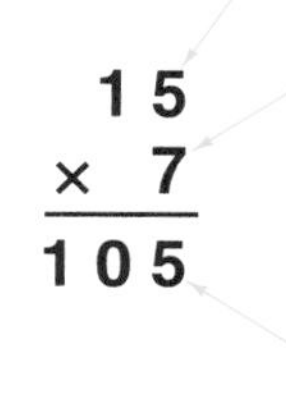

Multiple Choice

_______________ **1.** A ___ number is a number that can be divided an exact number of times only by itself and the number 1.

A. proper
B. period
C. product
D. prime

_______________ **2.** The most common operation in mathematics is ___.

A. addition
B. subtraction
C. multiplication
D. division

_______________ **3.** A(n) ___ fraction has a denominator larger than its numerator.

A. proper
B. improper
C. mixed
D. none of the above

_______________ **4.** The numerator is the ___-hand number of a fraction.

A. lower or right
B. lower or left
C. upper or right
D. upper or left

_______________ **5.** Fractional parts of an inch are always expressed in their ___ terms.

A. highest
B. most
C. lowest
D. none of the above

_______________ **6.** A ___ is a digit that indicates the precision of a measured value.

A. decimal
B. significant digit
C. leading zero
D. prime number

_______________ **7.** A(n) ___ triangle is a triangle that has no equal angles or equal sides.

A. scalene
B. isosceles
C. regular
D. none of the above

_______________ **8.** A ___ is the distance across a circle through its centerpoint.

A. circumference
B. radius
C. diameter
D. chord

____________ **9.** All fractions must have a common ___ before one can be subtracted from another.

A. sum
B. remainder
C. numerator
D. denominator

____________ **10.** ___ is a quantity that represents the size of a surface or two-dimensional shape.

A. Circumference
B. Hypotenuse
C. Volume
D. Area

Identification — Circle

____________ **1.** Diameter

____________ **2.** Sector

____________ **3.** Arc

____________ **4.** Segment

____________ **5.** Radius

____________ **6.** Tangent

____________ **7.** Centerpoint

____________ **8.** Circle

____________ **9.** Chord

____________ **10.** Quadrant

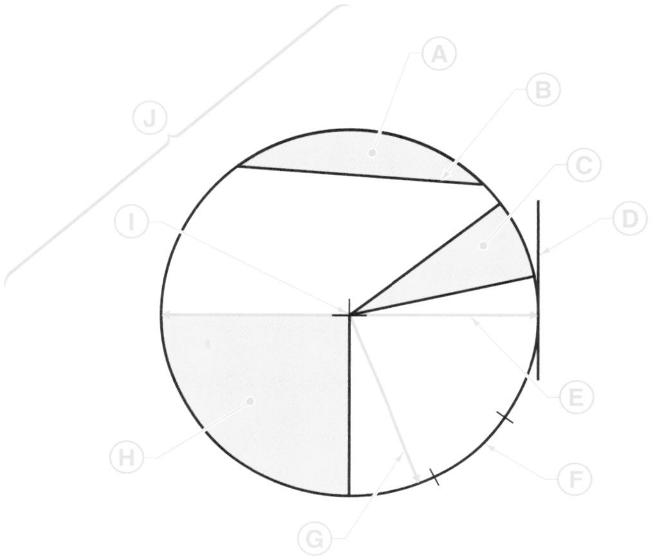

TRADE COMPETENCY TEST 4

Name ______________________ Date ______________

Word Problems

______ **1.** Four pieces (A, B, C, and D) were cut from the 96″ long piece of cold rolled steel. What is the length of the remaining piece? (Disregard the cutting waste.)

______ **2.** A machinist spends 3¼ hours cutting and grinding steel to shape and 1½ hours positioning, machining, and finishing parts to complete a job. How many hours did the job require?

______ **3.** A riveted assembly requires 32 rivets. How many rivets are required for 124 riveted assemblies?

______ **4.** How many 2¼″ × 16″ strips can be cut from the piece of 16″ by 48″ sheet metal? (Disregard the cutting waste.)

______ **5.** What is the volume of the tank in cubic feet?

______ **6.** What is the length of Side c of Triangle A?

______ **7.** What is the length of Side c of Triangle A in inches?

______ **8.** What is the area of Triangle B?

______ **9.** What is the circumference of the circle?

______ **10.** What is the volume of the rectangular solid?

16″ 21″ $3\frac{7}{8}$″ $3\frac{7}{8}$″ 96″
A B C D COLD ROLLED STEEL
48″ 16″ SHEET METAL
2′-0″ 6′-0″ TANK
TRIANGLE A 8′-0″ b c a 6′-0″
TRIANGLE B 13″ 10″
CIRCLE 30″
11″ 14″ 26″ RECTANGULAR SOLID
SPHERE 4′-0″
SQUARE $7\frac{1}{4}$″ $7\frac{1}{4}$″

______ **11.** What is the area of the circle?

______ **12.** What is the area of Triangle A?

______ **13.** What is the volume of the sphere?

______ **14.** What is the total length of all the sides of the square?

______ **15.** What is the area of the square?

Dowel Positioner

Refer to print on page 109.

______ **1.** The overall length of Part B is ___″.

______ **2.** Part ___ has rounded corners.

T F **3.** Four 3/8″ diameter holes are drilled in Part A.

______ **4.** The horizontal center-to-center distance between the holes in Part B is ___″.

______ **5.** The front surface area of Part B is ___ sq in. (Disregard the drilled holes.)

______ **6.** Part B has a volume of ___ cu in. (Disregard the drilled holes.)

______ **7.** The centerpoints of the two upper holes in Part B are ___″ from the top edge of Part A.

______ **8.** The holes in Part A are ___″ farther apart than the holes in Part B.

T F **9.** The overall height of the Dowel Positioner is 4 1/4″.

T F **10.** The Dowel Positioner is made from 1/2″ thick stock.

T F **11.** The holes in Part B are larger than the holes in Part A.

T F **12.** Part B is perpendicular to Part A.

______ **13.** The radius of each rounded corner is ___″.

______ **14.** All drilled holes are shown in the right side view with ___ lines.

______ **15.** The overall depth of the Dowel Positioner is ___″.

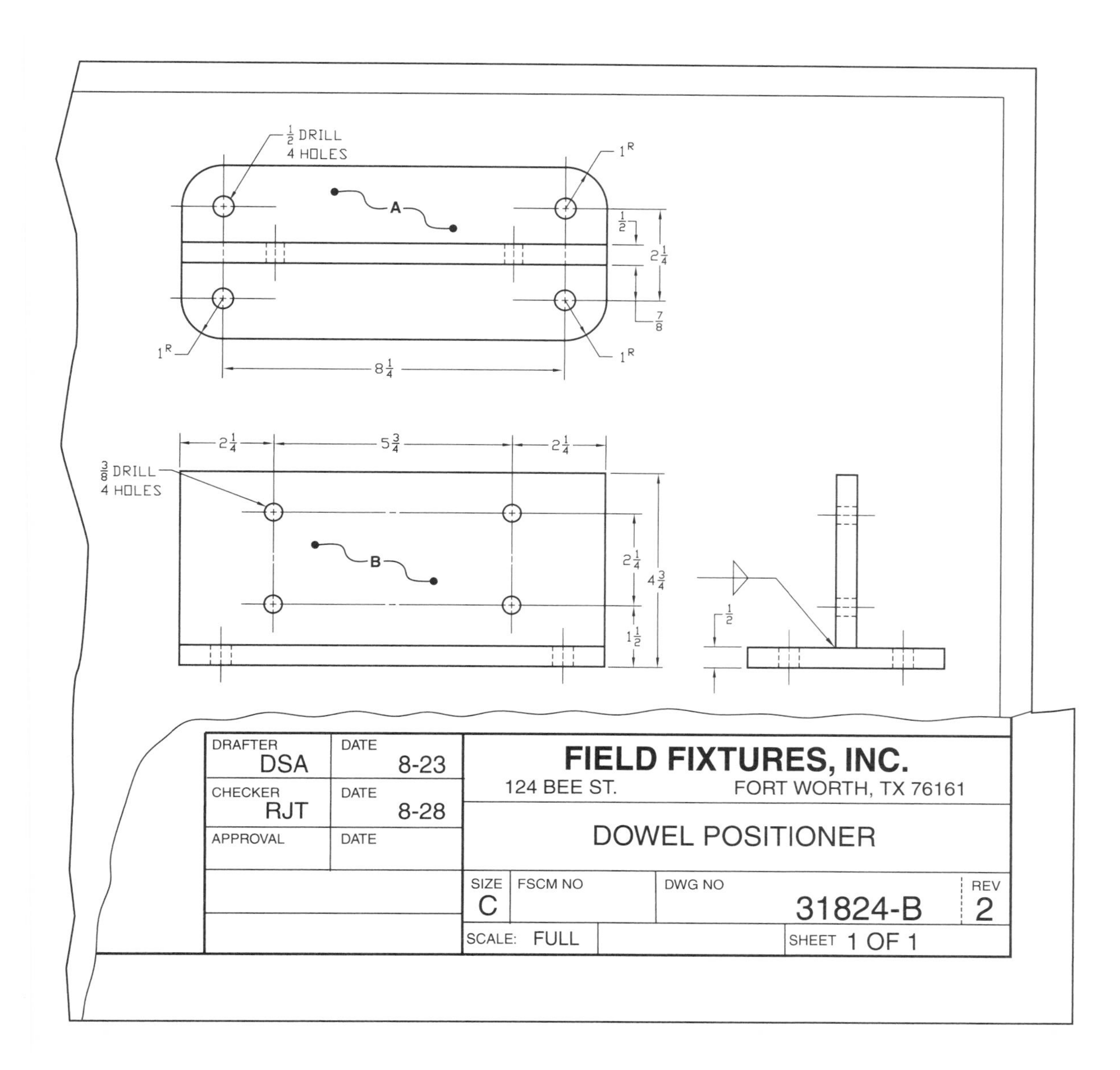

DOWEL POSITIONER

End Plate

Refer to print below.

This drawing uses a dimensioning method in which each centerline is given a coordinate based on its distance from a horizontal or vertical axis. The axes are marked with a circle at each end.

_______________ **1.** The distance between centerlines A and B is ___″. (decimal to four places)

_______________ **2.** The distance between centerlines B and C is ___″. (decimal to three places)

_______________ **3.** The distance between centerlines M and L is ___″. (decimal to three places)

_______________ **4.** The distance between centerlines F and E is ___″. (decimal to three places)

_______________ **5.** The distance between H and G is ___″. (fraction)

_______________ **6.** The distance between G and D is ___″. (decimal to three places)

_______________ **7.** The center-to-center distance between N and O is ___″. (fraction)

_______________ **8.** The center-to-center distance between A and K is ___″. (fraction)

_______________ **9.** The ream for the ∅3⁄16 drilled hole increases the diameter by ___″. (decimal to four places)

_______________ **10.** The decimal distance from I to the centerline at J is ___″. (decimal to four places)

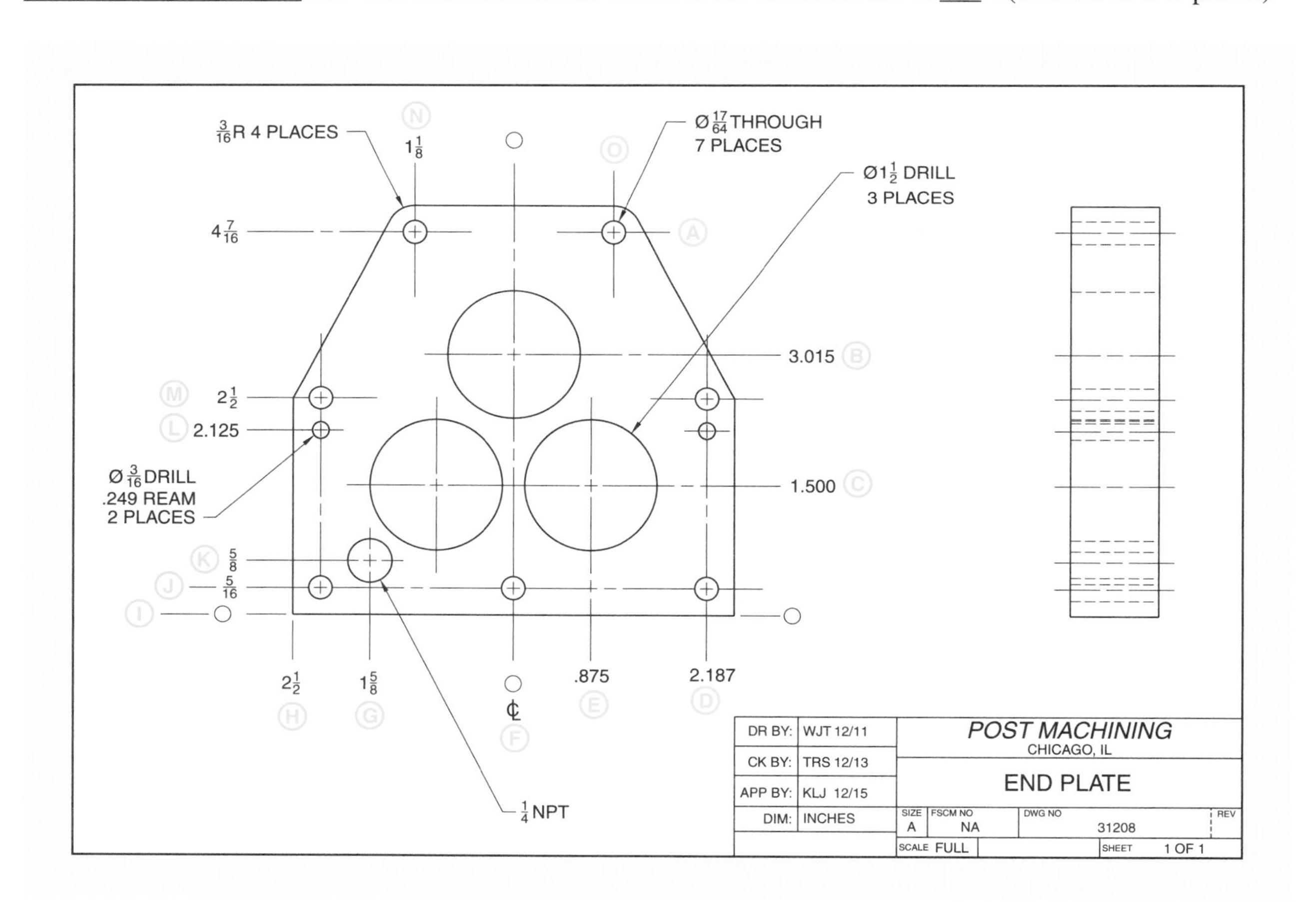

END PLATE

5 Measurement

The transfer of information from prints to manufactured parts requires proficiency with the measurement systems and units involved. Machine trades technicians must be able to interpret the quantities and units and be able to select and use the appropriate measuring tool. Measuring tools are used to set up machine tools and check manufactured parts to ensure that they meet the print specifications.

MEASUREMENT SYSTEMS

Prints quantify the sizes of part features with dimensions, which consist of a number and a unit. The type of units used on a print depends upon the measurement system. There are two common systems of measurement: the International System of Units (SI) and the U.S. customary system.

Both systems make use of base units and derived units. A *base unit* is a fundamental unit of measure that cannot be described using any other combination of units. Base units are defined in terms of physical constants, such as the speed of light or atomic mass. A *derived unit* is a unit of measure that can be defined as a combination of other units. For example, a unit for area is derived from the square of a unit of length.

International System of Units (SI)

The International System of Units (SI) is official in almost every country and used extensively in all others. The system was established and is maintained by the International Bureau of Weights and Measures. SI units are strictly defined and highly structured. The system consists of exactly seven base units, from which numerous derived units are also defined. **See Figure 5-1.**

A fundamental feature of SI is the use of prefixes for representing decimal multiples or submultiples of any unit. Prefixes are used to describe a new unit that is a power of ten of the original unit. **See Figure 5-2.** For example, the prefix "centi" is associated with 10^{-2} (0.01). Therefore, the centimeter (cm) is a unit equal to one hundredth of a meter (0.01 m). The prefix "kilo" is associated with 10^3 (1000), so the kilometer (km) is equal to one thousand meters (1000 m). Because of this prefix system, SI is also known as the metric system.

U.S. Customary System

The U.S. customary system is used mainly in the United States. This system is also known as the English system, since it evolved from the British Imperial system. Shop tools that are based on inch measurements may also be referred to as being SAE (Society of Automotive Engineers) type.

Unlike SI, the U.S. customary system is not administered by an authoritative organization, so there is no strictly defined set of base units. Therefore, there may be more than one unit of measurement for each type of quantity. For example, length is commonly measured in inches, feet, yards, or miles. The U.S. customary system also incorporates some units from SI.

INTERNATIONAL SYSTEM OF UNITS (SI) BASE UNITS		
QUANTITY	UNIT	SI SYMBOL
Length	meter	m
Mass	kilogram	kg
Time	second	s
Thermodynamic temperature	kelvin	K
Electric current	ampere	A
Luminous intensity	candela	cd
Amount of substance	mole	mol

SELECTED SI DERIVED UNITS				
QUANTITY	SPECIAL NAME	SYMBOL	EXPRESSED IN TERMS OF OTHER SI UNITS	
			OTHER DERIVED UNITS	BASE UNITS ONLY
Area	—	—	—	m^2
Volume	—	—	—	m^3
Velocity	—	—	—	m/s
Acceleration	—	—	—	m/s^2
Density	—	—	—	kg/m^3
Force	newton	N	—	$kg \cdot m/s^2$
Pressure	pascal	Pa	N/m^2	$kg \cdot m^{-1} \cdot s^{-2}$
Energy, work	joule	J	N·m	$kg \cdot m^2/s^2$
Power	watt	W	J/s	$kg \cdot m^2/s^3$
Electromotive force	volt	V	W/A	$kg \cdot m^2 \cdot s^{-3} \cdot A^{-1}$
Electric charge	coulomb	C	—	A·s

Figure 5-1. The International System of Units (SI) is a strictly defined list of seven base units, which are used to create any number of derived units.

INTERNATIONAL SYSTEM OF UNITS (SI) PREFIXES					
PREFIX	SYMBOL	MULTIPLE	PREFIX	SYMBOL	MULTIPLE
yotta	Y	10^{24}	yocto	y	10^{-24}
zetta	Z	10^{21}	zepto	z	10^{-21}
exa	E	10^{18}	atto	a	10^{-18}
peta	P	10^{15}	femto	f	10^{-15}
tera	T	10^{12}	pico	p	10^{-12}
giga	G	10^{9}	nano	n	10^{-9}
mega	M	10^{6}	micro	µ	10^{-6}
kilo	k	10^{3}	milli	m	10^{-3}
hecto	h	10^{2}	centi	c	10^{-2}
deca	da	10^{1}	deci	d	10^{-1}

Figure 5-2. The International System of Units (SI) uses a system of metric prefixes to create larger and smaller units.

U.S. customary units do not have a prefix system for denoting multiples or submultiples of units. Instead, entirely different units of larger or smaller sizes are usually used. The relationships between these units are typically in nondecimal multiples, such as 12 or 16. For example, there are 12 in. in 1 ft. This makes conversion between units generally more difficult than a decimal or prefix system.

Also, some measurements are represented in nondecimal fractions of a unit. For example, fractional inches are commonly expressed in eighths or sixteenths. Alternatively, many technical applications use decimal inches. This system allows for easier arithmetic of various values than if they were expressed as fractions.

UNITS AND UNIT CONVERSION

Conversion is possible between any two units of the same type of quantity, such as length, regardless of the system of measurement. The relationships between related units are provided in reference tables. For example, the relationship between the inch and the millimeter can be expressed as 1 in. = 25.4 mm and 1 mm = 0.03937 in.

Two methods of unit conversion use this type of relationship. **See Figure 5-3.** With the product method, the starting measurement is multiplied by the number of resulting units per starting units. With the ratio method, the starting measurement is multiplied by a ratio of 1 resulting unit to the equivalent value in starting units.

There are units in each measurement system for a wide variety of quantities, such as energy, power, and luminance. In printreading, the most common types of units are for length, area, volume, and angles. Specifications for materials may also commonly involve mass and force.

UNIT CONVERSION METHODS

CONVERT 200 mm INTO EQUIVALENT INCHES
GIVEN: 1 mm = 0.03937 in.
1 in. = 25.4 mm

200 mm × 0.03937 in./mm = **7.87 in.**

PRODUCT METHOD

$$200\text{ mm} \times \frac{1\text{ in.}}{25.4\text{ mm}} = \mathbf{7.87\text{ in.}}$$

RATIO METHOD

Figure 5-3. Quantities are converted from one unit to another using the product or ratio method.

The SI conventions for unit symbols dictate that the official symbol for the liter should be a lowercase "l". However, since this is easily confused with the numeral 1, the uppercase "L" is also officially accepted.

Units of Length

The SI base unit of length is the meter (m). Large and small lengths are typically measured in kilometers (km) and millimeters (mm), respectively. The U.S. customary system uses four primary units of length: the inch (in.), foot (ft), yard (yd), and mile (mi). **See Figure 5-4.**

Units of Area and Volume

In SI and the U.S. customary system, units of length are used to derive units of area and volume by raising the unit to the second or third power, respectively. **See Figure 5-5.** Additional special units are also used for liquid volume, such as liters (L), quarts (qt), and gallons (gal.). The liter is by definition equal to exactly 1 dm^3.

UNITS OF LENGTH

LENGTH	mm	m	km	in.	ft	yd	mi
1 mm	1	0.001	0.000001	0.03937	0.003281	0.001094	6.214×10^{-7}
1 m	1000	1	0.001	39.37	3.281	1.094	0.0006214
1 km	1,000,000	1000	1	39,370	3281	1094	0.6214
1 in.	25.4	0.0254	0.0000254	1	0.08333	0.02778	1.578×10^{-5}
1 ft	304.8	0.3048	0.0003048	12	1	0.3333	0.0001894
1 yd	914.4	0.9144	0.0009144	36	3	1	0.0005682
1 mi	1,609,344	1609.34	1.6093	63,360	5280	1760	1

Figure 5-4. Units of length are converted from one unit to another using conversion factors.

UNITS OF AREA					
AREA	cm^2	m^2	in^2	ft^2	yd^2
1 cm^2	1	0.0001	0.1550	0.001076	0.0001196
1 m^2	10000	1	1550	10.76	1.196
1 in^2	6.452	0.0006452	1	0.006944	0.0007716
1 ft^2	929.0	0.09290	144	1	0.1111
1 yd^2	8361	0.8361	1296	9	1

UNITS OF VOLUME							
VOLUME	cm^3	L	m^3	in^3	qt	gal.	ft^3
1 cm^2	1	0.001	0.000001	0.06102	0.001057	0.0002642	3.531×10^{-5}
1 L	1000	1	0.001	61.02	1.057	0.2642	0.03531
1 m^3	1,000,000	1000	1	61,024	1057	264.2	35.31
1 in^3	16.39	0.01639	1.639×10^{-5}	1	0.01732	0.004329	0.0005787
1 qt	946.4	0.9464	0.0009464	57.75	1	0.2500	0.03342
1 gal.	3785	3.785	0.003785	231.0	4	1	0.1337
1 ft^3	28,317	28.32	0.02832	1728	29.92	7.481	1

Figure 5-5. Units of area and volume are converted from one unit to another using conversion factors.

The circle has been divided into 360 equal parts since the second century BCE. Why the number 360 was chosen is uncertain, but theories involve ancient astronomy and Babylonian mathematics. Also, the number 360 is easily divisible by 24 whole-number divisors, including every number between 1 and 10, except 7.

Units of Angle

The most common unit of measure for angles (in both systems) is the degree (°), which is equal to 1⁄360 of a circle. Fractions of a degree can be represented with decimals or with the smaller units of minutes (′) and seconds (″). A minute is 1⁄60 of a degree, and a second is 1⁄60 of a minute. For example, the angle 42.125° equals 42°7′30″.

Plane angles are also measured in radians. **See Figure 5-6.** The radian (rad) is a plane (two-dimensional) angle with an arc length equal to its radius. Therefore, there are exactly 2π (approximately 6.28) radians in a circle. This means that a radian is equal to approximately 57.296°. The steradian (sr) is a solid (three-dimensional) angle with the area of the spherical surface equal to the square of its radius.

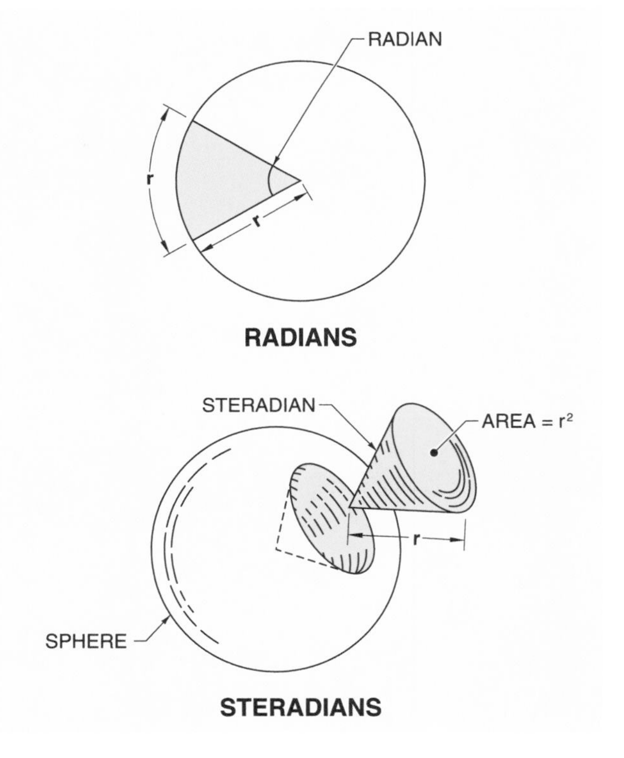

Figure 5-6. A radian is an alternative unit for measuring angles. A steradian is a three-dimensional version of a radian.

Units of Mass

The SI base unit of mass is the kilogram (kg). **See Figure 5-7.** Grams (g) are used for small mass measurements, and metric tons (t) are used for very large measurements. The U.S. customary unit for mass is the slug (sl). The pound (lb) is actually a unit of force, but in many applications, it can also be used for mass. To differentiate the way the pound is used, the suffixes "-force" or "-mass" may be added, as in "pound-mass" (lbm). Fractions of a pound are measured in ounces (oz). Large masses are measured in short tons or, less commonly, long tons. These two units do not have generally accepted unique abbreviations.

Units of Force

The SI unit of force is the newton (N), which is a derived unit equivalent to $kg \cdot m/s^2$. **See Figure 5-8.** Other units of force related to SI, but not officially part of the system, are the dyne (dyn) and the kilogram-force (kgf). The U.S. customary system primarily uses the pound (lb) for force. A related unit is the poundal (pdl).

MEASURING TOOLS

Measuring tools are an important part of machine trades. Measuring tools are used to set up and calibrate machine tools. They are also needed to check manufactured parts and ensure that they meet the print's specifications. The measuring tools used must be capable of precise and accurate measurements.

Common machine shop measuring tools include rules, protractors, squares, gauges, indicators, calipers, and micrometers. Each tool is designed for measuring a certain quantity. Many are adaptable for a variety of circumstances, though some are particularly suited for certain uses. All measuring tools must be used properly to ensure continued accuracy. Measurements should always be made with light contact of the tool to the surface being measured. Given the high precision of many measurements in machine trades, anything but light contact can affect the reading. Forcing a measuring tool onto a feature not only provides an inaccurate reading, but it may also damage the tool and the feature.

The difference between force and mass is often misunderstood, which leads to some of their units being used interchangeably. Most notably, the pound is often used as a unit of mass. While technically incorrect, this practice is still tolerated in some applications when the meaning is clear.

UNITS OF MASS

MASS	g	kg	t	sl	lbm	short ton
1 g	1	0.001	0.000001	6.852×10^{-5}	0.002205	0.0000011
1 kg	1000	1	0.001	0.06852	2.205	0.001102
1 t	1,000,000	1000	1	68.52	2205	1.102
1 sl	14,594	14.59	0.01459	1	32.17	0.01609
1 lbm	453.6	0.4536	0.0004536	0.03108	1	0.0005
1 short ton	907,185	907.2	0.9072	62.16	2000	1

Figure 5-7. Units of mass are converted from one unit to another using conversion factors.

UNITS OF FORCE

FORCE	dyn	N	kgf	lbf	pdl
1 dyn	1	0.00001	1.020×10^{-6}	2.248×10^{-6}	7.233×10^{-5}
1 N	100,000	1	0.1020	0.2248	7.233
1 kgf	980,665	9.807	1	2.205	70.93
1 lbf	444,822	4.448	0.4536	1	32.17
1 pdl	13,825	0.1383	0.01410	0.03108	1

Figure 5-8. Units of force are converted from one unit to another using conversion factors.

Measuring tools for U.S. customary units use inches as a unit of measurement. Metric measuring tools use centimeters and millimeters. Combination tools have both inch and metric graduations.

Rules and squares are common measuring tools in any machine shop.

Rules

A *rule* is a semiprecision measuring tool used for measuring length. **See Figure 5-9.** The rule is the most common measuring tool. The most common rules in the machine shop are steel rules. Steel rules are made of tempered steel, and both edges of both sides are marked with deeply etched graduations. Their length may be anywhere from 6″ to 36″.

U.S. customary unit steel rules are typically graduated in fractions of an inch, such as eighths, sixteenths, thirty-seconds, and sixty-fourths. Some are graduated in decimal inches. Metric rules are usually divided into millimeters or half millimeters.

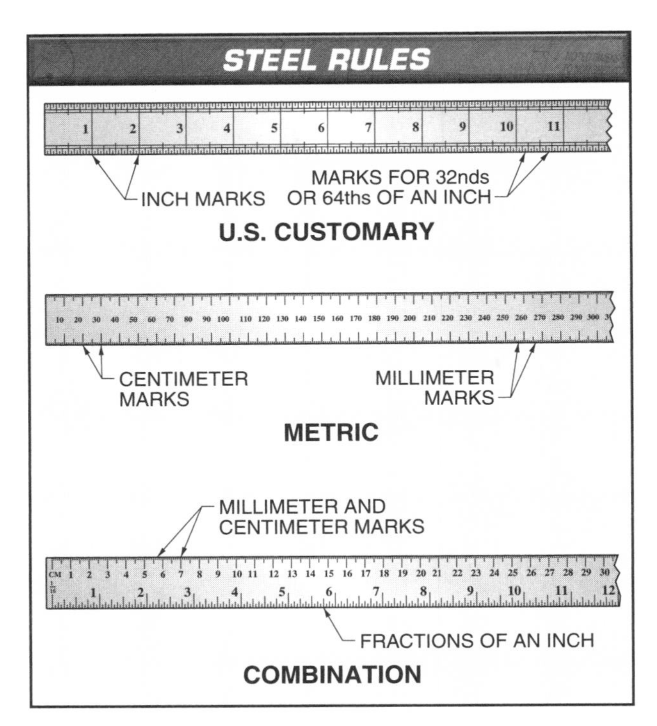

Figure 5-9. Steel rules are semiprecision measuring tools used to measure length.

Protractors

A *protractor* is a measuring tool for angles. **See Figure 5-10.** The baseline is placed on one leg of an angle with the centerpoint at the vertex, and the desired angle is measured or laid out.

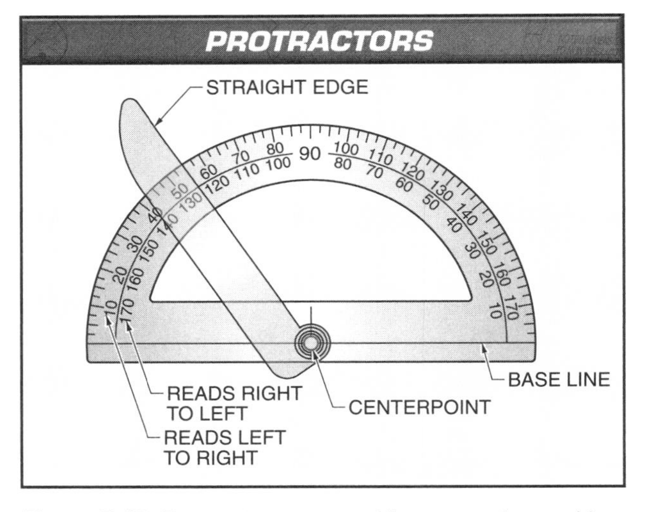

Figure 5-10. Protractors are used for measuring and laying out angles.

Squares

A *square* is a tool for laying out and checking right angles. **See Figure 5-11.** Various types of squares are available for different trades. Squares intended for use in the machine trades typically have no graduations, though squares that do have graduations are also used.

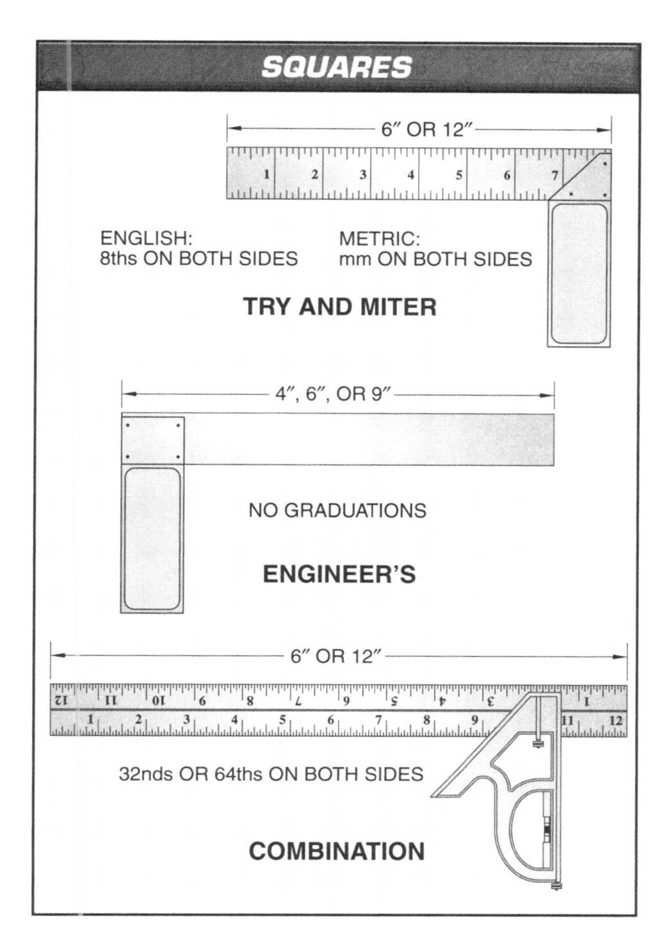

Figure 5-11. Squares are used primarily to check right angles.

The combination square is a two-piece square that consists of a head that slides along a special ruled blade. The head engages with the blade such that it can be removed only at the ends. These squares are typically short, either 6″ or 12″ blades. The standard head allows a combination square to be used to check both right and 45° angles. This type of square is also used to check flatness, to find the centers of round features, to measure depths, and for other layout tasks. An alternative protractor head allows the combination square to measure and lay out other angles.

Gauges

A *gauge* is a standard that is used as a tool for checking sizes and pitches. A *standard* is an object that is created to exactly match a certain defined characteristic. Standards are then used to evaluate whether other objects meet the definition. Common gauges in a machine shop include thread pitch, radius, thickness, and drill and wire size. **See Figure 5-12.** For example, a thread pitch gauge is machined with notches that match the spacing and size of a certain thread pitch. By holding the gauge up against a thread, it is apparent whether the thread meets the specification.

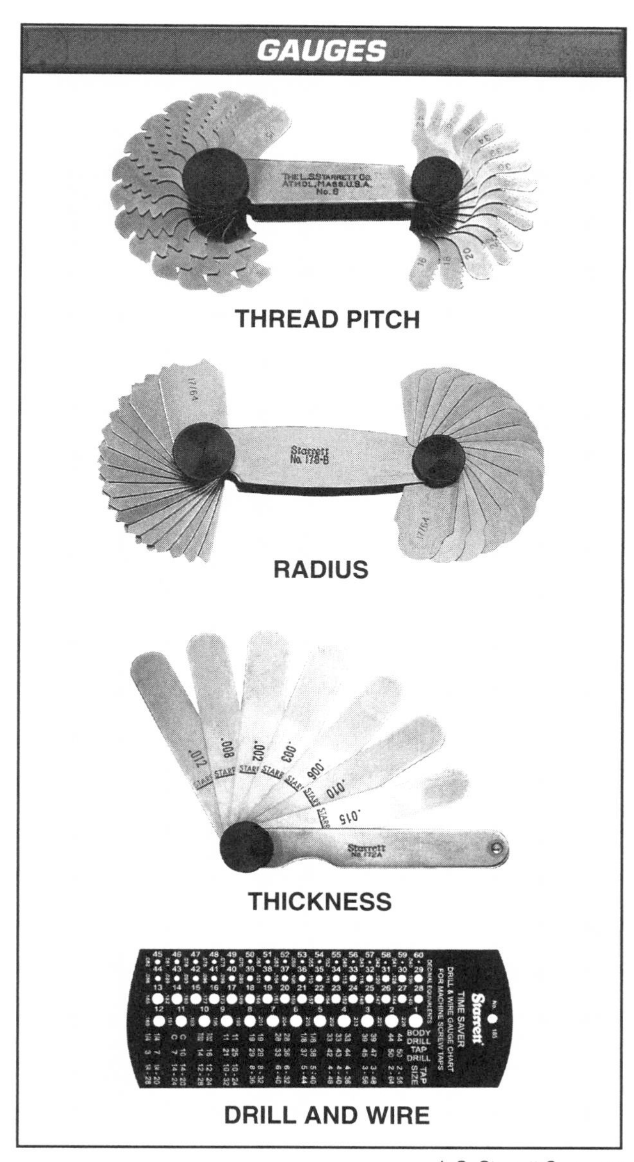

L.S. Starrett Company

Figure 5-12. Gauges are used to compare features against standard sizes, thicknesses, and pitches.

Combination squares can be used to check or mark miter joints, flatness, squareness, levelness, and plumb.

Indicators

An *indicator* is a measuring tool for very small linear displacements. **See Figure 5-13.** Indicators are attached to stable mounts with their moveable plungers touching the surface to be tested. Any longitudinal movement in the plunger is quantified on the calibrated dial. A common application of indicators is to check for out-of-roundness in cylindrical parts. As the surface is rotated around its axis, the plunger moves and the indicator displays the surface variations.

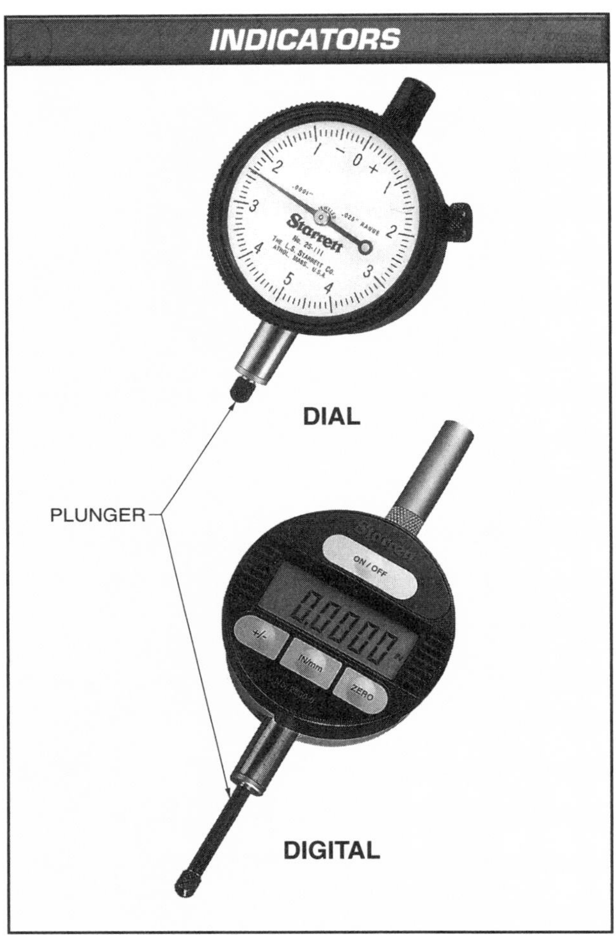

L.S. Starrett Company

Figure 5-13. An indicator measures very small linear displacement by movement of its plunger.

The terms "accuracy" and "precision" have different meanings. Accuracy relates to how closely a measurement matches the true value. Precision relates to how much variation there is among measurements that are taken multiple times.

Calipers

A *caliper* is a measuring tool for inside and outside dimensions. There are two basic caliper designs: reading and nonreading. **See Figure 5-14.** Both calipers consist of two parts that are attached together, but the parts can be moved in and out so that each contacts one side of the feature to be measured.

The reading design consists of a graduated rule with one half of a jaw at the end. Another part slides along the rule and completes the jaw. The jaw is placed around the feature to be measured and the reading is indicated from the marks on the rule.

The nonreading design consists of two long pieces attached at one end to form a pivot. This design is not graduated for direct measurements. Instead, these calipers are used to transfer sizes.

Micrometers

A *micrometer* is a measuring tool for relatively small but extremely precise linear measurements. Most micrometers are a type of caliper that is specialized for small distances, usually up to a few inches. **See Figure 5-15.** A micrometer uses the same principle as the screw to amplify extremely fine linear motion to graduated marks on a sleeve and rotating thimble. Micrometer measurements are made between the fixed anvil and the movable spindle. Variations on the micrometer design are intended for inside, outside, and depth measurements.

PRECISION MEASUREMENTS

Most measuring tools are read by counting the unit divisions marked along one edge of the tool. Alternatively, some tools use dial scales, which indicate measurements by the position of a needle on a graduated and calibrated dial. Digital measuring tools include digital readouts that directly display measurements to the maximum precision. A special feature available on most digital measuring tools is the ability to automatically convert between units with the press of a button.

The precision of the measurement tool is equal to the smallest division marked on its scale. Measurements can be read to additional places by interpolating the position of the measured quantity between the divisions. However, this method is not as accurate as using measuring tools with greater precision. Handheld precision measuring tools, particularly calipers and micrometers, use special scales and methods for indicating the measured quantity.

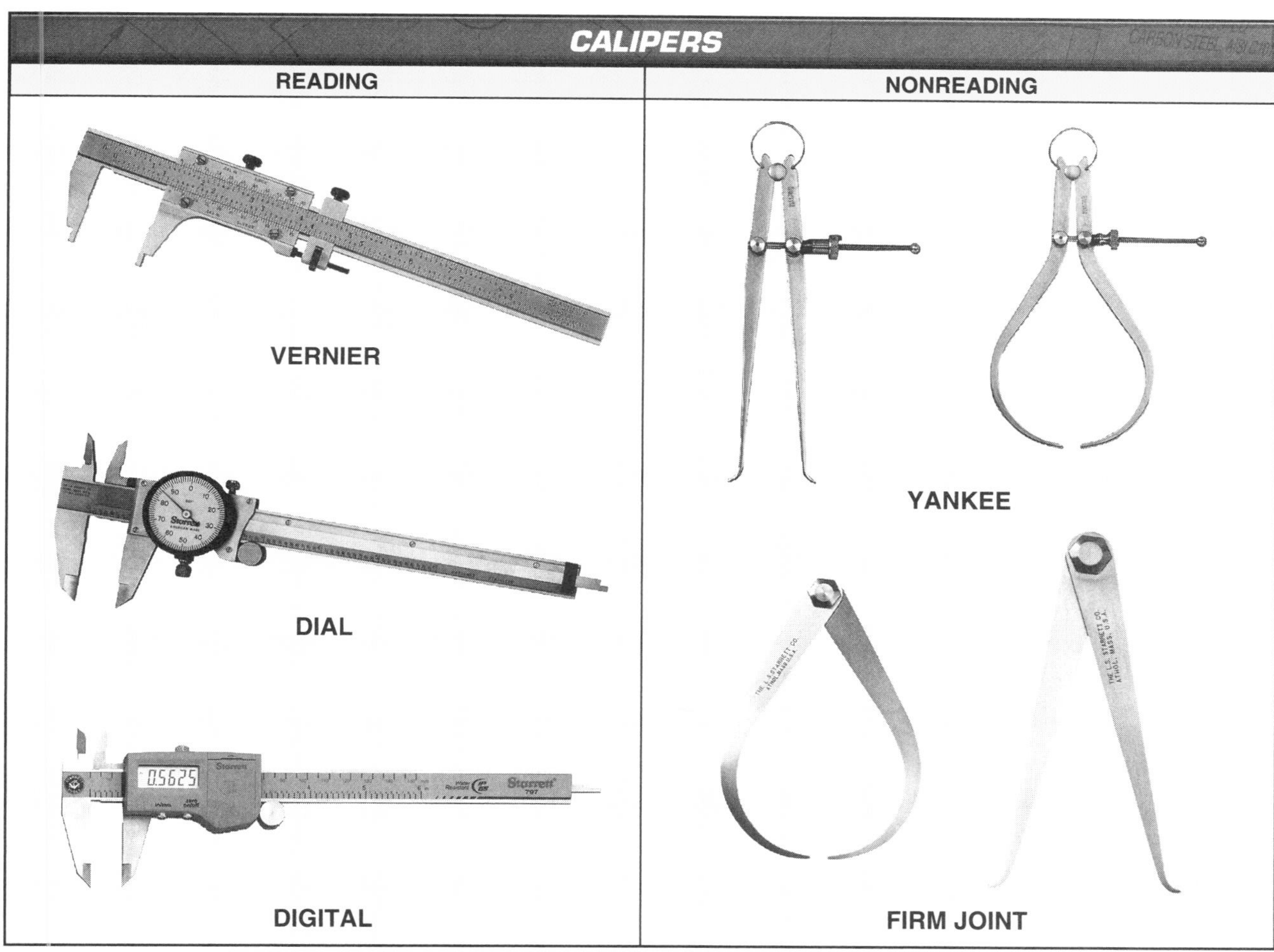

L.S. Starrett Company

Figure 5-14. Calipers include both reading and nonreading types. Both types are used to measure or compare inside and outside dimensions.

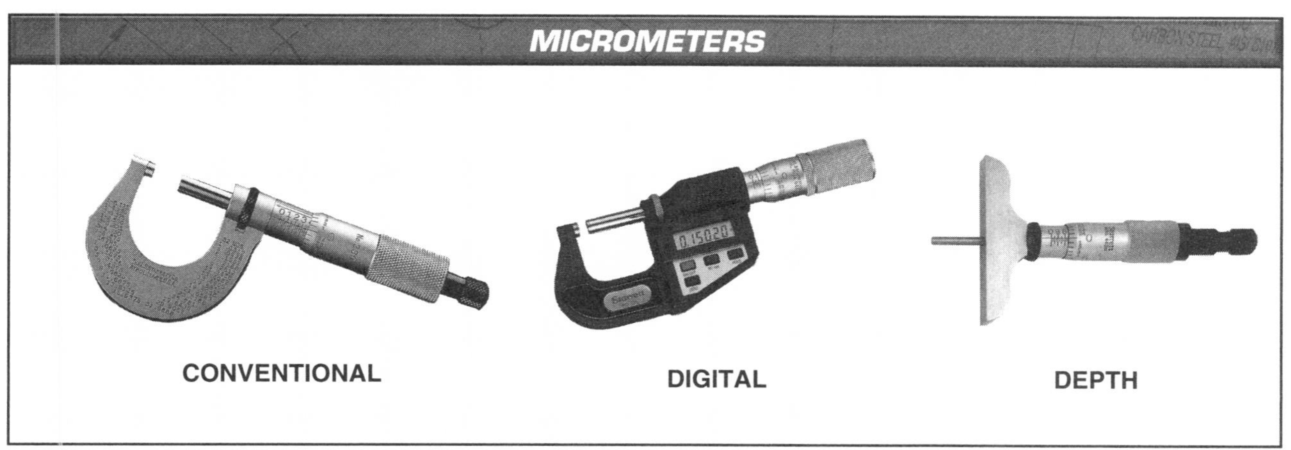

L.S. Starrett Company

Figure 5-15. Micrometers are usually limited to measuring relatively small dimensions precisely.

Caliper Measurements

A reading caliper consists of a long beam marked with graduated divisions and a fixed jaw at one end. **See Figure 5-16.** A movable jaw with a zero-line mark slides along the beam. The quantity to be measured is placed between the jaws, and a measurement on the beam's main scale is indicated by the zero-line mark.

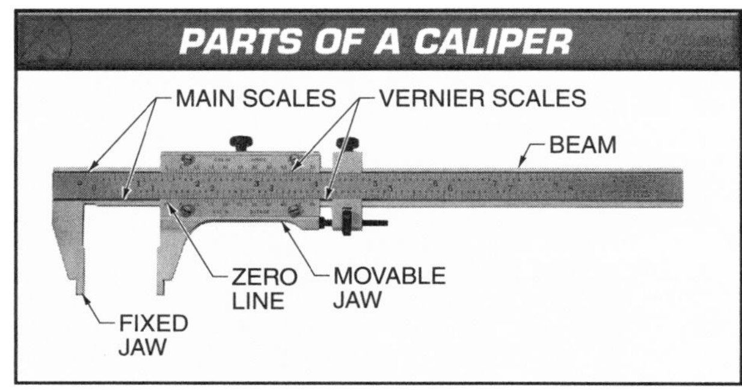

L.S. Starrett Company

Figure 5-16. Calipers used in machine shops follow a basic design in which a jaw moves along a graduated beam.

> Using a vernier scale is a simple and versatile way to make high-precision measurements. It can be added to the design of any type of adjustable measuring tool that measures the space between two points.

Standard Calipers. Standard calipers include only a fixed main scale and a zero-line mark on the movable jaw. The measurement is precise to the smallest division on the main scale. However, standard calipers are relatively uncommon compared to vernier calipers, which offer greater precision.

Vernier Calipers. A vernier caliper is similar to a standard caliper, but it includes a vernier scale in addition to its main scale. A *vernier scale* is an additional scale on a measuring tool that is used to precisely determine measurements that lie between the smallest marks on the main scale. **See Figure 5-17.** The marks on the main scale and the vernier scale have slightly different spacing, so the marks align at only one position at a time. Since it is relatively easy for the human eye to discern which marks are aligned, this arrangement is used to precisely measure fractions of the smallest marked division.

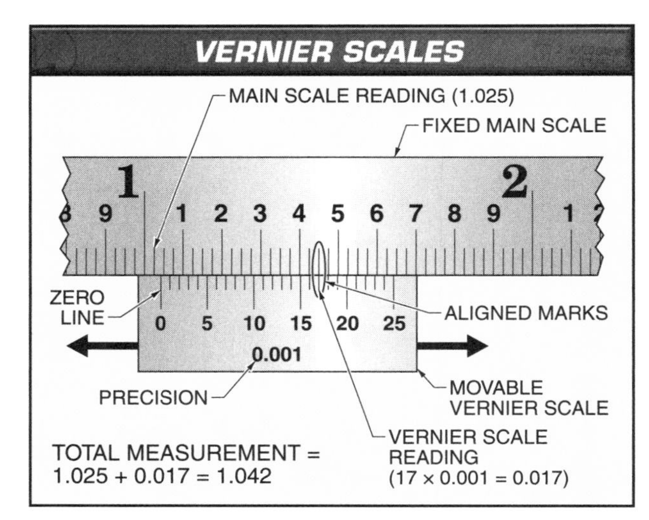

Figure 5-17. A vernier scale is a secondary scale that, when used properly, increases the precision of a measuring tool.

Vernier scales can be incorporated into several different types of measuring tools, all using the same principle. **See Figure 5-18.** In general, the procedure for using a vernier scale is as follows:

1. Position the movable portion of the measuring tool as appropriate.
2. Note the reading on the main scale to the next smallest whole marked division as indicated by the zero line on the vernier scale.
3. Determine which vernier scale mark is aligned with a mark on the main scale.
4. Note the precision of the vernier scale and the number associated with the aligned vernier mark. The vernier scale reading is the unit number multiplied by the precision.
5. Add the vernier scale reading to the main scale reading. The sum is the total measurement.

If the vernier scale's zero line is the mark that best aligns with a graduation of the main scale, then the measurement is precise to exactly that whole unit on the main scale.

Vernier calipers may include 10, 25, or 50 divisions on their vernier scales. In combination with the main scale divisions, the vernier scale divisions determine the measurement precision. Most inch-based vernier calipers have a precision of 0.001 in., and most metric calipers have a precision of 0.02 mm. Calipers may also have two sets of main and vernier scales, which can be used to provide inside and outside measurements or measurements in two different units, such as inches and centimeters.

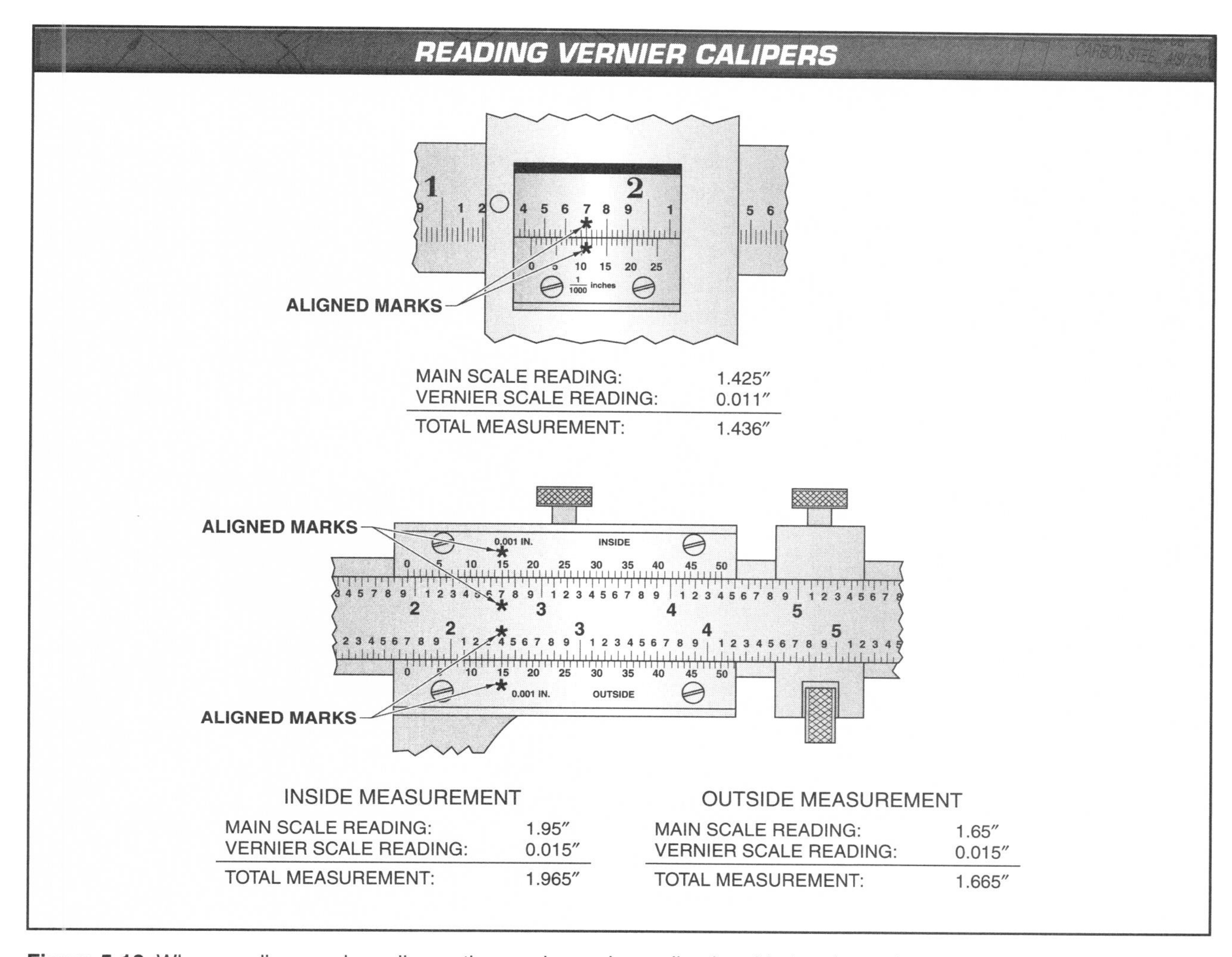

Figure 5-18. When reading vernier calipers, the vernier scale reading is added to the main scale reading.

Micrometer Measurements

A micrometer uses a main scale that is marked on the sleeve. **See Figure 5-19.** As the thimble is turned, the spindle moves towards or away from the anvil, and the thimble covers or uncovers divisions on the main scale. The largest reading exposed on the sleeve is the beginning of the measurement. Fractions of the final main scale division are then determined by using a secondary scale marked around the circumference of the thimble. The micrometer may include a vernier scale in addition to the standard secondary scale.

Standard Micrometers. A micrometer's standard secondary scale provides a direct reading of the fraction of the main scale division. This is the number associated with the mark that is aligned with the centerline of the main scale.

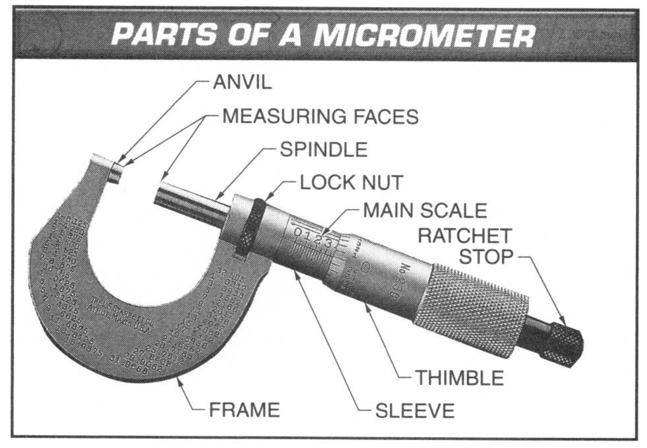

L.S. Starrett Company

Figure 5-19. Micrometers use the mechanical advantage of rotating screws to amplify very tiny changes in the space between the measuring faces.

The total measurement from a standard micrometer is equal to the sum of the main scale reading from the sleeve and secondary scale reading from the thimble. **See Figure 5-20.** If the zero mark on the thimble aligns with the centerline of the main scale, the measurement is precise to exactly the whole unit indicated on the main scale.

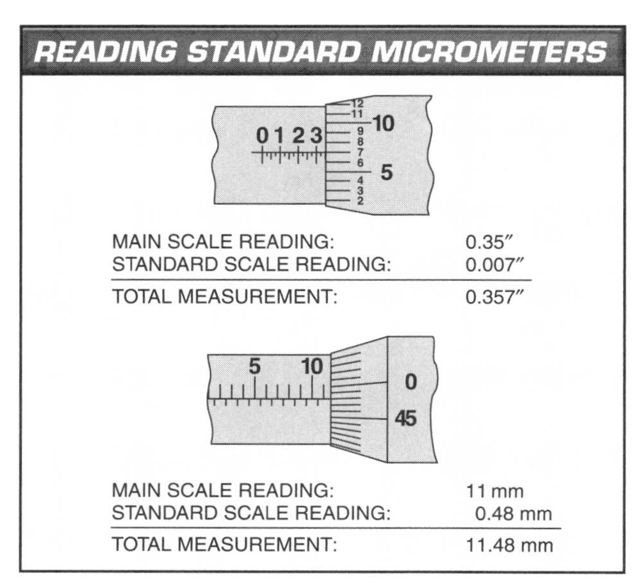

Figure 5-20. When reading standard micrometers, the standard secondary scale reading is added to the main scale reading.

On most inch-based standard micrometers, one complete thimble revolution moves the spindle 0.025 in. The circumference of the thimble is marked with 25 divisions, so each division corresponds to 0.001 in. Therefore, the measurement precision is 0.001 in. On most metric standard micrometers, one revolution moves the spindle 0.5 mm. The thimble is marked with 50 divisions, providing a measurement precision of 0.01 mm.

Vernier Micrometers. A vernier micrometer is similar to a standard micrometer, but it has a set of long marks along the length of its sleeve. The total measurement is determined using the same procedure as a standard micrometer, but a third reading from the vernier scale is added, further increasing the measurement precision. Most inch-based vernier micrometers measure to a precision of 0.0001 in. Most metric vernier micrometers measure to a precision of 0.001 mm.

The vernier scale measurement is the number associated with the vernier scale mark that aligns with one of the standard scale thimble marks. **See Figure 5-21.** After noting the measurement visible on the sleeve's main scale, both the standard scale and the vernier scale measurements are added.

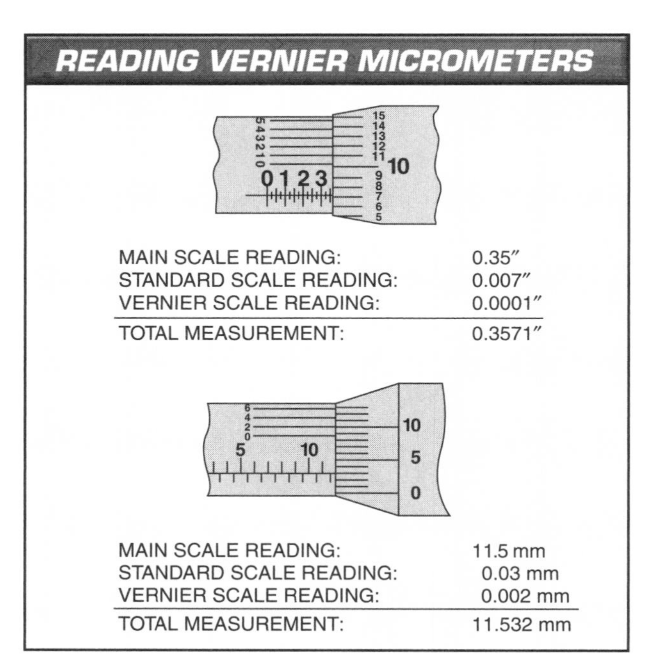

Figure 5-21. When reading vernier micrometers, both the standard scale and the vernier scale readings are added to the main scale reading.

Vernier Protractor Measurements

The vernier protractor is a precision instrument used to measure angles. The vernier scale on the protractor consists of 12 divisions, thus making measurements with a precision of $\frac{1}{12}°$ or 5′ (5 min) possible. **See Figure 5-22.** The total reading of a vernier protractor is the number of whole degrees indicated on the main scale plus the number of minutes indicated on the vernier scale where it is aligned with the main scale.

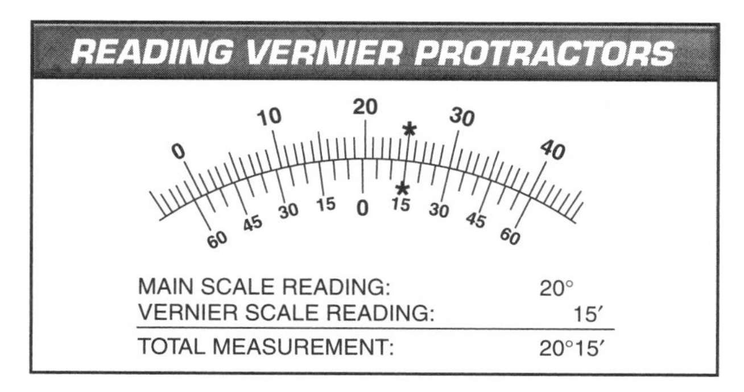

Figure 5-22. When reading vernier protractors, the vernier scale reading is added to the main scale reading.

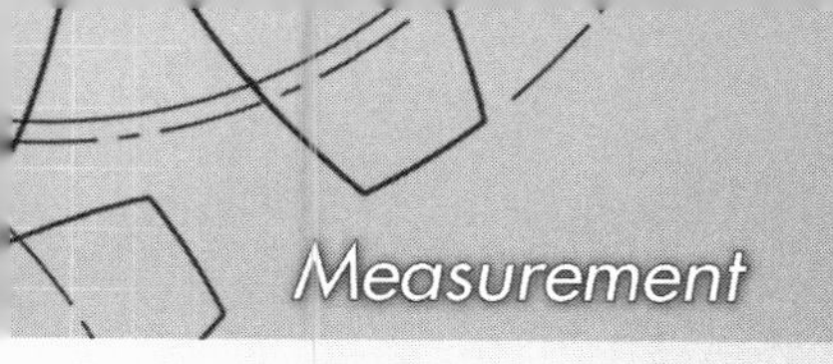

REVIEW QUESTIONS 5

Name ______________________________ Date ____________________

True False

T F **1.** A derived unit is a unit of measure that can be defined as a combination of other units.

T F **2.** The International System of Units (SI) is official in almost every country and used extensively in all others.

T F **3.** The U.S. customary system has a strictly defined set of base units.

T F **4.** Many technical applications use decimal inches.

T F **5.** A second is $\frac{1}{60}$ of a degree.

T F **6.** A protractor is a semiprecision measuring tool used for measuring length.

T F **7.** Combination tools have both inch and metric graduations.

T F **8.** Squares intended for use in the machine trades typically do not have graduations.

T F **9.** Vernier scales can be incorporated into only calipers.

T F **10.** Vernier calipers may include only 10 divisions on their vernier scales.

T F **11.** A micrometer may include a vernier scale in addition to the standard secondary scale.

T F **12.** Most inch-based vernier micrometers measure to a precision of 0.0001 in.

T F **13.** The alignment of marks on a main scale and a vernier scale is used to measure fractions of the smallest marked division precisely.

T F **14.** SI prefixes are used to describe a new unit that is a power of 12 of the original unit.

T F **15.** Metric rules are typically graduated in fractions, such as eighths, sixteenths, thirty-seconds, and sixty-fourths.

Matching – Metrics

______	**1.** hecto-	**A.**	thousandth
______	**2.** m	**B.**	mega-
______	**3.** g	**C.**	second (time)
______	**4.** A	**D.**	kilo-
______	**5.** milli-	**E.**	centi-
______	**6.** M	**F.**	meter (length)
______	**7.** hundredth	**G.**	deci-
______	**8.** thousand	**H.**	ampere
______	**9.** tenth	**I.**	hundred
______	**10.** s	**J.**	gram (mass)

Completion

______ **1.** A(n) ___ unit is a fundamental unit of measure that cannot be described using any other units.

______ **2.** A fundamental feature of SI is the use of ___ for representing decimal multiples or submultiples of any unit.

______ **3.** Units of ___ are used to derive units of area and volume by raising the unit to the second or third power, respectively.

______ **4.** ___ is possible between any two units of the same type of quantity, regardless of the system of measurement.

______ **5.** There are exactly ___ radians in a circle.

______ **6.** A(n) ___ is a standard that is used as a tool for checking sizes and pitches.

______ **7.** A feature available on most ___ measuring tools is the ability to convert between units automatically with the press of a button.

______ **8.** A(n) ___ caliper is not graduated for direct measurements.

______ **9.** The ___ of a measurement tool is equal to the smallest division marked on its scale.

______ **10.** A(n) ___ scale is an additional scale on a measuring tool that is used to determine measurements that lie between the smallest marks on the main scale.

______ **11.** In combination with the main scale divisions, the vernier scale divisions determine the measurement ___.

______ **12.** A micrometer's standard secondary scale provides a direct reading of the fraction of the ___ scale division.

__________ **13.** A micrometer uses a(n) ___ scale that is marked on the sleeve.

__________ **14.** Standard calipers are relatively uncommon compared to ___ calipers, which offer greater precision.

__________ **15.** Because of its prefix system, SI is also known as the ___ system.

Multiple Choice

__________ **1.** There are two common systems of measurement: the ___ and the U.S. customary system.
- A. SAE system
- B. International System of Units (SI)
- C. English system
- D. British Imperial system

__________ **2.** The U.S. customary system uses the ___ as a unit of length.
- A. inch
- B. foot
- C. yard
- D. all of the above

__________ **3.** The ___ is a plane (two-dimensional) angle with an arc length equal to its radius.
- A. radian
- B. steradian
- C. minute
- D. degree

__________ **4.** The SI unit of force is the ___.
- A. poundal
- B. newton
- C. dyne
- D. kilogram

__________ **5.** Fractions of a pound are measured in ___.
- A. poundals
- B. grams
- C. dynes
- D. ounces

__________ **6.** A ___ is a tool for laying out and checking right angles.
- A. square
- B. rule
- C. gauge
- D. compass

______ **7.** A(n) ___ square is a two-piece square that consists of a head that slides along a special ruled blade.

A. try and miter
B. engineer's
C. combination
D. vernier

______ **8.** A ___ uses the same principle as the screw to amplify extremely fine linear motion to graduated marks on a sleeve and rotating thimble.

A. micrometer
B. caliper
C. compass
D. protractor

______ **9.** A common application of ___ is to check for out-of-roundness in cylindrical parts.

A. rules
B. protractors
C. indicators
D. all of the above

______ **10.** Most inch-based vernier calipers have a precision of ___ in.

A. 1
B. 0.1
C. 0.01
D. 0.001

Measurements

Determine the measurements indicated by the following tool scales.

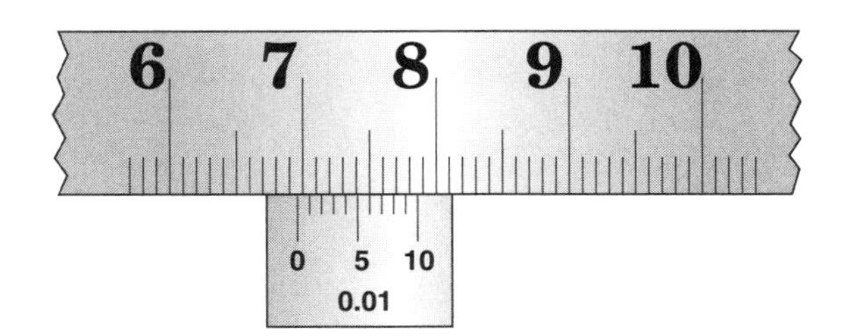

1. ______

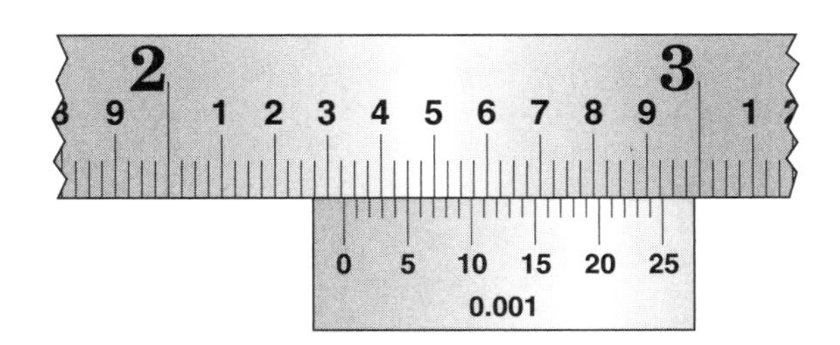

2. ______

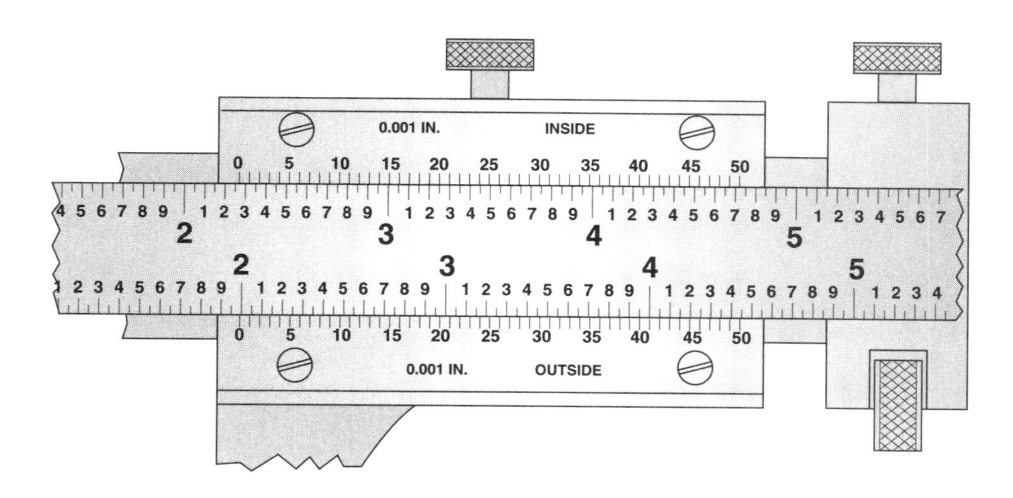

3. ____________ (inside dimension)

4. ____________ (outside dimension)

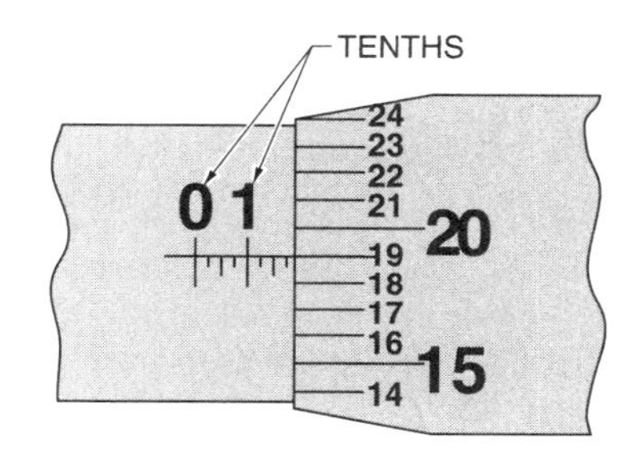

5. ____________

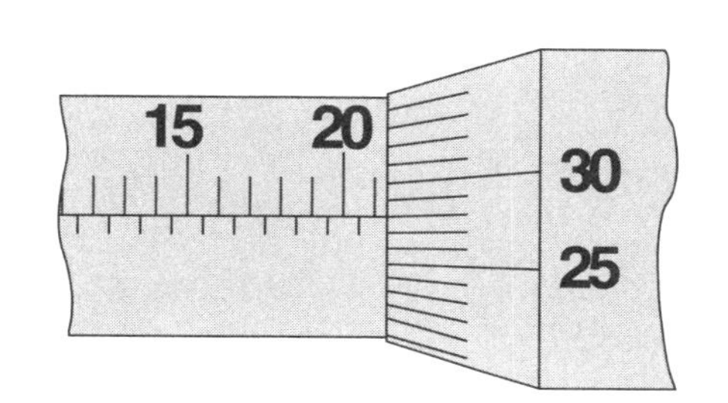

6. ____________

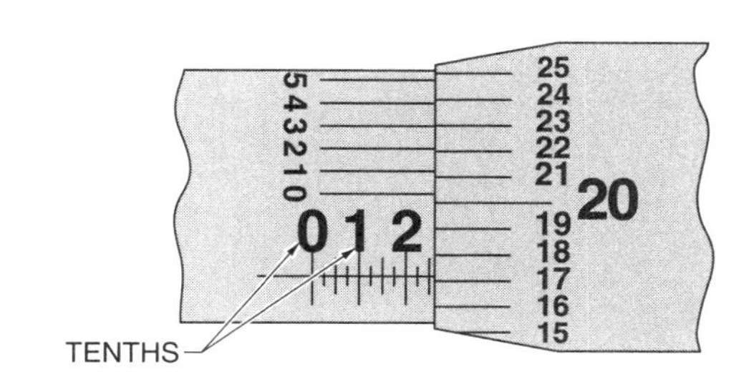

7. ____________

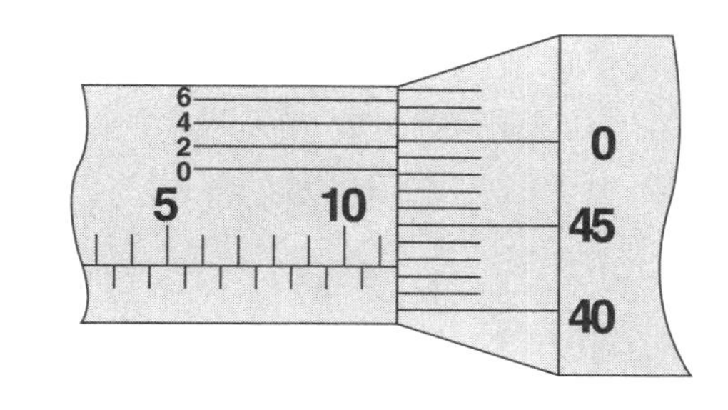

8. ____________

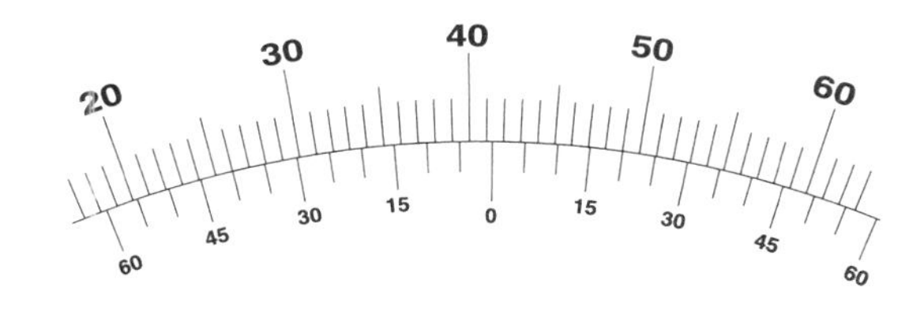

9. ____________

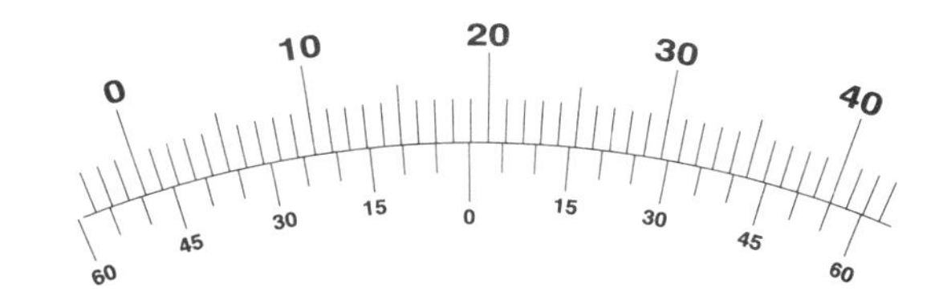

10. ____________

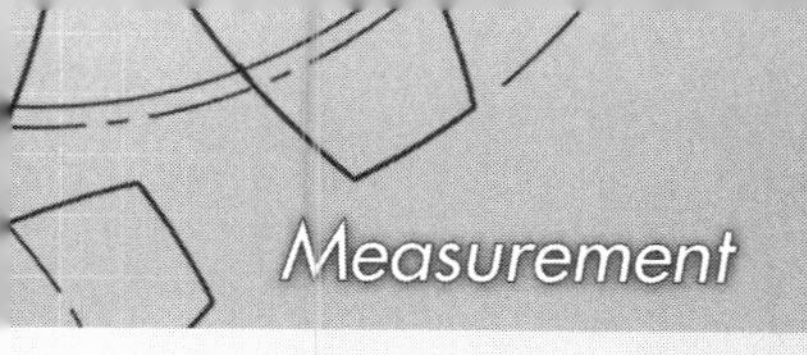

TRADE COMPETENCY TEST 5

Name ______________________________ **Date** ______________

Parallel Set

Refer to print on page 130.

T F **1.** The length of both pieces is 6.000″.

__________ **2.** All decimal dimensions have significant digits to the ___ place.
- A. tenths
- B. hundredths
- C. thousandths
- D. ten-thousandths

__________ **3.** The center-to-center spacing of both pairs of holes is ___″.

__________ **4.** The depth of the slots is ___″.

T F **5.** The decimal equivalent of the diameter of hole A is .75″.

T F **6.** All linear dimensions have the same tolerance.

__________ **7.** The maximum height of the bottom piece is ___″.
- A. 1.250
- B. 1.2505
- C. 1.255
- D. 1.2655

__________ **8.** The dimension at B is ___″.

__________ **9.** The width of the slots is ___″.

__________ **10.** The length of the slots is ___″.

T F **11.** All holes are ∅3/4″ in diameter.

T F **12.** A fractional equivalent of dimension C is 3/8″.

__________ **13.** All holes are to be ___.
- A. blind
- B. chamfered
- C. counterbored
- D. all of the above

__________ **14.** The drawing was drawn by ___.

T F **15.** There are multiple pages for this drawing.

T F **16.** The hole size tolerance is ±.005″.

____________________ **17.** The minimum thickness is ___″.

T F **18.** The drawing uses all decimal numbers.

T F **19.** The material is cold rolled steel (CRS).

____________________ **20.** The largest hole diameter is ___″.

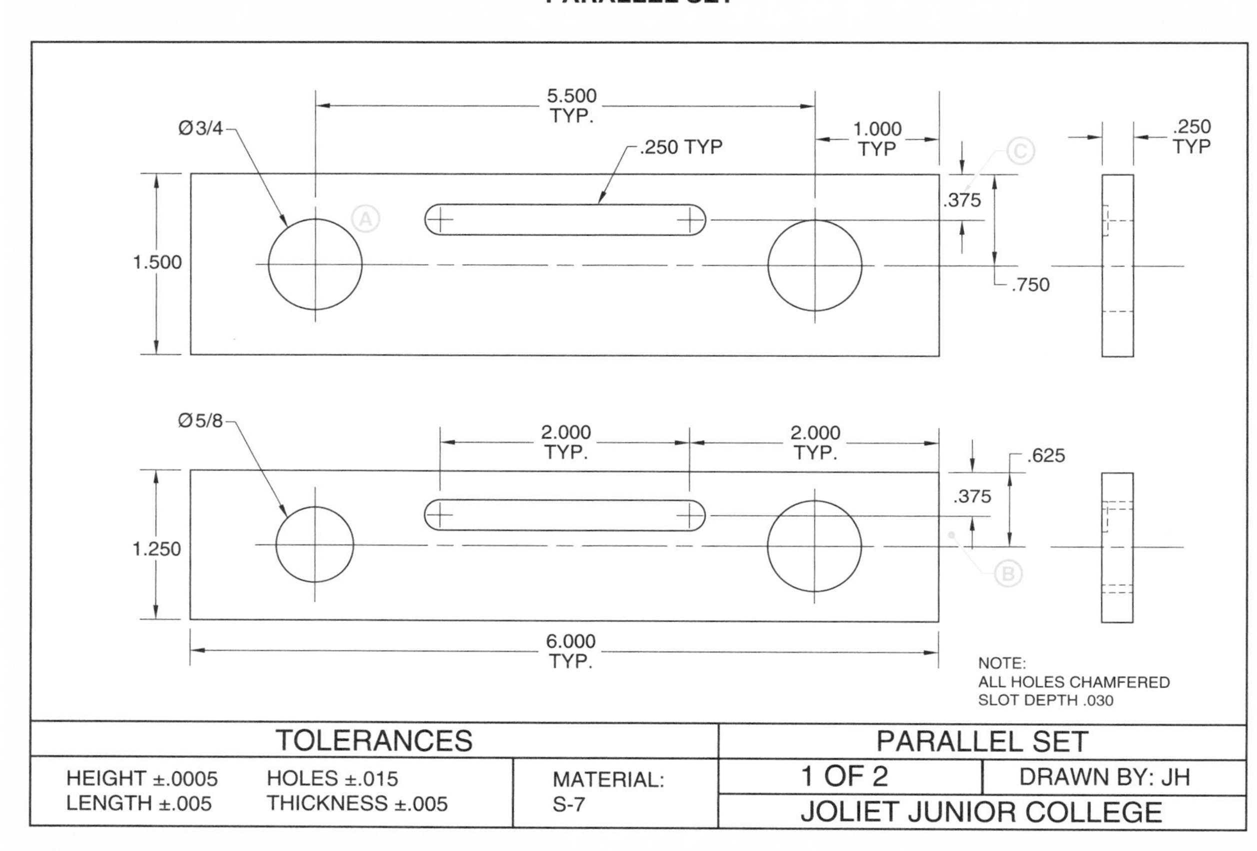

Geometric Dimensioning and Tolerancing

6

Geometric dimensioning and tolerancing is a standardized method of dimensioning the form and location of features in a drawing. Advances in machining techniques have permitted greater control and accuracy in parts manufacturing. Specific dimensioning and tolerancing methods have been developed to take advantage of precision machining and produce highly accurate parts. These methods are detailed in ANSI and ISO standards.

DIMENSIONING

Geometric dimensioning and tolerancing is a method of specifying the size, shape, and location of features on manufacturing prints, as well as the allowable variations. The accepted symbols and practices are described in ANSI/ASME Y14.5, *Dimensioning and Tolerancing.* The dimensioning section of the standard covers dimension types and methods.

Geometric dimensioning and tolerancing is covered in ISO Standards 1101, 2692, 5459, and others.

Dimensions

Dimensioning is a method of identifying and quantifying the size of features. Dimensions consist of dimension and extension lines that designate features and the numbers that quantify them.

Most dimensions shown on a print are ordinary dimensions that are subject to general or specific tolerances. However, basic and reference dimensions are two types of dimensions that have special uses and do not include tolerances.

Basic Dimensions. A *basic dimension* is a numerical value used to describe a theoretical exact size, shape, or location of a feature or datum. **See Figure 6-1.** It serves as the basis for establishing permissible tolerances on other dimensions. A basic dimension is identified by enclosing it in a rectangular frame.

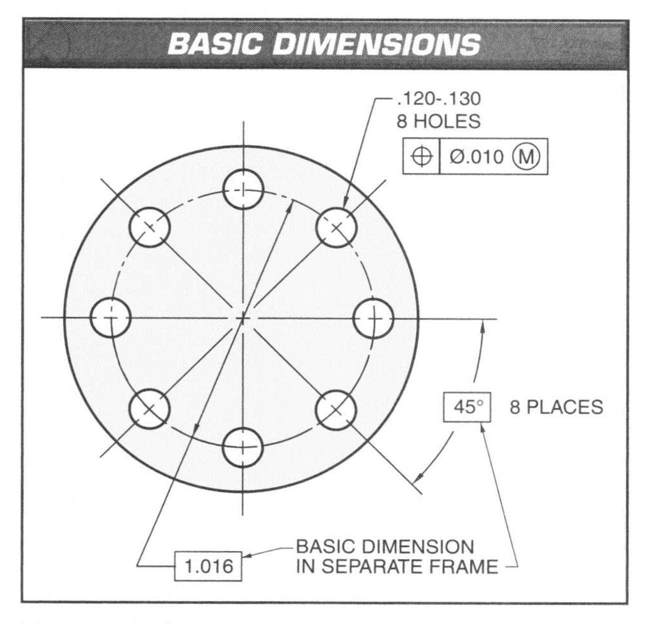

Figure 6-1. A basic dimension is a theoretically exact dimension.

Reference Dimensions. A *reference dimension* is a dimension that is used for informational purposes only and is not intended to govern manufacturing operations. **See Figure 6-2.** Reference dimensions are enclosed in parentheses or are followed by the abbreviation REF. A reference dimension is a repeat of a dimension or is derived from other values on the drawing.

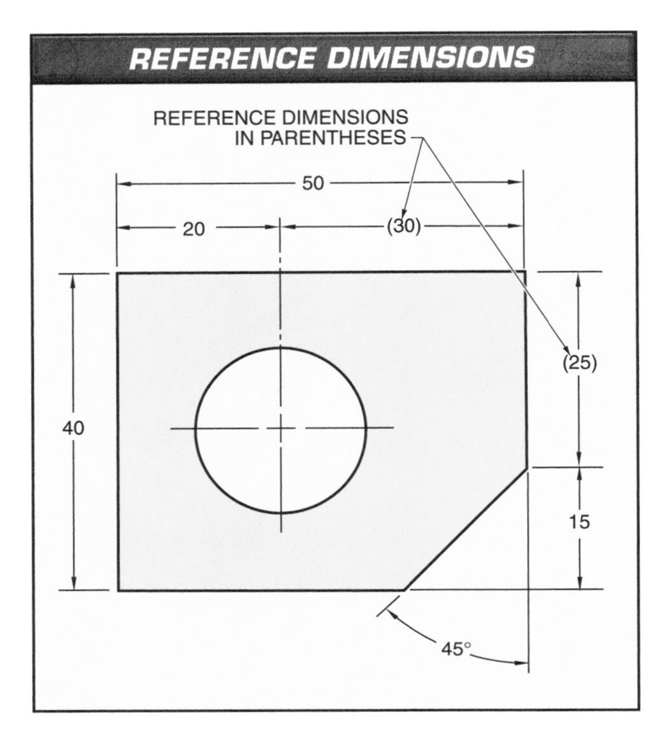

Figure 6-2. Reference dimensions are used for informational purposes only.

Either U.S. customary or metric tools can be used to measure dual-dimensioned parts.

Units. Units are usually not included with numerical dimensions because a general note on the print specifies the measuring system. If some dimensions must be in a different system than the rest of the print, they may include a unit abbreviation, such as in. or mm.

Dual dimensioning is the practice of showing both inch and millimeter dimensions together. **See Figure 6-3.** The first number in a dual dimension pair is the design dimension, which is in the unit used to originally design the feature. The second number is the conversion dimension, which is the calculated equivalent of the first. The two numbers are either separated by a slash ("/") or the second number is enclosed in parentheses or brackets. A general print note indicates the order of the units.

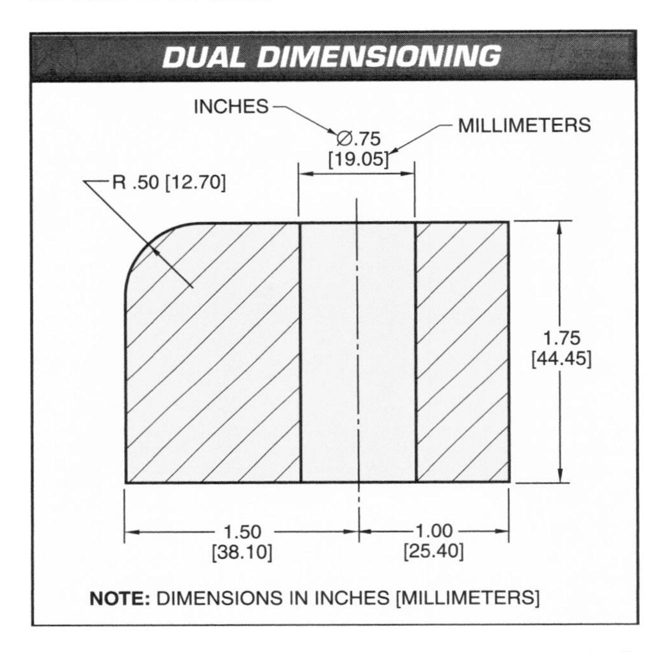

Figure 6-3. Dual dimensioning shows both inch and millimeter dimensions.

Leading Zeros. There are two rules regarding leading zeros for dimensions less than one unit. When the dimension is less than 1 in., a zero is not used before the decimal point. However, a zero is always used before the decimal point when the dimension is less than 1 mm.

Dimensioning Methods

There are a number of different ways of placing identical dimensioning information on a drawing. Any method is acceptable as long as the lines and symbols remain consistent and the method clearly shows the required information without unnecessary clutter.

The four primary methods of applying dimensions include point-to-point, rectangular coordinate, tabular, and polar coordinate dimensioning. The point-to-point and rectangular coordinate methods, and a mixture of the two, are the most common, but the other methods may be particularly useful for certain types of drawings.

Point-to-Point Dimensioning. *Point-to-point dimensioning* is a dimensioning method that specifies sizes along an object from one feature to another. **See Figure 6-4.** The dimensions are only as long as the space between features.

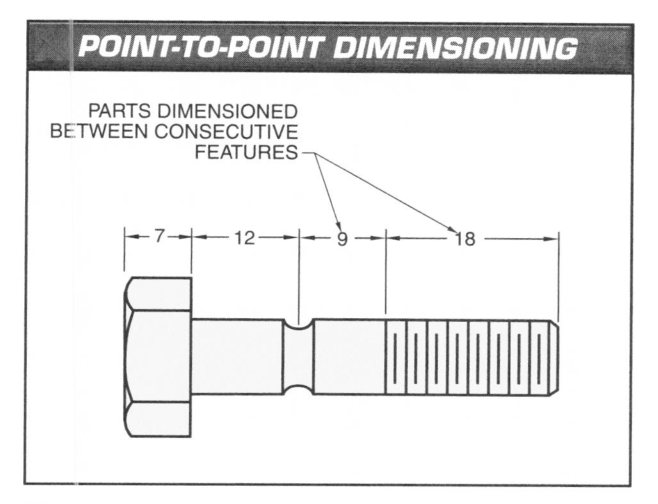

Figure 6-4. Point-to-point dimensioning shows only the dimensions between features.

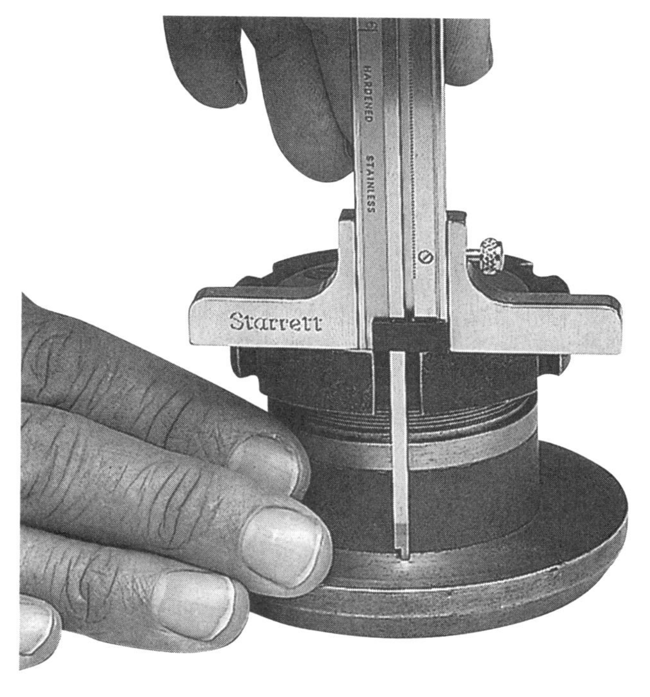

L.S. Starrett Company

Rectangular coordinate dimensioning involves measuring all features from one surface.

Rectangular Coordinate Dimensioning. *Rectangular coordinate dimensioning* is a dimensioning method that specifies all dimensions from a baseline or datum. **See Figure 6-5.** This method usually takes up a lot of space on a print.

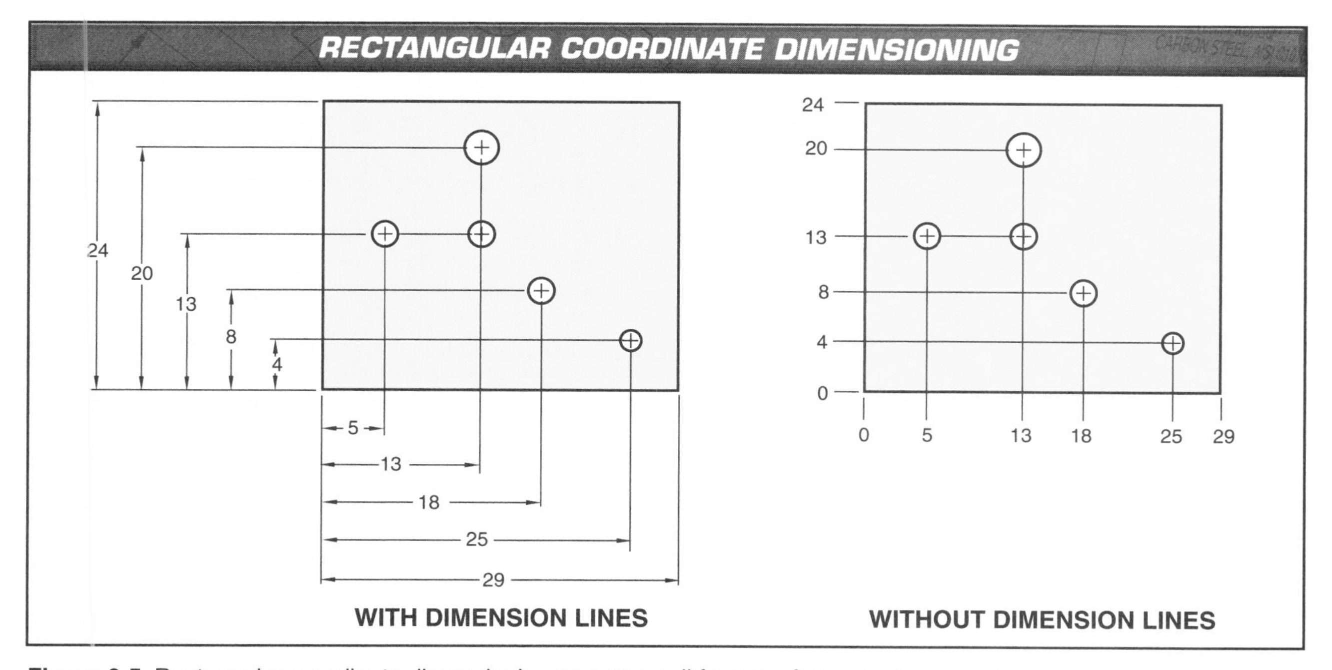

Figure 6-5. Rectangular coordinate dimensioning measures all features from common baselines.

A less cluttered alternative to rectangular coordinate dimensioning removes the dimension lines and places the numerical dimensions at the ends of the extension lines. Each baseline is labeled as zero and each extension line is dimensioned from the zero baseline.

Tabular Dimensioning. *Tabular dimensioning* is a dimensioning method that assembles all the numerical feature dimensions in table form. **See Figure 6-6.** Features are identified using letters or symbols, and all dimensions are determined from zero baselines using the rectangular coordinate method. The feature symbols and the associated dimensions are listed together in table form. Only overall dimensions and baseline indicators are typically included on the drawing.

Polar Coordinate Dimensioning. *Polar coordinate dimensioning* is a dimensioning method that determines location with angular and radius dimensions. **See Figure 6-7.** This method is particularly useful for parts with features arranged in a circle, when it is usually more convenient to indicate the circle's radius or diameter and then the angular distance of each feature from the circle's centerlines.

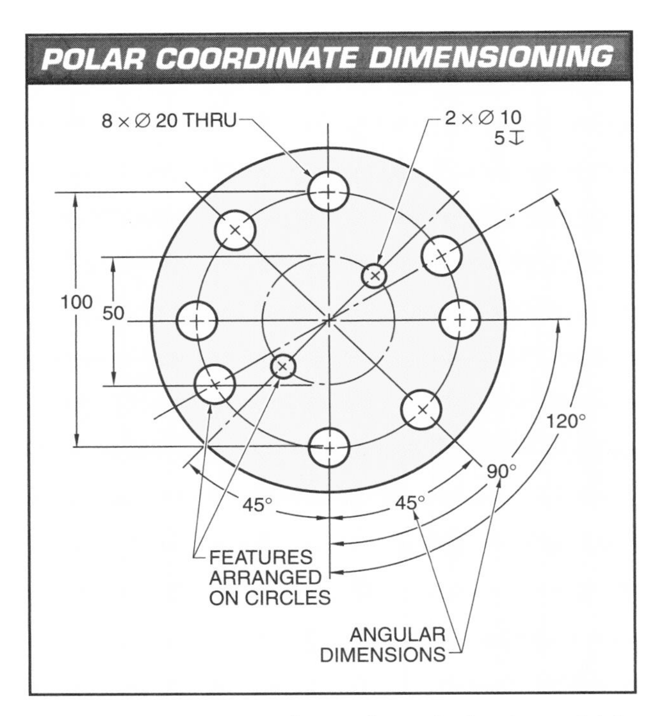

Figure 6-7. Polar coordinate dimensioning is useful for features arranged in circles.

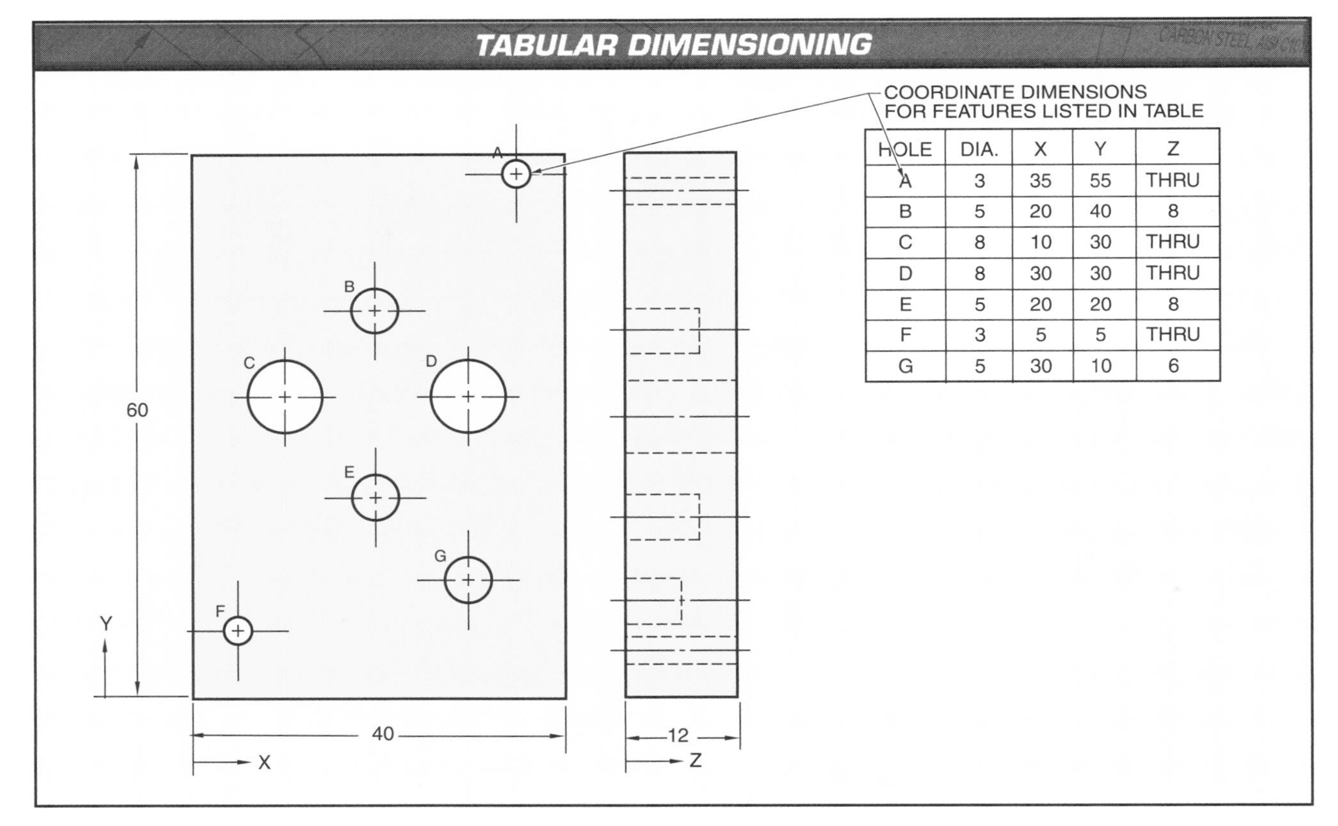

HOLE	DIA.	X	Y	Z
A	3	35	55	THRU
B	5	20	40	8
C	8	10	30	THRU
D	8	30	30	THRU
E	5	20	20	8
F	3	5	5	THRU
G	5	30	10	6

Figure 6-6. Coordinate dimensions can be arranged in a tabular form.

TOLERANCING

There is always some variation in size and shape among manufactured parts, and this variation affects the interchangeability of the parts. Well-designed assemblies accept some degree of inevitable variation while retaining full functionality. *Tolerancing* is a method of specifying the allowable variations of a feature.

Tolerances should not require a precision greater than is necessary for the fit and function of the part. Also, the precision capability of the manufacturing or machining process must be considered. The type of process and the condition of the tooling determine the range of possible tolerances. **See Figure 6-8.**

The combination of optimization and tolerancing techniques in manufacturing is the driving force of statistical process control (SPC). SPC is the application of statistical methods to monitor and control a production process for more efficient operation.

Direct Tolerancing

Direct tolerancing is the practice of specifying a dimension's permissible range directly within the dimensioning lines. The two common methods of direct tolerancing are limit dimensioning and plus and minus tolerancing. **See Figure 6-9.**

Limit Dimensioning. *Limit dimensioning* is the practice of including only the maximum and minimum values of a dimension. The two limits can be expressed in a single line, with low limit and high limit separated by a dash, or in two lines, with the high limit above the low limit.

The high and low limits should be shown to the same number of decimal places for uniformity, adding zeros if necessary.

Plus and Minus Tolerancing. *Plus and minus tolerancing* is the practice of providing an ideal dimension along with its allowable deviations in the positive and negative directions. The set of tolerances follows the dimension, with the positive tolerance always shown above the negative tolerance. The dimension and tolerances should be shown to the same number of decimal places, adding zeros for uniformity if necessary.

When the numerical tolerance is the same in both directions, the plus and minus values can be combined into a single expression using the ± symbol, such as 15.0 ±.1.

General Tolerances

Often, most or all of the dimensions on a drawing require the same degree of tolerance. To save space on the print, a general note may indicate a general tolerance. A *general tolerance* is a specified tolerance that is common for all dimensions on a drawing that are not otherwise toleranced. This replaces the inclusion of tolerances on every dimension separately.

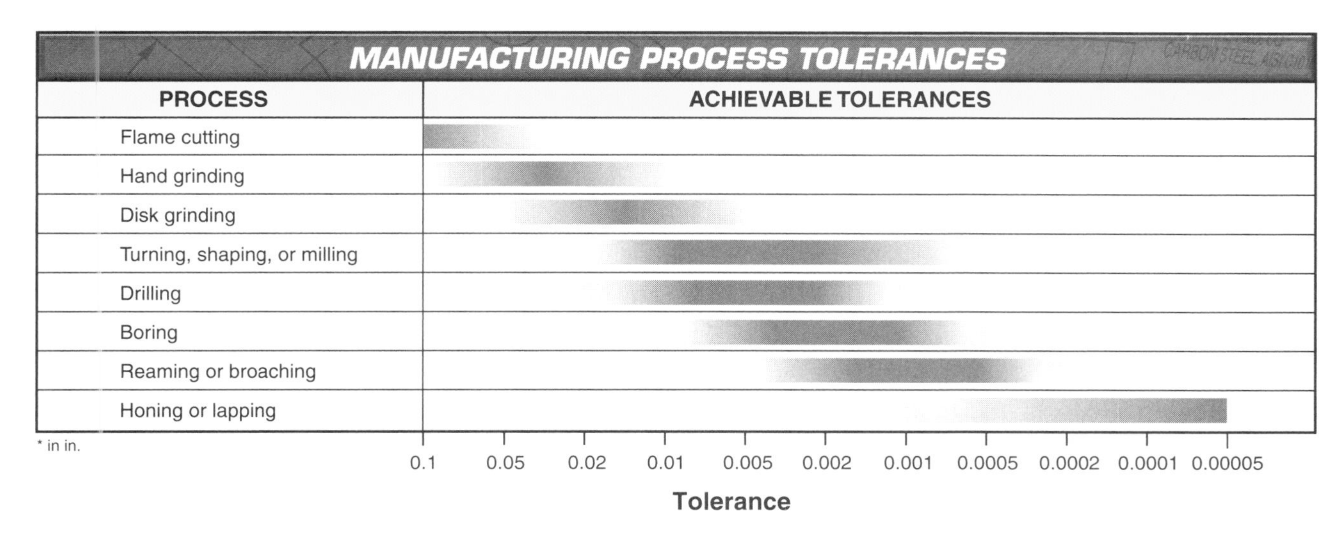

Figure 6-8. Specified tolerances should not be tighter than is necessary for the application or than is achievable with the intended manufacturing process.

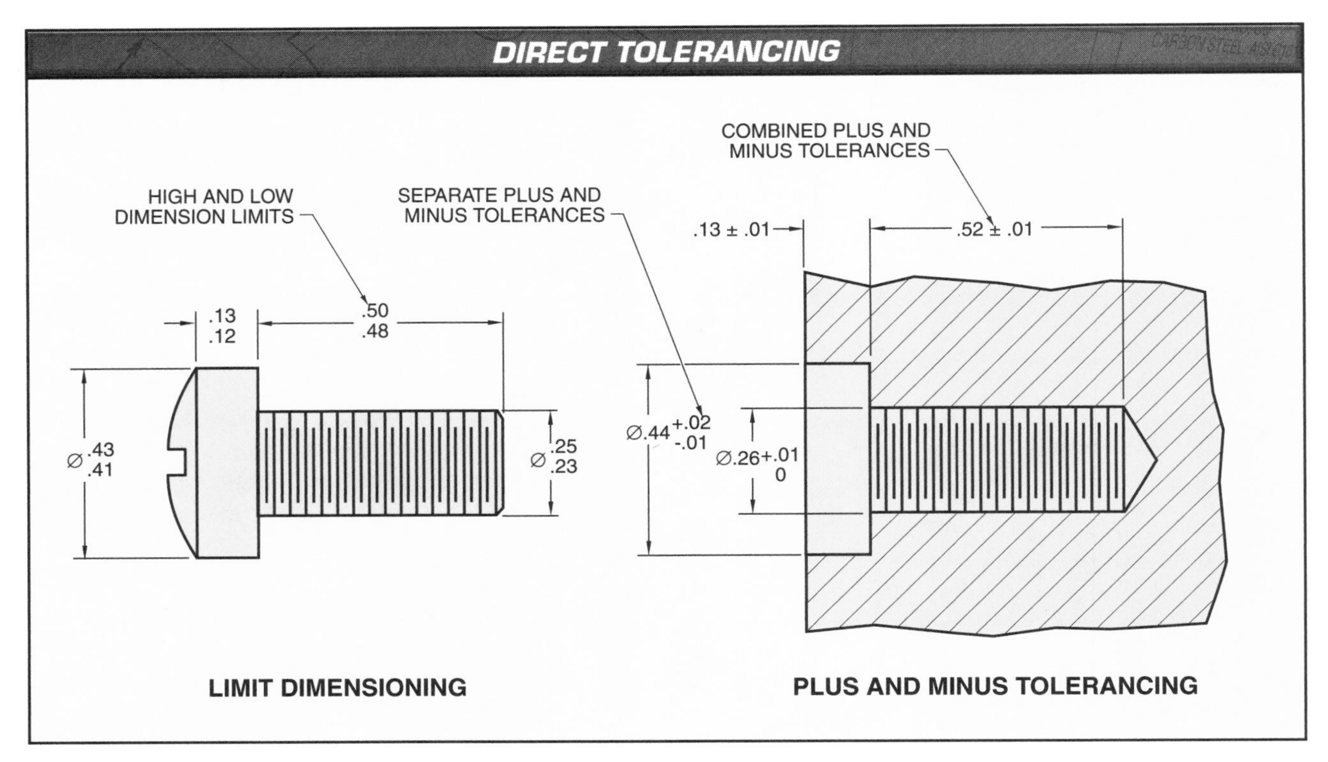

Figure 6-9. Direct tolerancing applies tolerances to each dimension separately.

Multiple general tolerances may be specified in order to cover dimensions of different precisions. For example, dimensions specified to hundredths of a unit may have a general tolerance of ±.01 while dimensions specified to thousandths have a tolerance of ±.002. Any special tolerances that must differ from the general tolerance can still be included. These are noted on the drawing at the dimension.

Tolerance Stack

A disadvantage of point-to-point dimensioning when tolerancing is that since each dimension is based on the one before it, tolerances for each successive dimension add up. **See Figure 6-10.** *Tolerance stack* is the accumulation of excessive tolerance due to the referencing of dimensions to features with their own tolerance. For example, if one feature is dimensioned as 50 ±2 from the baseline, and a second feature is dimensioned as 15 ±1 from the first feature, the second feature is equivalent to 65 ±3 from the baseline. The tolerances continue to increase for the third feature, fourth feature, and so on. This stacking of tolerances is typically undesirable. To avoid tolerance stack, a method based on a common datum, such as rectangular coordinate dimensioning, is recommended.

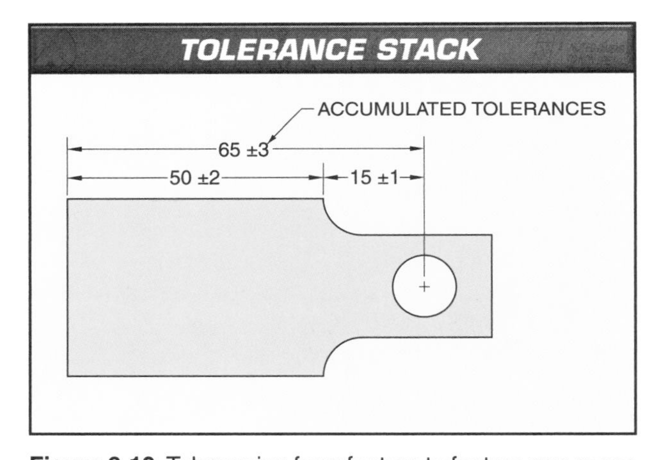

Figure 6-10. Tolerancing from feature to feature can cause an undesirable accumulation of tolerance for the overall part.

Datums

A *datum* is a theoretically exact point, line, axis, or surface that serves as the origin for dimensions and a reference for tolerances. Datums are identified in the primary orthographic views or any other view that is needed to establish the required relationships. The two types of datums are datum features and datum targets.

Datum Features. A *datum feature* is a datum that is an actual line or surface on a part. **See Figure 6-11.** Datums are identified using uppercase letters enclosed in a square frame. The frame is connected to the datum feature with a leader that is terminated with a triangle, which may be filled or unfilled. Datum frames can be connected directly to the datum's visible line, or extension line, or to a feature control frame already in place. (Previous versions of the ANSI standard define this symbol as a frame enclosing an uppercase letter between long dashes. This older symbol was placed directly adjacent to extension lines or other control frames.)

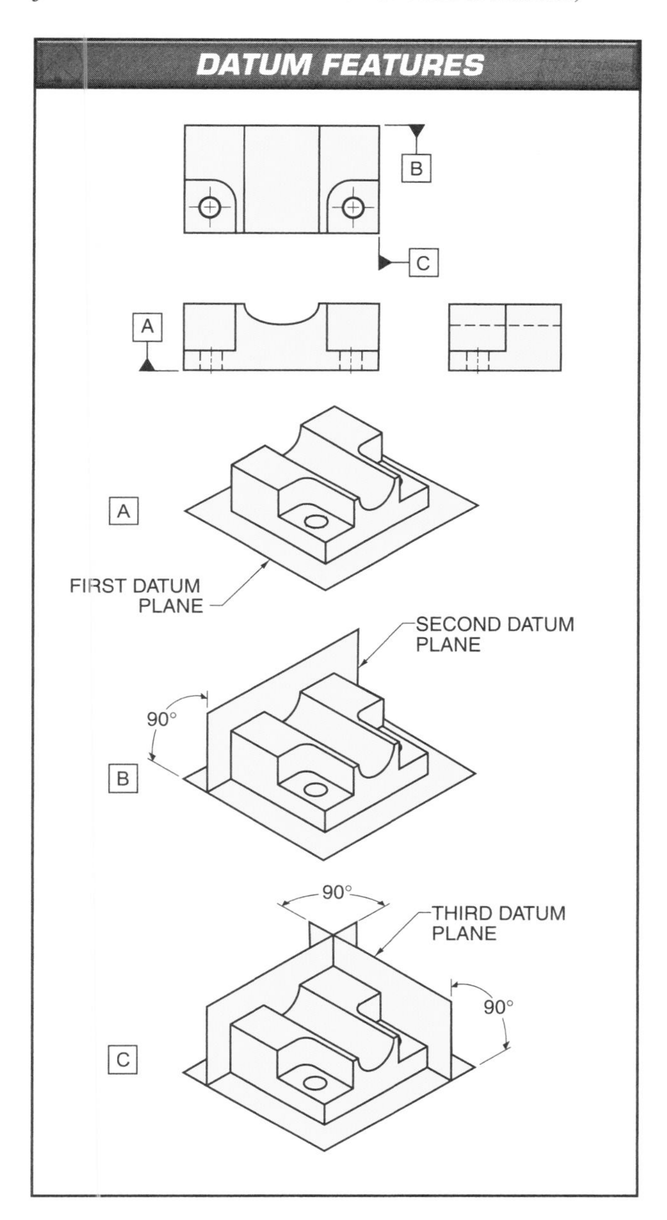

Figure 6-11. Surfaces are usually used as datum features.

The letters identifying datum features are assigned beginning with A and continuing through the alphabet, except I, O, and Q, which are too easily confused with similar numerals. If more than 23 unique datums are required, letter combinations such as AA through AZ, BA through BZ, and so on are used.

Datum Targets. A *datum target* is a datum that is a point, line, or area that is designated on a part but may not be a physical feature. **See Figure 6-12.** Points are marked with large × symbols. Lines are drawn with phantom lines in some views and appear as points when from the edge in other views. Areas are filled with general purpose section lining. The location of a target is given with basic dimensions.

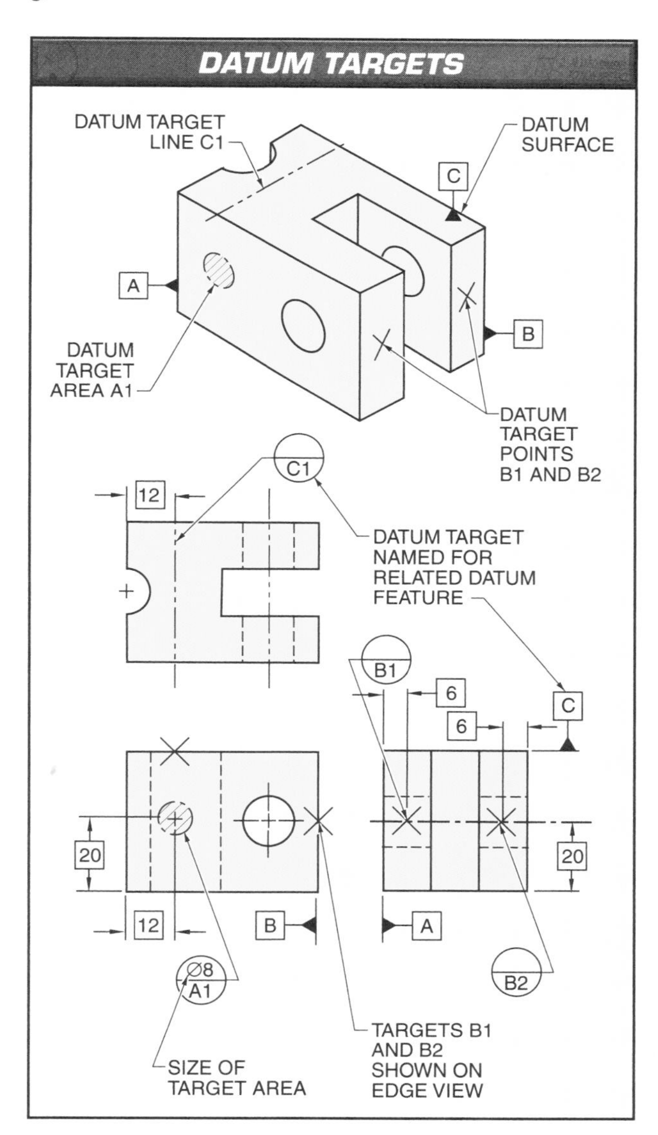

Figure 6-12. Datum targets can be points, lines, or areas.

A datum target symbol uses a circular frame divided into two halves. The bottom contains a letter identifying the associated datum feature and a serial number. The top half contains the size of the target, which is only required when the target is an area. Otherwise, the top half is empty.

Feature Control

Feature control is a tolerancing method that addresses special characteristics and can account for the interrelationships of multiple features. This method is typically used when manufacturing conditions require complex tolerances. A *feature* is any surface, angle, hole, round, or other characteristic on a part that can be dimensioned and controlled. A feature's relationships to datums determine whether it is individual or related.

Individual Features. An *individual feature* is a feature that is an independent characteristic of a part and does not relate to any datum. Tolerances on individual features stipulate how much the part may vary from a theoretical geometric form. Individual feature characteristics include straightness, flatness, circularity, and cylindricity. For example, a piece of square bar stock may have a straightness tolerance of .002″.

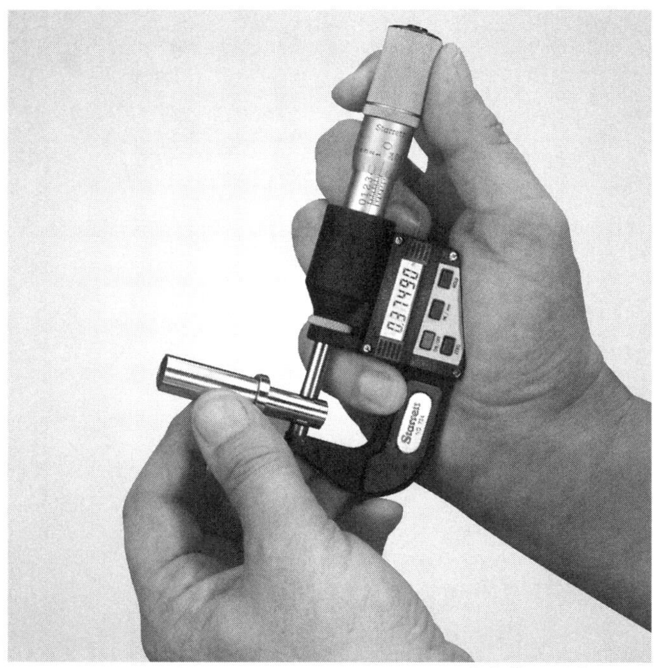

L.S. Starrett Company

Tolerancing specifies the allowable range of dimensions for a part feature.

Related Features. A *related feature* is a feature that is associated with one or more specific datums. Tolerances on related features specify the allowable deviation from a relationship with one or more datums. The relationships include position, concentricity, symmetry, angularity, perpendicularity, parallelism, circular runout, and total runout. For example, a hole may have a position tolerance of 1 mm in relation to the part's edge.

Feature Control Frames

Feature tolerance information is enclosed within a feature control frame. A *feature control frame* is a rectangular frame divided into sections that encloses characteristic symbols, numerical tolerances, datum references, and modifiers. **See Figure 6-13.**

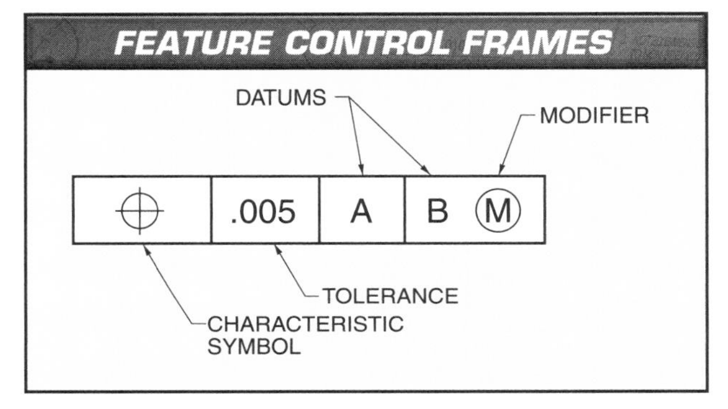

Figure 6-13. Feature control frames contain information for applying very specific tolerances to features.

A *characteristic symbol* is a symbol that represents a feature characteristic. **See Figure 6-14.** Tolerance in the frame is applied to the characteristic represented by the symbol. For example, a drawing may require that a surface be flat to a tolerance of 0.08. The feature control frame for that surface would include the symbol for flatness and the tolerance of 0.08.

For related features, datums are included in the feature control frame. Multiple datums can be referenced and are typically added individually in their order of importance for the feature. **See Figure 6-15.** There may be primary, secondary, and tertiary datums. In addition, multiple datums may be identified as primary if they are equally necessary to determine a feature relationship.

FEATURE CHARACTERISTIC SYMBOLS			
	TYPE OF FEATURE CONTROL	**CHARACTERISTIC**	**SYMBOL**
FOR INDIVIDUAL FEATURES	Form	Straightness	—
		Flatness	⏥
		Circularity	○
		Cylindricity	⌭
FOR INDIVIDUAL OR RELATED FEATURES	Profile	Profile of a line	⌒
		Profile of a surface	⌓
FOR RELATED FEATURES	Location	Position	⌖
		Concentricity	◎
		Symmetry	⌯
	Orientation	Angularity	∠
		Perpendicularity	⊥
		Parallelism	//
	Runout	Circular runout	↗
		Total runout	⌰

Figure 6-14. Many different feature characteristics can be controlled.

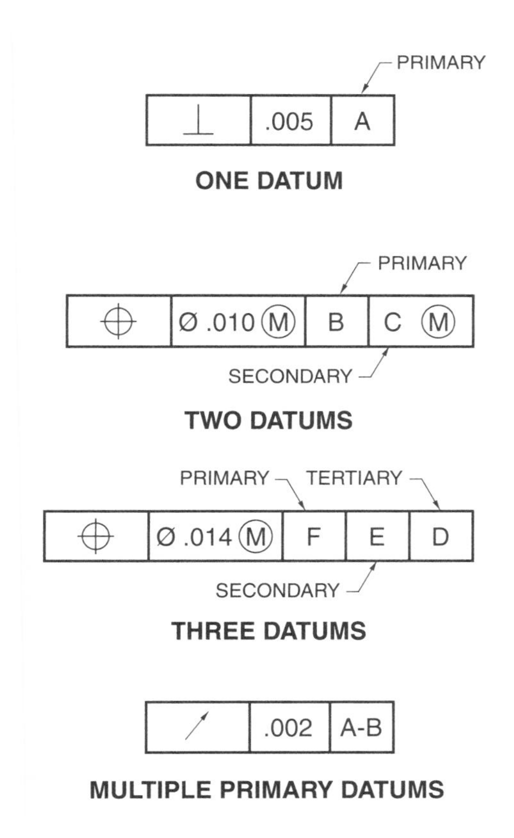

Figure 6-15. Feature control frames may include more than one datum reference.

A *modifier symbol* is a symbol that represents a specific condition of the part or feature when the tolerance is applied. For instance, one tolerance may require that it be applied when another feature is at a maximum tolerance condition. Modifiers may not be necessary at all or they may appear multiple times within a feature control frame.

Tolerance Zones

A *tolerance zone* is an area or volume that defines the space within which a feature may acceptably vary. Parts with features outside their respective tolerance zones may be rejected. The smaller the tolerance zone is, the more accurately the feature must be manufactured. **See Figure 6-16.** The method used to define a feature's acceptable variations determines the size and shape of the tolerance zone. Tolerance zones can be created by direct tolerancing or by using feature control.

FEATURE CHARACTERISTICS

There are five categories of feature characteristic control: form, profile, location, orientation, and runout. Each category includes two or more characteristics that can be directly controlled through feature control frames. A *characteristic* is an aspect of a feature that can be toleranced.

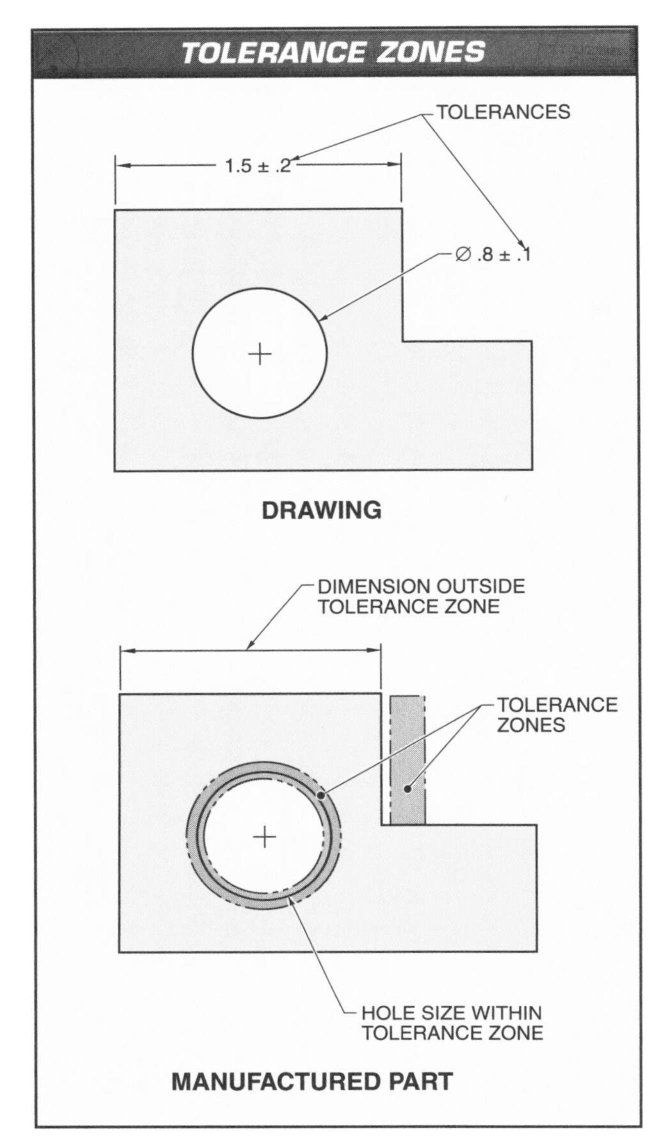

Figure 6-16. The area or volume that a feature may vary within is the tolerance zone.

Form

Tolerances of form refer to how well a feature matches a theoretically perfect feature. Form characteristics are not related to any datum, so the referenced features are independent. These tolerances specify the maximum permissible variation of desired surface conditions for straightness, flatness, circularity, or cylindricity.

Straightness. *Straightness* is the condition of a line feature where all points on the feature are in perfect alignment. **See Figure 6-17.** Straightness tolerance can be applied to a surface in one dimension or to an axis. The tolerance creates a tolerance zone between two parallel lines. The distance between the lines is equal to the numerical tolerance. The actual surface or axis must lie between the lines.

Flatness. *Flatness* is the condition of a surface feature with no surface variations. **See Figure 6-18.** Flatness tolerance applies only to flat surfaces. The specified tolerance provides a distance between two parallel planes. The actual surface must lie between the planes.

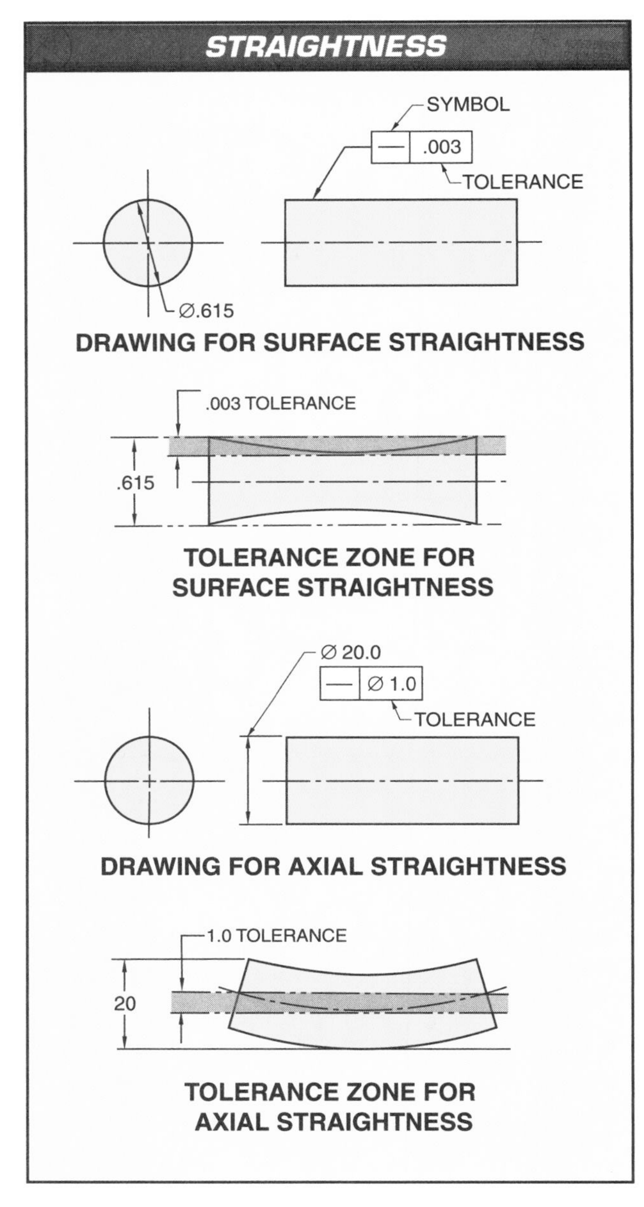

Figure 6-17. A straightness tolerance applies to surfaces or axes.

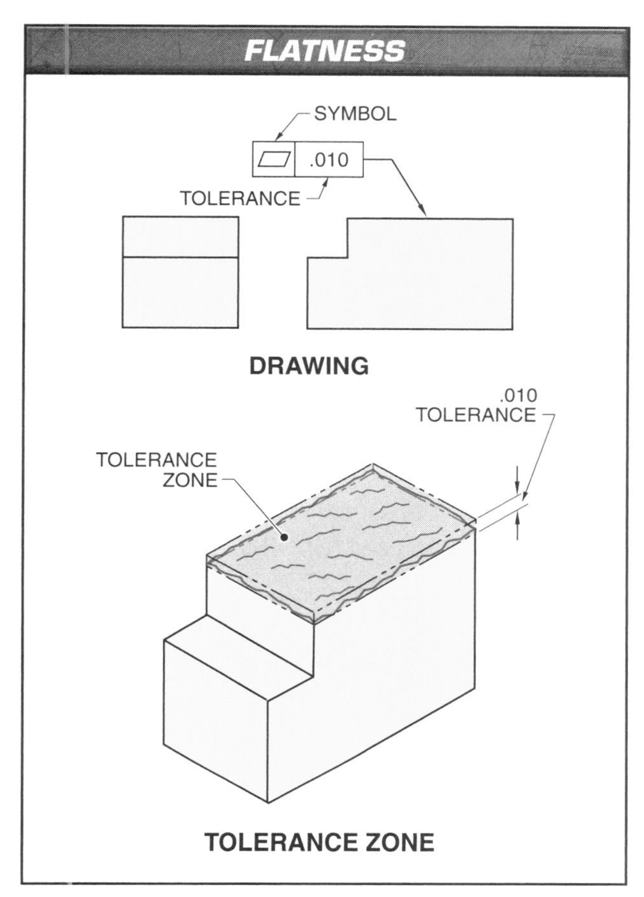

Figure 6-18. A flatness tolerance provides a tolerance zone between two parallel planes.

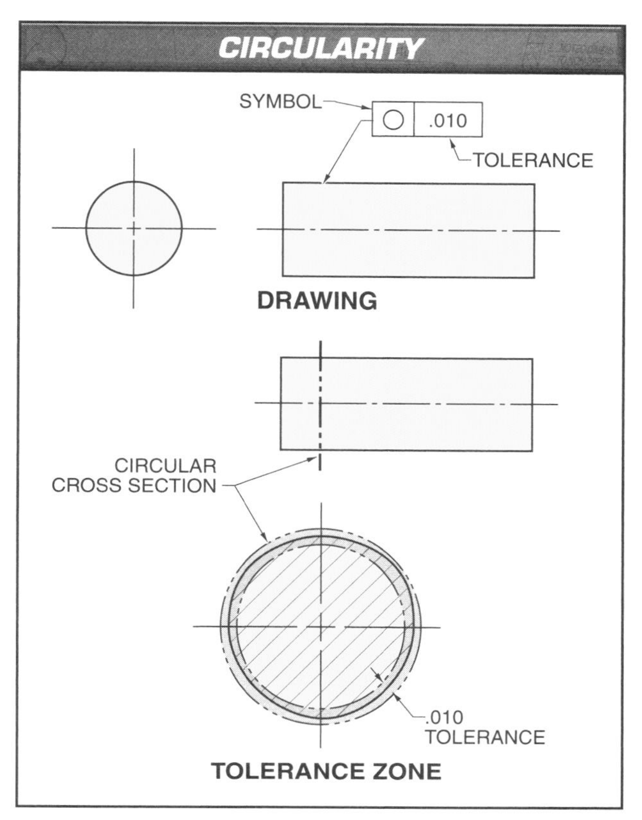

Figure 6-19. A circularity tolerance applies to circular cross sections of an object.

Circularity. *Circularity* is the condition of a circular feature where all points are equally distant from a centerpoint. **See Figure 6-19.** Circularity tolerance creates a ring-shaped tolerance zone around the centerpoint formed by two concentric circles. The distance between the circles is equal to the numerical tolerance. Circularity tolerance is applied to features with circular cross sections, such as cylinders and spheres. The cross section of the feature shall be in the tolerance zone. There is no consideration given to the surface of the object except where the section is taken.

Cylindricity. *Cylindricity* is the condition of a cylindrical feature where all points are equally distant from a common axis. **See Figure 6-20.** Tolerance for cylindricity creates a tolerance zone around the axis in the form of two concentric cylinders. The distance between the cylinder walls is equal to the numerical tolerance. The surface of the cylindrical feature must lie within the tolerance zone.

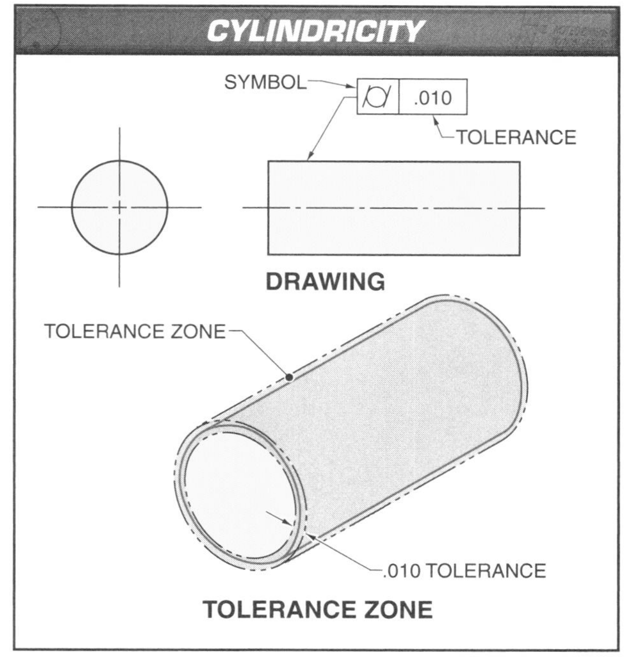

Figure 6-20. A cylindricity tolerance creates a tolerance zone from two concentric cylinders.

Profile

A *profile* is an outline of an object in a given plane. Profiles are formed by orthographic projection or cross section. Elements that make up a profile include lines, arcs, and other curved lines. Theoretical profiles are specified on a print with lines, angles, and arc radii in basic dimensions. Toleranced profiles can be all around a part or only around a section whose end points are identified on the drawing with letters. Profile tolerances include profile of a line and profile of a surface.

Profile of a Line. A *profile of a line* is a mathematically defined two-dimensional shape. **See Figure 6-21.** The profile of a line characteristic specifies how closely the actual part matches the profile defined on an orthographic view or cross section. The tolerance zone follows the shape of the profile and is as wide as the numerical tolerance.

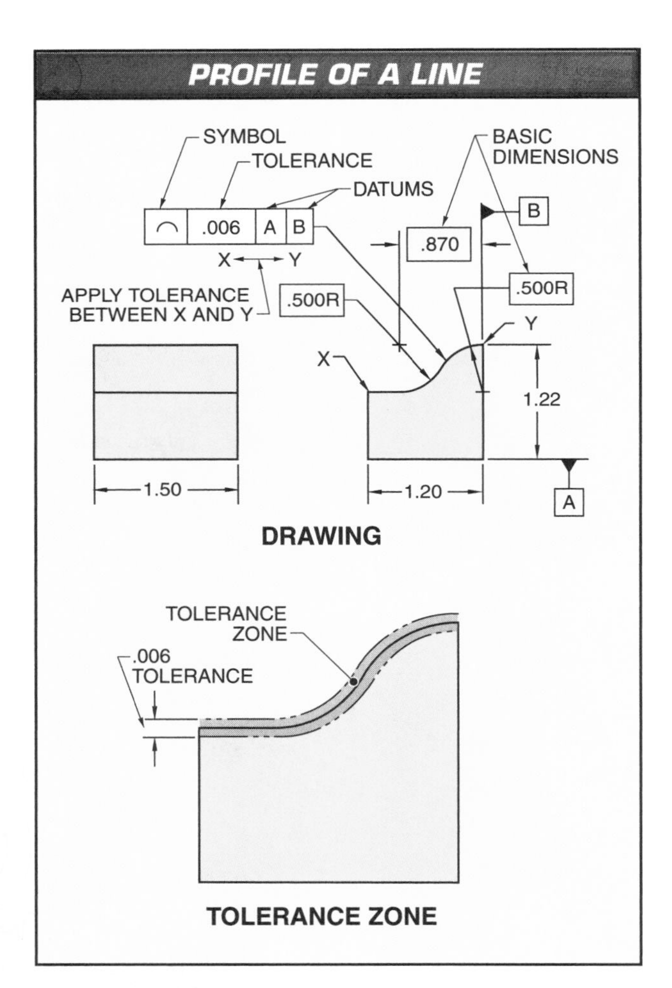

Figure 6-21. A profile of a line tolerance is applied around a cross section.

This type of tolerance usually references one or more datums, but datums may not be required if the profile is taken cross section by cross section, as in a continuous extrusion.

Profile of a Surface. A *profile of a surface* is a mathematically defined three-dimensional surface. **See Figure 6-22.** A profile of a surface tolerance is applied in the same way as a profile of a line tolerance, except that it controls a three-dimensional surface. The tolerance zone follows the theoretical surface and is as thick as the numerical tolerance. A profile of a surface tolerance usually references datums.

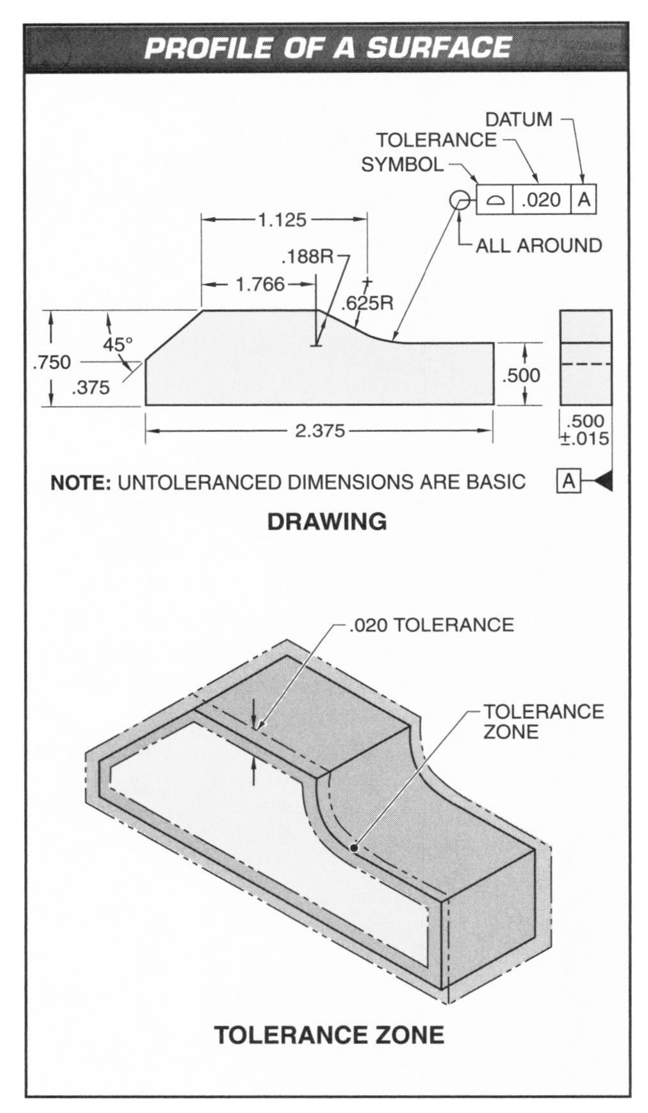

Figure 6-22. A profile of a surface tolerance is applied around an entire area.

Location

Location characteristics define tolerances for the arrangement of features in relation to one or more datums. Basic dimensions establish the nominal positions.

Position. *Position* is the condition of a feature's center, axis, or center plane being at a nominal distance from one or more datums. **See Figure 6-23.** Position tolerances may be referenced to only one datum, but they are more commonly referenced to two or three. This indicates the number of directions that are important to the tolerance.

Positional tolerances are usually given as a diameter, which creates a circular tolerance zone. The toleranced centerpoint or axis can vary anywhere within the circular tolerance zone.

Concentricity. *Concentricity* is the condition in which a circular feature's axis exactly matches a datum axis. **See Figure 6-24.** Circular features include circles, cylinders, spheres, and cones. Concentricity is usually referenced to one datum. Tolerances are given as a diameter that, when projected along the axis of the datum, creates a cylindrical tolerance zone. Axes of toleranced features must remain within the tolerance zone.

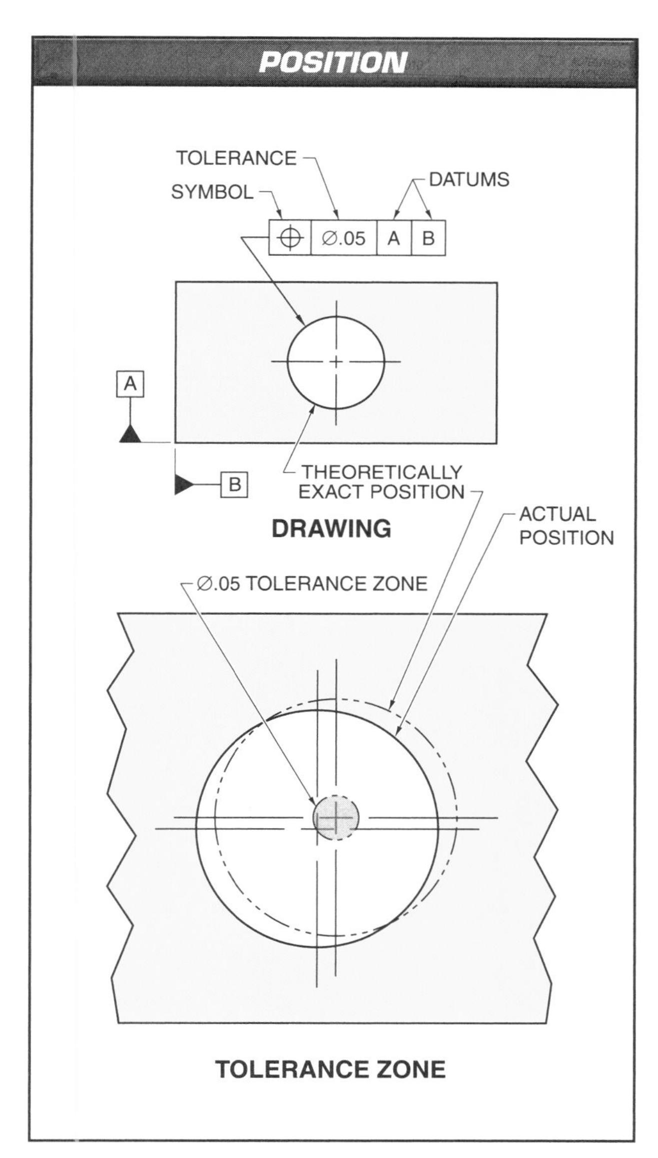

Figure 6-23. A positional tolerance provides a clear description of the location of a feature.

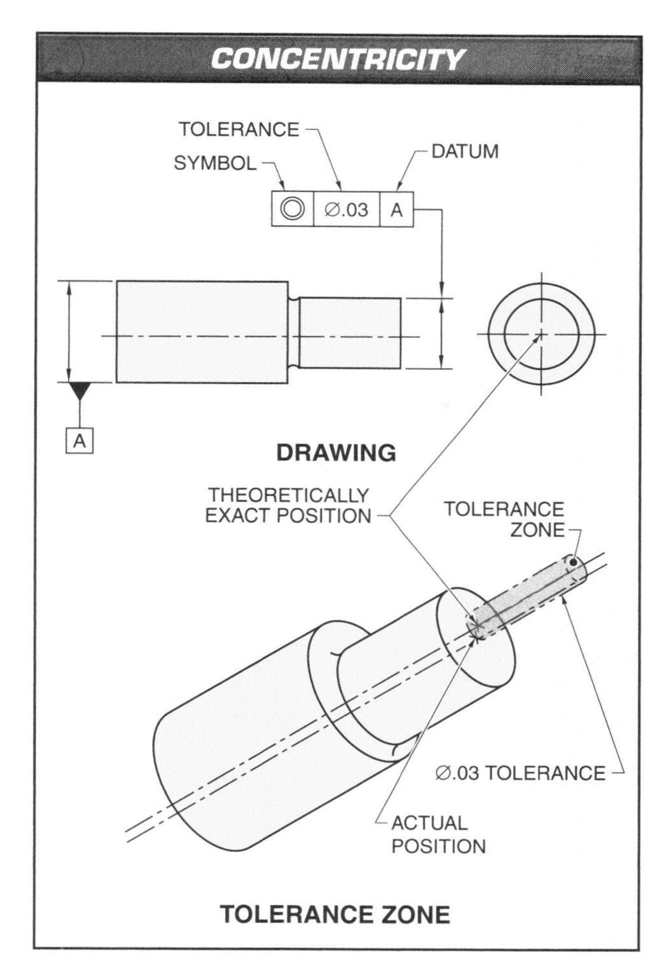

Figure 6-24. A concentricity tolerance specifies the relationship between a datum axis and circular features.

Concentricity is most commonly used when tolerancing cylinders, but can also be applied to spherical features.

Symmetrical relationships can be toleranced in multiple ways, which can involve position, profile, or symmetry feature characteristics. For example, position may be used to define the relationship of a center plane (between the two symmetrical features) to a datum plane.

Symmetry. *Symmetry* is the condition where the median points of elements on opposing features lie exactly on the part's centerline. **See Figure 6-25.** On a part that is not perfectly symmetrical, the median points will vary away from the centerline. The tolerance for symmetry creates a tolerance zone of two parallel planes. Median points between all symmetrically opposed features must lie within the tolerance zone.

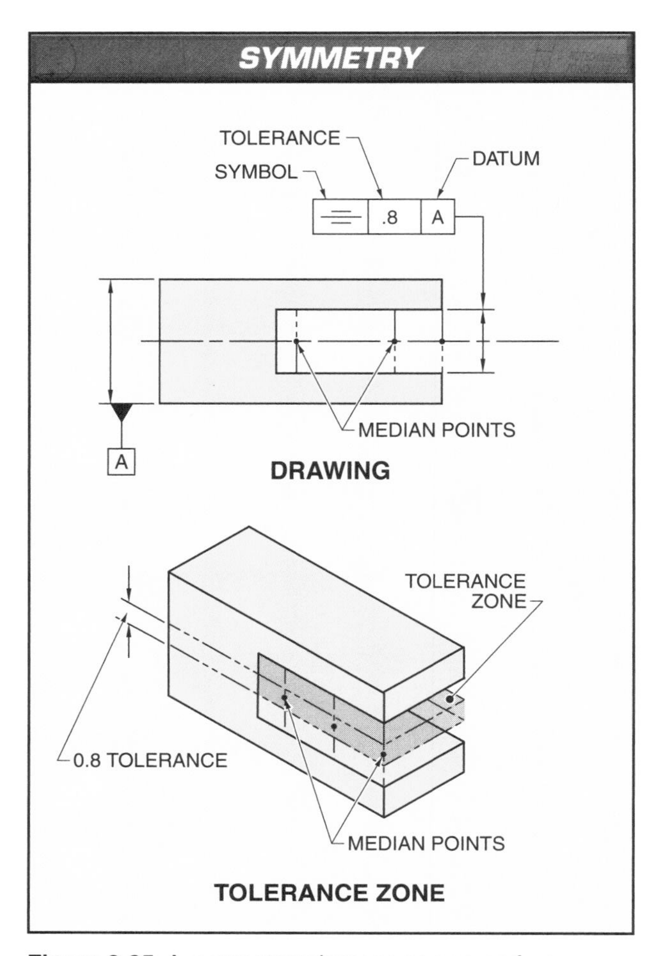

Figure 6-25. A symmetry tolerance compares features on either side of a symmetry line.

Orientation

Orientation tolerancing for related features involves the control of the relationship between two or more lines or surfaces. Orientation tolerances control angularity, perpendicularity, and parallelism.

L.S. Starrett Company

Indicators are used to check cylindricity and runout.

Angularity. *Angularity* is the condition of a feature's axis or surface being at a specified angle (other than 90°) from a datum. **See Figure 6-26.** When the angular tolerance is given as a numerical value only, a tolerance zone is formed by two parallel planes whose distance apart is equal to the numerical tolerance. An angular tolerance specified as a diameter (a numerical value with a diameter symbol) forms a cylindrical tolerance zone. The toleranced line or axis may vary in angularity as long as it remains inside the tolerance zone.

Perpendicularity. *Perpendicularity* is the condition of a feature's surface or axis being at a right angle (exactly 90°) to one or two datums. **See Figure 6-27.** Tolerancing perpendicularity functions in a similar way to angularity, except that the angle is always 90°. One difference, however, is that perpendicularity can reference two datums at once so that the surface must be perpendicular to both, within the tolerance zone.

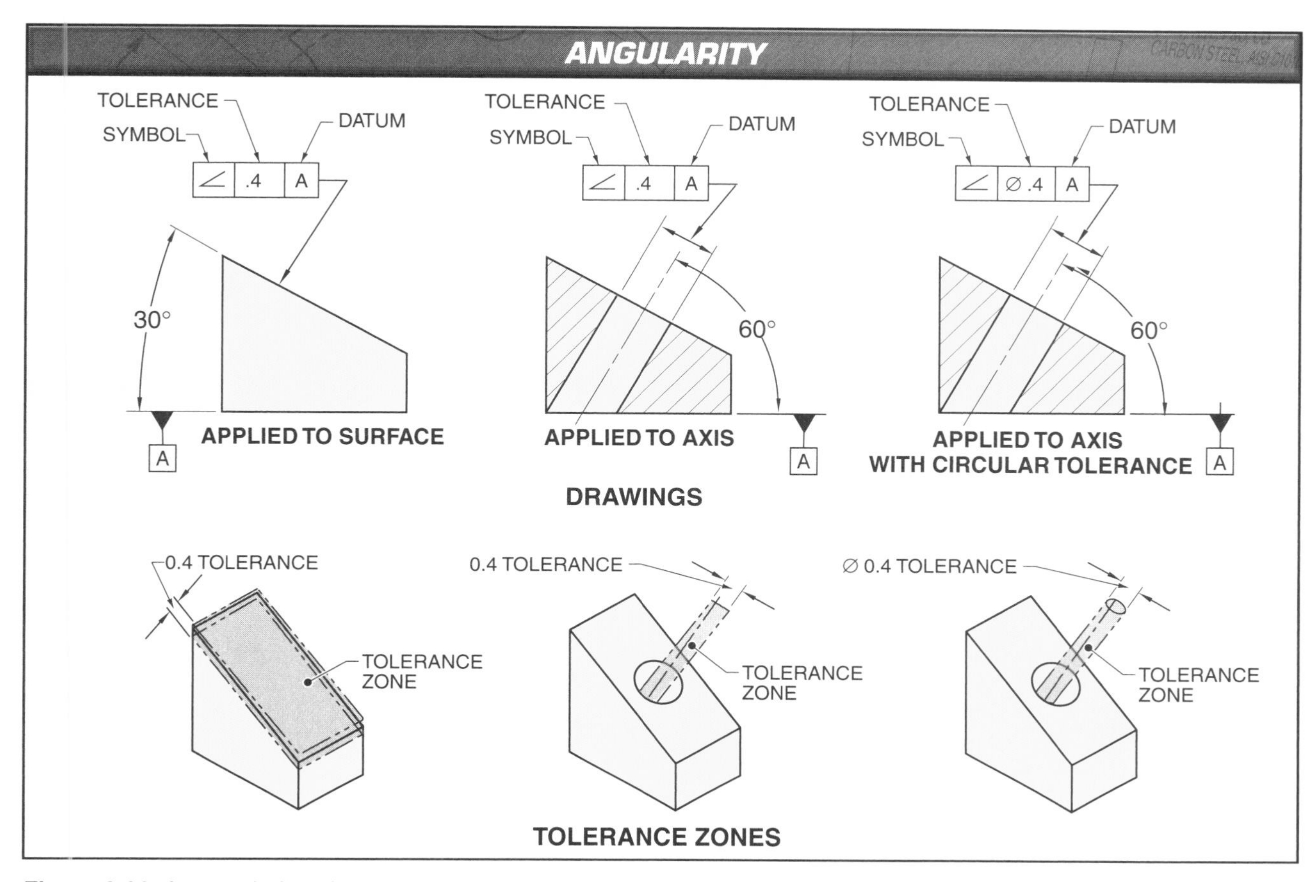

Figure 6-26. An angularity tolerance relates a feature at a specified angle to a datum.

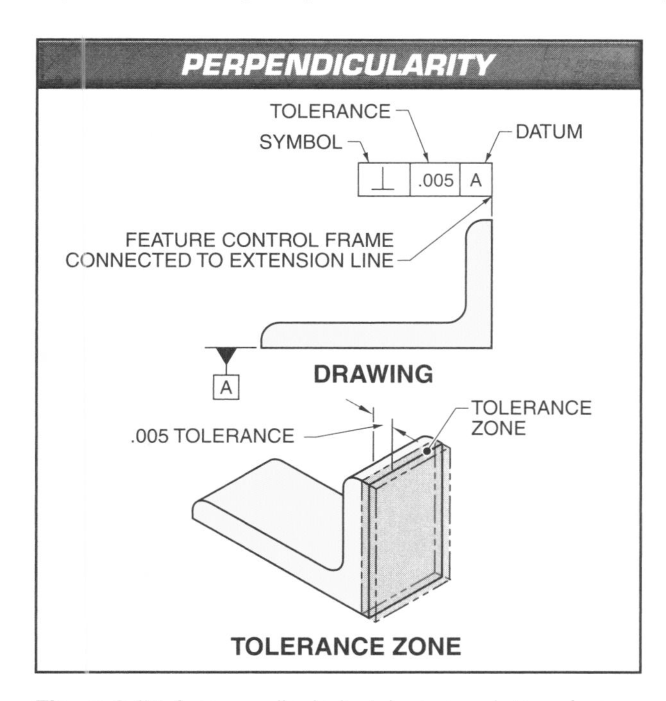

Figure 6-27. A perpendicularity tolerance relates a feature at a right angle to a datum.

Parallelism. *Parallelism* is the condition of a feature's axis or surface where all the points are equidistant from a datum line or plane. **See Figure 6-28.** Tolerancing parallelism is also similar to angularity because parallelism can be considered to be an angularity of 0° to the datum.

Runout

Runout is the measurement of the relationship of circular features to a datum axis. Runout is measured with an indicator placed perpendicular to the surface and fixed in relation to the axis. The indicator measures variations in the surface height as the part is rotated for a full revolution about its axis. The full indicator movement (FIM) must fall within the tolerance zone for the part to be acceptable.

Circular Runout. *Circular runout* is the maximum variation between high and low spots on the edge of a circular feature in relation to its centerpoint. **See Figure 6-29.** Actual circular runout cannot be greater than the specified tolerance.

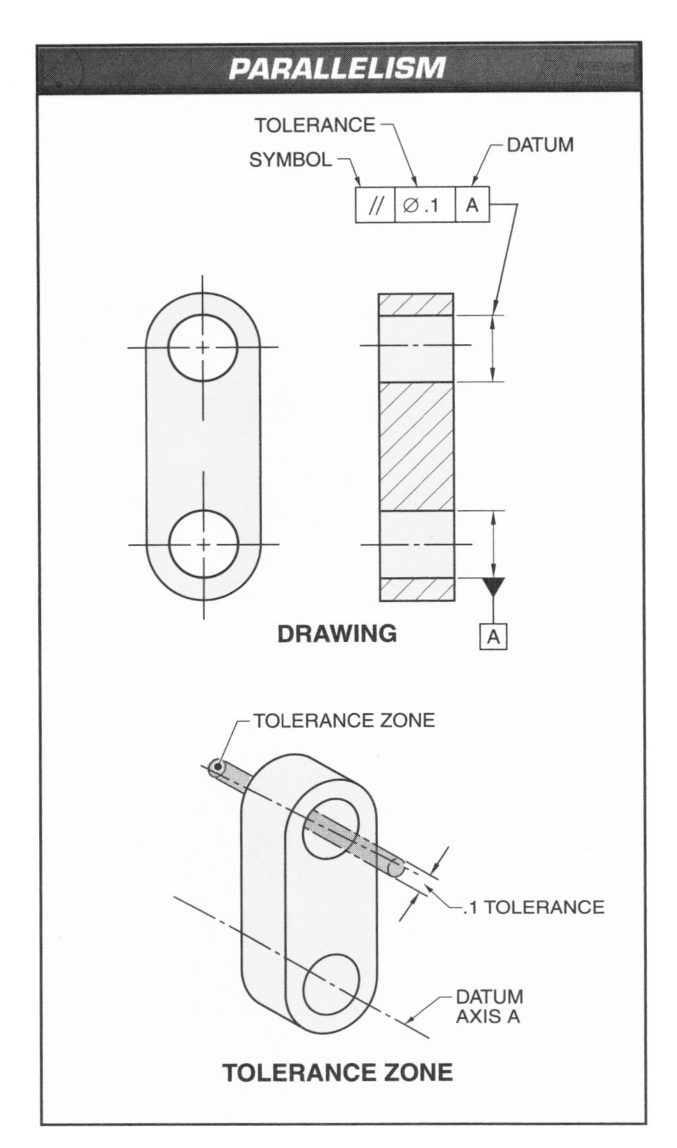

Figure 6-28. A parallelism tolerance applies to features equidistant from a datum.

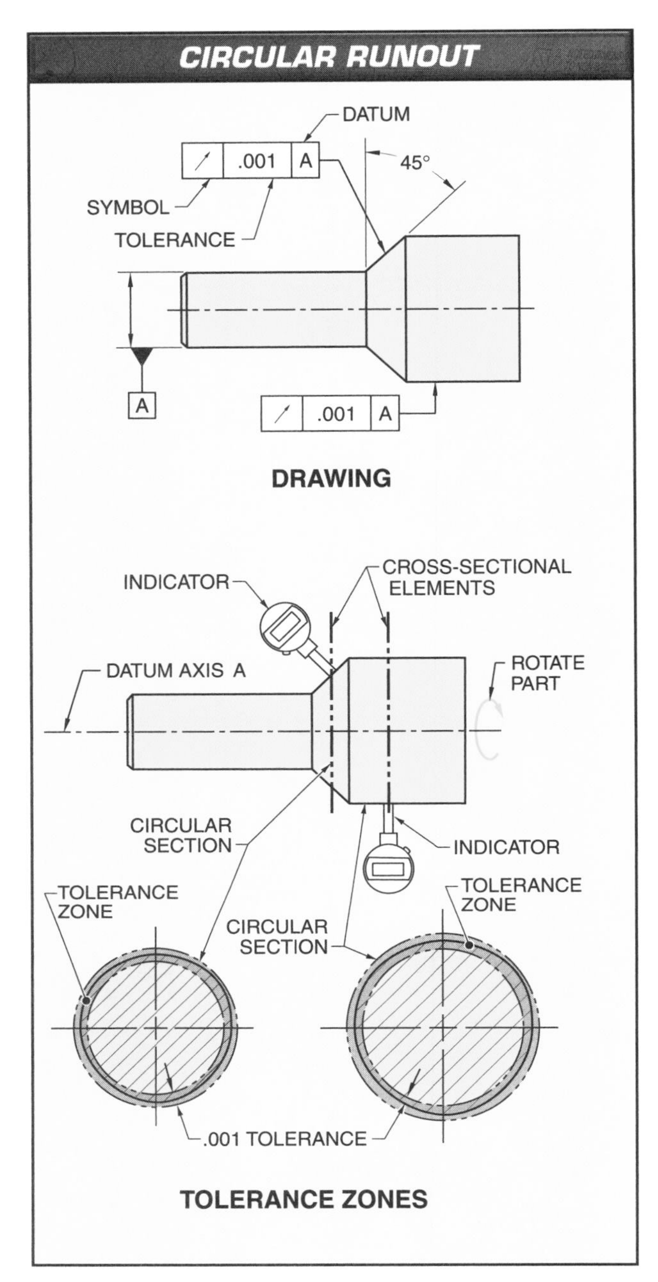

Figure 6-29. A circular runout tolerance provides maximum allowable variation around circular cross sections.

Datum targets are not actual features on the part and are usually determined by how the part mates with other parts in an assembly. For example, a drawing may include a datum target point because a pin must contact the part at a certain location. A datum target line may indicate where a part must rest on an edge or the side of a shaft. A datum target area may be necessary when surfaces of parts must contact each other precisely. The areas are usually circular, but may be any shape clearly described and dimensioned.

Total Runout. *Total runout* is the maximum variation between high and low spots on a cylindrical surface feature in relation to its axis. **See Figure 6-30.** The indicator is moved back and forth along the length of the cylinder, which is rotated many times, in order to sufficiently test the entire surface. Actual total runout cannot be greater than the specified tolerance.

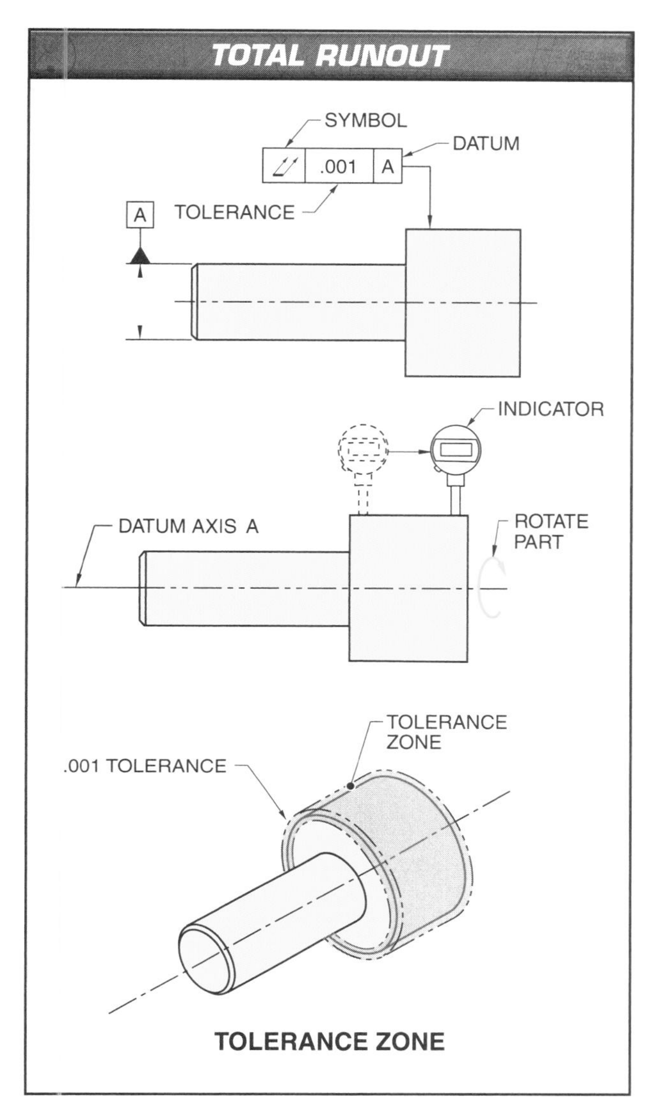

Figure 6-30. A total runout tolerance provides maximum allowable variation to a cylindrical surface area.

MODIFIERS

A *modifier* is a specification for a special condition of a feature or tolerance zone when a tolerance is applied. This allows tolerances to vary for different conditions. Modifiers are particularly important for parts that mate in assemblies. This is because when two parts must fit together, the actual manufactured size of one part affects the required size of the other part.

Modifiers are specified with symbols, which are placed in the feature control frame. The placement of the modifier identifies the aspect of the tolerance that is being modified. Multiple modifiers can be used together in a feature control frame to specify complex conditions.

Common Modifiers

There are several possible modifiers defined in the ANSI/ASME Y14.5 standard. **See Figure 6-31.** However, some are rarely used. The most commonly applied modifiers are regardless of feature size, maximum material condition, and least material condition.

MODIFIER SYMBOLS

MODIFIER	SYMBOL
Regardless of feature size	No symbol
Maximum material condition	Ⓜ
Least material condition	Ⓛ
Translation	▷
Projected tolerance zone	Ⓟ
Free state	Ⓕ
Tangent plane	Ⓣ
Unequally disposed profile	Ⓤ
Independency	Ⓘ
Statistical tolerance	⟨ST⟩
Continuous feature	⟨CF⟩

Figure 6-31. Modifier symbols are included in feature control frames to define special tolerancing conditions.

Regardless of Feature Size. *Regardless of feature size* is the condition where the characteristic tolerances are not affected by the size tolerances of the feature. As noted in the current standard, this state is assumed when no modifier symbols are included in the feature control frame. When regardless of feature size is explicitly identified, it is designated with the abbreviation RFS or the letter S in a circle. These symbols are holdovers from previous standards when they were included among the modifier symbols.

Maximum Material Condition. *Maximum material condition (MMC)* is the condition of a feature having the maximum amount of material permitted by the tolerance zone. For internal features such as holes and slots, this is the minimum allowable size. For external features, this is the maximum allowable size. This modifier may be designated with the circle M symbol or the abbreviation MMC.

Multiple modifiers and datums can be combined within a feature control frame for complex tolerancing.

Least Material Condition. *Least material condition (LMC)* is the condition of a feature having the minimum amount of material permitted by the tolerance zone. For internal features, this is the maximum allowable size for the feature. For external features, this is the minimum permissible size. Although this is the opposite of MMC, it works in the same way to increase tolerances as the feature size changes. This modifier may be designated with the circle L symbol or the abbreviation LMC.

Modifying Features

The size of a feature may affect other characteristics, particularly position. For example, consider an assembly in which a shaft in a particular location must fit into a hole on an adjacent part. The hole has a range of acceptable diameters. **See Figure 6-32.** At its smallest permissible size, its position must be very precise for the shaft to fit. Therefore, the hole's feature control frame specifies a positional tolerance based on its maximum material condition (smallest size). If the hole's actual size is larger, the position of the hole may vary more than its original tolerance and still fit onto the shaft. In that case, the positional tolerance zone is larger.

An example applying the concept of least material condition is an assembly in which a boss (protruding surface) must contact a specific area on an adjacent part. **See Figure 6-33.** Controlling the size and position of the boss is a way to adequately contact the required area under all conditions. At the boss's smallest size (least material condition), the position tolerance is the smallest. As the size of the boss increases, there is greater permissible variance in position, so the tolerance zone also increases.

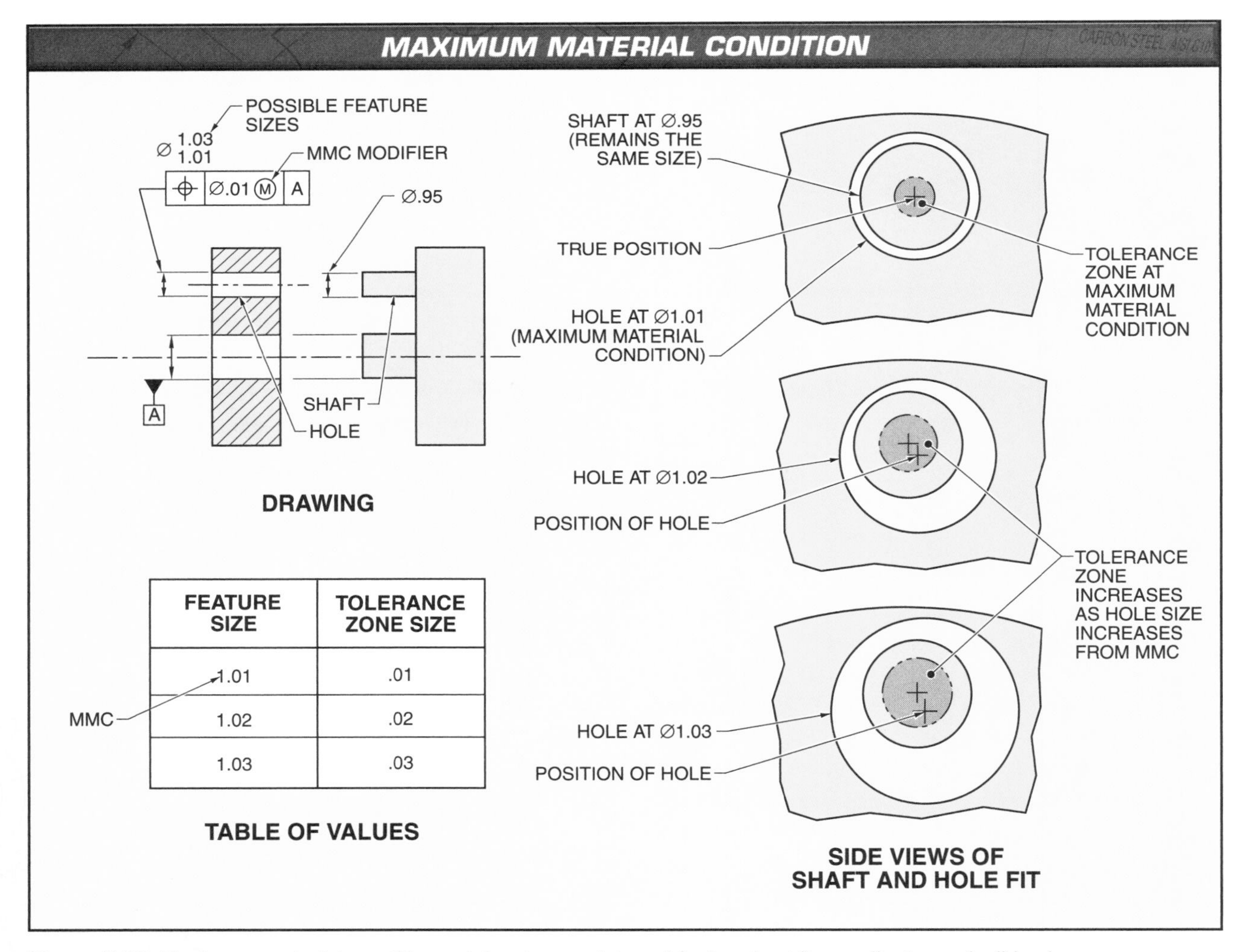

Figure 6-32. Maximum material condition exists when an internal feature is at its smallest permissible size.

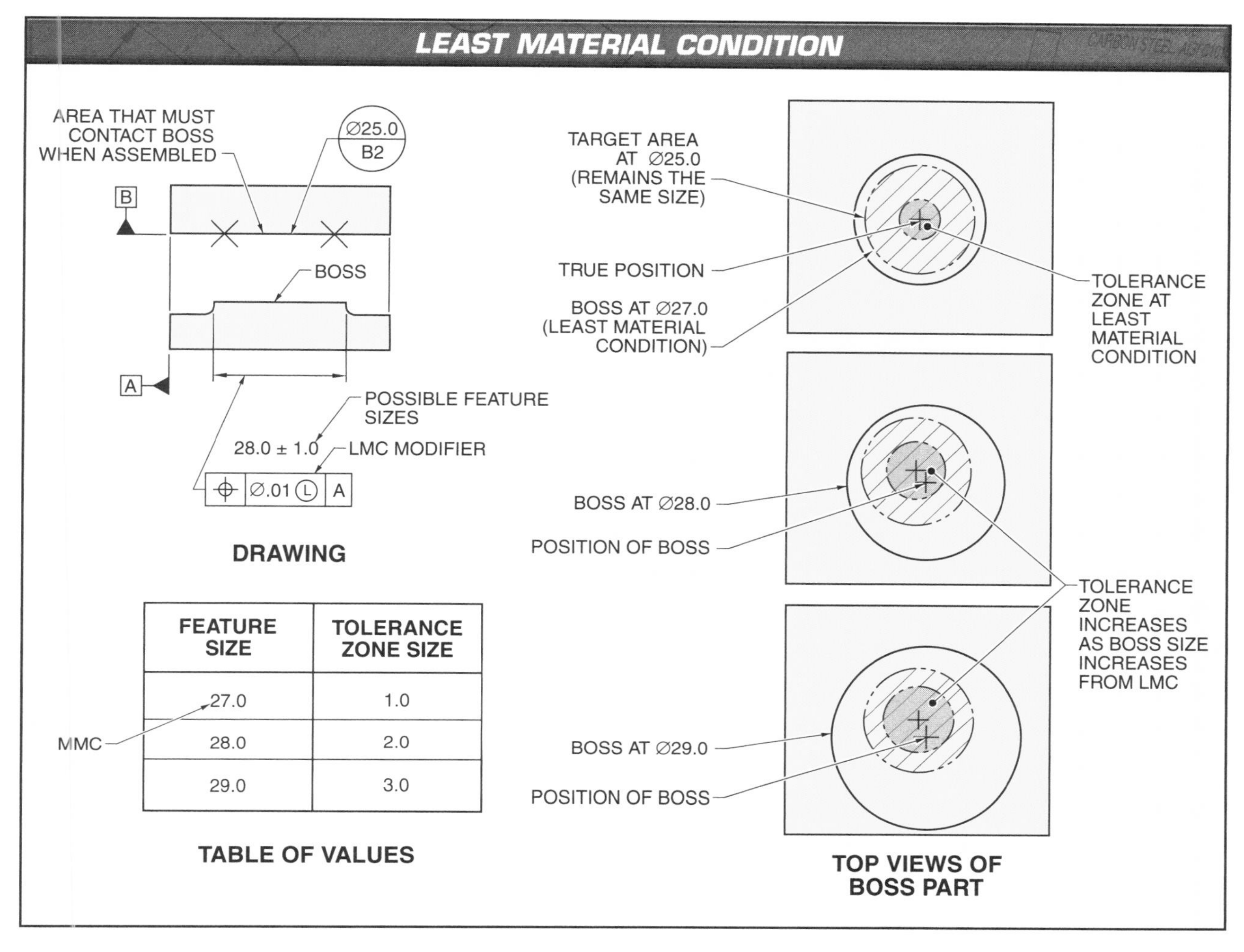

FEATURE SIZE	TOLERANCE ZONE SIZE
27.0	1.0
28.0	2.0
29.0	3.0

Figure 6-33. Least material condition defines tolerances for external features at their smallest permissible size.

The statistical tolerance modifier involves the statistical probability of mass-produced components fitting together in an assembly.

Modifying Datums

Modifiers can also be applied to datums with dimensional tolerances. In this case, the modifier symbol is included in the same frame section as the datum letter. The tolerance in the feature control frame applies when the datum is at a modifier condition, and the tolerance changes as the datum varies within its tolerance zone. When both a tolerance and a datum are modified, tolerance change is affected by both the feature and datum sizes. **See Figure 6-34.**

L.S. Starrett Company
Indicators are used to measure runout in reference to the datum axis.

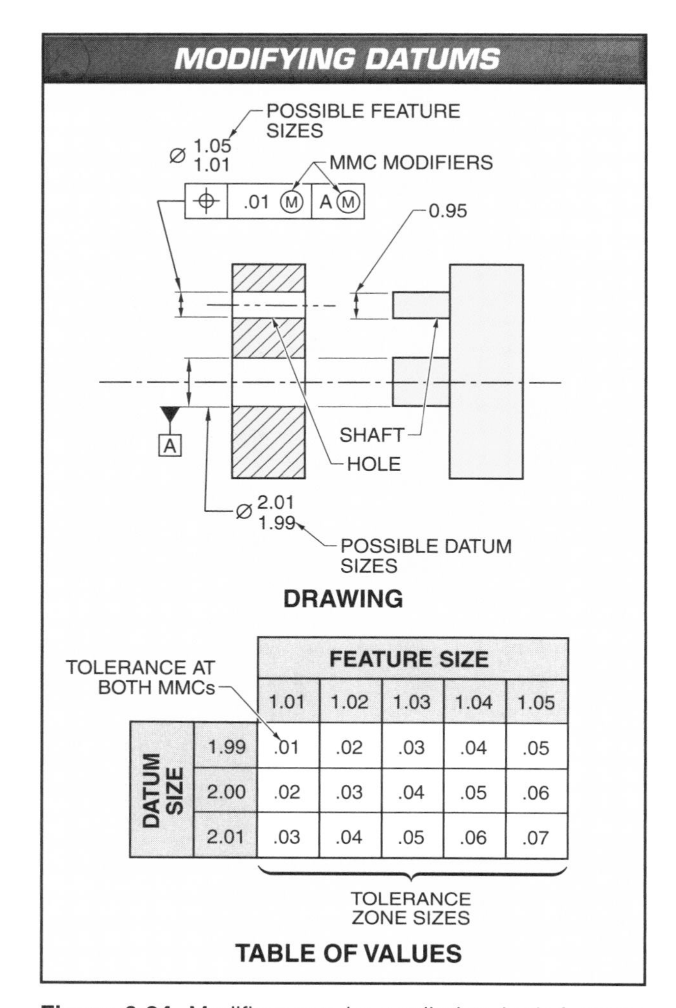

DATUM SIZE	FEATURE SIZE 1.01	1.02	1.03	1.04	1.05
1.99	.01	.02	.03	.04	.05
2.00	.02	.03	.04	.05	.06
2.01	.03	.04	.05	.06	.07

Figure 6-34. Modifiers can be applied to both features and datums.

L.S. Starrett Company

The specification of engineering fits is critical to tolerancing features in an assembly.

STANDARD TOLERANCING

Both ANSI and the ISO have established standards for the specification of manufacturing tolerances. Drawings or specifications referencing one of these standards can identify the appropriate general tolerances by including only a grade number. This does not preclude the use of other tolerances for special applications, but it provides a consistent system of tolerance designations that are appropriate for most applications.

> The projected tolerance zone modifier specifies the diameter and height of a tolerance zone volume that extends away from a feature, usually an axis. An extension of the toleranced feature must remain within the tolerance zone.

Standard ANSI Tolerances

ANSI lists tolerance grades in the standard ANSI/ASME B4.1, *Preferred Limits and Fits for Cylindrical Parts.* The tolerances are based on the part's basic dimension and the desired degree of precision. The grades range from 4 (tightest) to 13 (loosest). **See Figure 6-35.** The most commonly used tolerances are between 5 and 7.

Standard ISO Tolerances

The ISO provides information on standard tolerances in the standard ISO 286, *Geometrical Tolerance Specifications.* This group of documents lists 20 international tolerance (IT) grades: IT01 through IT18. Grades IT01 through IT16 are the most common. **See Figure 6-36.** The general tolerance for a particular part is determined by the part's basic dimension and the degree of tolerance needed for the design.

ENGINEERING FITS

An *engineering fit* is the specification of the appropriate tolerances and allowances for the way two parts are intended to mate together. *Allowance* is the amount of space between mating parts when at their actual size. For example, a common application of fit specification is a shaft that must be assembled into a hole in another part. **See Figure 6-37.** Tolerances for the shaft and hole can be designed so that the shaft may move freely in the hole (positive allowance) or is locked into place after being forced into the hole (zero or negative allowance).

STANDARD ANSI TOLERANCES*										
Nominal Size	Tolerance Grade									
	4	5	6	7	8	9	10	11	12	13
0 – 0.12	0.00012	0.00015	0.00025	0.0004	0.0006	0.0010	0.0016	0.0025	0.004	0.006
0.12 – 0.24	0.00015	0.00020	0.0003	0.0005	0.0007	0.0012	0.0018	0.0030	0.005	0.007
0.24 – 0.40	0.00015	0.00025	0.0004	0.0006	0.0009	0.0014	0.0022	0.0035	0.006	0.009
0.40 – 0.71	0.00020	0.0003	0.0004	0.0007	0.0010	0.0016	0.0028	0.004	0.007	0.010
0.71 – 1.19	0.00025	0.0004	0.0005	0.0008	0.0012	0.0020	0.0035	0.005	0.008	0.012
1.19 – 1.97	0.0003	0.0004	0.0006	0.0010	0.0016	0.0025	0.0040	0.006	0.010	0.016
1.97 – 3.15	0.0003	0.0005	0.0007	0.0012	0.0018	0.0030	0.0045	0.007	0.012	0.018
3.15 – 4.73	0.0004	0.0006	0.0009	0.0014	0.0022	0.0035	0.005	0.009	0.014	0.022
4.73 – 7.09	0.0005	0.0007	0.0010	0.0016	0.0025	0.0040	0.006	0.010	0.016	0.025
7.09 – 9.85	0.0006	0.0008	0.0012	0.0018	0.0028	0.0045	0.007	0.012	0.018	0.028
9.85 – 12.41	0.0006	0.0009	0.0012	0.0020	0.0030	0.005	0.008	0.012	0.020	0.030
12.41 – 15.75	0.0007	0.0010	0.0014	0.0022	0.0035	0.006	0.009	0.014	0.022	0.035
15.75 – 19.69	0.0008	0.0010	0.0016	0.0025	0.004	0.006	0.010	0.016	0.025	0.04
19.69 – 30.09	0.0009	0.0012	0.0020	0.003	0.005	0.008	0.012	0.020	0.03	0.05
30.09 – 41.49	0.0010	0.0016	0.0025	0.004	0.006	0.010	0.016	0.025	0.04	0.06
41.49 – 56.19	0.0012	0.0020	0.003	0.005	0.008	0.012	0.020	0.03	0.05	0.08
56.19 – 76.39	0.0016	0.0025	0.004	0.006	0.010	0.016	0.025	0.04	0.06	0.10
76.39 – 100.9	0.0020	0.003	0.005	0.008	0.012	0.020	0.03	0.05	0.08	0.125
100.9 – 131.9	0.0025	0.004	0.006	0.010	0.016	0.025	0.04	0.06	0.10	0.16
131.9 – 171.9	0.003	0.005	0.008	0.012	0.020	0.03	0.05	0.08	0.125	0.20
171.9 – 200	0.004	0.006	0.010	0.016	0.025	0.04	0.06	0.10	0.16	0.25

* in in.

Figure 6-35. Standard tolerances are defined by ANSI for ranges of nominal sizes. Grade numbers indicate the tightness of the tolerance.

An engineering fit is determined by two specified values: tolerance and fundamental deviation. The tolerance is the permissible variation in a feature's dimension. The *fundamental deviation* is the difference between a basic dimension and the closest limit of the tolerance zone. While it is common for a tolerance zone to be centered on a basic dimension, it does not have to be, especially when describing fits. The fundamental deviation describes the amount of offset of the tolerance zone from the basic size. The combination of tolerance and fundamental deviation determines the clearance (positive allowance) or interference (negative allowance) between two mating parts.

ISO standard tolerances are calculated from a formula based on the basic size and IT grade.

Fit Types

Fit types are characterized as clearance, interference, or transitional. A *clearance fit* is a fit design for two mating parts that always have space between them, as long as both parts are within their tolerance zones. For example, the smallest possible size of the hole is still greater than, or at least equal to, the largest possible size of the shaft. An *interference fit* is a fit design for two mating parts that always has some overlap between the parts, regardless of their actual sizes within their tolerance zones. The largest possible size of the hole is smaller than, or at least equal to, the smallest possible size of the shaft. A *transitional fit* is a fit design for two mating parts that may result in either clearance or interference, depending on the actual sizes of the parts within their tolerance zones. Therefore, the tolerance zones of the hole and shaft partly or completely overlap.

STANDARD ISO TOLERANCES*

NOM. SIZE	TOLERANCE GRADE																
	01	0	1	2	3	4	5	6	7	8	9	10	11	12	13	14	15	16
0 – 3	0.0003	0.0005	0.0080	0.0012	0.0020	0.003	0.004	0.006	0.010	0.014	0.025	0.040	0.060	0.10	0.14	0.25	0.40	0.60
3 – 6	0.0004	0.0006	0.0010	0.0015	0.0025	0.004	0.005	0.008	0.012	0.018	0.030	0.048	0.075	0.12	0.18	0.30	0.48	0.75
6 – 10	0.0004	0.0006	0.0010	0.0015	0.0025	0.004	0.006	0.009	0.015	0.022	0.036	0.058	0.090	0.15	0.22	0.35	0.58	0.90
10 – 18	0.0005	0.0008	0.0012	0.0020	0.003	0.005	0.008	0.011	0.018	0.027	0.043	0.070	0.11	0.18	0.27	0.43	0.70	1.1
18 – 30	0.0006	0.0010	0.0016	0.0025	0.004	0.006	0.009	0.013	0.021	0.033	0.052	0.084	0.13	0.21	0.33	0.52	0.84	1.3
30 – 50	0.0008	0.0010	0.0016	0.0028	0.004	0.007	0.011	0.016	0.025	0.030	0.062	0.100	0.16	0.25	0.39	0.62	1.00	1.6
50 – 80	0.0008	0.0012	0.0020	0.003	0.005	0.008	0.013	0.019	0.030	0.045	0.074	0.120	0.19	0.30	0.46	0.74	1.20	1.9
80 – 120	0.0010	0.0015	0.0025	0.004	0.006	0.010	0.016	0.022	0.036	0.054	0.087	0.140	0.22	0.36	0.54	0.87	1.40	2.2
120 – 180	0.0012	0.002	0.0035	0.005	0.080	0.012	0.018	0.025	0.040	0.063	0.100	0.160	0.25	0.40	0.63	1.00	1.60	2.5
180 – 250	0.0020	0.003	0.0045	0.007	0.010	0.014	0.020	0.029	0.046	0.072	0.115	0.186	0.29	0.46	0.72	1.15	1.85	2.9
250 – 315	0.0025	0.004	0.006	0.008	0.012	0.016	0.022	0.032	0.052	0.081	0.130	0.21	0.32	0.52	0.81	1.30	2.1	3.2
315 – 400	0.0030	0.005	0.007	0.009	0.013	0.018	0.025	0.036	0.057	0.089	0.140	0.23	0.36	0.57	0.89	1.40	2.3	3.6
400 – 500	0.0040	0.006	0.008	0.010	0.016	0.020	0.027	0.040	0.063	0.097	0.165	0.25	0.40	0.63	0.97	1.55	2.5	4.0
500 – 630	0.0045	0.006	0.009	0.011	0.016	0.022	0.030	0.044	0.070	0.110	0.175	0.28	0.44	0.70	1.10	1.75	2.8	4.4
630 – 800	0.0050	0.007	0.010	0.013	0.018	0.026	0.035	0.050	0.080	0.125	0.20	0.32	0.50	0.80	1.25	2.0	3.2	5.0
800 – 1000	0.0055	0.008	0.011	0.015	0.021	0.029	0.040	0.056	0.090	0.140	0.23	0.38	0.56	0.90	1.40	2.3	3.6	5.5
1000 – 1250	0.0065	0.009	0.013	0.018	0.024	0.034	0.046	0.066	0.105	0.165	0.26	0.42	0.68	1.06	1.65	2.6	4.2	6.5
1250 – 1600	0.008	0.011	0.016	0.021	0.019	0.040	0.054	0.078	0.126	0.198	0.31	0.50	0.78	1.26	1.98	3.1	5.0	7.5
1600 – 2000	0.009	0.013	0.018	0.026	0.035	0.048	0.065	0.092	0.150	0.23	0.37	0.60	0.92	1.50	2.3	3.7	6.0	9.2
2000 – 2500	0.011	0.016	0.022	0.030	0.041	0.067	0.077	0.110	0.175	0.26	0.44	0.70	1.10	1.75	2.8	4.4	7.0	11.0
2500 – 3160	0.013	0.018	0.026	0.036	0.050	0.069	0.093	0.135	0.210	0.33	0.54	0.88	1.36	2.10	3.3	5.4	8.0	13.5

* in mm

Figure 6-36. The ISO publishes standard tolerances for ranges of nominal feature sizes. Grade numbers indicate the tightness of the tolerance.

Boston Gear

Engineering fits are critical for bearings and bushings, which are usually pressed into housings.

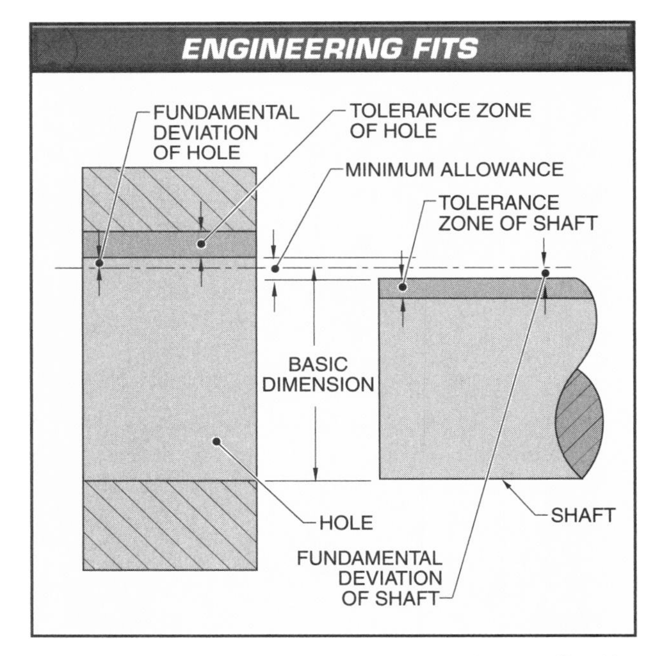

Figure 6-37. The fit of two parts in an assembly is defined by the basic dimension, fundamental deviation, and tolerance.

Engineering fits are represented graphically as tolerance zones drawn in relation to the basic dimension, represented by a baseline. **See Figure 6-38.** The height of the tolerance zone is based on the standard tolerance grade or other specified tolerance. The position of the tolerance zone is based on the fundamental deviation. Tolerance zones for internal features (such as holes) and external features (such as shafts) are drawn on opposite sides of the zero line. For clearance fits, the distance between the two tolerance zones represents the open clearance space. For transitional fits, the tolerance zones overlap on the zero line. For interference fits, the tolerance zones switch relative positions and the distance between them represents the interference.

Standard fundamental deviations are identified by letter designations. Uppercase letters are used for internal features (holes), while lowercase letters are used for external features (shafts). **See Figure 6-39.** Positive fundamental deviations are above the baseline, and those that are negative are below the baseline.

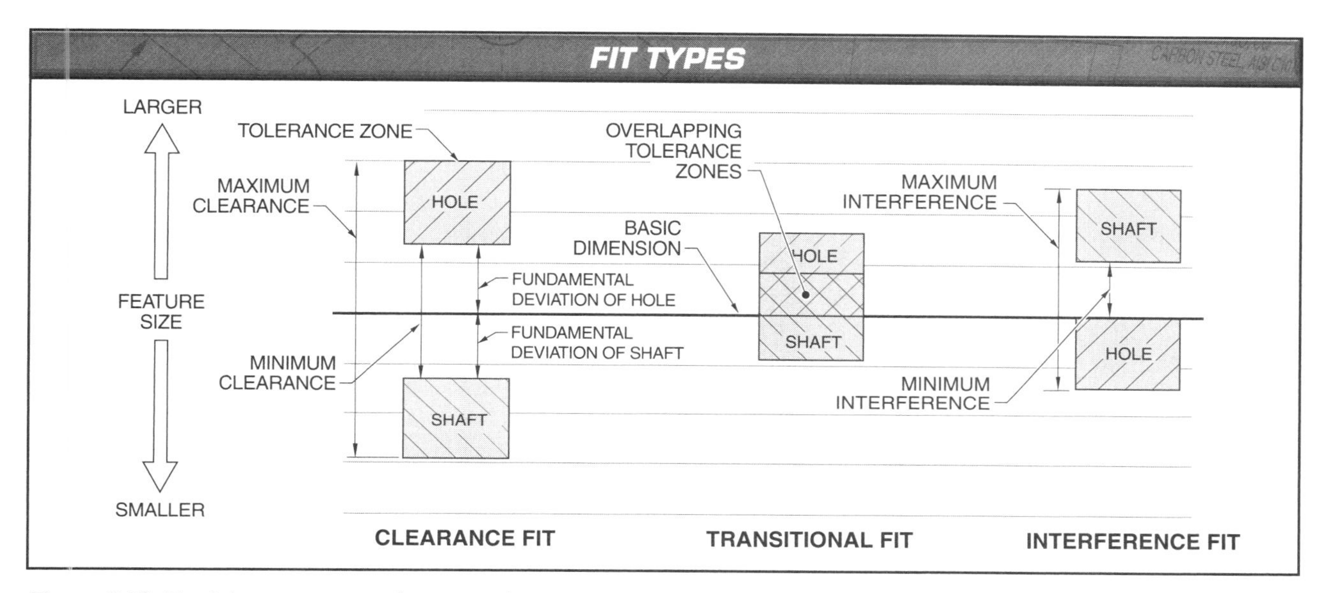

Figure 6-38. The tolerance zones of two parts in an engineering fit can be represented graphically on a chart. This illustrates the difference between clearance, transitional, and interference fits.

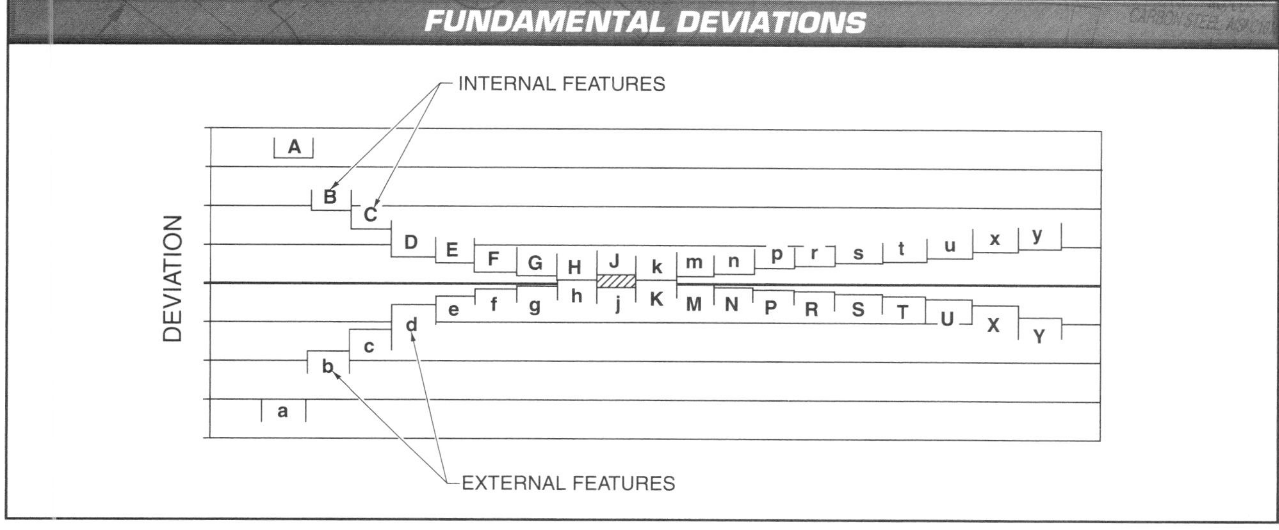

Figure 6-39. Fundamental deviations are standard offsets that are identified by letters. Uppercase letters are used for internal features and lowercase for external features.

Fit Specifications

When the fit of an assembly must be specified, notes are included on the drawing with fundamental deviation and tolerance information. This usually consists of a set of two number-letter combinations, such as H5-g4. The "H" and "g" specify the fundamental deviations of the hole and shaft, respectively. Likewise, the "5" and "4" specify their tolerances. Depending upon the origin of the drawing, either ANSI or ISO tolerance grades may be used.

With the basic dimension of the fit, the numerical tolerance limits can be determined. For example, if the basic dimension of an H5-g4 fit is 2 in., then the actual hole size must be between 2.0000 in. and 2.0005 in. (a tolerance of 0.0005 in.), and the actual shaft size must be between 1.9996 in. and 1.9993 in. (a tolerance of 0.0003 in.).

Standard ANSI Fits

The specification of an engineering fit may list any combination of tolerance and fundamental deviation. However, ANSI lists standard types of fits and their specifications in ANSI/ASME B4.1, *Preferred Limits and Fits for Cylindrical Parts*. This document identifies three general categories of fits: running (or sliding) clearance, locational, and force (or shrink) fits. The specification of an ANSI standard fit need only include the basic dimension and one of these standard fit designations.

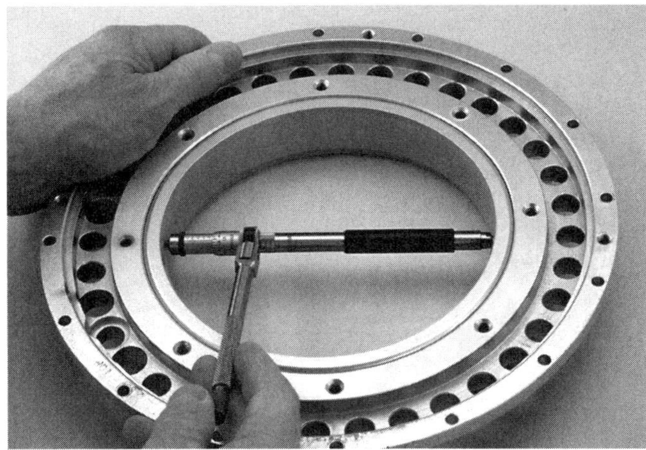

L.S. Starrett Company

Inside diameter micrometers can be used to check circularity, concentricity, and runout.

Running Clearance Fits. Running (or sliding) clearance (RC) fits allow free motion, with suitable lubrication, between the mated parts. RC fits include a range of allowances, which are further broken down into individual specifications and designated as RC1 (tightest) through RC9 (loosest). For example, RC1 is H5-g4 and RC8 is H10-c9.

Locational Fits. Locational fits are intended for parts that remain stationary relative to each other but may require disassembly and reassembly. These fits range from loose to snug and are further categorized by letter and number designations. There are three types of locational fits:

- Locational clearance (LC) fits are the loosest of the locational fits and range from LC1 through LC11.
- Locational transition (LT) fits allow small amounts of either clearance or interference between the mating parts. LT fits range from LT1 through LT6.
- Locational interference (LN) fits allow a certain amount of interference, so the mating parts must be assembled by force. LN fits range from LN1 through LN3.

Force Fits. Force (or shrink) (FN) fits provide highly stressed assemblies that must be forced together. FN fits require the internal feature part to be heated, which enlarges it slightly, before it can be assembled. When it cools, the assembly becomes tightly locked together.

Standard ISO Fits

The standard ISO 286, *Geometrical Tolerance Specifications*, provides a system of designations for specifying engineering fits that involves the same fit components of basic dimension, fundamental deviation, and tolerance as the ANSI standard. However, the ISO system does not categorize certain combinations like the ANSI system. The fit is simply specified with these three pieces of information. For example, 40H9-d9 is an ISO designation for a hole and shaft assembly. The basic dimension for the assembly is 40 mm. The IT grade for both the hole and shaft is 9, which for this size indicates a tolerance zone of 0.062 mm. The fundamental deviations of H for the hole and d for the shaft mean that this is a clearance fit.

REVIEW QUESTIONS 6

Name ______________________ Date ______________

True-False

T F **1.** The ± symbol is used when limit dimensioning.

T F **2.** A feature is any surface, angle, hole, round, or other characteristic on a part that can be dimensioned and controlled.

T F **3.** Dimensioning is a method of identifying and quantifying the size of features.

T F **4.** When the dimension is less than 1 in., a zero is not used before the decimal point.

T F **5.** Parallelism can be considered to be an angularity of 0° to the datum.

T F **6.** Tolerances of location refer to angularity, perpendicularity, and parallelism.

T F **7.** Runout is measured with an indicator.

T F **8.** Runout may be either circular or rectangular.

T F **9.** Tolerance is the amount of space between mating parts when at their actual size.

T F **10.** A hole at MMC is as large as possible with a given tolerance.

Completion

______________ **1.** ___ is a method of specifying the allowable variations of a feature.

______________ **2.** A(n) ___ dimension is used for informational purposes only.

______________ **3.** When the ___ of an assembly must be specified, notes are included on the drawing with fundamental deviation and tolerance information.

______________ **4.** ___ material condition is the condition of a feature having the minimum amount of material permitted by the tolerance zone.

______________ **5.** Tolerances of ___ refer to how well a feature matches a theoretically perfect feature.

______________ **6.** ___ are identified using uppercase letters enclosed in a square frame.

______________ **7.** ___ is the condition of a feature's surface being at a right angle (exactly 90°) to one or two datums.

______________ **8.** A(n) ___ is an outline of a surface in a given plane.

______________ **9.** A(n) ___ is a specification for a special condition of a feature or tolerance zone when a tolerance is applied.

______________ **10.** When the fit of an assembly must be specified, notes are included on the drawing with the ___ and tolerance information.

________ **11.** A(n) ___ is an area or volume that defines the space within which a feature may acceptably vary.

________ **12.** A(n) ___ fit is a fit design for two mating parts that always has some overlap between the parts.

________ **13.** A datum ___ symbol uses a circular frame.

________ **14.** A(n) ___ feature is a feature that is an independent characteristic of a part and does not relate to any datum.

________ **15.** A related feature is associated with one or more specific ___.

Identification—Feature Characteristic Symbols

________ **1.** Angularity

________ **2.** Parallelism

________ **3.** Flatness

________ **4.** Symmetry

________ **5.** Concentricity

________ **6.** Straightness

________ **7.** Perpendicularity

________ **8.** Total runout

________ **9.** Circularity

________ **10.** Profile of a surface

________ **11.** Circular runout

________ **12.** Position

________ **13.** Profile of a line

________ **14.** Cylindricity

A B C D

E F G H

I J K L

M N

Identification—Dimensioning and Tolerancing

________ **1.** Reference dimension

________ **2.** Basic dimension

________ **3.** Least material condition

________ **4.** Datum feature

________ **5.** Maximum material condition

________ **6.** Limit dimensioning

________ **7.** Datum target

________ **8.** Plus and minus tolerancing

M (A) L (B) 4.501 4.499 (C) Ø10 B2 (D)

(4.500) (E) C (F) 4.500 (G) 4.500 ±.001 (H)

TRADE COMPETENCY TEST 6

Armature

Refer to print on page 159.

______________	**1.**	All dimensions specified with three decimal places have a tolerance of ±___″.
______________	**2.**	The Armature is made of ___ steel.
T F	**3.**	Datum B is the breakout side of the stamped part.
______________	**4.**	The minimum width of slot A is ___″.
______________	**5.**	All material conditions are either RFS or ___.
______________	**6.**	The Armature has a maximum thickness of ___″.
______________	**7.**	Datum A is positioned with reference to datum ___.
______________	**8.**	Datum B is ___ within a .001 tolerance zone.
______________	**9.**	The center hole may be ___ on the breakout side.
______________	**10.**	There are ___ slots around the circumference of the Armature.
T F	**11.**	The breakout side of the part may be larger than the ∅4.650 dimension.
______________	**12.**	All position tolerance zones are given as a ∅___″.
______________	**13.**	Surface B is ___ to datum B.
______________	**14.**	The size of datum C at MMC is ___″.
______________	**15.**	This drawing is made in compliance with ANSI ___.
T F	**16.**	The breakout side of the part is cleaned using glass bead blasting.
______________	**17.**	The corners of the slots are rounded to a(n) ___ radius.
T F	**18.**	Surface B is a related feature.
T F	**19.**	The slots are located around a circle with a dimension of ∅4.500.
T F	**20.**	Two-place decimals are toleranced to a zone .02″ wide.
______________	**21.**	Material specification is found in Note ___.
T F	**22.**	A black chromate dip is used to zinc flash the Armature.
______________	**23.**	The true position of C is toleranced to datum ___.

_______________ **24.** The slots are separated by an angle of ___°.

_______________ **25.** The slots are located using ___ dimensioning.

A. point-to-point
B. rectangular coordinate
C. tabular
D. polar coordinate

_______________ **26.** Finishing requirements are specified in Note(s) ___.

A. 3
B. 4
C. 5
D. 4 and 5

_______________ **27.** Angles are accurate to a tolerance of ± ___.

A. ¼°
B. 1°
C. 0°30′
D. 2°

T F **28.** Datum A is toleranced for concentricity to datum C.

_______________ **29.** The Armature is used in assembly ___.

T F **30.** The angle 45° is a reference dimension.

_______________ **31.** The drawing is made to ___ scale.

T F **32.** The drawing may be scaled to obtain measurements.

T F **33.** Datum B is a related feature.

_______________ **34.** The ___ tolerancing method is used on this drawing.

A. general note
B. plus and minus
C. feature control
D. all of the above

T F **35.** This drawing uses datum targets.

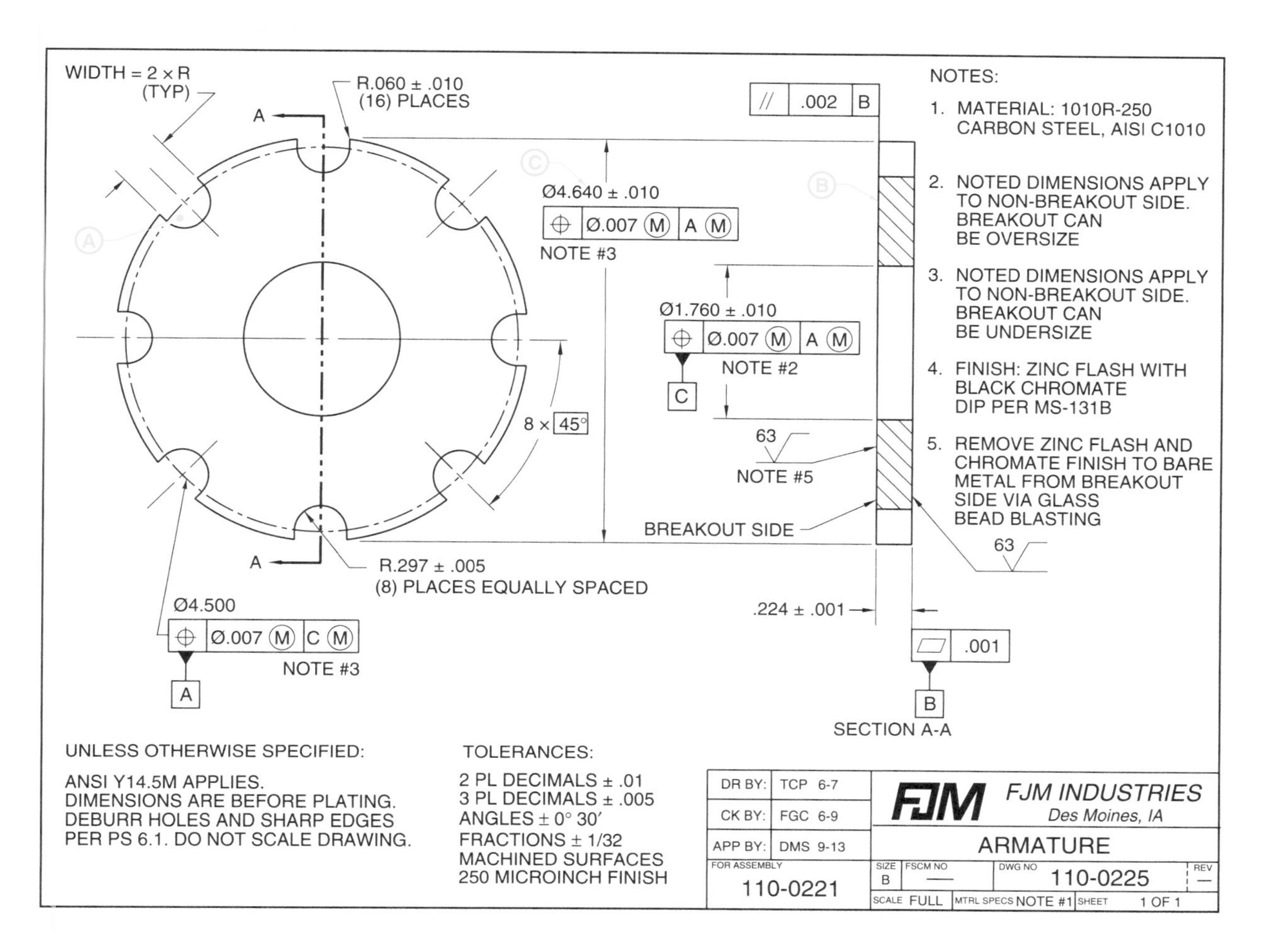

WIDTH = 2 × R (TYP)
R.060 ± .010 (16) PLACES
A
Ø4.640 ± .010
Ø.007 Ⓜ A Ⓜ
NOTE #3
Ø1.760 ± .010
Ø.007 Ⓜ A Ⓜ
NOTE #2
C
8 × 45°
63
NOTE #5
BREAKOUT SIDE
R.297 ± .005 (8) PLACES EQUALLY SPACED
Ø4.500
Ø.007 Ⓜ C Ⓜ
NOTE #3
A
.002 B
.224 ± .001
.001
B
SECTION A-A
63
NOTES:
1. MATERIAL: 1010R-250 CARBON STEEL, AISI C1010
2. NOTED DIMENSIONS APPLY TO NON-BREAKOUT SIDE. BREAKOUT CAN BE OVERSIZE
3. NOTED DIMENSIONS APPLY TO NON-BREAKOUT SIDE. BREAKOUT CAN BE UNDERSIZE
4. FINISH: ZINC FLASH WITH BLACK CHROMATE DIP PER MS-131B
5. REMOVE ZINC FLASH AND CHROMATE FINISH TO BARE METAL FROM BREAKOUT SIDE VIA GLASS BEAD BLASTING
UNLESS OTHERWISE SPECIFIED:
ANSI Y14.5M APPLIES.
DIMENSIONS ARE BEFORE PLATING.
DEBURR HOLES AND SHARP EDGES PER PS 6.1. DO NOT SCALE DRAWING.
TOLERANCES:
2 PL DECIMALS ± .01
3 PL DECIMALS ± .005
ANGLES ± 0° 30′
FRACTIONS ± 1/32
MACHINED SURFACES 250 MICROINCH FINISH
DR BY: TCP 6-7
CK BY: FGC 6-9
APP BY: DMS 9-13
FOR ASSEMBLY 110-0221
FJM
FJM INDUSTRIES
Des Moines, IA
ARMATURE
SIZE B
FSCM NO —
DWG NO 110-0225
REV —
SCALE FULL
MTRL SPECS NOTE #1
SHEET 1 OF 1

ARMATURE

7 Detail and Assembly Prints

Detail prints include special views, close-ups, or any other extra information necessary to completely describe a part for production. While simple parts may not involve detail prints, complex parts may require several detail prints and may refer to important drawings or notes on their detail prints. Assembly drawings are produced to describe how two or more parts are oriented in relation to each other and how they fit together.

DETAIL PRINTS

A *detail print* is a supplemental print that provides certain types of additional information needed to produce a part. Detail prints accompany general or assembly prints and may be used for different types of information. Together, the set of prints includes all the necessary manufacturing instructions.

Many detail prints consist of a set of detail views, which focus on a part's small or complex features. If there is adequate space, detail views may be included on regular orthographic view prints. However, if several detail views are needed, they may be grouped together on a separate detail print. **See Figure 7-1.** In this case, the same general manufacturing process (usually machining) is involved throughout the set, but more detail is necessary.

Alternatively, detail prints are also used to convey special information about additional manufacturing processes required for a part. For example, a part may be roughly shaped with one process and then machined to precise dimensions with another. A detail print would include the additional information needed for the roughing process. The most common supplemental processes requiring detail prints are patternmaking, forging, welding, stamping, and machining. **See Figure 7-2.**

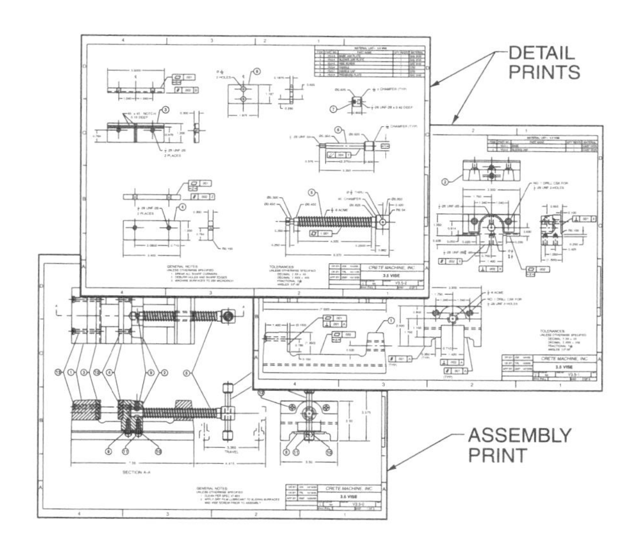

Figure 7-1. Detail prints are required for each part of an assembly.

Cast parts are usually made to be slightly larger than the intended finished product so that surfaces can be machined down to the exact dimensions required.

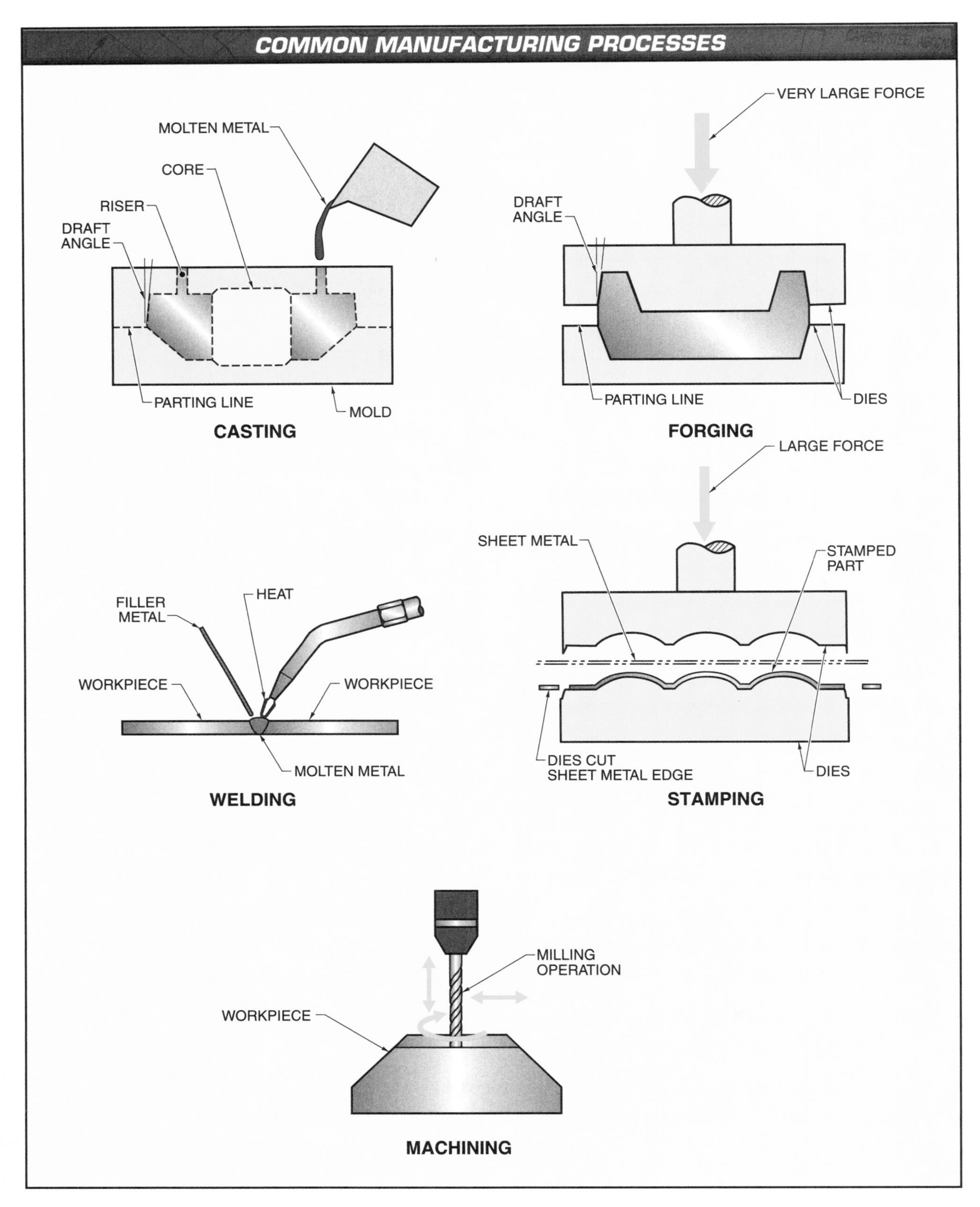

Figure 7-2. Some manufacturing processes require extra detail prints.

Patternmaking (Casting) Prints

A *patternmaking (casting) print* is a print that details the information needed to make the mold for a cast part. *Casting* is a manufacturing process where a shaped cavity is filled with molten metal, which takes on the shape of the cavity as it cools and solidifies. The cavity mold may be made from sand or metal.

Casting is a relatively simple procedure for producing parts quickly, but the design of the mold must account for several unique factors. These factors include the shrinkage of the metal while cooling, extra material needed to facilitate later machining, and the parting line for removing the part from the mold. Therefore, patternmaking drawings provide the size and shape information for the mold, which may differ from the size and shape of the finished part. **See Figure 7-3.** A separate drawing for the finished part with any machining detail drawings may also be required.

Special information included on patternmaking prints includes parting lines, draft angles, core size and shape, and risers. A *parting line* is a boundary dividing two parts of a mold. Reusable molds are opened along parting lines to remove the cast part. Some casting molds are used only once and destroyed in order to release the part.

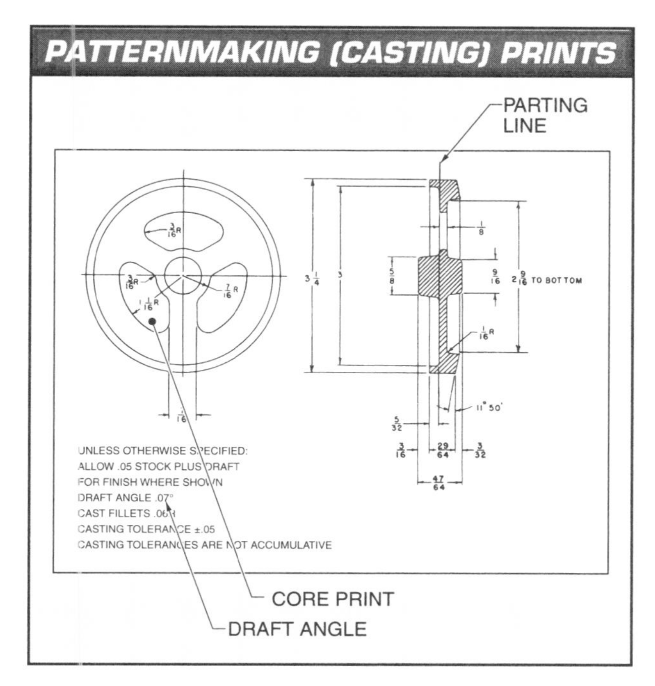

Figure 7-3. Patternmaking (casting) prints show features such as core, draft angle, and parting line information.

Injection molding dies have many features similar to casting molds.

A *draft angle* is the small angle between a mold or die surface and an imaginary surface that is perpendicular to the parting line. A typical draft angle is between 0.5° and 2°. This angle helps in removing the part from the mold without damage.

A *core* is a separate, internal part of a casting mold that is used to form a void or open area in a part. For example, cores are used to produce cup or ring shapes. Information on how to produce cores for casting is specified on a core print.

A *riser* is an internal cavity in a casting mold that funnels molten metal into the part cavity and serves as a reservoir for extra molten metal after the part cavity is filled. As the desired part cools and shrinks slightly, this extra metal keeps the cavity filled. Metal solidified in risers is later removed from the part and recycled. Since the size and placement of risers in a mold affects the quality of the casting, a patternmaking print may include information on risers.

Forging Prints

A *forging print* is a print that details the information needed to forge a part. *Forging* is a manufacturing process in which a workpiece is deformed into the desired shape with high compressive forces, usually between two dies. A *die* is a very strong piece of metal with specially shaped cavities that is used to form shapes when pressed against softer materials. Forging can be done with or without heat.

Forging prints are drawn much like patternmaking prints, since both involve first producing a negative of the desired part. The mold cavity is the negative of a cast part and the die cavity is the negative of a forged part. Forging prints must also take into account shrinkage of the metal and extra material for machining operations, so a die cavity is usually slightly larger than the finished part.

Information on draft angles and parting lines is also important to forging operations and is noted on forging prints. **See Figure 7-4.** Since forged parts may not meet final tolerance requirements, separate machining processes may also be specified. Additional machining drawings of the finished part would then be included.

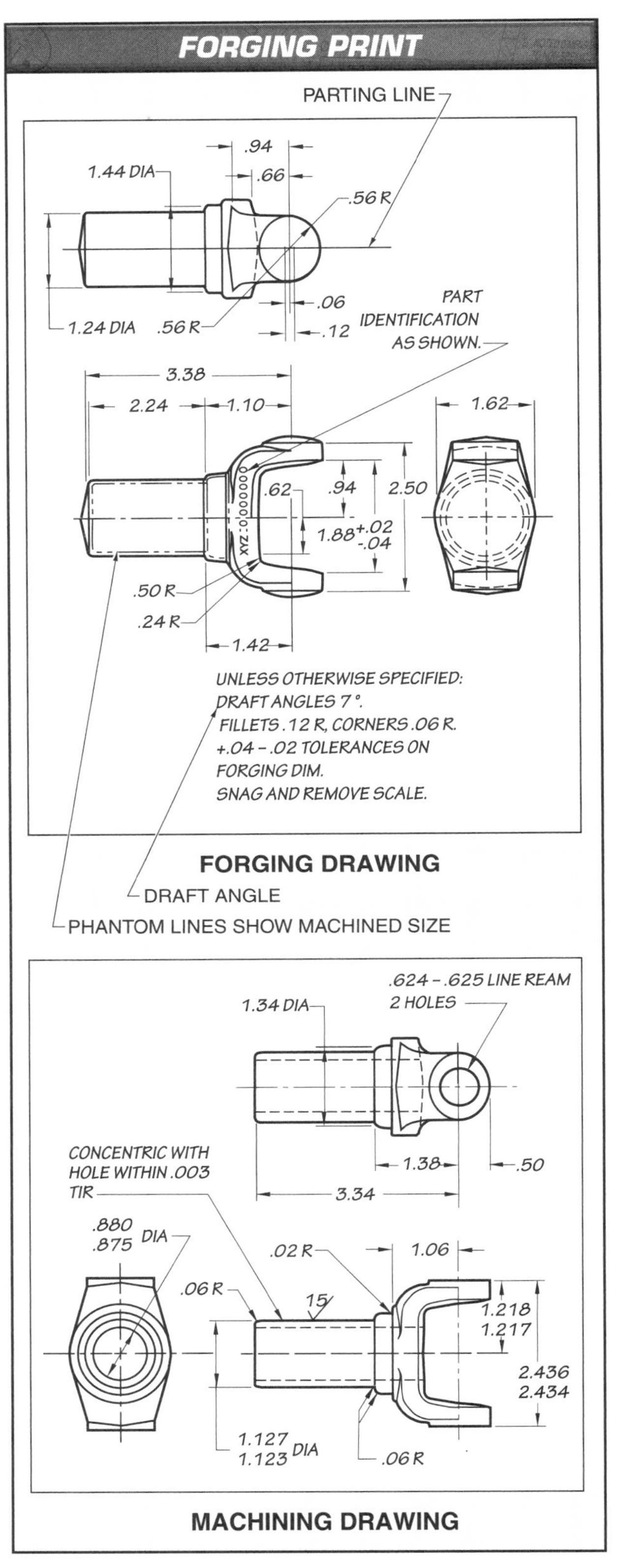

Figure 7-4. Composite forging prints show rough forged features, with machining drawings showing finished detail.

Welding Prints

A *welding print* is a print that details the information needed to weld assemblies together. *Welding* is the process of joining metal parts by heating them until molten and allowing the molten metals to merge, usually adding extra filler metal to the joint as well. When cool, the welded bond can be as strong as the base material itself.

Each part to be welded is dimensioned separately and may also be drawn separately in other detail views. A parts list is included to list all individual components needed to make the final object. **See Figure 7-5.**

Miller Electric Manufacturing Company

Welding processes are commonly used to fabricate assemblies.

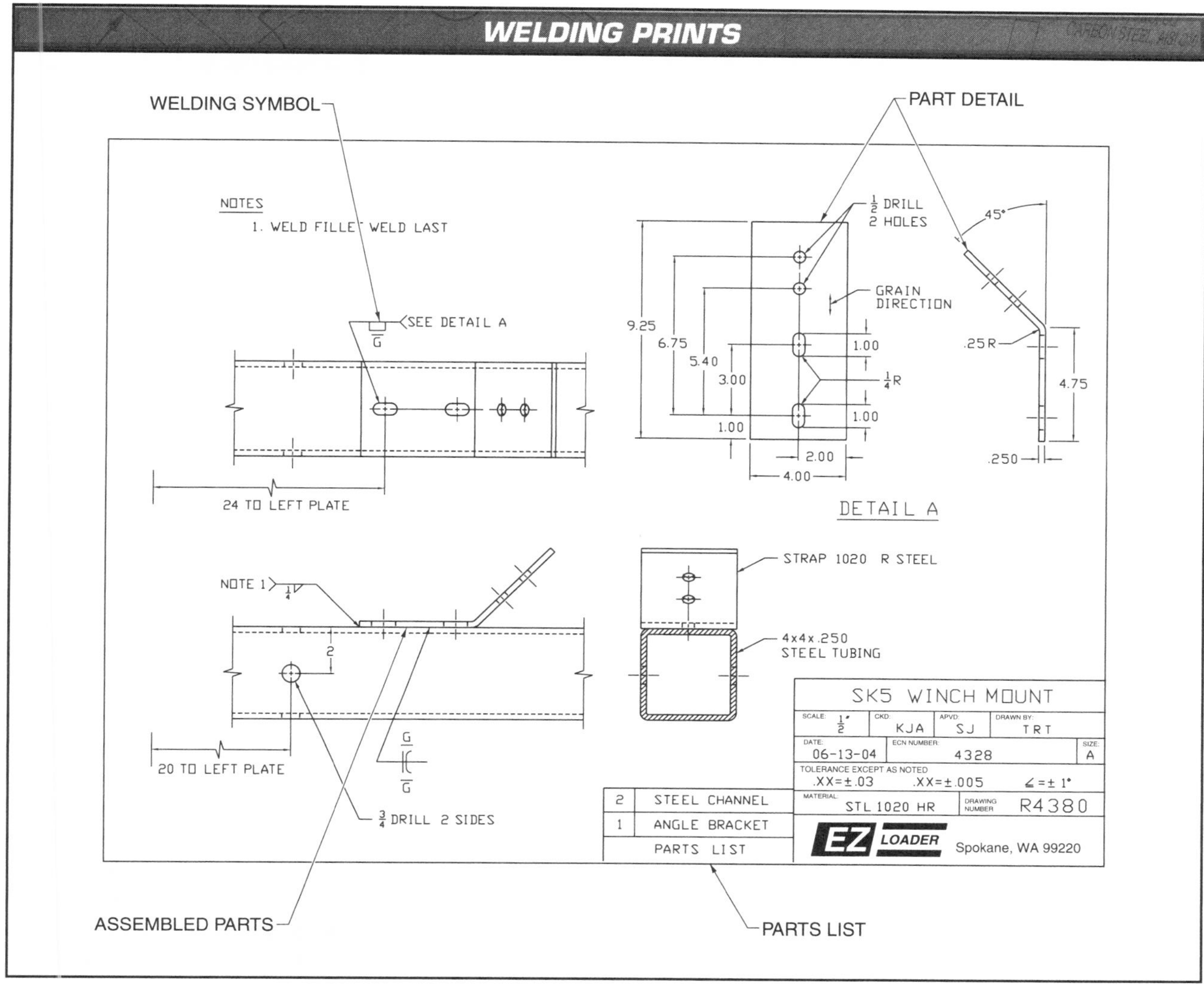

Figure 7-5. Welding prints show assembled parts and individual part details.

Welding prints use symbols to specify the characteristics of a weld in a compact form. **See Figure 7-6.** These symbols have been standardized by the American Welding Society (AWS) in ANSI/AWS A2.4, *Symbols for Welding, Brazing, and Nondestructive Examination* and adopted by ANSI as a national standard.

The standard allows for uniform practices in conveying weld joint information such as the process, weld site, edge preparation, weld length, root opening, or weld contour. Simple welds may require only a few specifications, while complex welds may include information for each characteristic provided for on the welding symbol. **See Appendix.**

Stamping Prints

A *stamping print* is a print that details the information needed to stamp thin material parts into the desired final size and shape. *Stamping* is a manufacturing process that involves forming sheet metal between dies with pressure while cutting the part from the sheet metal with the edge of the die.

The stamped parts in detail drawings appear essentially the same as machined parts. The primary difference is that stamped parts generally allow for greater tolerances on the breakout side of the stamping. **See Figure 7-7.** The *breakout side* is the side of a stamped part that is opposite the die that breaks through the surface.

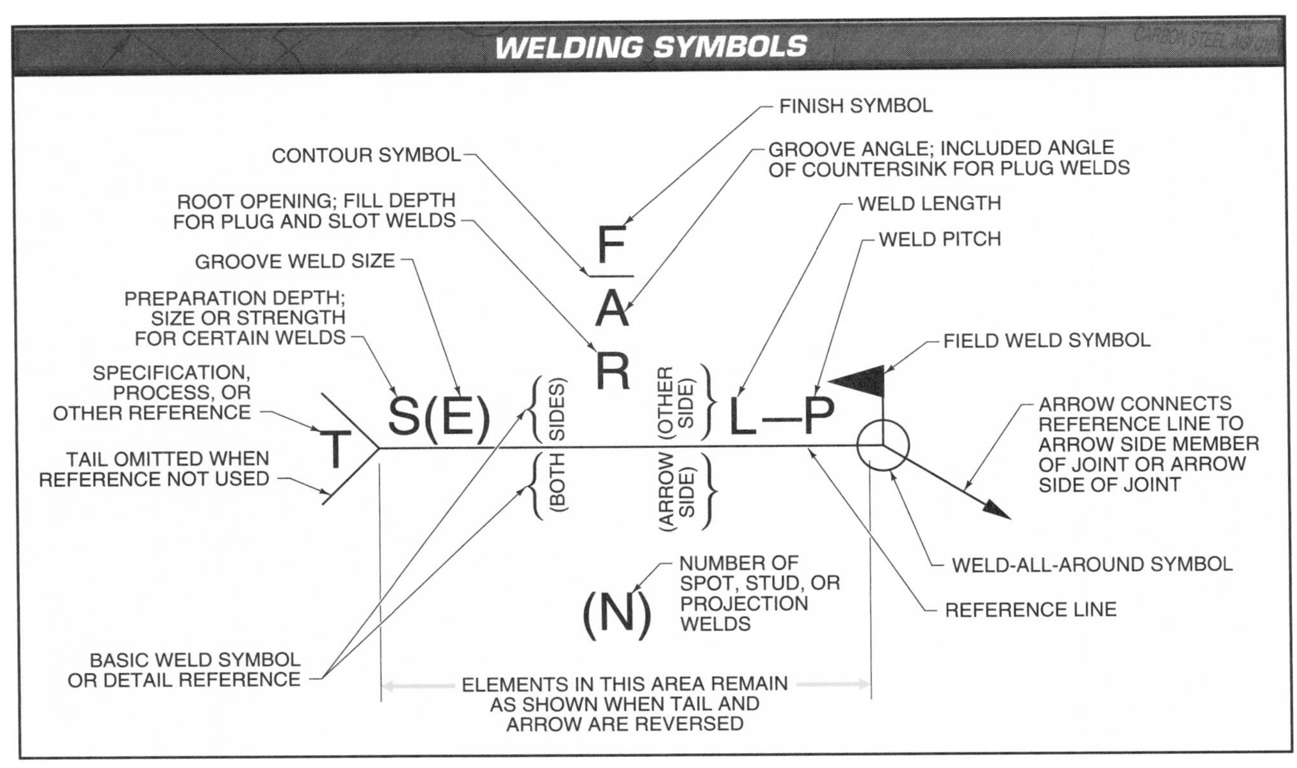

Figure 7-6. Welding symbols provide a standardized system for presenting weld information.

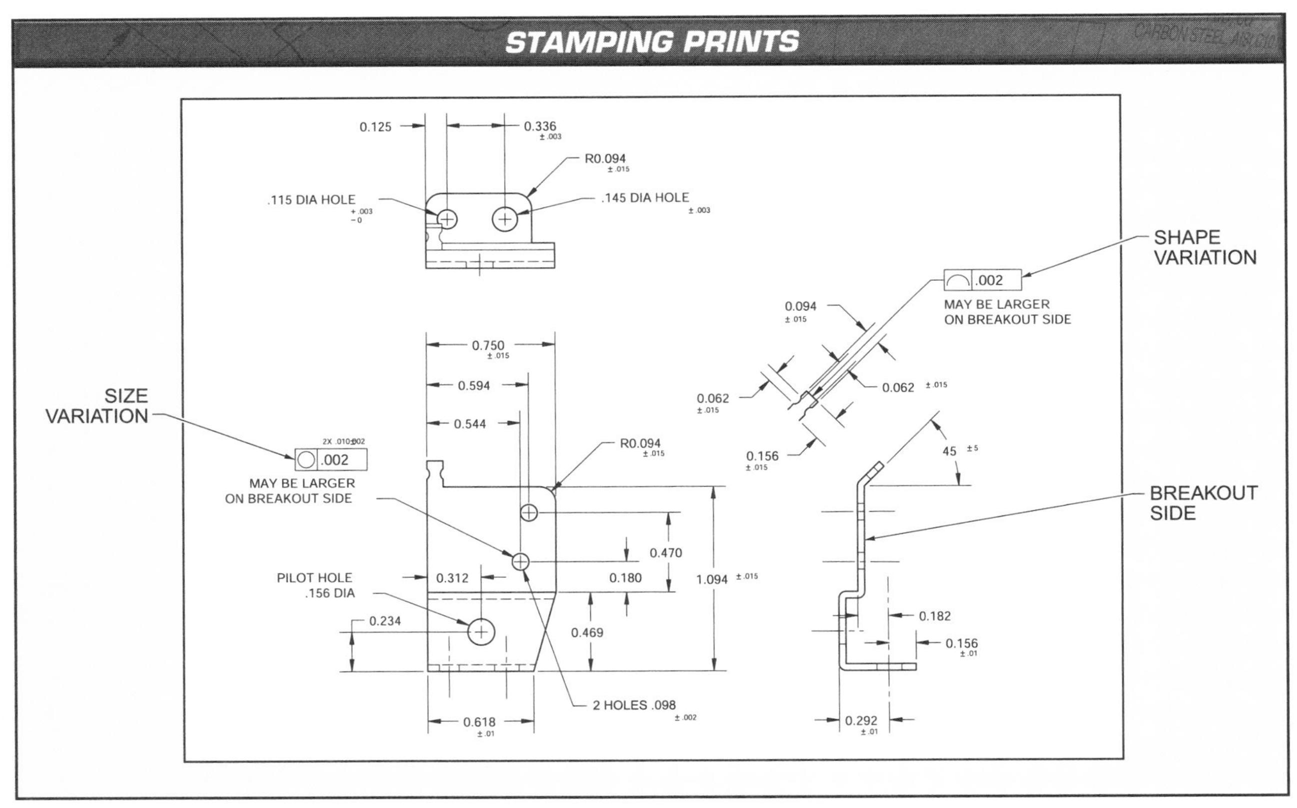

Figure 7-7. Stamping prints allow for tolerance variations on the breakout side.

Machining Prints

A *machining print* is a print that details the information needed to machine a part. Machining covers a broad range of processes relating to the removal of material from a workpiece, including drilling, boring, milling, turning, shaping, and grinding. These processes produce parts with the tightest tolerances.

Machining prints show the greatest amount of information about the size and shape of an object because the manufacturing of a part generally ends with the machining stage. **See Figure 7-8.** Machining detail prints specify more information about surface finish, tolerances, dimensions, and other critical information than other types of detail drawings.

Machining detail prints can sometimes be simplified to make them easier to understand. For example, repetitive views may be eliminated. When this is done, notes and symbols are placed on the drawing to relay any necessary information the extra view would have contained. These notes contain information about dimensions or describe a geometric shape. Also, symbols are occasionally used to identify repetitive features, such as holes of a common diameter, without using multiple leaders. **See Figure 7-9.**

Stock items or purchased hardware is usually not drawn. Instead, notes include the necessary specifications. Simplified drawings convey all the critical information needed to make a part while reducing extraneous lines.

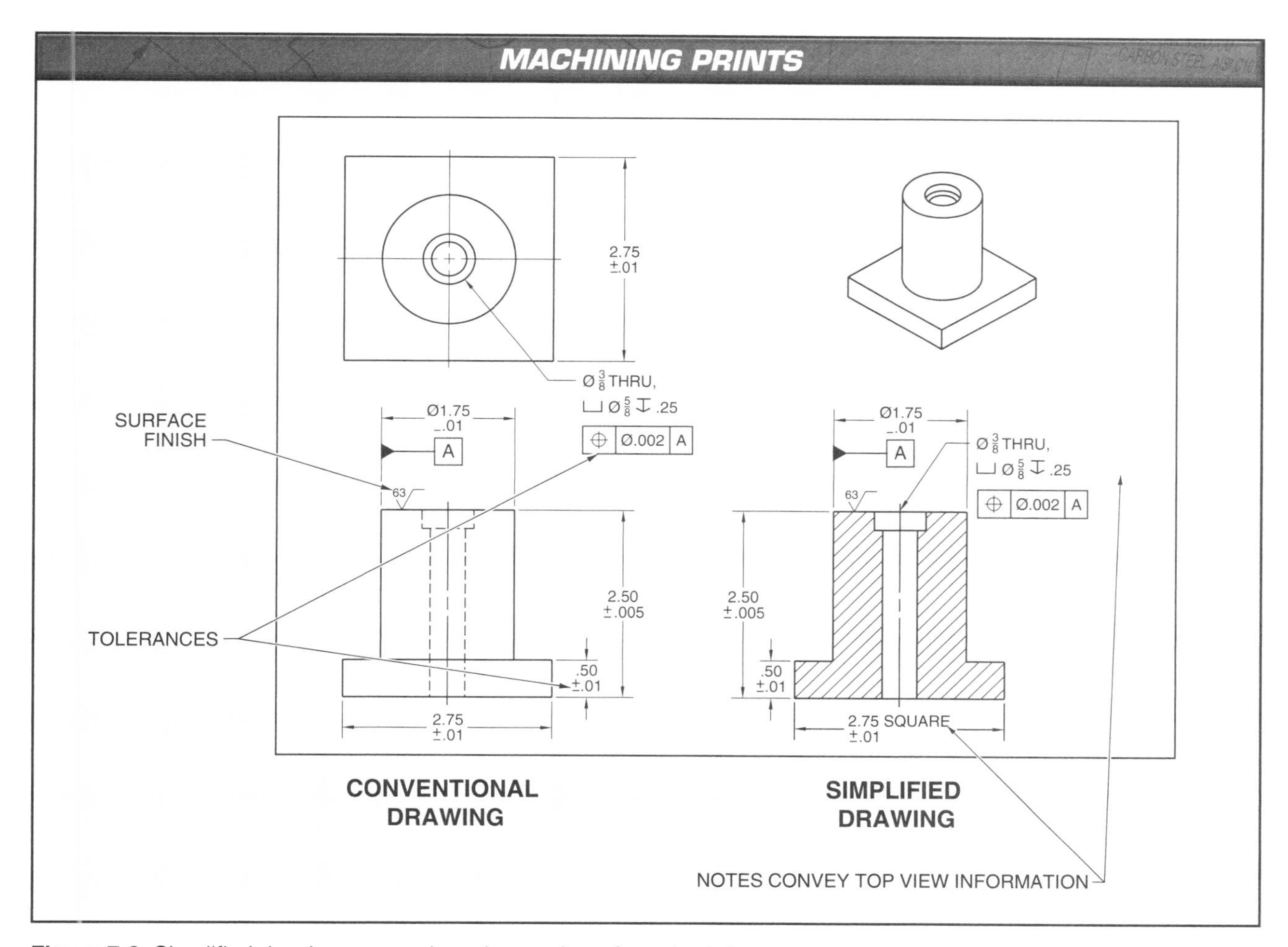

Figure 7-8. Simplified drawings can reduce the number of required views.

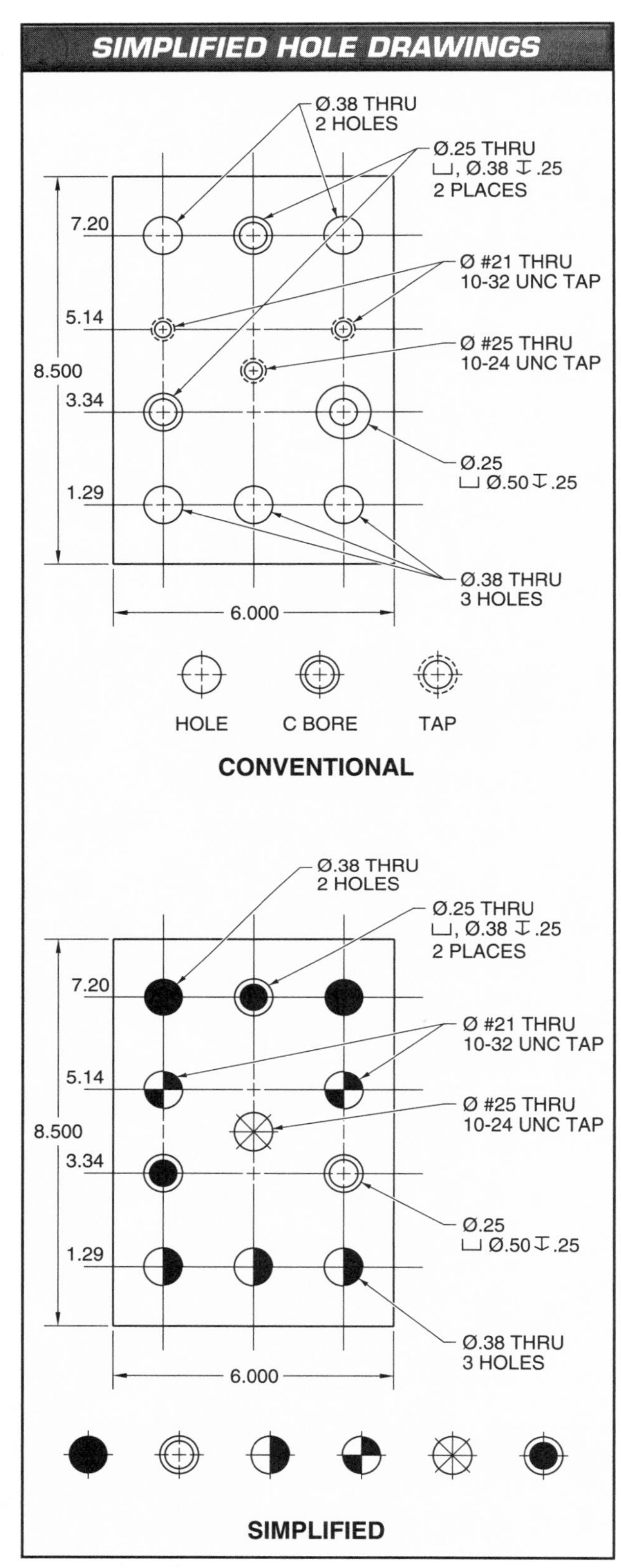

Figure 7-9. Symbols for repetitive features may be used to simplify drawings.

Composite Prints

If a casting or forging detail is not complex, a single composite print may be used to show both the casting or forging operation and any required machining details. These drawings show the casting or forging detail using phantom lines and the machining detail using visible lines. The finished part is dimensioned, and the casting or forging sizes are adjusted to allow for shrinkage.

ASSEMBLY PRINTS

An *assembly print* is a print that illustrates how two or more parts fit together. Assembly prints identify all parts required for the assembly, including both fabricated parts and purchased parts such as bolts, set screws, or keys. A part is identified on an assembly print using an encircled number, which relates to the accompanying parts list. Assembly prints also include overall dimensions or any other specifications needed to understand how the parts fit together.

Depending upon the drawing layout and the complexity of the product, assembly drawings may be included on the same print as the part manufacturing drawings. These detail assembly prints show both the manufacturing and assembly information on a single sheet. **See Figure 7-10.** However, this arrangement is only practical for very simple products. Many assembly prints are separate sheets that accompany the manufacturing prints.

Assembly prints use either orthographic or pictorial views. Schematic and installation drawings also use similar views to illustrate how larger systems are assembled. The type of drawing selected is based on its intended use.

Orthographic Assembly Prints

Orthographic assembly prints include views in one or more of the primary planes of projection. Orthographic assembly prints are used primarily as working assembly drawings in a manufacturing setting. The fewest number of views that completely describe the assembly are shown. Generally, one or two views will suffice. The views are shown either as conventional views or as sectional views. **See Figure 7-11.**

Conventional Views. A *conventional view* is an exterior orthographic view used for assembly drawings. Hidden lines are omitted unless they are required to understand the assembly, though phantom lines are common for identifying related assemblies or parts. These views are used for simple assemblies with few internal features.

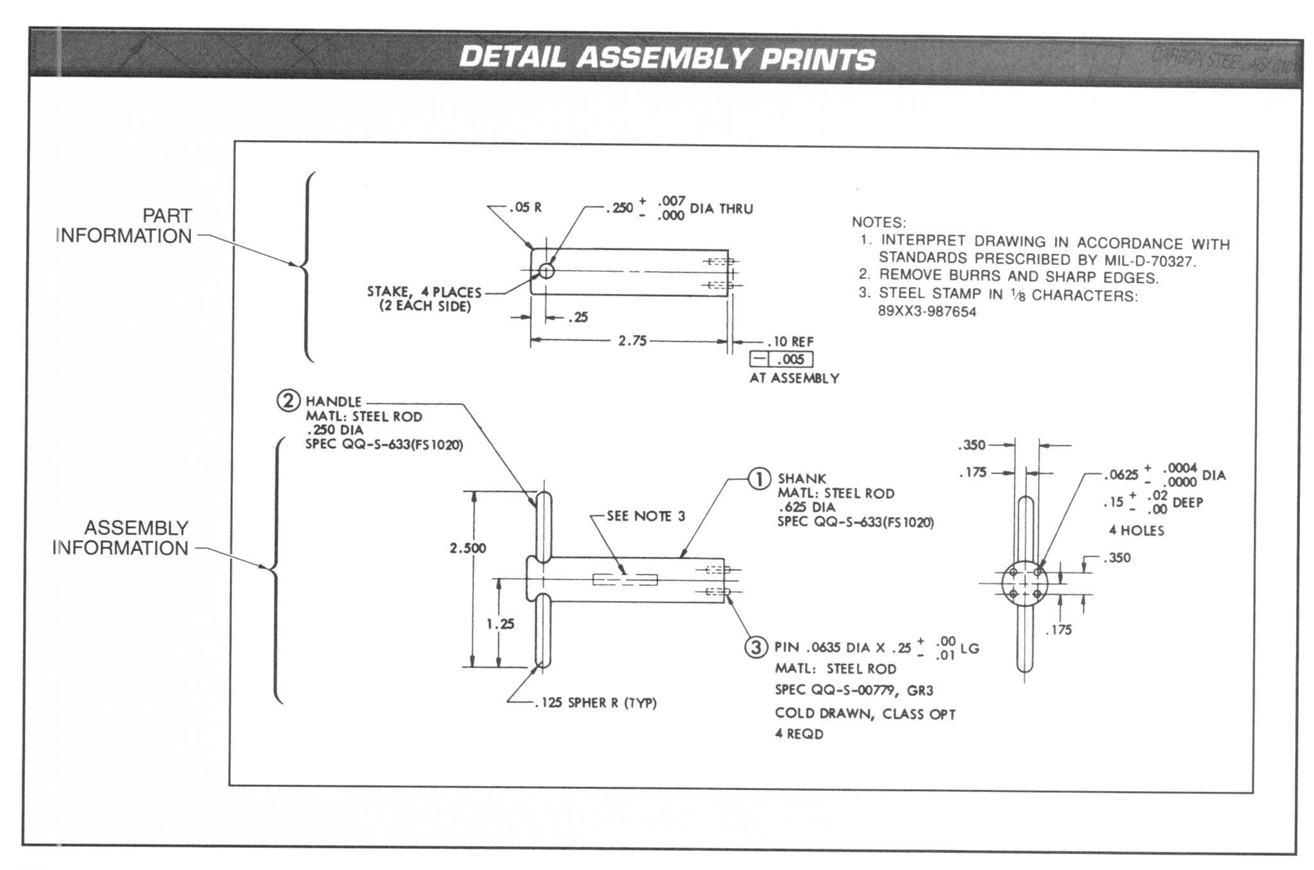

Figure 7-10. Detail assembly prints show both assembly and part manufacturing information.

Sectional Views. Sectional views are orthographic views that show the internal features of an assembly. The section lines are drawn at various angles to differentiate the parts in the assembly. These drawings are often needed to illustrate more complex assemblies, especially those with internal features.

Pictorial Assembly Prints

Pictorial assembly prints represent the assembly as it appears in three dimensions. These are generally drawn as isometric or perspective representations. Pictorial assembly drawings are used in parts catalogs, service manuals, and technical manuals to provide a clear description of how parts are arranged in an assembly. Sectioned parts may be included to show internal features. The most common type of pictorial assembly drawing is an exploded view.

A pictorial view shows all three dimensions of an object such that it resembles a photo of the object.

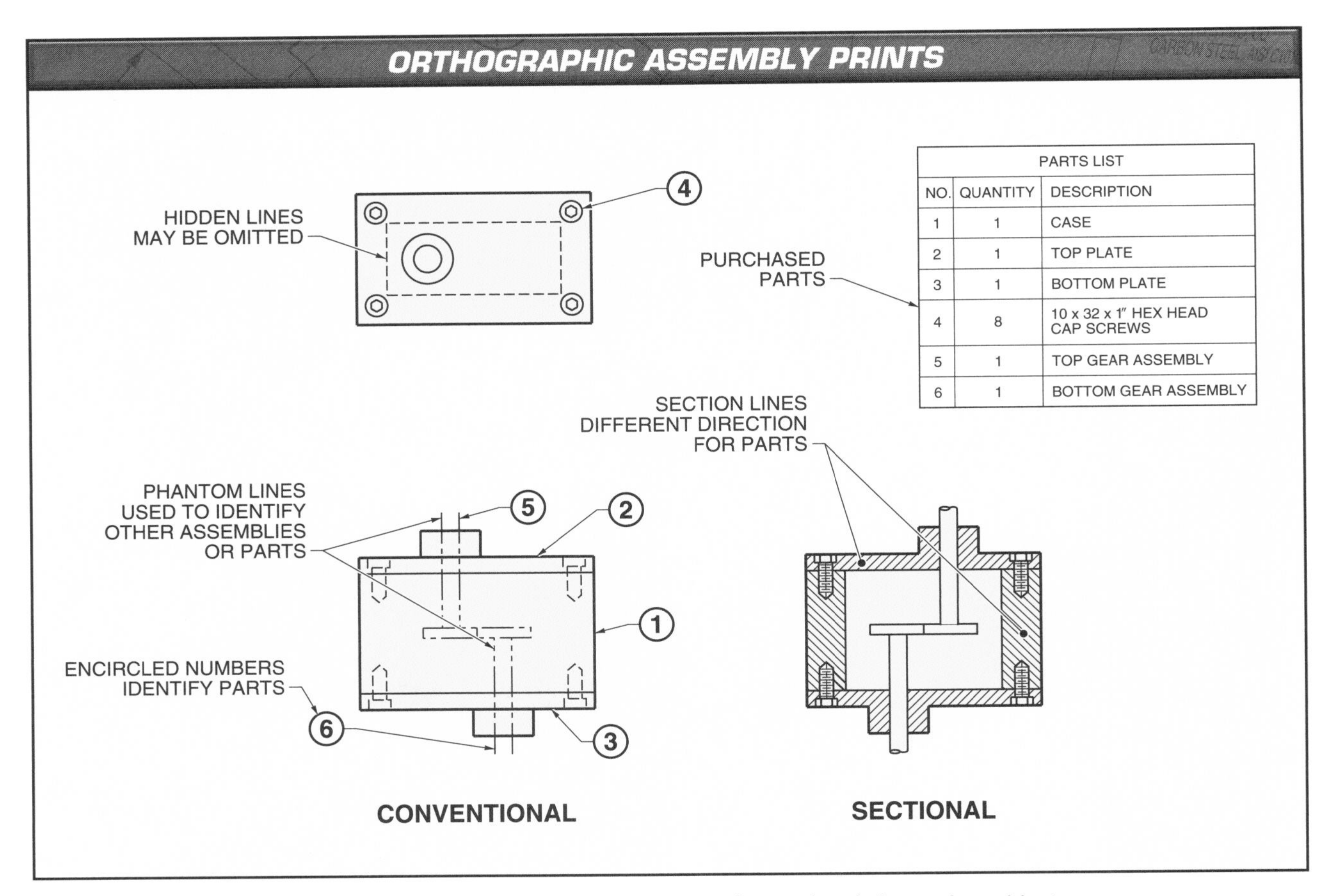

PARTS LIST		
NO.	QUANTITY	DESCRIPTION
1	1	CASE
2	1	TOP PLATE
3	1	BOTTOM PLATE
4	8	10 x 32 x 1″ HEX HEAD CAP SCREWS
5	1	TOP GEAR ASSEMBLY
6	1	BOTTOM GEAR ASSEMBLY

Figure 7-11. Orthographic assembly prints show conventional and/or sectional views of an object.

Exploded Views. An *exploded view* is a drawing that separates all the components of an assembly, but retains their alignment and orientation for reassembly. Dashed lines are usually included to aid in visualizing the reassembly process. **See Figure 7-12.** Exploded views are particularly useful because they can individually identify each component in a complicated assembly and illustrate how the components fit together in one drawing.

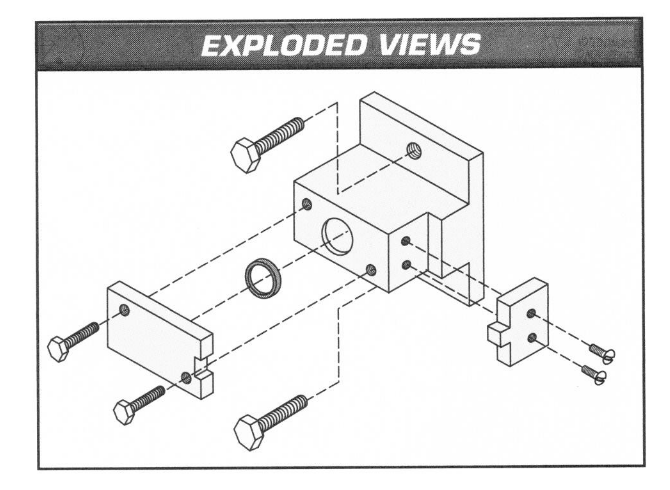

Figure 7-12. Exploded views are pictorial assembly drawings that show how parts in an assembly relate to each other.

Exploded views are commonly found in service manuals, technical manuals, and parts catalogs.

Schematic Assembly Prints

Special assembly prints are used to show system schematics or installations. A *schematic assembly print* is a print that illustrates in pictorial or plan view the relative locations and connections of equipment within a system. **See Figure 7-13.** These drawings are not made to scale. They may show an electrical or hydraulic circuit, a piping layout, a ventilation system, or another system. The only dimensions indicated on these drawings are distances between critical points needed for installation.

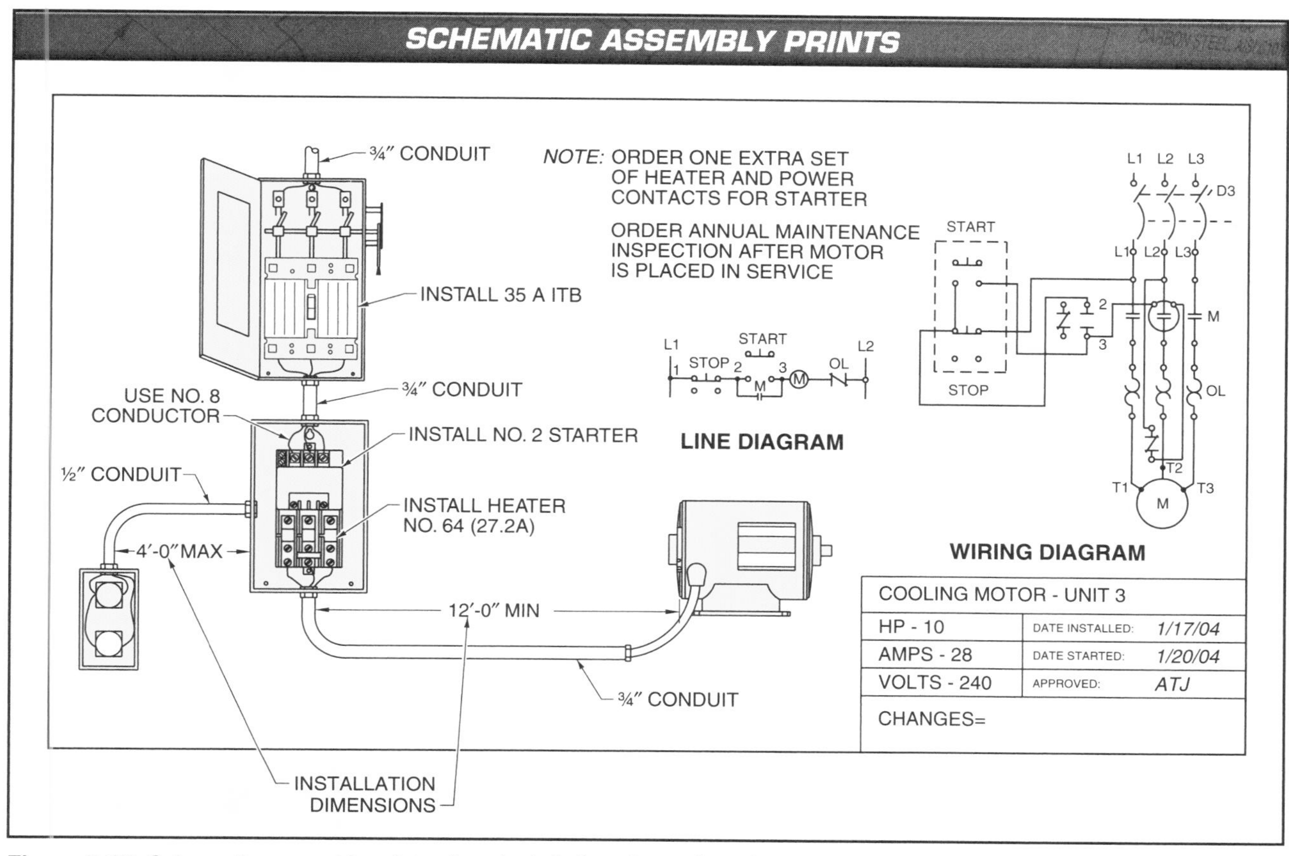

Figure 7-13. Schematic assembly prints show installation dimensions for a system.

Installation Prints

An *installation print* is a print that illustrates the general configuration and information needed to install a specific piece of equipment. **See Figure 7-14.** Mounting dimensions, outline dimensions, clearance requirements, and feature information may be included.

Mounting Dimensions. A *mounting dimension* is a dimension used to locate fastening points on equipment. This information is useful for preparing areas for equipment installation before the equipment arrives.

Outline Dimensions. An *outline dimension* is a dimension for the minimum space required to install the piece of equipment. This information shows the overall size and contour of the equipment and the surfaces related to the mounting dimensions.

> Mounting and outline dimensions are important for preparing an area for equipment installation.

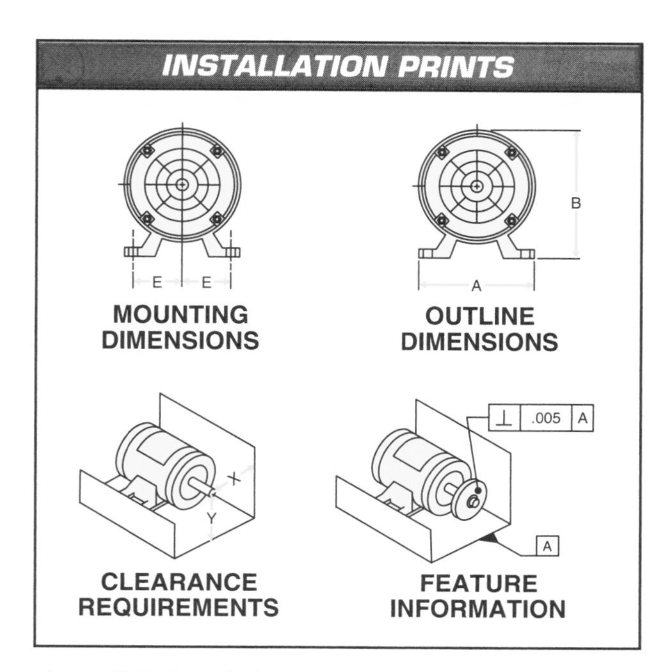

Figure 7-14. Installation prints give information required to install a specific piece of equipment.

Clearance Requirements. A *clearance requirement* is a specification for the empty space needed around the outside surface of a piece of equipment. Clearances may be function or service related. Functional clearances cover all extremes of extended and retracted positions, angles of operation, or moving components. Service clearances are given for all areas such as brushes, covers, or panels that may require service.

Feature Information. *Feature information* is installation information about the relationships between equipment components and equipment or facility features. For instance, it may be essential for a component to be perpendicular to the floor following installation, within a certain tolerance requirement. These relationships can be conveyed with the same feature control frames, symbols, and datums used in manufacturing prints.

REVIEW QUESTIONS 7

Name ______________________________ Date ______________

Completion

______________ 1. A(n) ___ is a very strong piece of metal with specially shaped cavities that is used to form shapes when pressed against softer materials.

______________ 2. ___ are occasionally used to identify repetitive features.

______________ 3. Machining prints show the ___ amount of detail about an object.

______________ 4. A(n)___ print illustrates how two or more parts fit together.

______________ 5. Parts on an assembly print are each identified with a(n) ___ number.

______________ 6. Orthographic assembly prints are shown as conventional or ___ views.

______________ 7. A(n) ___ requirement is a specification for the empty space needed around the outside surface of a piece of equipment.

______________ 8. Pictorial assembly prints are generally drawn as isometric or ___ representations.

______________ 9. A(n) ___ print is a print that illustrates the general configuration and information needed to install a specific piece of equipment.

______________ 10. ___ assembly prints may show electrical or hydraulic circuits, piping layouts, or other plan drawings.

Multiple Choice

______________ 1. A ___ is an internal cavity in a casting mold that funnels molten metal into the part cavity and serves as a reservoir for extra molten metal after the part cavity is filled.

A. draft angle
B. core
C. riser
D. none of the above

______________ 2. A patternmaking print details the information needed to make a mold for a ___ part.

A. forged
B. cast
C. welded
D. machined

______________ 3. ___ is a manufacturing process in which a workpiece is deformed into the desired shape with high compressive forces, usually between two dies.
A. Casting
B. Forging
C. Patternmaking
D. Welding

______________ 4. Patternmaking prints show ___.
A. finished sizes
B. machined features
C. draft angles
D. thread sizes

______________ 5. A(n) ___ view is a drawing that separates all the components of an assembly, but retains their alignment and orientation for reassembly.
A. conventional
B. sectional
C. exploded
D. none of the above

______________ 6. Stock items are ___ in a simplified drawing.
A. not drawn
B. shown in outline only
C. used in the assemblies
D. none of the above

______________ 7. Machining detail prints show information about ___.
A. surface finish
B. tolerances
C. dimensions
D. all of the above

______________ 8. Orthographic assembly prints are used primarily in ___.
A. product catalogs
B. manufacturing
C. consumer manuals
D. exploded view drawings

______________ 9. ___ dimensions are used to locate fastening points on equipment.
A. Clearance
B. Mounting
C. Outline
D. Feature

______________ 10. A(n) ___ dimension is the minimum space required to install the piece of equipment.
A. clearance
B. mounting
C. outline
D. feature

True-False

T F **1.** Composite prints shows the casting or forging detail using phantom lines.

T F **2.** A welding print contains a parts list.

T F **3.** Stamping prints generally allow for greater tolerances on the breakout side.

T F **4.** A draft line is a boundary dividing two parts of a mold.

T F **5.** Forging is a manufacturing process that involves forming sheet metal between dies.

T F **6.** Welding symbols have been standardized by the AWS and adopted by ANSI as a national standard.

T F **7.** Detail assembly prints show both the manufacturing and assembly information on a single sheet.

T F **8.** Sectional views show internal features.

T F **9.** Sectioned parts are never used in pictorial assemblies.

T F **10.** Schematic assembly prints are never drawn to scale.

TRADE COMPETENCY TEST 7

Name ______________________________ Date ____________________

Wire EDM

Refer to print on pages 179-180.

T F **1.** The assembly was drawn before the detail was completed.

__________ **2.** The Top V-Block has a maximum height of ___″.

__________ **3.** The final assembly contains ___ different parts.
- A. three
- B. four
- C. five
- D. 10

T F **4.** The drawing number is etched into the surface of the Top V-Block.

__________ **5.** The Top V-Block may be ___″ thick.
- A. .370 to .375
- B. .370 to .380
- C. .3725 to .3775
- D. .375 to .380

__________ **6.** The through slot has a nominal width of ___″.

__________ **7.** The position for the dowel pin holes is accurate to a tolerance of ±___″.
- A. .0005
- B. .005
- C. .01
- D. .03

__________ **8.** The distance of A is ___″.

T F **9.** Part 2 in the assembly is press fit into Part 1.

__________ **10.** All nuts, screws, and bolts are made of ___.

T F **11.** Part 3 has a metric thread.

T F **12.** The slot at C is centered on an axis 2.00″ from the bottom of the Top V-Block.

__________ **13.** The dimension at B cannot be larger than ___″.

__________ **14.** Angle E can be within a range of ___.
- A. 118½° to 120½°
- B. 119° to 121°
- C. 119°30′ to 120°30′
- D. 120° to 121°

_______________ **15.** The gauge at D has a diameter of ___″.

A. 1.338 ± .000

B. 1.338 ± .005

C. 1.338 ± .03

D. not shown on print

_______________ **16.** The gauge at D locates the center of ___ holes.

A. one

B. two

C. three

D. four

_______________ **17.** Part G is ___″ long.

_______________ **18.** Part H is ___ mm long.

_______________ **19.** The radius of the corner at F is ___″.

A. .125

B. .25

C. .38

D. not shown on print

T F **20.** All holes in the detail drawing are shown using an exact representation of their appearance.

T F **21.** The assembly drawing is referenced on the detail drawing.

T F **22.** All surfaces are ground to a final finish.

_______________ **23.** The Top V-Block is drawn at ___ scale.

T F **24.** All hidden lines are shown in the right side view of the Top V-Block.

T F **25.** There are two HEX HD. SCR. #10-24 × 1⅛ LG. required for the assembly.

_______________ **26.** The drawings were completed by ___.

_______________ **27.** Three-place decimal dimensions are accurate to ±___″.

T F **28.** The assembly print has the same tolerances as the detail print.

T F **29.** The assembly drawing number is B-803-28531.

_______________ **30.** The minimum width of the slot at C is ___″.

_______________ **31.** The detail was completed on ___.

_______________ **32.** The in-line location tolerance for the holes in the Top V-Block to locate the dowel pins is ±___″.

_______________ **33.** The Top V-Block is made from ___.

T F **34.** Angles are accurate to ±½°.

_______________ **35.** Part 1 references Mecatool #___.

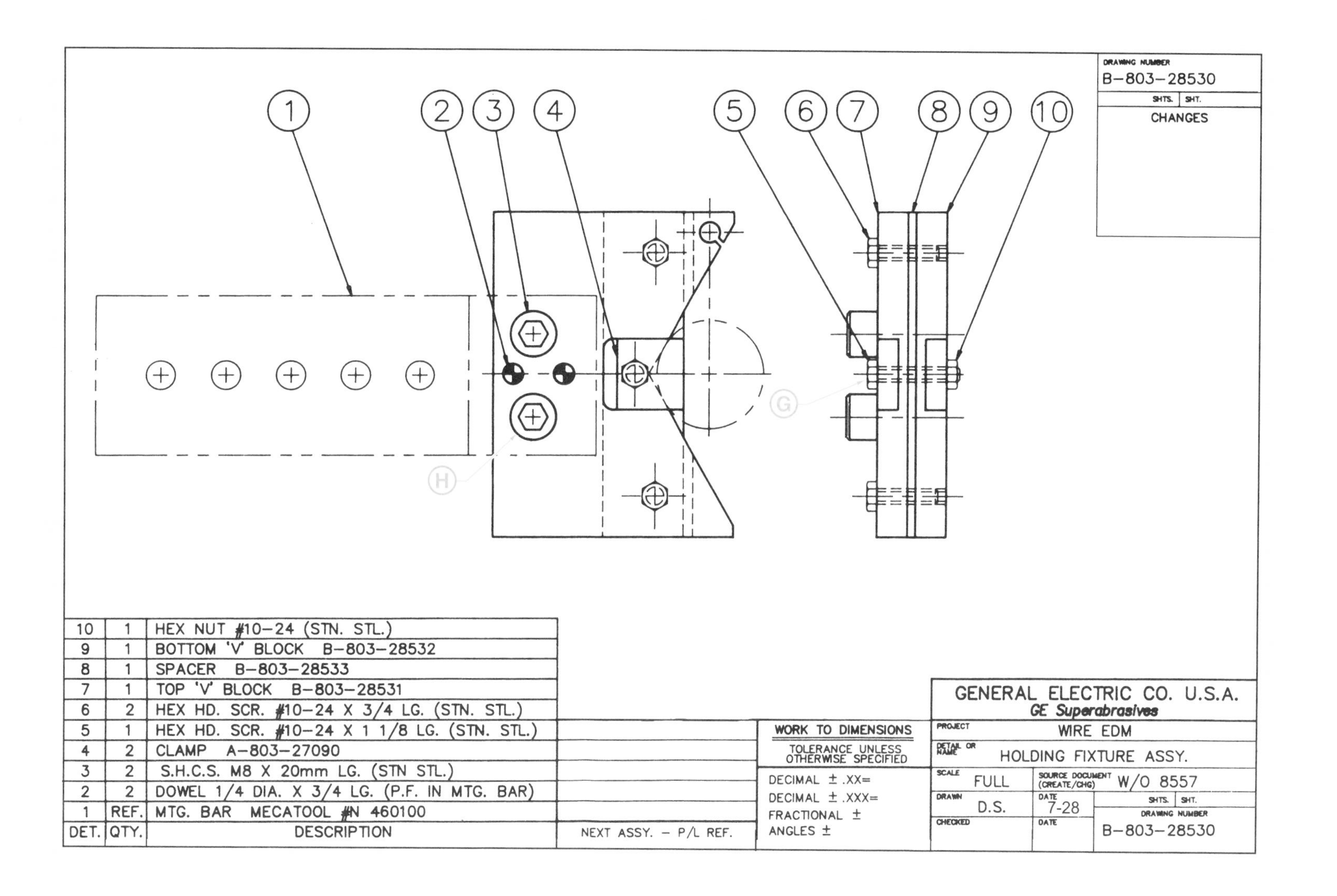

DRAWING NUMBER
B-803-28530
SHTS. SHT.
CHANGES
1
2
3
4
5
6
7
8
9
10
G
H
10 1 HEX NUT #10-24 (STN. STL.)
9 1 BOTTOM 'V' BLOCK B-803-28532
8 1 SPACER B-803-28533
7 1 TOP 'V' BLOCK B-803-28531
6 2 HEX HD. SCR. #10-24 X 3/4 LG. (STN. STL.)
5 1 HEX HD. SCR. #10-24 X 1 1/8 LG. (STN. STL.)
4 2 CLAMP A-803-27090
3 2 S.H.C.S. M8 X 20mm LG. (STN STL.)
2 2 DOWEL 1/4 DIA. X 3/4 LG. (P.F. IN MTG. BAR)
1 REF. MTG. BAR MECATOOL #N 460100
DET. QTY. DESCRIPTION
NEXT ASSY. - P/L REF.
WORK TO DIMENSIONS
TOLERANCE UNLESS OTHERWISE SPECIFIED
DECIMAL ± .XX=
DECIMAL ± .XXX=
FRACTIONAL ±
ANGLES ±
GENERAL ELECTRIC CO. U.S.A.
GE Superabrasives
PROJECT WIRE EDM
DETAIL OR NAME HOLDING FIXTURE ASSY.
SCALE FULL
SOURCE DOCUMENT (CREATE/CHG) W/O 8557
DRAWN D.S.
DATE 7-28
SHTS. SHT.
CHECKED
DATE
DRAWING NUMBER
B-803-28530

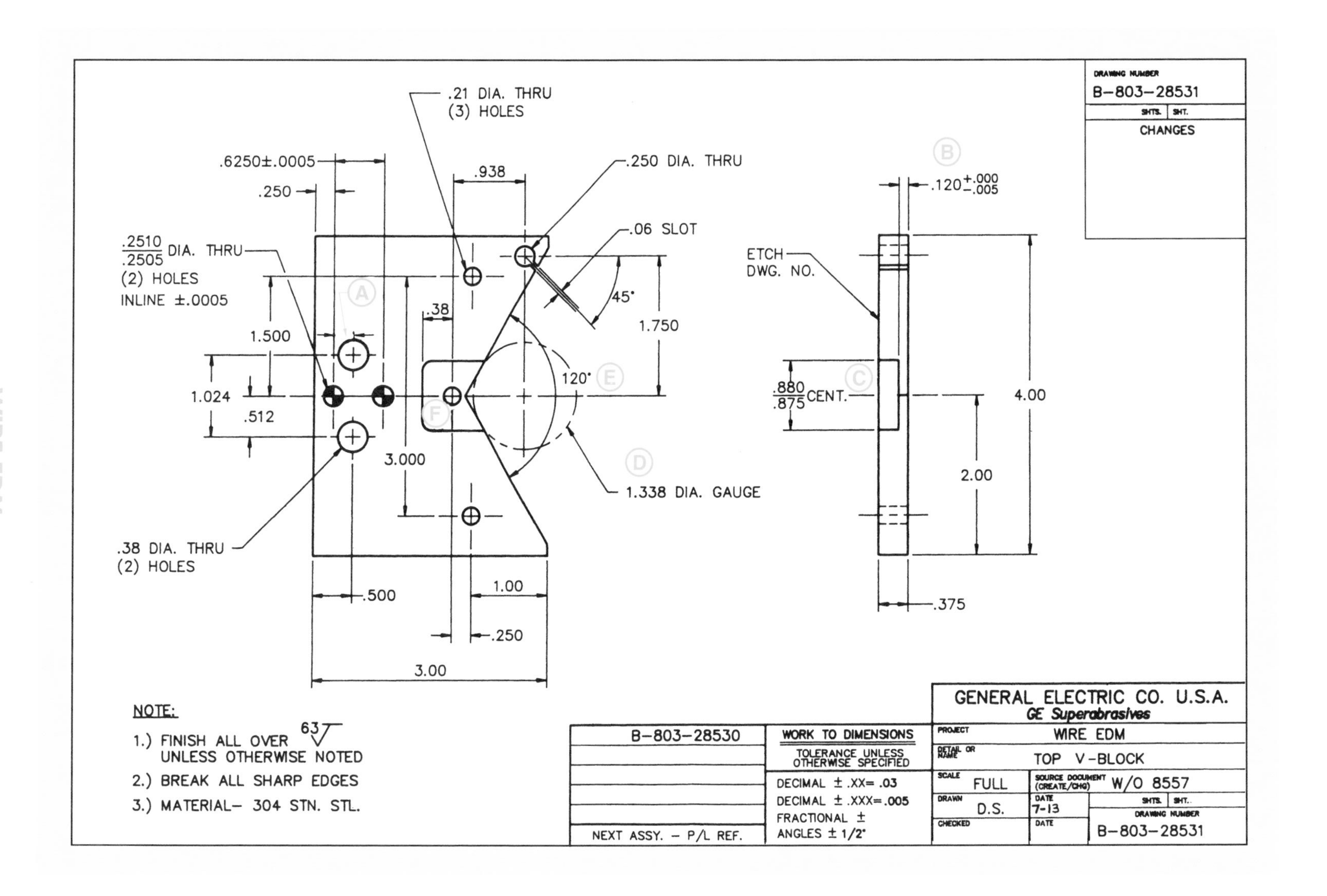

.21 DIA. THRU
(3) HOLES
.6250±.0005
.250
.938
.250 DIA. THRU
.06 SLOT
.2510
.2505
DIA. THRU
(2) HOLES
INLINE ±.0005
A
.38
45°
1.750
1.500
1.024
.512
120°
E
F
3.000
D
1.338 DIA. GAUGE
.38 DIA. THRU
(2) HOLES
.500
1.00
.250
3.00
B
.120+.000 -.005
ETCH
DWG. NO.
.880
.875
CENT.
C
4.00
2.00
.375
DRAWING NUMBER
B-803-28531
SHTS. SHT.
CHANGES
NOTE:
1.) FINISH ALL OVER 63
UNLESS OTHERWISE NOTED
2.) BREAK ALL SHARP EDGES
3.) MATERIAL- 304 STN. STL.
B-803-28530
NEXT ASSY. - P/L REF.
WORK TO DIMENSIONS
TOLERANCE UNLESS
OTHERWISE SPECIFIED
DECIMAL ± .XX= .03
DECIMAL ± .XXX= .005
FRACTIONAL ±
ANGLES ± 1/2°
GENERAL ELECTRIC CO. U.S.A.
GE Superabrasives
PROJECT
WIRE EDM
DETAIL OR NAME
TOP V-BLOCK
SCALE
FULL
SOURCE DOCUMENT (CREATE/CHG)
W/O 8557
DRAWN
D.S.
DATE
7-13
SHTS. SHT.
CHECKED
DATE
DRAWING NUMBER
B-803-28531

Materials and Machining

8

Materials and their properties greatly affect not only the design of a finished part, but also the process used to manufacture the part. Design features are specified on the print with dimensions, notes, and details. All of this information is interpreted and used to manufacture the part using appropriate machining processes. Machining processes are selected based on design requirements, material properties, the forming process, machinability of the material, and production efficiency.

MATERIALS

Material selection is an important part of component design. Materials and their properties determine the size, shape, and thickness a component needs for its intended function. For this reason, design engineers note the materials to be used to make components on the manufacturing prints. Material specifications may be noted next to the part or listed in the bill of materials on a print, sometimes called a parts list. Materials may be specified as notes on drawings of simple parts or assemblies with few components. General notes in or near the title block can also give material information.

The bill of materials provides a listing of all materials required to produce the product specified on the print. Information commonly found in the bill of materials includes the item number, quantity, part number or name, material specifications, and weight. **See Figure 8-1.** Materials commonly used in the machine trades include metals and plastics.

A *metal* is a material consisting of one or more chemical elements having a crystalline structure, high thermal and electrical conductivity, the ability to be deformed when heated, and high reflectivity. A metal is either a pure metal or an alloy. A *pure metal* is a metal that consists of one chemical element. Pure metals, which are usually soft and have relatively low strength, have very limited usage in engineering applications. An *alloy* is a metal that consists of two or more chemical elements.

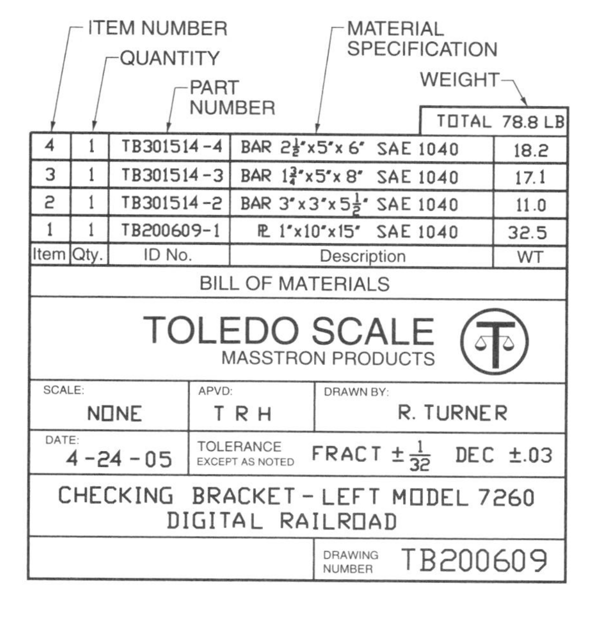

			TOTAL	78.8 LB
4	1	TB301514-4	BAR 2½"x5"x6" SAE 1040	18.2
3	1	TB301514-3	BAR 1¾"x5"x8" SAE 1040	17.1
2	1	TB301514-2	BAR 3"x3"x5½" SAE 1040	11.0
1	1	TB200609-1	PL 1"x10"x15" SAE 1040	32.5
Item	Qty.	ID No.	Description	WT

BILL OF MATERIALS

TOLEDO SCALE
MASSTRON PRODUCTS

SCALE: NONE	APVD: T R H	DRAWN BY: R. TURNER
DATE: 4-24-05	TOLERANCE EXCEPT AS NOTED	FRACT $\pm\frac{1}{32}$ DEC ±.03

CHECKING BRACKET - LEFT MODEL 7260
DIGITAL RAILROAD

DRAWING NUMBER TB200609

Figure 8-1. The bill of materials lists specifications of parts included on an assembly print.

Ferrous Metals

A *ferrous metal* is a metal that has iron as the major alloying element. Ferrous metals are usually magnetic. Pure iron is very soft, extremely ductile, and melts at a low temperature. Iron is commonly alloyed with carbon and other elements to form cast iron and carbon steel.

Cast Iron. Cast iron is an alloy of iron and carbon containing 1.70% to 4.50% carbon. Cast iron is commonly classified as gray, white, or malleable based on composition and heat treatment. Gray cast iron has a gray-colored fracture and is the most common cast iron. It can be easily machined and is widely used for engine blocks, machine tools, and pipe. White cast iron has a silvery-white fracture and is very hard and brittle. It is used in applications where high resistance to abrasion is desired. Malleable cast iron is white cast iron with additives and/or heat treatment to improve strength, ductility, and machinability.

Carbon Steel. Carbon steel has a higher tensile strength than pure iron and is commonly used in fabrication and manufacturing. Carbon steel is broadly grouped into low-, medium-, and high-carbon steel classifications depending on the percentage of carbon. **See Figure 8-2.**

Low-carbon steel contains 0.30% or less of carbon. Low-carbon steel is the weakest of the carbon steels and cannot be hardened. It is easily welded and is commonly used for machine parts where soft steel is required. Low-carbon steel is sometimes called mild steel.

Medium-carbon steel contains approximately 0.30% to 0.70% carbon and is stronger than low-carbon steel. Medium-carbon steel can be hardened and is used where high tensile strength is required on parts such as hammers, wrenches, and screwdrivers.

High-carbon steel contains more than 0.70% carbon. High-carbon steel can be hardened to obtain high strength for use in cutting tools, machine parts, and drills. The application determines the carbon content of the steel required. In addition to carbon, other elements such as nickel, chromium, and tungsten can be added to change the properties of the steel.

Carbon and other alloy steels are classified by the American Iron and Steel Institute (AISI) and the Society of Automotive Engineers (SAE) designation system. The AISI-SAE designation system uses four digits for classification. **See Appendix.**

The first digit indicates the family to which the steel belongs. For example, the number 1 indicates a carbon steel, 2 indicates a nickel steel, and 3 indicates a nickel-chromium steel. **See Figure 8-3.** The numbers continue depending on the family. The second digit indicates the subfamily or approximate percentage of the principal alloying element. The third and fourth digits typically indicate the carbon content in points (hundredths of a percent). For example, 1045 carbon steel has 45 points, or 0.45% carbon content. In the case of 2340 nickel steel, there is approximately 3% nickel and 0.40% carbon.

CARBON STEEL

CLASSIFICATION	% CARBON	CHARACTERISTICS AND TYPICAL APPLICATIONS
Low-Carbon	0.05 to 0.12	Chain, stampings, rivets, nails, wire, pipe, welding stock requiring very soft plastic steel
	0.10 to 0.20	Soft, tough steel; structural steels, machine parts, for case-hardened machine parts, screws
	0.20 to 0.30	Better grade of machine and structural steel; gears, shafting, bars, bases, levers, etc.
Medium-Carbon	0.30 to 0.40	Responds to heat treatment; connecting rods, shafting, crane hooks, machine parts, axles
	0.40 to 0.50	Crankshafts, gears, axles, shafts, and heat-treated machine parts
	0.50 to 0.60	High hardness; springs, parts machined from bar stock, small forgings
	0.60 to 0.70	Good shock strength, drop hammer dies, set screws, locomotive wheels, screwdrivers
High-Carbon	0.70 to 0.80	Tough and hard steel; anvil faces, band saws, hammers, wrenches, cable wires, etc.
	0.80 to 0.90	Punches for metal, rock drills, shear blades, cold chisels, rivet sets, and many hand tools
	0.90 to 1.00	Used for hardness and high tensile strength springs; high-tensile wire, knives, axes, dies

Figure 8-2. Recommended applications vary for steel of various carbon contents.

AISI-SAE STEEL DESIGNATION SYSTEM

Carbon Steels
10xx – Plain Carbon (1% Mn max)
11xx – Resulfurized
12xx – Resulfurized and Rephosphorized
15xx – Plain Carbon (1.00% Mn to 1.65% Mn Max)

Manganese Steels
13xx – 1.75% Mn

Nickel Steels
23xx – 3.5% Ni
25xx – 5% Ni

Nickel-Chromium Steels
31xx – 1.25% Ni; 0.65% Cr and 0.80% Cr

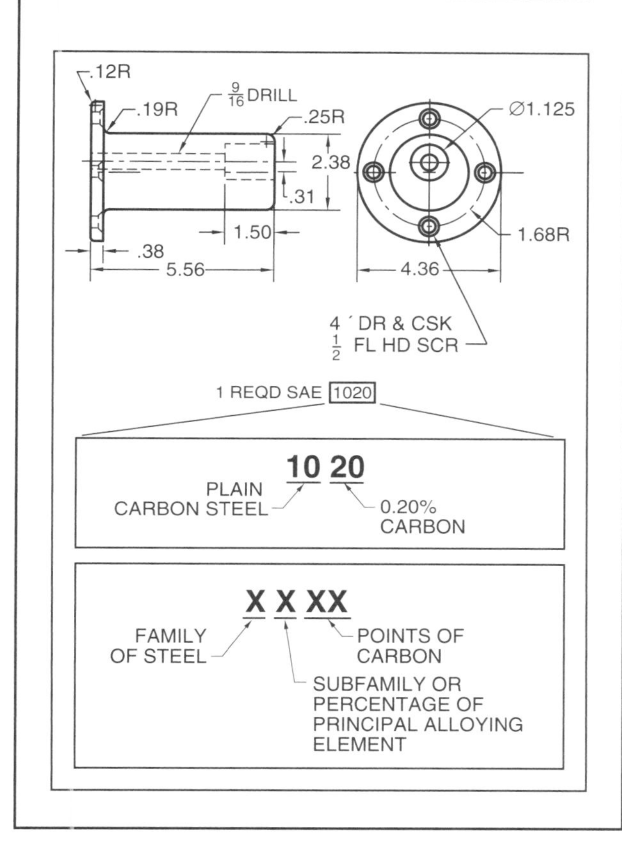

Figure 8-3. The AISI-SAE designation system indicates the alloy content of steels.

> **Nickel and its alloys are used in special** applications, usually **applications requiring a specific degree** of corrosion **resistance or high strength at high temperature.**

The Unified Number System (UNS) was developed through a joint effort of the SAE and ASTM International (formerly the American Society for Testing and Materials). The UNS is useful for correlating metals denoted by different designation systems. The UNS uses an uppercase letter followed by five numbers. For ferrous metals, the UNS designation consists of D, G, H, or K followed by five numbers. The letter indicates the grouping of the metal. The numbers indicate the composition of the metal as indicated by the AISI-SAE designation. For example, the AISI-SAE designation of 1045 has a UNS number of G10450. **See Appendix.**

Nonferrous Metals

A *nonferrous metal* is a metal that does not contain iron. Nonferrous metals are typically softer than ferrous metals and have distinctive color differences. Nonferrous metals commonly used in manufacturing include aluminum, copper, brass, bronze, and magnesium.

Aluminum. Aluminum is the most commonly used nonferrous metal. It has good machinability, weldability, and resistance to corrosion. Pure aluminum is too soft for most applications and is commonly alloyed with silicon, magnesium, or copper-silicon. Aluminum alloys are used extensively in the aircraft industry.

Copper. Copper is a heavy, soft metal with high electrical and heat conductivity and good corrosion resistance. Copper is easily processed and is used in a variety of applications such as electrical components, ammunition cartridges, plumbing pipe and fittings, and cooking utensils.

Copper is commonly alloyed to increase its strength. Copper alloyed with zinc forms brass, and copper alloyed with tin forms bronze. Brass and bronze were some of the earliest alloys developed.

Magnesium. Magnesium is a light structural metal with good machinability and strength-to-weight characteristics. It is commonly used as an alloy with aluminum in castings requiring high strength with minimum weight.

Nonferrous Alloy Classifications. Aluminum alloys are classified by the Aluminum Association using a four-digit designation system similar to the AISI-SAE designation system. **See Figure 8-4.** The first digit identifies the type based on the principal alloying elements. The second digit indicates the degree of control on individual impurities. Digits three and four are identifying serial numbers for alloys within a group, except for group

10XX. Series 10XX indicates aluminum with 99.00% or greater purity. In this group, the third and forth digits indicate the hundredths of a percent of aluminum above 99.00%. Therefore, designation 1030 indicates 99.30% aluminum. Temper designations are also added after the four-digit identifiers, separated by a hyphen.

UNS designation numbers for aluminum alloys include the existing numbers developed by the Aluminum Association, adding the letter A and a fifth number at the beginning. **See Appendix.** Thus, UNS A96061 is the same as 6061 aluminum.

The original designation for copper alloys used a three-digit number preceded by the letters CA. These designations have now been made part of the UNS system by expanding the numbers to five digits, preceded by the letter C. The original three-digit alloy numbers are embedded in the UNS designations.

Magnesium alloy designations begin with two letters representing the two principal alloying elements. The letters are followed by two numbers representing their respective composition percentages rounded to whole numbers, and one serial letter, indicating variations of additional alloying elements. Tempering designations, similar to those used for aluminum, are added after a hyphen. UNS designations for magnesium alloys begin with the letter M, followed by five digits.

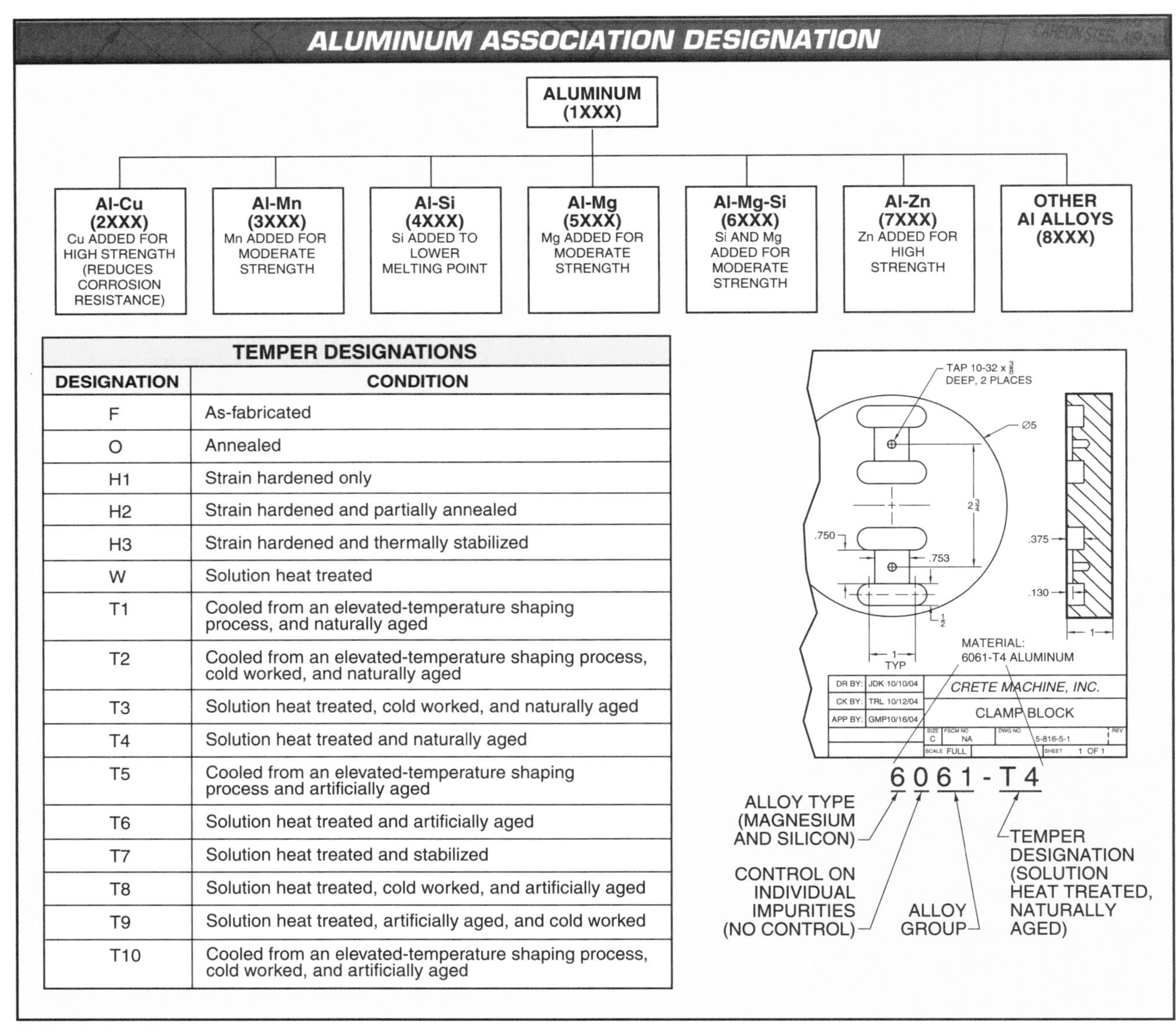

TEMPER DESIGNATIONS	
DESIGNATION	**CONDITION**
F	As-fabricated
O	Annealed
H1	Strain hardened only
H2	Strain hardened and partially annealed
H3	Strain hardened and thermally stabilized
W	Solution heat treated
T1	Cooled from an elevated-temperature shaping process, and naturally aged
T2	Cooled from an elevated-temperature shaping process, cold worked, and naturally aged
T3	Solution heat treated, cold worked, and naturally aged
T4	Solution heat treated and naturally aged
T5	Cooled from an elevated-temperature shaping process and artificially aged
T6	Solution heat treated and artificially aged
T7	Solution heat treated and stabilized
T8	Solution heat treated, cold worked, and artificially aged
T9	Solution heat treated, artificially aged, and cold worked
T10	Cooled from an elevated-temperature shaping process, cold worked, and artificially aged

Figure 8-4. Aluminum alloy type, impurity control, alloy group, and temper designation are included in the Aluminum Association designation number.

Plastics

A *plastic* is a material made up of repeating groups of atoms or molecules linked in long chains called polymers. The lengths of the chains, the bonds, and the elements involved determine the mechanical and physical properties of the materials, such as machinability, resistance to stretching (tensile strength), and resistance to compression (compressive strength). **See Figure 8-5.** Plastics can be broadly grouped into thermoplastic and thermosetting plastics.

A *thermoplastic* is a plastic that softens when heat is applied and reforms into a solid when cooled. This material is recyclable. Common thermoplastic plastics include acrylics, polystyrenes, polyethylenes, and polyvinyl chlorides.

A *thermoset* is a plastic that is chemically changed during initial processing and does not soften with subsequent application of heat. Since heating a thermoset degrades the material, these materials usually cannot be recycled. Common thermosetting plastics include epoxies, silicones, and polyesters.

An *elastomer* is a flexible material that can be stretched up to twice its length and return to its original length when released. Elastomers may be either thermoplastic or thermosetting materials.

MATERIAL PROPERTIES

The characteristics of materials are classified as mechanical, physical, or chemical properties. Machinability is a material property based on the material's mechanical, physical, and chemical properties. Machinability determines the ease of machining and affects the choice of machining process.

Mechanical Properties

A *mechanical property* is the response of a material under applied loads. In machining, the most important mechanical properties are strength, ductility, brittleness, hardness, toughness, and malleability. **See Figure 8-6.**

PLASTICS

CLASSIFICATION	POLYMER	MACHINABILITY			TENSILE STRENGTH*	COMPRESSIVE STRENGTH*
		FAIR	GOOD	EXC		
THERMOPLASTIC	Acetal			✓	8800	13,000
	Acrylic		✓		9000	16,000
	Acrylonitrile butadiene styrene (ABS)		✓		7000	10,000
	Cellulose acetate			✓	8000	28,000
	Nylon			✓	15,000	13,000
	Poly (amide-imide)			✓	20,000	35,000
	Polycarbonate			✓	9500	12,500
	Polyethylene, low density		✓		2000	2300
	Polyethylene, high density			✓	5000	3200
	Polyphenylene oxide			✓	9600	16,000
	Polypropylene		✓		5300	7000
	Polystyrene		✓		7000	15,000
	Polytetrafluorethylene (PTFE)			✓	4500	2180
	Polyvinyl chloride (PVC)			✓	4800	11,000
THERMOSET	Epoxy	✓			17,000	35,000
	Phenoliz			✓	7000	30,000
	Polyester			✓	30,000	25,000
	Polyurethane			✓	6000	20,000
	Silicone	✓			28,000	15,000

* in psi

Figure 8-5. Plastics are classified as thermoplastic or thermosetting plastics and have distinct material properties.

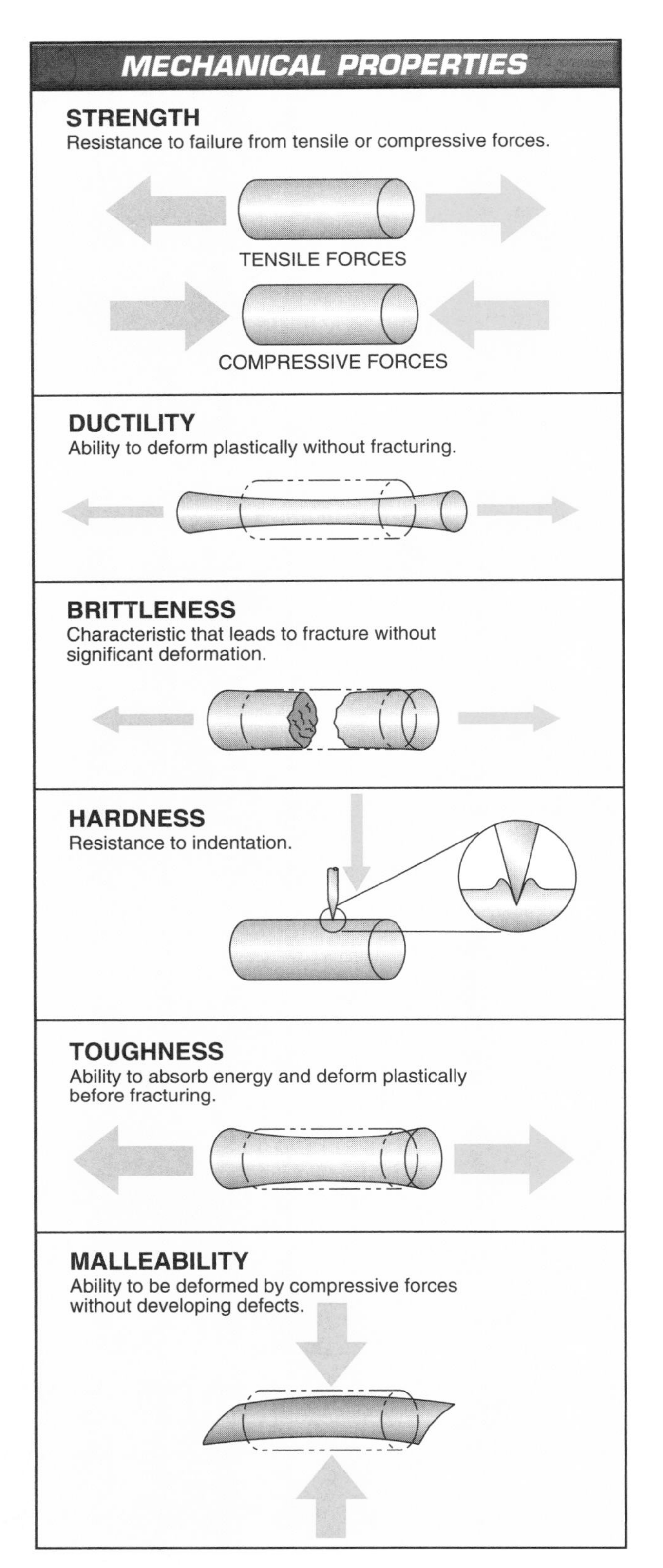

Figure 8-6. Mechanical properties of materials have the greatest effect on the machinability of the part.

Strength is the pressure that a material can withstand before it fails. Depending upon the desired behavior of the material, failure can mean yielding (the material permanently deforms) or fracture (the material breaks). Strength is denoted in pounds per square inch (psi) or pascals (Pa). *Tensile strength* is strength under tensile forces, which pull on the material. *Compressive strength* is strength under compressive forces, which squeeze the material.

Ductility is the ability of a material to deform under tensile forces without breaking or cracking. For example, high-ductility metals deform and fail gradually. This characteristic allows metals such as copper to be drawn into thin wires without fracturing.

Brittleness is the tendency of a material to fracture under pressure without significant deformation. Brittleness is a lack of ductility in a material. Brittle materials may still withstand very high loads, but fracture soon after beginning to deform. This contrasts with ductile materials, which deform greatly before failing.

Hardness is the ability of a material to resist permanent deformation, usually by indentation. Greater hardness means a greater resistance to wear and abrasion.

Toughness is the ability of a material to withstand fracture when stressed. Toughness may be considered as a combination of strength and ductility.

Malleability is the ability of a material to deform under compressive forces without developing defects. A malleable material is one that can be stamped, hammered, forged, pressed, or rolled into thin sheets.

Machinability

Machinability is the ease with which a material can be acceptably machined. A material with good machinability does not shorten the life of the cutting tool, requires low force and power to cut, and leaves a good surface finish. The machinability of a material is most affected by its hardness. Harder materials are difficult to machine. Machinability is a factor in selecting machining operations.

Machinability is quantified approximately with a rating. The steel alloy 1112 is defined as having a machinability rating of 100%. Machinability ratings of less than 100% indicate difficulty in machining. For example, 1045 carbon steel has a machinability rating of 55%. Machinability ratings of more than 100% indicate ease in machining. Aluminum is relatively soft and has a low tensile strength, so the machinability ratings for aluminum alloys range from 300% to 2000%.

Physical Properties

Physical properties are the thermal, electrical, optical, magnetic, and other general properties of a material. Of the physical properties, thermal properties affect machining operations the most.

Thermal expansion is the elongation of a material when subjected to heat. The amount of thermal expansion is expressed as the coefficient of thermal expansion. A coefficient of thermal expansion defines a material's change in length per unit of temperature change. **See Figure 8-7.**

Thermal expansion in all dimensions occurs when metal is exposed to heat. Different metals have different coefficients of thermal expansion.

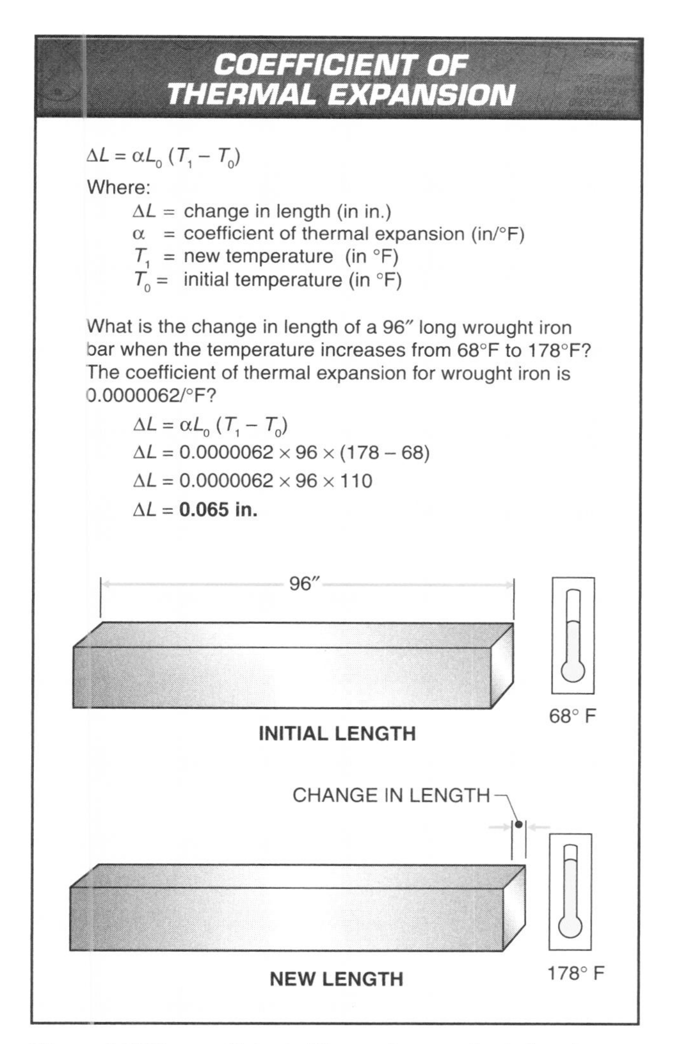

Figure 8-7. The coefficient of thermal expansion is the change in the length of a material per unit of temperature change.

Chemical Properties

Chemical properties are the properties of a material such as corrosion and oxidation pertaining to chemical reactivity of the material. Machining operations are not affected by the chemical properties of most common materials. Machining exotic materials may require special safety precautions.

MATERIAL FORMING

There are many methods of shaping materials into useful forms. These forming processes may be involved in the manufacture of finished products or in the wholesale production of standard shapes. Standard shapes can then be further modified into final parts.

Forming Processes

A *forming process* is a process that shapes a part prior to final machining operations. Common forming processes include casting, forging, stamping, rolling, drawing, and extruding. **See Figure 8-8.**

Casting is a manufacturing process where a shaped cavity is filled with molten metal, which takes on the shape of the cavity as it cools and solidifies. The design of cast parts maximizes the strength of the parts by eliminating sharp corners. Rounded corners of intersecting planes also provide better release from the casting mold.

Forging is a manufacturing process in which a workpiece is deformed into the desired shape with high compressive forces, usually between two dies. Forging can be accomplished with or without heat applied to the material.

Stamping is a manufacturing process that involves forming sheet metal between dies with pressure while cutting the part from the sheet metal with the edge of the die. As pressure is applied, the metal is bent in several directions in one operation. Generally, stamping is used when a uniform thickness is desired throughout the part.

Rolling is a manufacturing process in which material is squeezed between two revolving rolls to obtain the desired thickness. The rolls are placed slightly closer together than the starting thickness of the material. Several passes with the rolls successively closer may be needed to achieve the desired thickness.

Drawing is a manufacturing process in which material is pulled through a die in order to shape the material to final size and shape. Drawing may require several steps to reach final specifications.

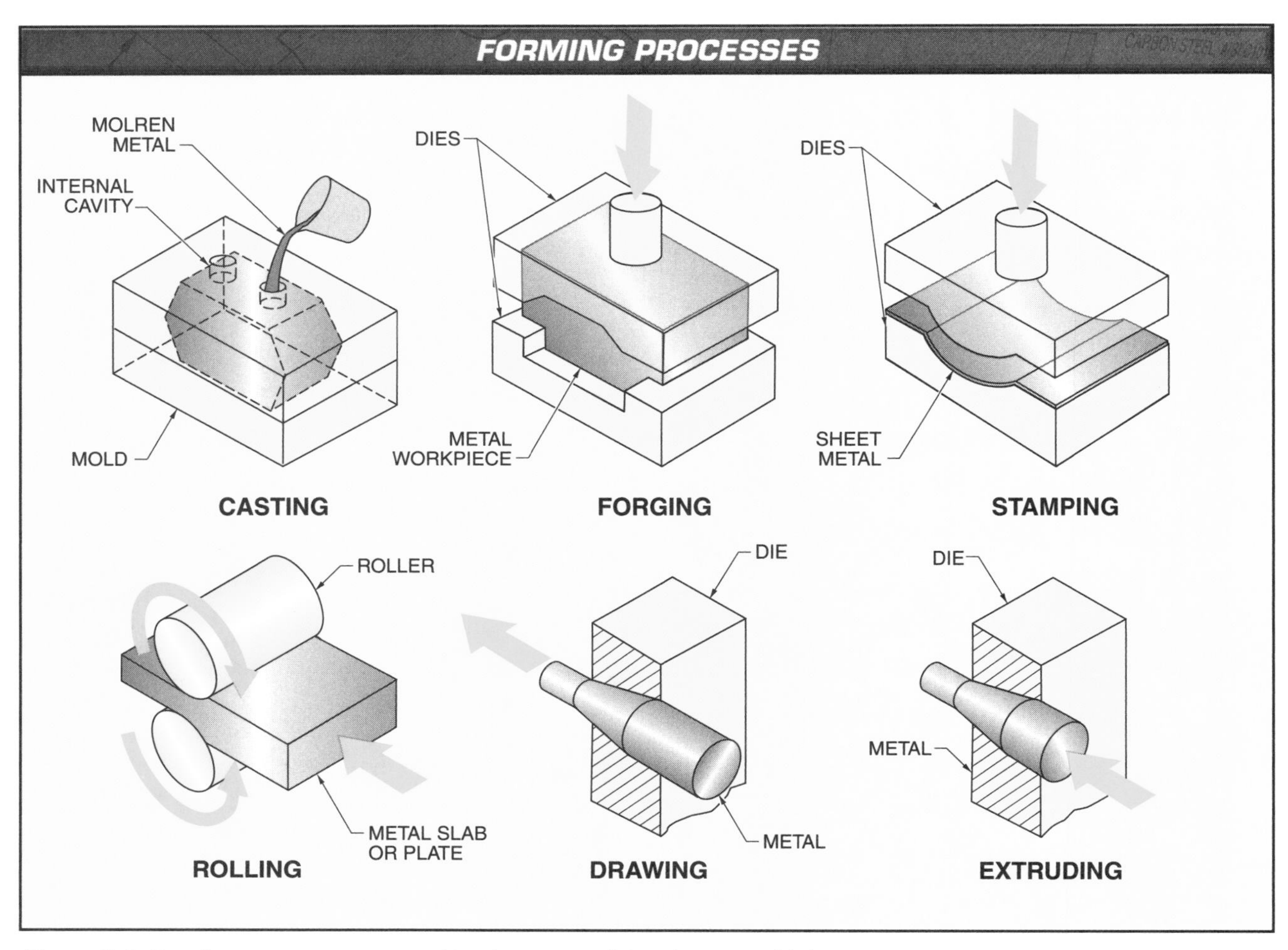

Figure 8-8. Forming processes are used to shape materials prior to machining.

Extruding is a manufacturing process in which a material is pushed through a die in order to obtain the desired shape. Several steps may be required to reach the desired size and shape.

Standard Shapes

Forming processes produce commonly available standard shapes and sizes for many materials. These shapes are sometimes called structural or mill shapes. **See Figure 8-9.** Manufacturers prefer to use standard shapes as the bases for new parts because of the economy and availability of standard shapes. Parts are designed to take advantage of standard shapes whenever possible, saving the manufacturer the need for unnecessary machining and waste. For instance, an equipment bracket may begin as a standard angle shape of a common size, and then be machined to include the necessary holes and slots. Standard shapes include beams, channels, angles, tees, tubing, bars, plate, sheet, and pipe.

HEAT TREATMENT

Some components require heat treatment. *Heat treatment* is the application of heat to change the properties of a metal without changing its size and shape.

Annealing

Stresses that build up in the part during the manufacturing process can be removed by annealing. *Annealing* is the process of heating metal until the metal's crystalline structure changes, and then allowing it to cool very slowly. This alters the metal's microstructure to relieve internal stresses and obtain the desired mechanical and physical properties.

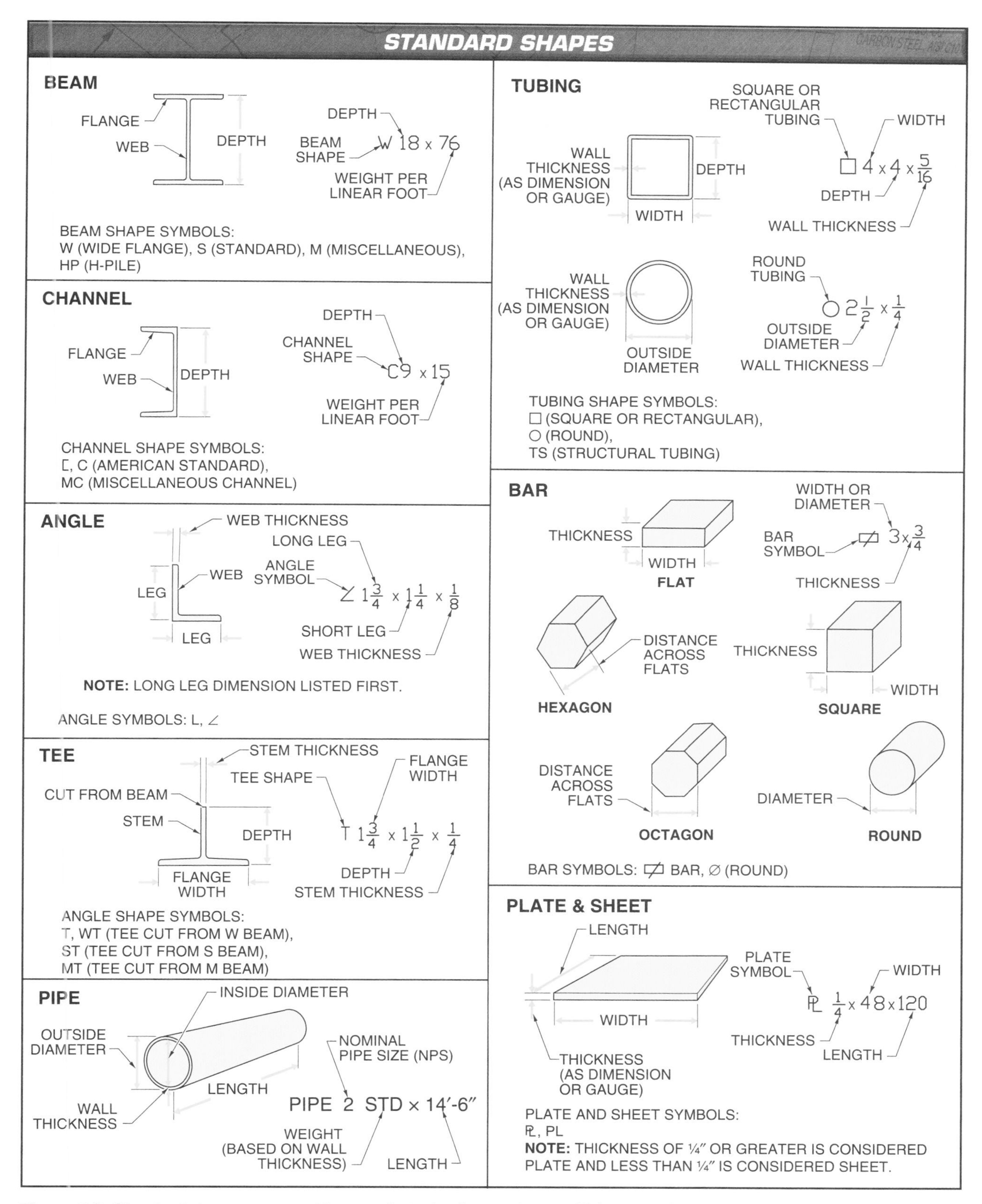

Figure 8-9. Standard shapes are used in manufacturing for maximum efficiency and economy.

Hardening and Tempering

Some parts require additional wear resistance and strength obtained through hardening and tempering. *Hardening* is the process of heating metal followed by quenching in oil, water, or another cooling medium to bring the temperature down quickly. Parts may require further heat treatment to retain the desired mechanical properties.

Tempering is the process of heating metal followed by controlled cooling at a specific rate. During tempering, the temperature of the metal does not rise to the crystalline transformation point. Tempering increases toughness and ductility.

Case hardening is the process of increasing the hardness of a metal surface without changing the mechanical properties of the core. Case hardening is commonly performed by carburizing, induction hardening, or flame hardening. *Carburizing* is a case hardening process in which carbon is introduced into a solid iron-base alloy heated above a certain temperature. Case hardening can also be obtained by induction hardening and flame hardening.

MACHINING PROCESSES

Machining is the process of removing material from a workpiece with cutting tools in order to achieve a desired size and shape. The shape and dimensions of the part and the machinability of the material determine the machining process required.

Machining operations consider depth of cut, cutting speed, and feed. **See Figure 8-10.** *Depth of cut* is the penetration of a cutting tool for each pass. *Cutting speed* is the speed of the surface of a cutting tool. The cutting speed can also indicate the movement of the workpiece. Cutting speed is expressed in feet per minute (ft/min) or meters per minute (m/min). *Feed* is the rate at which the cutting tool advances into the workpiece. In drilling, turning, and boring operations, feed is expressed in inches per revolution (in./rev) or millimeters per revolution (mm/rev). In shaping operations, feed is expressed in inches per stroke (in./st) or millimeters per stroke (mm/st). In milling operations, feed is expressed in inches per tooth (ipt) or millimeters per tooth (mmpt).

The machining process selected is based on factors such as cutting speed, depth of cut, cutting tool design, and the cutting tool characteristics. Machining processes commonly used include milling, shaping, turning, grinding, drilling, and boring. **See Figure 8-11.**

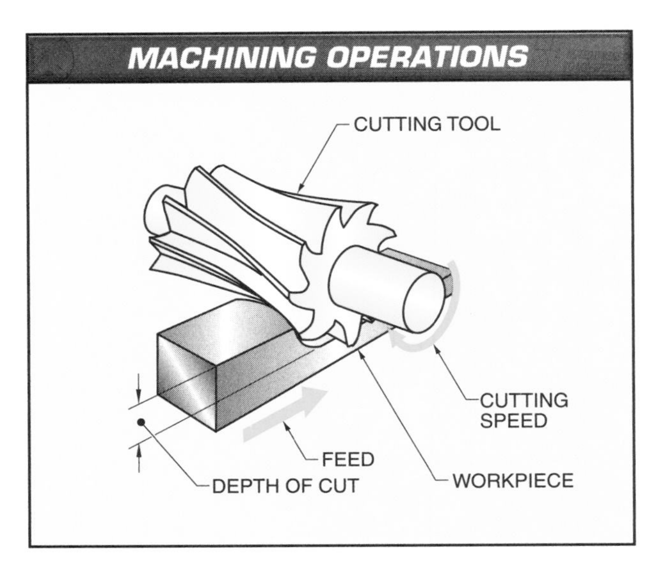

Figure 8-10. The material to be machined determines the depth of cut, cutting speed, and feed required.

Milling

Milling is a cutting operation that combines the rotation of a cutting tool and the feeding of the workpiece into the path of the cutter. The milling machine cutting tool is a cylindrical tool with multiple teeth. Each tooth removes a portion of the workpiece as it rotates on its axis, and the workpiece is fed into the path of the cutting tool by the movement of the milling machine's table. The cutting tool and the workpiece are positioned and controlled by x, y, and z coordinates. A milling machine can be used for a variety of industrial operations including drilling, boring, slotting, machining flat and irregular surfaces, and producing gears.

Milling machines are commonly classified as horizontal or vertical milling machines. On a horizontal milling machine, the cutting tool is mounted on a horizontal arbor or spindle. The arbor or spindle axis is parallel to the table. On a vertical milling machine, the cutting tool is mounted in a spindle that is perpendicular to the table.

Shaping

Shaping is a cutting operation performed by the reciprocating motion of a cutting tool. The cutting tool moves back and forth parallel to the table. The table controls the feeding path of the workpiece into the cutting tool. Shapers are commonly used for machining flat surfaces, external and internal keyseats, and T-slots.

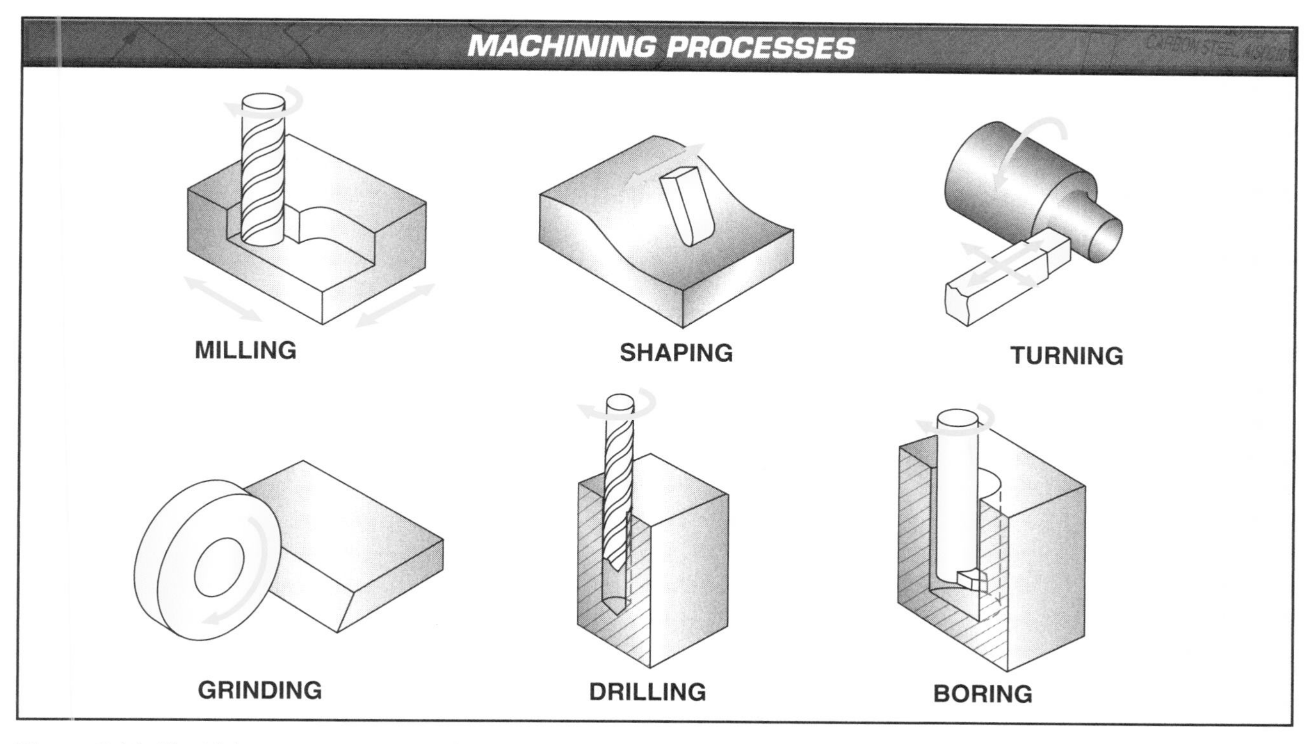

Figure 8-11. Machining processes are selected for efficiency and according to part requirements.

Turning

Turning is a cutting operation performed with the workpiece rotating and a cutting tool fed into or across the workpiece. Turning operations are commonly performed on a lathe. The lathe is one of the most versatile machine tools. In addition to producing cylindrical parts, the lathe is used for cutting threads, tapers, and other machined features.

The cutting tool is mounted in a tool post fastened to the compound rest. The compound rest can be rotated on the cross slide to position the cutting tool at various angles to the work. The cross slide moves the cutting tool in and out at 90° to the axis of the work.

The compound rest and cross slide are mounted on the carriage which allows the cutting tool to travel parallel or perpendicular to the axis of the work. Cutting tools on the lathe are designed for obtaining the desired cut and finish. Different cutting tools and cutting operations require specific speeds and feeds.

Grinding

Grinding is the process of removing material with an abrasive. This is done to produce the desired dimensional and surface characteristics. Abrasives used for machining are adhered to wheels mounted on rotating horizontal or vertical arbors. The work is fed into the path of the grinding wheel. Abrasives can also be bonded to paper or cloth or used loose in special operations. The abrasive grains act as miniature cutting tools that remove a controlled amount of material. Grinding can be performed with very light pressure and can remove material to close tolerances.

Grinding wheels are classified according to the abrasive material, the grit number, and type of bond used to join the grains. The grinding wheel required for an operation is determined by the workpiece material properties and the desired surface finish.

Common abrasive materials include aluminum oxide, silicon carbide, and diamond. Grit number is determined by the grain size. Grains are sorted through successively finer mesh screens until they become trapped. Grit number relates to the size of the mesh that confines grains of a certain size. For example, an 80-grit grinding disk has abrasives that will not fall through a screen with more than 80 openings per linear inch. Larger grit numbers represent smaller grain sizes and therefore finer grinding, less material removal, and a smoother surface finish.

The four main types of bond materials are vitrified (glass-like), resinoid (thermosetting plastic), rubber, and metal. Some bond materials are only used with some abrasives. For instance, diamond is only set into metal bonded wheels. Bond type affects the application, cost, and useful life of the wheel.

Grinding may be required after other machining operations to obtain the desired surface texture. Surface texture is determined by a material's roughness, waviness, lay, and flaws. **See Figure 8-12.**

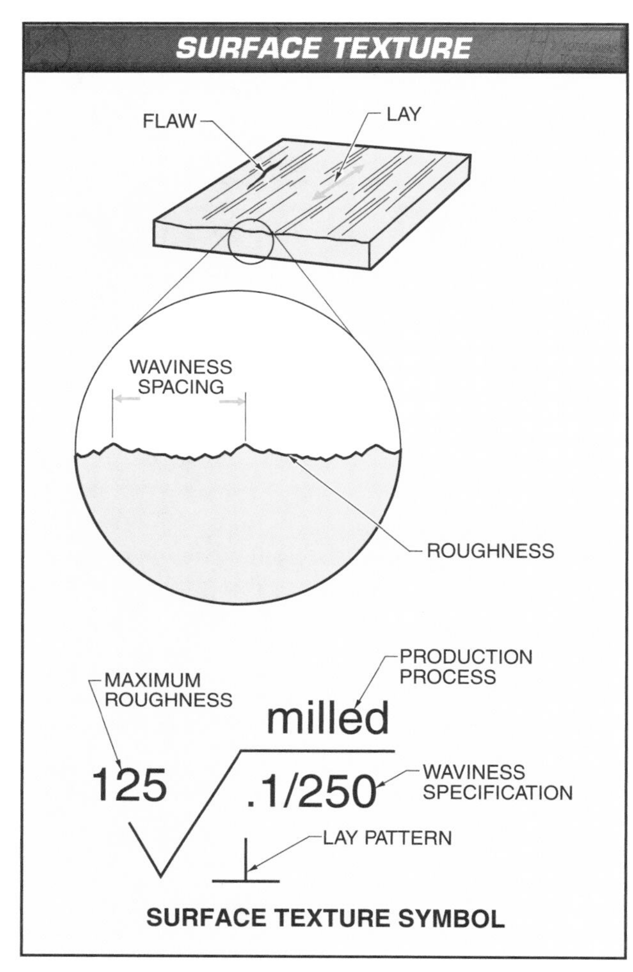

Figure 8-12. Elements of surface texture include roughness, waviness, lay, and flaws and some or all may be specified for a part on a print.

Speed, feed rate, grit size, bonding material, wear, and lubrication all affect where a surface roughness falls within the range for each manufacturing process.

Roughness is the degree of irregularity in the smoothness of a surface. Some roughness is produced by machining processes. Roughness is measured by subtracting the theoretical ideal surface from an average of the tiny peaks and valleys of an actual surface. When a manufacturing print specifies the surface texture, the symbol √ is used, and the desired roughness is given in microinches or micrometers. The symbol can be expanded to $\sqrt{}$ when additional information is specified. Symbols and notations for surface texture are based on the conventions in ANSI/ASME Y14.36M, *Surface Texture Symbols.*

Surface roughness varies between different forming and machining operations. Therefore, the desired surface roughness determines the finishing operation. **See Figure 8-13. See Appendix.**

Waviness is a widely spaced surface texture pattern. Waviness results from machining variations such as deflections, vibration, or chatter. *Lay* is the direction of the dominant surface texture pattern. A *flaw* is an unintentional interruption in the characteristic texture of a surface. Allowable waviness, lay, and flaws may or may not be specified on a surface texture symbol along with the surface roughness.

Drilling

Drilling is the cutting of round holes in material with a rotating twist drill. Holes are made or enlarged by drilling. The drill, the workpiece, or both may be held in rotating spindles during the drilling process. The drilling process may be done on a drill press, lathe, or milling machine.

Symbols are used on prints to show the specific drilling operation. Drill symbols include counterbore or spotface, countersink, and depth. A size note accompanies the drill symbol. **See Figure 8-14.** Drill symbols are based on the conventions in ANSI/ASME Y14.5, *Dimensioning and Tolerancing.*

Twist drills (also known as drills) are sized by diameter. Sizes are designated by numbers 1 through 80, letters A through Z, fractions 1⁄64″ through 3½″, and metric sizes 1 mm through 13 mm. **See Appendix.** Numbered drills range from 0.2280″ (No. 1) to 0.0135″ (No. 80). Letter drills range from 0.234″ (A) to 0.413″ (Z). Fractional drills range from 1⁄64″ to 1¾″ and are available in 1⁄64″ increments. From 1¾″ to 2¼″, they are available in 1⁄32″ increments, and from 2¼″ to 3½″, they are available in 1⁄16″ increments. Drills in metric sizes range from 1 mm to 13 mm, typically in 0.5 mm increments.

ROUGHNESS PER MANUFACTURING PROCESS

PROCESS	2000* (50)†	1000 (25)	500 (12.5)	250 (6.3)	125 (3.2)	63 (1.6)	32 (0.80)	16 (0.40)	8 (0.20)	4 (0.10)	2 (0.05)	1 (0.025)	0.5 (0.012)
Planing, shaping													
Drilling													
Milling													
Boring, turning													
Grinding													
Sand casting													
Forging													
Cold rolling, drawing													

* in μin.
† in μm

Figure 8-13. Production and machining processes generate varying surface texture roughness.

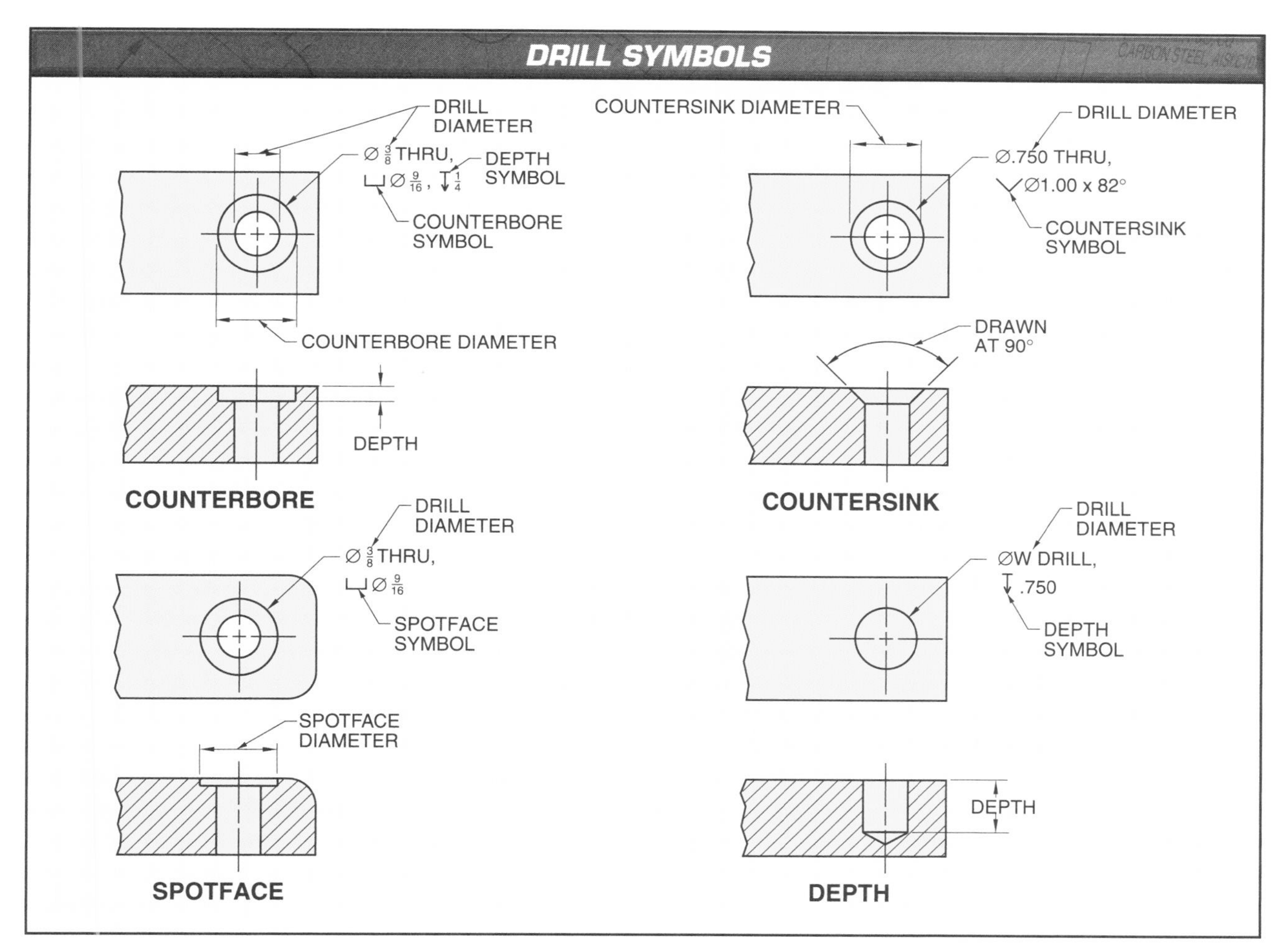

Figure 8-14. Symbols are used on prints to show drilling operations.

Radial surface finishing produces a circular lay pattern.

Jobber-length drills are used for general purpose drilling. Taper-length drills are used for general purpose drilling where a longer length is required. Screw machine-length drills are used for thin metals where a shorter length is required.

Drills may have round, hex, or tapered shafts. They have a variety of cutting point designs adapted to their application.

Boring and Reaming

Boring is the process of enlarging an existing hole or circular internal shape with a rotating cutting tool. The purpose of the boring operation is to enlarge holes to specified tolerances, clean a drilled hole, or remove the eccentricity of a hole.

Boring is most commonly performed using a single-point cutting tool, but can also be performed with a multiple-edge tool. Boring can be performed on a drill press, lathe, or vertical or horizontal milling machine.

Reaming is the process of enlarging an existing hole slightly in order to improve its dimensional accuracy and surface quality to tighter tolerances. A reamer is a multiple-edge cutting tool that removes very little material. The most accurate holes are produced by drilling, boring, and reaming in sequential steps. Each step makes the hole slightly larger than the previous step.

MACHINED FEATURES

Special features are machined into a part in order to improve the functionality or facilitate the assembly of components. Common machined features are grooves, slots, spotfaces, tapers, chamfers, and knurls.

Grooves

A *groove* is a shallow channel machined into a surface. Grooves may be on external or internal sections of a cylindrical part or on flat surfaces. Grooves can be used to provide relief when cutting threads or when performing knurling operations. Grooves are commonly produced using a milling machine, shaper, or lathe. In some instances, grooves can be produced with a grinding operation. Groove shapes are classified as square grooves, V-grooves, or round grooves. **See Figure 8-15.**

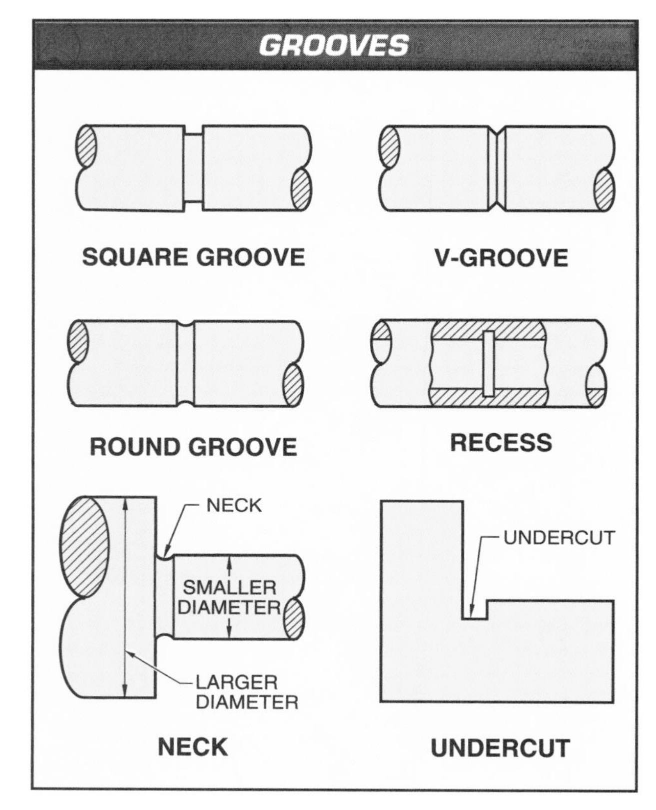

Figure 8-15. Grooves are machined below the surface of the material to provide clearance and bearing surfaces.

Square grooves are cut into the work and have shoulders that are 90° to the axis of the work. This provides a bearing surface for seats of mating parts. Square grooves are also used with retaining rings and clips.

V-grooves are cut into the work and have shoulders that are at an angle. The angled surface is commonly used as a bearing surface for drive belts. The angled surface can also be used as a finished surface before cutting the completed part to final size.

Round grooves have shoulders that provide greater strength than square grooves. Round grooves are commonly used for seating springs, washers, and seals.

A *recess* is a groove cut into the internal diameter of a cylinder. A recess can be used for the same functions as other groove types.

A *neck* is a groove cut into a cylindrical part to provide a space where the diameter on a cylinder changes. The neck ensures that a pulley installed on the shaft would fit in full contact against the shoulder of the larger diameter.

An *undercut* is a groove machined at the intersection of two perpendicular planes and runs the length of the part. Like a neck on a cylindrical part, the undercut provides clearance to allow mating parts to fit in full contact with one another.

Slots

A *slot* is an elongated hole machined either through a part or to a specified depth. **See Figure 8-16.** Slots are produced with a milling machine or a shaper. Internal slots begin and end inside the perimeter of the part. Other slots can run from one edge to another.

Slots are specified by hole diameter, length, and depth. Wide slots are specified by length and width and include a radius dimension for rounded corners. Centerpoints and centerlines can be used to indicate the slot size and location.

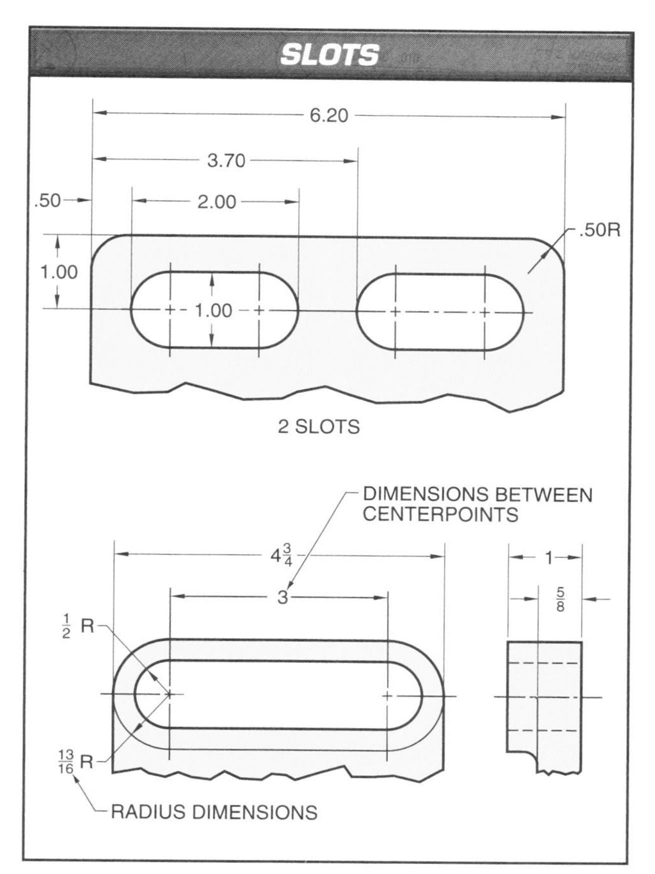

Figure 8-16. Slots are commonly machined to specifications using a milling machine.

Keyseats

A *keyseat* is a rectangular groove along the axis of a shaft or hub that mates with a key. A *key* is a removable fastener that provides a positive means of transmitting torque between a shaft and a hub when mounted in a keyseat. **See Figure 8-17.** This maintains consistent rotation speed between a shaft and attached parts such as pulleys, cranks, or gears.

The keyseat specified is based on the strength requirements and the key selected. Keys commonly used include the parallel, taper, and Woodruff. Dimensions for keys are standardized to fit in specific dimensions.

Keyseats on a shaft are dimensioned by the thickness and shape of the mating key and the distance from the opposite side of the shaft or hole. Keyseats on a hub are dimensioned by thickness, depth, and length. The distance from the opposite side of the shaft or hole is also given. Keyseats are commonly machined into the part using the milling process.

Spotfaces and Bosses

A *spotface* is a flat surface machined at a right angle to a drilled hole. A spotface can be used on recessed or raised surfaces of the part to provide a bearing surface for a mating part. On a recessed surface, spotfacing is a shallow counterboring operation. On a raised surface, spotfacing is used to finish a boss on the part. **See Figure 8-18.** Spotfaces are produced using the drilling or milling process.

A keyed joint consists of a key and two keyseat grooves. The keyseats, also known as keyways, are machined into the interior surface of a hub and the exterior surface of a shaft. The assembly keeps the hub and shaft engaged together under rotation, transmitting torque between them. Commonly keyed components include gears, pulleys, and couplings.

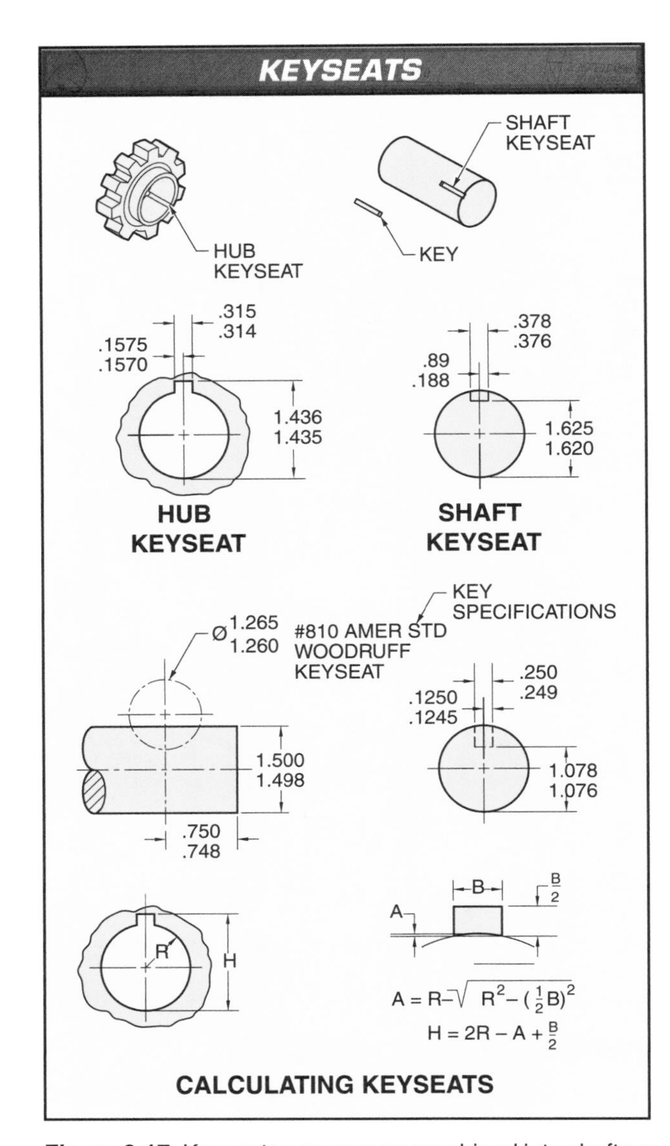

Figure 8-17. Keyseats are grooves machined into shafts or hubs that match key specifications.

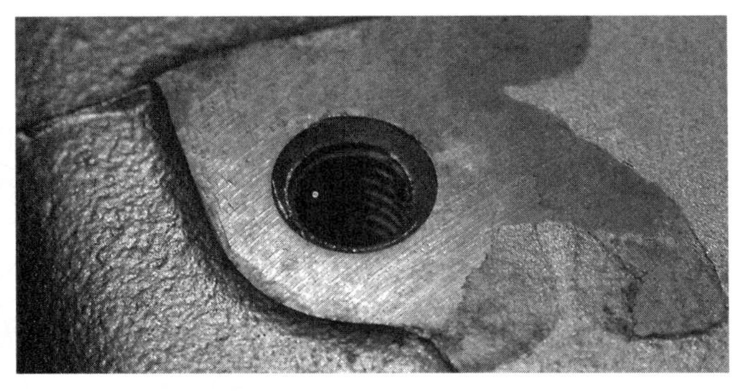

A boss is a feature that projects from another surface and is machined flat.

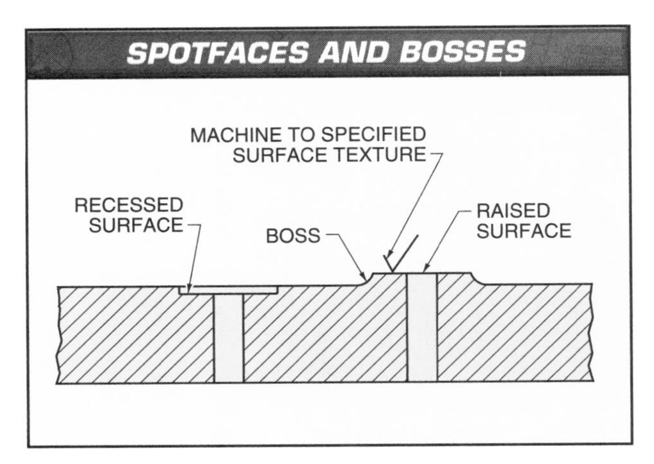

Figure 8-18. Spotfaces and bosses provide level bearing surfaces for load transfer.

A *boss* is a short projection with a finished surface that extends above the surface of a part. The top surface of the projection is machined to remove any surface irregularities and to match the bearing surface of a mating part. A boss increases the strength in load-bearing areas of a casting.

Tapers

A *taper* is a solid or hollow cylinder in which the diameter changes uniformly from one end to the other. Taper specifications are determined by the accuracy required. For noncritical applications, the taper may be specified by the large diameter, the small diameter, and the length of the axis of the taper, using the necessary tolerances. **See Figure 8-19.** Taper can also be specified using the included angle desired.

For greater accuracy, the amount of taper per linear unit on the diameter is specified. For example, tapers can be specified as taper per inch or taper per foot. Taper per inch is equal to the difference in inches between the diameters, divided by the length in inches of the taper. Taper per foot uses the length in feet as the linear unit.

Standard tapers used in industry include the Morse taper, Brown and Sharpe taper, American National Standard taper, and Jarno taper. Standard tapers are commonly specified for the shank of twist drills, end mills, and lathe centers. Standard tapers are also commonly specified for tapered holes used on drill, milling machine, and lathe spindles. Tapers are produced by turning operations, grinding operations, or both.

Chamfers

A *chamfer* is a beveled edge. Chamfers are specified on the print by angle and linear dimension or by two linear dimensions. **See Figure 8-20.** Chamfers are commonly produced using a milling machine for square shapes or a lathe for cylindrical shapes. Chamfers can be external or internal.

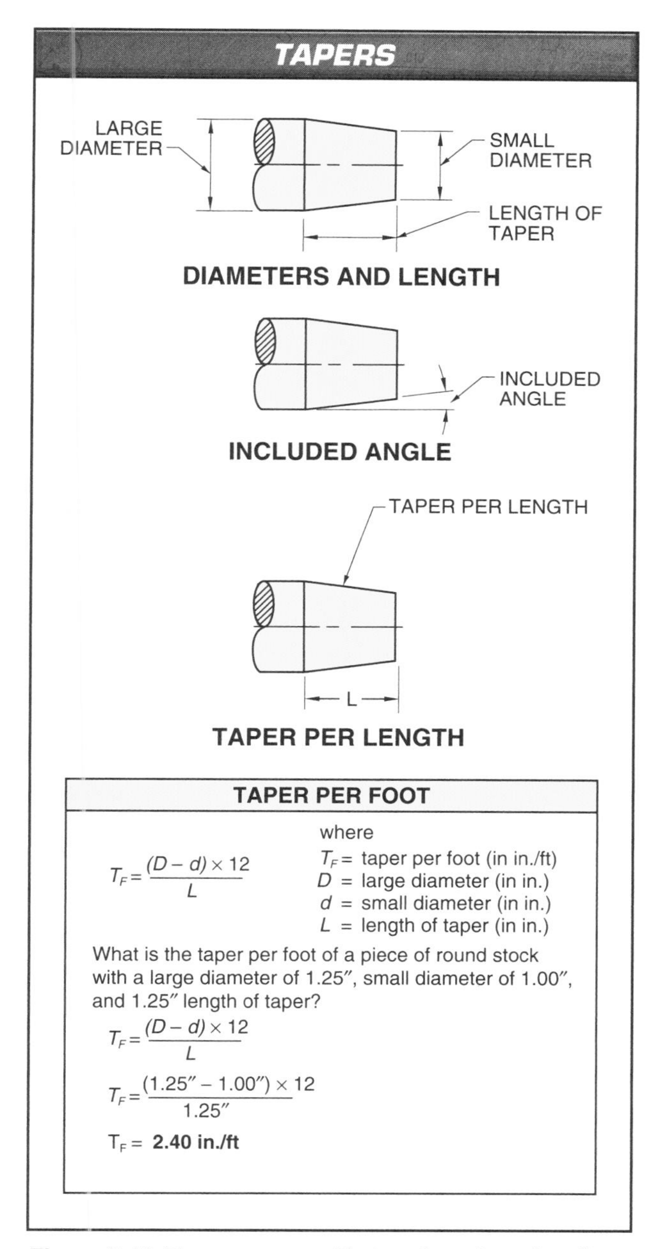

Figure 8-19. Tapers are specified on the print using diameters, included angle, and taper per length specifications.

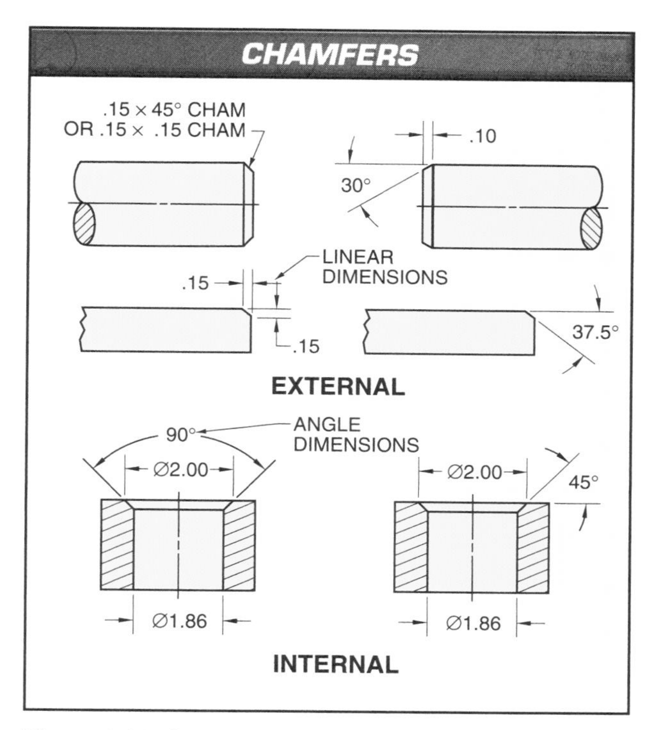

Figure 8-20. Chamfers are specified with linear or angle dimensions.

Knurls

A *knurl* is a raised pattern formed on a material for improving the grip. Knurling changes a smooth surface into uniform ridges and projections. The knurl pattern is formed with uniform, localized pressure from the rolls of a knurling tool on a lathe. The rolls remain in contact with the work until the knurling operation is complete. The knurling tool rolls have straight, diagonal, or diamond-shaped teeth. **See Figure 8-21.**

Straight tooth patterns are commonly used for fastener heads. Diagonal knurls can be used to serrate the surface of parts to be press-fit or locked together. The diamond patterns are most commonly used for grips on handles. The diamond pattern can also be formed by two operations of rolls having diagonal teeth.

Knurled areas are specified by type, diametral pitch, and length of the knurled area. *Diametral pitch* is the number of raised projections per unit of pitch diameter. Surfaces with noncritical dimensions such as a gripping surface may not require diameter dimensions. Critical dimensions, such as on parts to be press-fit, require diameters to be specified before and after the knurling operation.

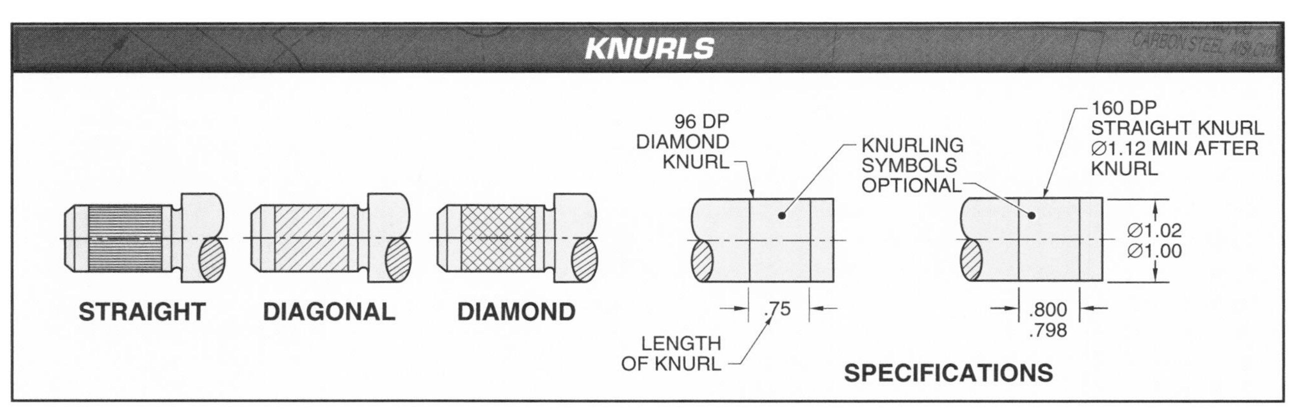

Figure 8-21. Knurls are specified by type, diametral pitch, and length of knurled area.

Materials and Machining

REVIEW QUESTIONS 8

Name ______________________________ **Date** ______________

Completion

______________ **1.** ___-carbon steel is used for cutting tools and drills.

______________ **2.** ___ is a manufacturing process in which material is squeezed between two revolving rolls to obtain the desired thickness.

______________ **3.** Copper alloyed with ___ forms brass.

______________ **4.** Hardening is the process of heating metal followed by ___ in oil or water.

______________ **5.** Machinability is quantified approximately with a(n) ___.

______________ **6.** A(n) ___ is a raised pattern formed on a material for improving the grip.

______________ **7.** A key provides a positive means of transmitting torque between a shaft and hub when mounted in a(n) ___.

______________ **8.** The coefficient of ___ expansion defines a material's change in length per unit of temperature change.

______________ **9.** A(n) ___ is a short projection with a finished surface that extends above the surface of a part.

______________ **10.** ___ is the process of removing material with an abrasive.

True-False

T F **1.** Shaping is a cutting operation that uses a circular motion of the cutting tool.

T F **2.** Turning operations are commonly performed on a lathe.

T F **3.** Feed is the rate at which the cutting tool advances into the workpiece.

T F **4.** Nonferrous metal contains iron.

T F **5.** Carburizing is a case hardening process.

T F **6.** A thermoset is a plastic that does not soften with subsequent application of heat.

T F **7.** Chemical properties are the thermal, electrical, and magnetic properties of a material.

T F **8.** Drawing is a manufacturing process in which material is pulled through a die.

T F **9.** A knurl is a sloped edge of an object running from surface to side.

T F **10.** A taper may be specified by the large diameter, the small diameter, and the length of the axis of the taper.

Identification—Mechanical Properties

_____________ **1.** Ductility

_____________ **2.** Hardness

_____________ **3.** Brittleness

_____________ **4.** Toughness

_____________ **5.** Strength

_____________ **6.** Malleability

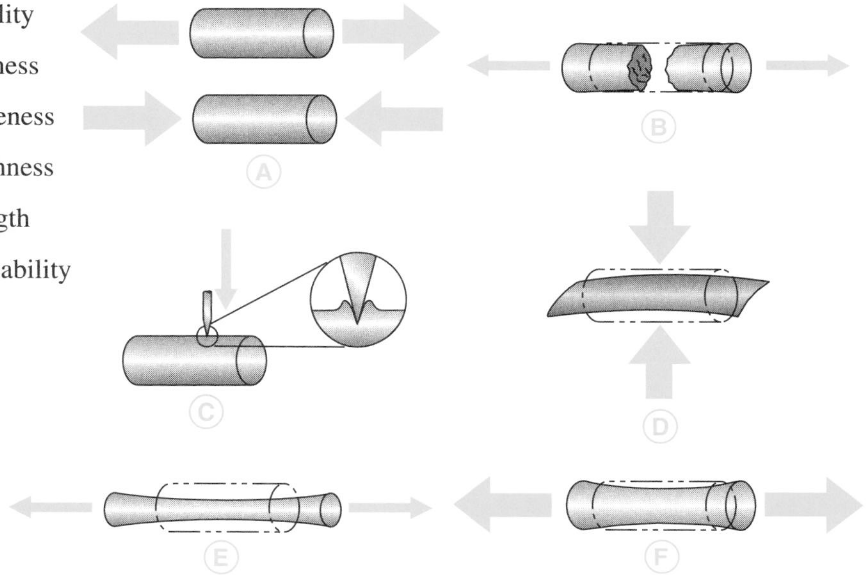

A B C D E F

Identification—Machined Features

_____________ **1.** Square groove

_____________ **2.** Slots

_____________ **3.** Neck

_____________ **4.** Keyseat

_____________ **5.** Taper

_____________ **6.** Chamfer

_____________ **7.** Spotface

_____________ **8.** Knurl

_____________ **9.** V-groove

_____________ **10.** Recess

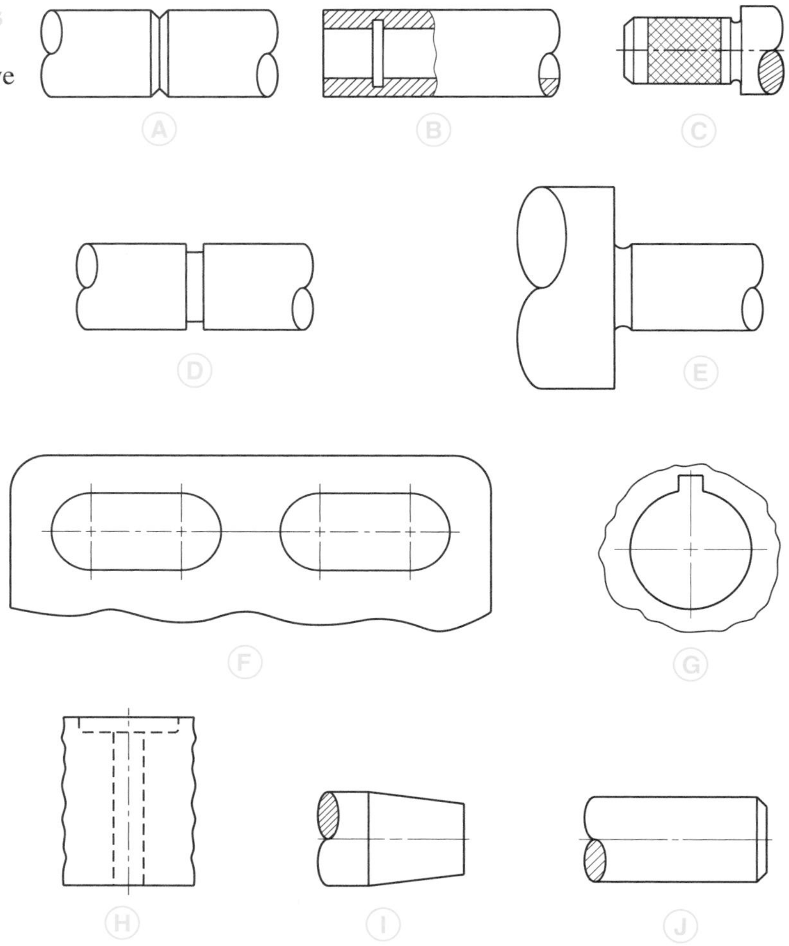

A B C D E F G H I J

Identification—Machining Processes

____________________ **1.** Grinding

____________________ **2.** Boring

____________________ **3.** Turning

____________________ **4.** Milling

____________________ **5.** Shaping

____________________ **6.** Drilling

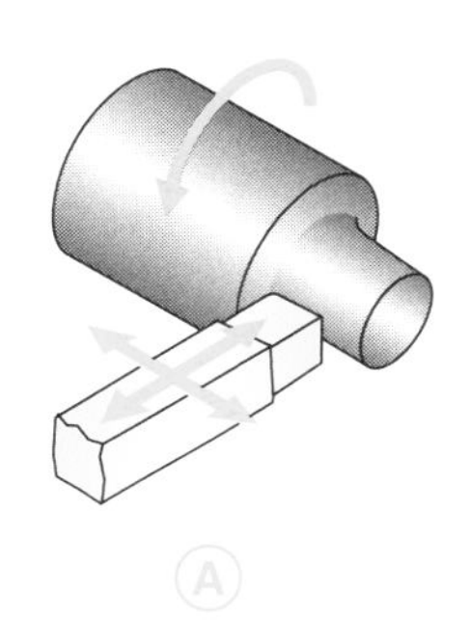
A

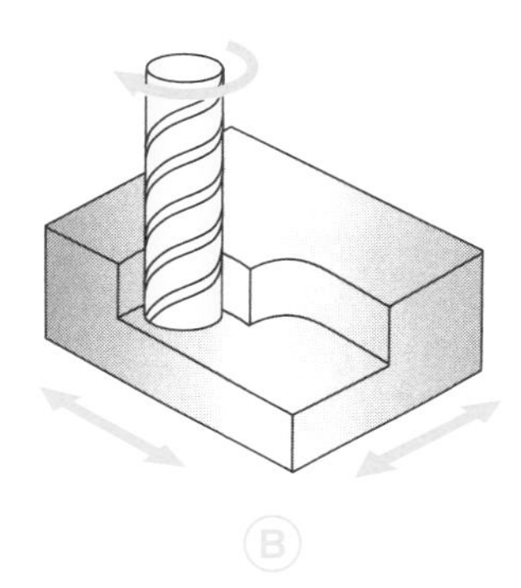
B

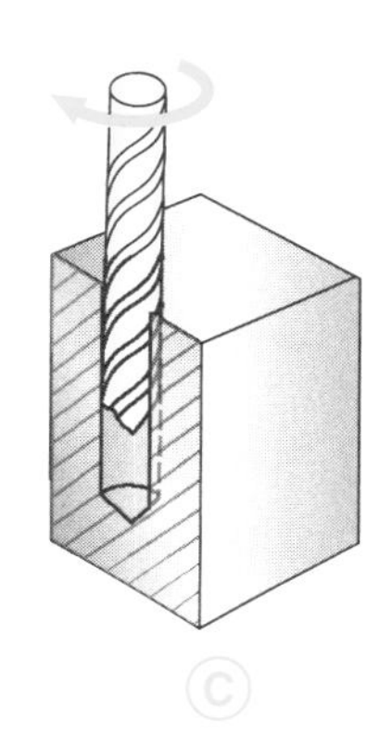
C

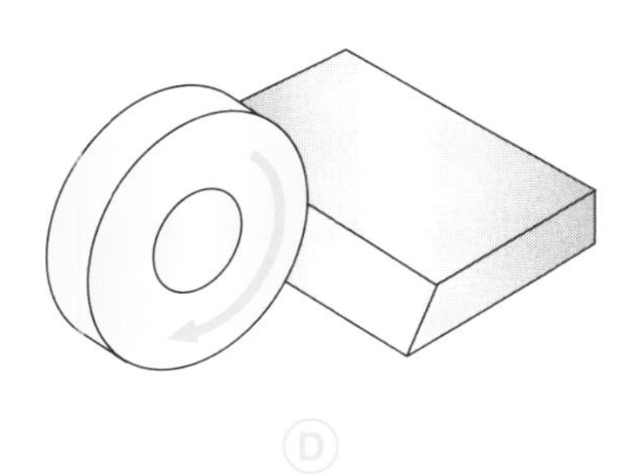
D

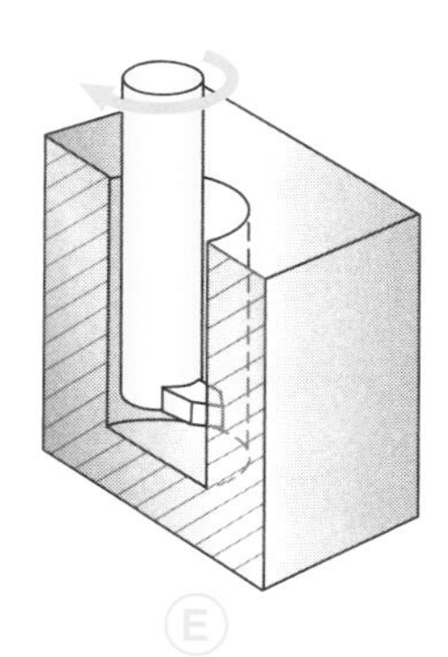
E

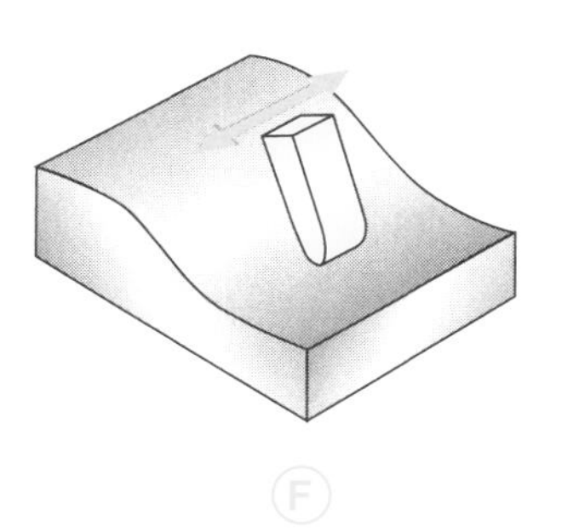
F

Identification—Forming Processes

______________________ **1.** Forging

______________________ **2.** Rolling

______________________ **3.** Casting

______________________ **4.** Stamping

______________________ **5.** Extruding

______________________ **6.** Drawing

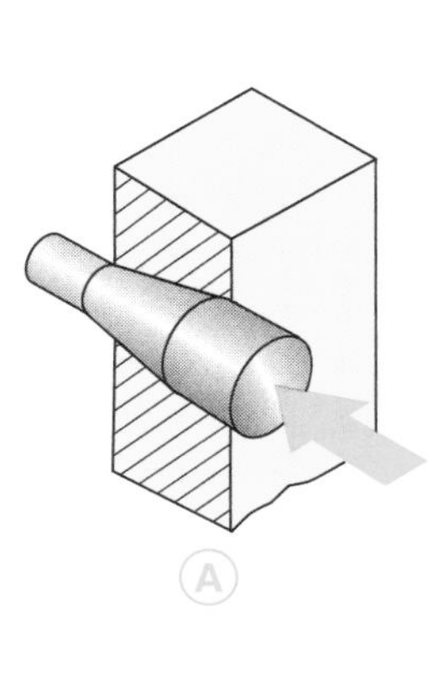

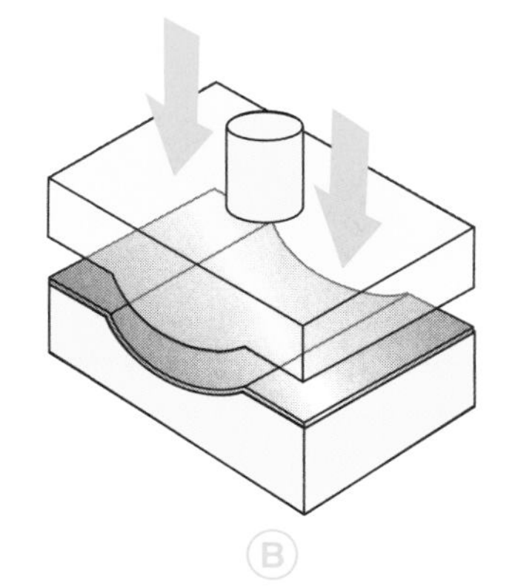

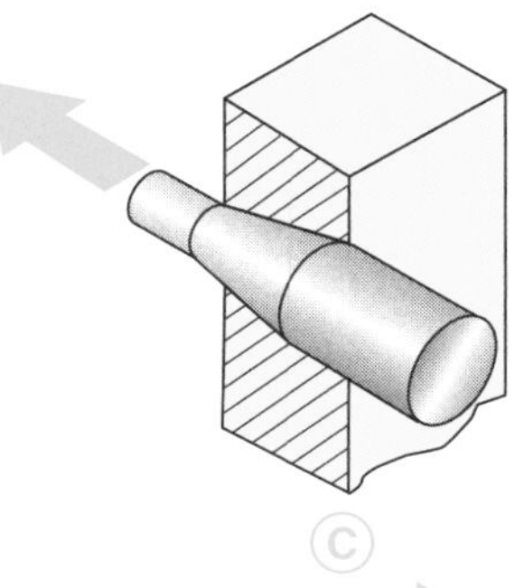

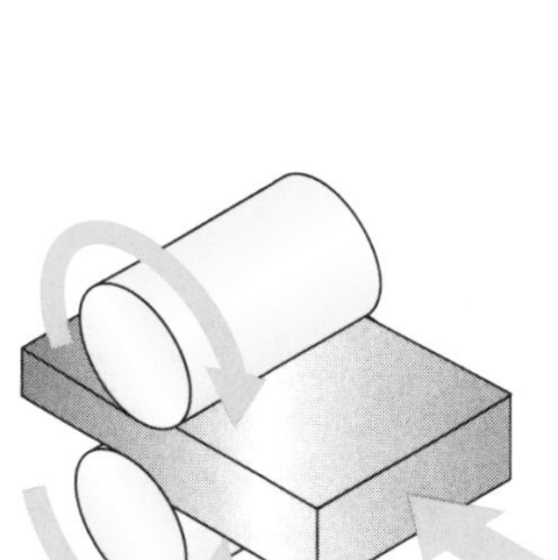

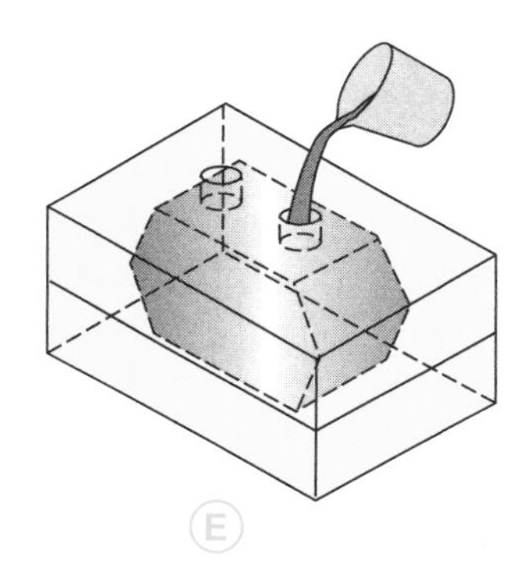

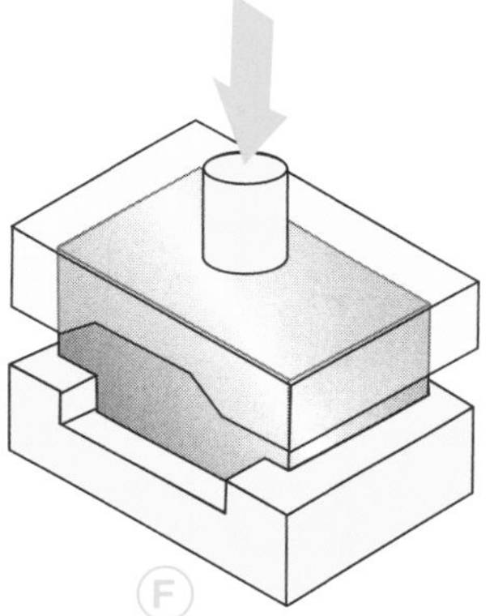

Multiple Choice

______ **1.** A ferrous metal is a metal that has ___ as the major alloying element.
A. aluminum
B. lead
C. copper
D. iron

______ **2.** A ___ property is the response of a material under applied loads.
A. chemical
B. physical
C. mechanical
D. thermal

______ **3.** ___ is a manufacturing process where a shaped cavity is filled with molten metal.
A. Stamping
B. Forging
C. Rolling
D. Casting

______ **4.** Stresses that build up in a part during the manufacturing process can be removed by ___.
A. tempering
B. hardening
C. annealing
D. carburizing

______ **5.** A(n) ___ is an elongated hole machined into the part.
A. neck
B. undercut
C. groove
D. slot

______ **6.** Cast iron is an alloy of iron and ___.
A. aluminum
B. carbon
C. chromium
D. zinc

______ **7.** A ___ is a plastic that softens when heat is applied and reforms into a solid when cooled.
A. thermoplastic
B. thermoset
C. hardened alloy
D. carburized material

______ **8.** Corrosion is a ___ property of a material.
A. mechanical
B. thermal
C. physical
D. chemical

_____________________ **9.** ___ is the ability of a material to deform under tensile forces without breaking or cracking.

A. Ductility
B. Toughness
C. Malleability
D. Strength

_____________________ **10.** ___ is the process of enlarging an existing hole slightly in order to improve its dimensional accuracy and surface quality to tighter tolerances.

A. Reaming
B. Knurling
C. Tapping
D. Drilling

TRADE COMPETENCY TEST 8

Name ______________________________ Date ____________________

Armature Shaft

Refer to print on page 207.

__________ **1.** The maximum end diameter of the Armature Shaft is ___.

__________ **2.** The knurled section of the Armature Shaft is ___″ long.

__________ **3.** All chamfers are specified for a(n) ___° angle.

__________ **4.** The total length of the Armature Shaft is ___″.

__________ **5.** The minimum diameter of the Armature Shaft at the small end is ___″.

__________ **6.** Chamfers are specified for ___″ from the ends of the shaft.

__________ **7.** The distance from the knurl to the small end is ___″.

__________ **8.** The maximum diameter of the knurl is ___″.

T F **9.** Both ends of the Armature Shaft must be drilled.

T F **10.** Note A_4 specifies material requirements.

T F **11.** End chamfers have been increased since the original drawing was made.

T F **12.** The minimum diameter of the large end is .5000″.

T F **13.** A diamond tooth knurl pattern is specified on the print.

__________ **14.** The knurl is specified for ___ TPI.

__________ **15.** The distance from the large end to the knurl is ___″.

__________ **16.** The Armature Shaft is drawing number ___.

T F **17.** The Armature Shaft requires heat treatment.

__________ **18.** The drawing was drawn to ___ scale.

__________ **19.** The gear specified requires a(n) ___″ full tooth.

__________ **20.** The minimum diameter for the knurl is ___″.

Center-Die

Refer to print on page 208.

______ **1.** Dimension E is ___″.

______ **2.** Radius K has a depth of ___″.

______ **3.** Chamfer I specifies a(n) ___° angle.

______ **4.** The maximum diameter of the Center-Die is ___″.

______ **5.** Radius K specifies a(n) ___″ radius.

______ **6.** The symbol at G specifies ___.

______ **7.** The datum located at ___ is the axis of the overall diameter of the Center-Die.

______ **8.** Radius M is ___″.

______ **9.** The minimum diameter of F is ___″.

______ **10.** Datum A must be ___ to datum B.

______ **11.** The number at D specifies a(n) ___ of .0005″.

______ **12.** Letter ___ is located at the centerpoint for radius M.

______ **13.** The depth of chamfer I is ___″.

______ **14.** The symbol at H specifies ___.

T F **15.** Datum B is used to specify the length of the Center-Die.

T F **16.** The small hole diameter is .802″.

T F **17.** The angle specified at L is 35°.

T F **18.** The depth of the small hole is .753″.

______ **19.** The symbol at A specifies ___.

______ **20.** Roughness height is specified for ___ microinches unless specified otherwise.

T F **21.** The symbol at B specifies a datum surface.

T F **22.** The surface of the large hole is finished to a roughness of 32 microinches.

______ **23.** Chamfer J has a specified dimension of ___.

______ **24.** The drawing was checked on ___.

______ **25.** The Center-Die is hardened to ___.

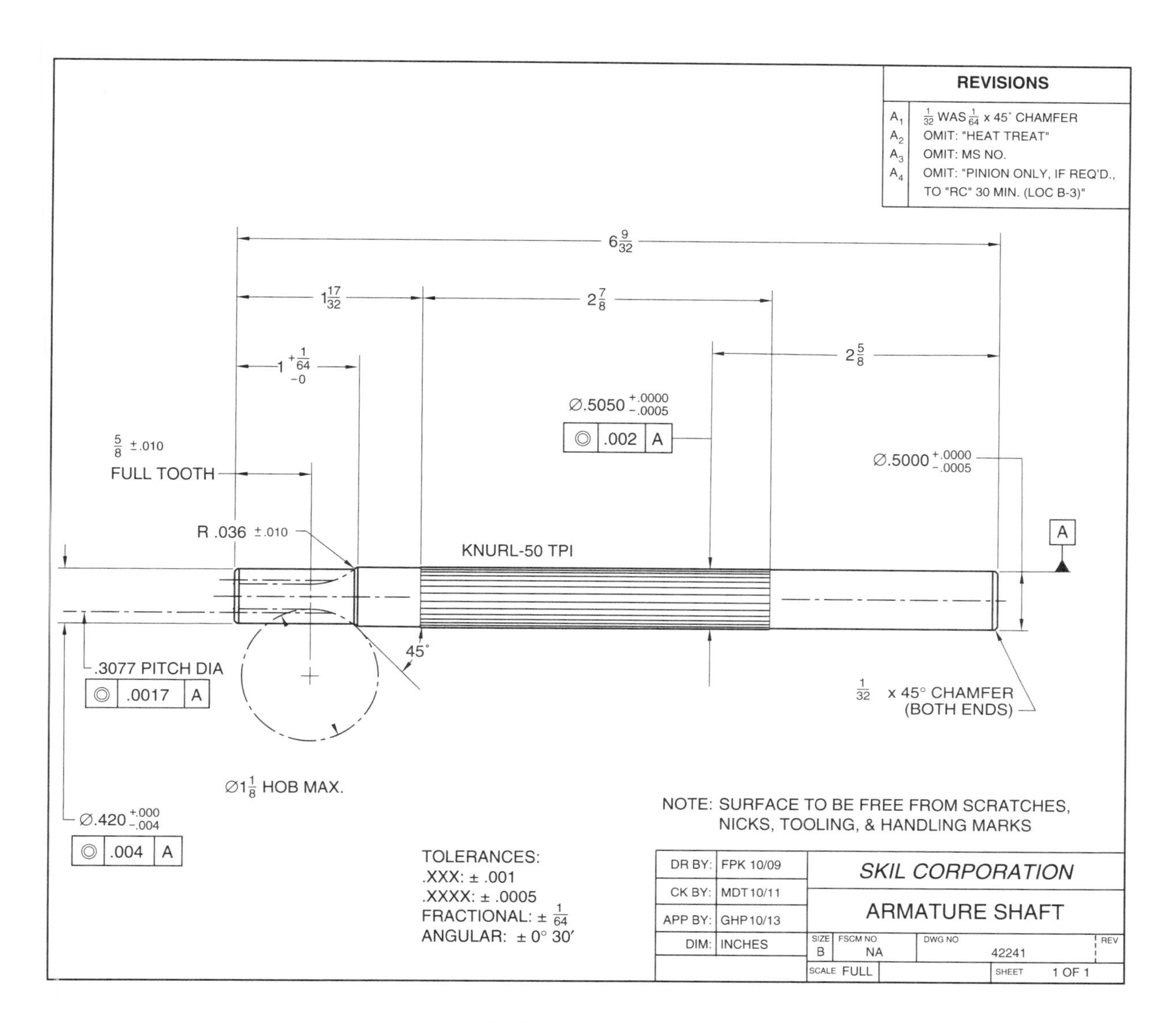

ARMATURE SHAFT

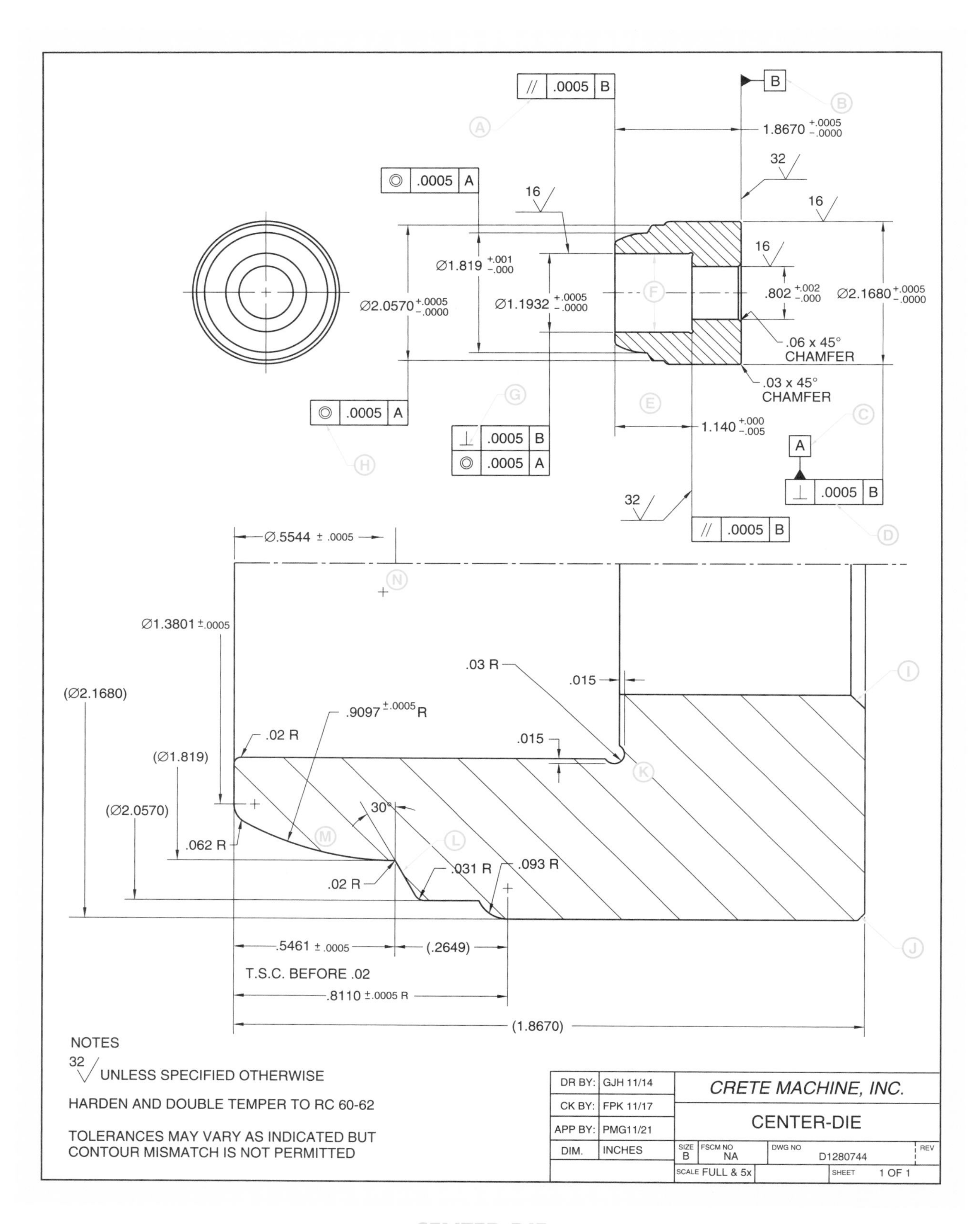

NOTES
32 UNLESS SPECIFIED OTHERWISE
HARDEN AND DOUBLE TEMPER TO RC 60-62
TOLERANCES MAY VARY AS INDICATED BUT
CONTOUR MISMATCH IS NOT PERMITTED
DR BY: GJH 11/14
CK BY: FPK 11/17
APP BY: PMG11/21
DIM. INCHES
CRETE MACHINE, INC.
CENTER-DIE
SIZE B
FSCM NO NA
DWG NO D1280744
REV
SCALE FULL & 5x
SHEET 1 OF 1
T.S.C. BEFORE .02
.06 x 45° CHAMFER
.03 x 45° CHAMFER

CENTER-DIE

Machine Trades
PRINTREADING

9 Threads and Fasteners

Threads are used for fastening, adjusting, and transmitting power. Fasteners are devices that join or fasten parts together. They may be threaded or nonthreaded. Threaded fasteners are easily installed or removed. Nonthreaded fasteners are more permanent. Threaded and nonthreaded fasteners must meet ANSI standards. Manufacturing prints may include specifications for the type, form, size, thread pitch, and material of the fasteners to be used in assemblies.

THREADED FASTENERS

A *threaded fastener* is a device that uses threads to join parts together. The most common types of threaded fasteners are bolts and screws. Threaded fasteners have several advantages. For example, threaded fasteners are commercially available in a variety of sizes, styles, strengths, and materials and are capable of joining similar or dissimilar materials. They are easily installed in the shop or the field with hand or power tools and are easily removed and replaced.

Threads

Threaded fasteners are based on the principle of an inclined plane wrapped around a cylinder. **See Figure 9-1.** A *thread* is a ridge of uniform section that forms a spiral on the internal or external surface of a cylinder or cone. A thread formed on a cylinder is a straight or parallel thread. A *taper thread* is a thread formed on a cone.

An *external thread* is a thread on the external surface of a cylinder or cone. Threads on bolts are external. An *internal thread* is a thread on the internal surface of a hollow cylinder or cone. Threads on nuts are internal. **See Figure 9-2.**

A *right-hand thread* is a thread that, when viewed axially, winds clockwise when receding. Viewed from the side, a right-hand thread always slopes up to the right. Right-hand threads are the most common. Threads are assumed to be right-handed unless otherwise specified.

A *left-hand thread* is a thread that, when viewed axially, winds counterclockwise when receding. Viewed from the side, a left-hand thread always slopes up to the left. Left-hand threads are generally only used in applications where a right-hand thread would be subject to spin off and become disengaged. All left-hand threads are designated LH.

The most common application for threads is fastening parts together. Another use is for power transmission and linear displacement. Threads used in this manner are called power screws. A rotating threaded rod will cause internally threaded parts such as nuts to move along the rod's length. Milling tables and lead screws on lathes use threads in this way.

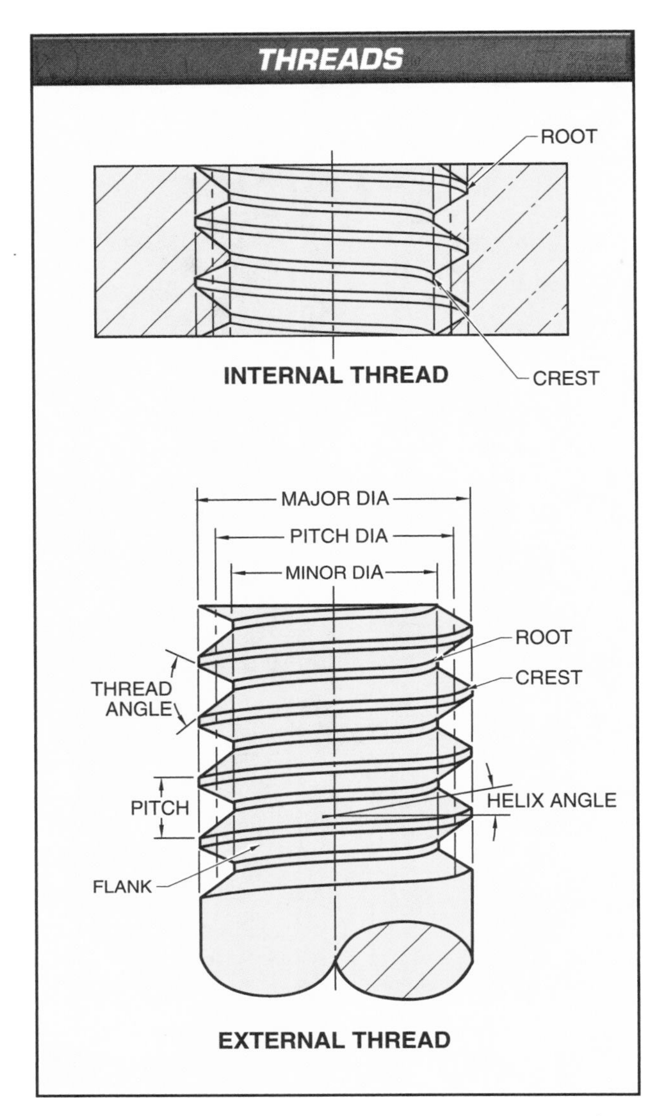

Figure 9-1. A thread is a ridge of uniform section in the form of a helix on the internal or external surface of a cylinder.

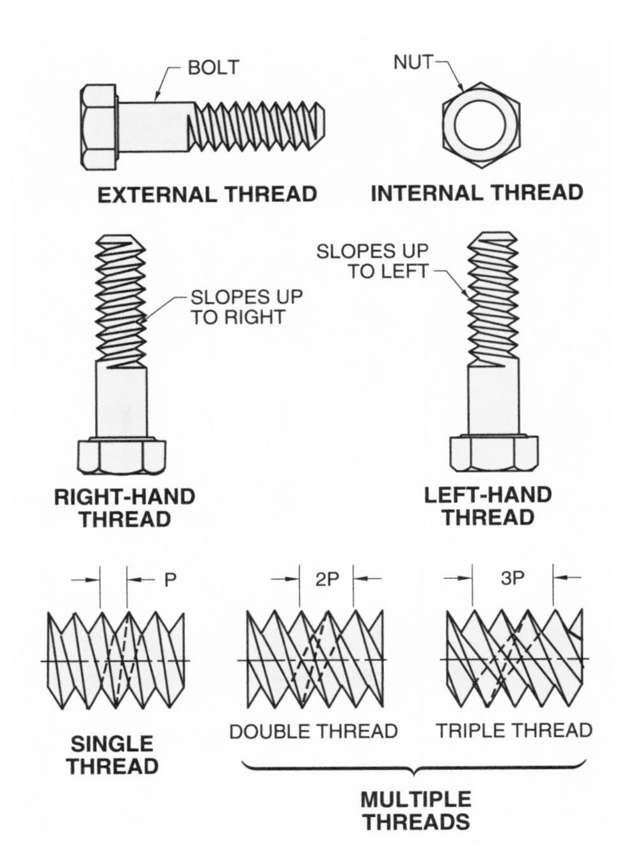

Figure 9-2. Threads are external or internal, right- or left-handed, and single or multiple.

A fastener with a single thread has only one thread ridge wrapped around the cylinder or cone. With each revolution, single-threaded fasteners advance a distance equal to the thread spacing. A fastener with multiple threads has two or more thread ridges wrapped around the cylinder or cone. With each revolution, multiple-threaded fasteners advance a distance equal to an integer multiple of the thread spacing. For example, double-threaded fasteners advance two threads per revolution and triple-threaded fasteners advance three threads per revolution. The *lead* is the distance a threaded fastener advances per revolution.

Thread Dimensions

The dimensions of screw threads are based on the pitch and the diameter. Pitch determines the number of threads per inch. The diameter determines the width of the threaded form. **See Figure 9-3.**

Pitch is the distance between corresponding points on adjacent raised projections. The reciprocal of pitch (when measured in inches) is threads per inch (TPI). The *thread angle* is the angle between the flanks of a thread, measured in an axial plane.

The *major diameter* is the diameter of the imaginary coaxial cylinder that bounds the crest of an external thread or the root of an internal thread. On a taper thread, it is the diameter of the major cone at a given position on the thread axis.

The *minor diameter* is the diameter of the imaginary coaxial cylinder that bounds the root of an external thread or the crest of an internal thread. On a taper thread, it is the diameter of the minor cone at a given position on the thread axis.

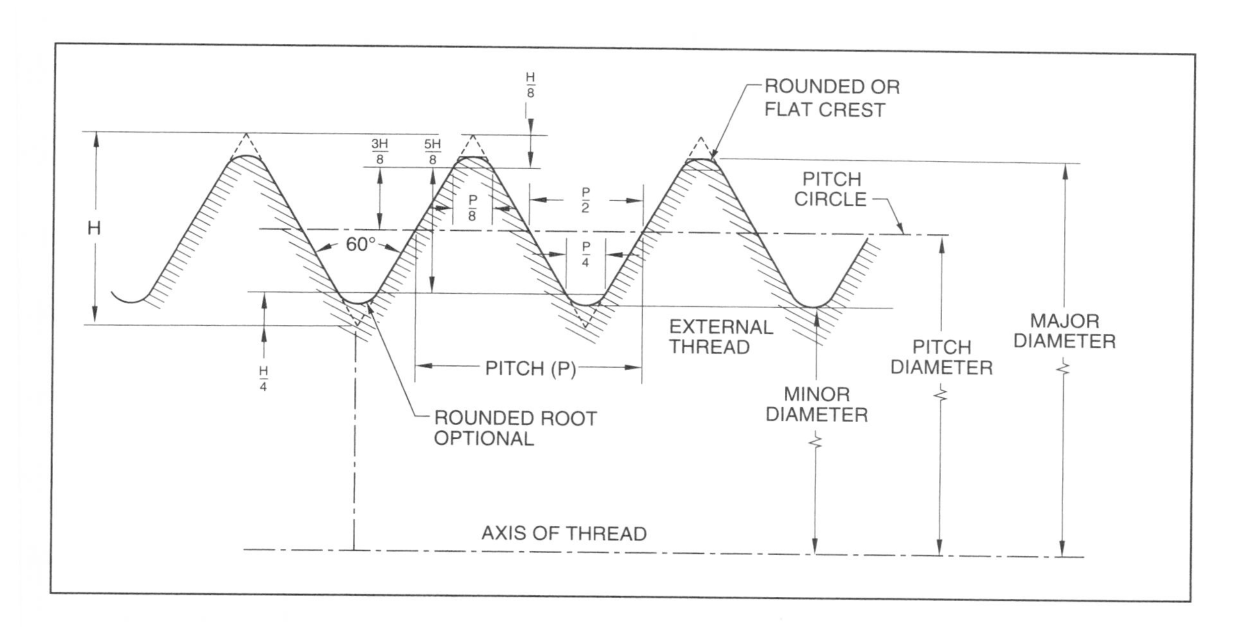

Figure 9-3. Thread profiles define the shapes and dimensions of the thread forms.

The *pitch circle* is the imaginary circle located approximately halfway between the tops and roots of raised projections. The *pitch diameter* is the diameter of the pitch circle. On a taper thread, it is the diameter of the pitch cone at a given position on the thread axis.

In a thread assembly, the length of thread engagement is the distance between the extreme points of contact on the pitch cylinders or cones of two mating threads measured parallel to the axis. *Crest clearance* is the distance, measured perpendicular to the axis, between the crest of a thread and the root of its mating thread.

Thread Forms

A *form* is a thread's shape profile in an axial plane. The *flank* is either surface of a thread that connects its crest with its root. The *crest* is the surface joining the flanks that is farthest from the thread's axis. The *root* is the surface joining the flanks that is identical in position with, or immediately adjacent to, the cylinder or cone from which the thread projects. Crests and roots are comparable to the peaks and valleys of mountain ranges, with flanks as the sloping sides.

A *complete (full) thread* is a thread having full form at both crest and root. An *incomplete thread* is a thread that is not fully formed. This occurs when the thread intersects the edge or chamfer of a workpiece. A *blunt start* is the removal of the partial thread at the starting end of the thread. This is a feature of threaded parts that are repeatedly assembled by hand, such as hose couplings and thread plug gauges. It also prevents cutting hands and crossing threads.

Threaded fasteners rely on uniform and standardized sizes and thread forms in order to be interchangeable. The process of thread standardization has resulted in several lasting thread forms. The most common thread forms are the unified, metric (M), Acme, buttress, and knuckle forms. **See Figure 9-4.** The application determines which thread profile is most appropriate. Many other forms exist, but are limited to special applications.

Internal and external threads must be of the same size, pitch, and form to be assembled together.

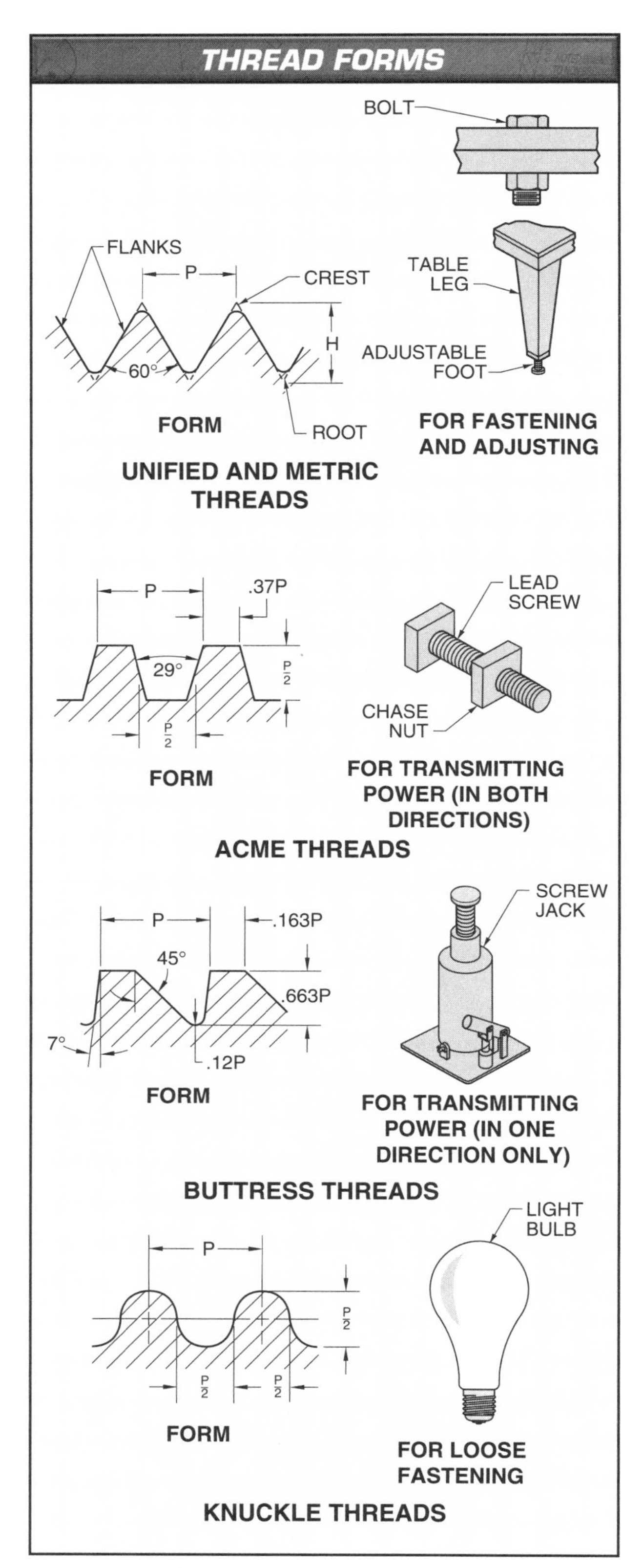

Figure 9-4. Thread forms may be unified, metric, Acme, buttress, or knuckle.

Unified Threads. The Unified Screw Thread Standard was developed by committees from the United States, Great Britain, and Canada based on a 60° standard thread angle. Subsequent developments led to the Unified National Inch Screw Thread Form, which includes the profiles of Unified National (UN) and Unified National Rounded (UNR) threads. These two are the same, except that the roots and crests of UNR threads are rounded. **See Appendix.** Unified threads are used for a variety of general-purpose fastening and adjusting.

Metric Threads. Metric threads share the same basic profile as UN threads, but differ in diameter and thread pitch, which are based on the metric system. Metric threads are also for general-purpose use. Metric threads are the primary system in most countries besides the United States and are defined in ISO 68-1, *Metric Screw Threads*. With metric threads becoming more common in American manufacturing, ASME International has developed the American National Standard B1.13M, *Metric Screw Threads,* which closely follows the ISO standard.

Acme Threads. Acme threads were developed from a square thread form, which is particularly well suited to move or translate heavy loads. The square thread is more efficient than the Acme form for this purpose, but is very difficult to cut with its parallel flanks. Acme threads are slightly less efficient, but are slightly stronger, easier to manufacture, and can be adjusted for wear.

Buttress Threads. Buttress threads combine the efficient heavy-load capacity of a square thread with the advantages of the Acme thread. This results in an asymmetrical thread form, so the buttress thread can only provide force in one direction.

Knuckle Threads. Knuckle threads are used for relatively loose fastening of parts such as lightbulb sockets and glass jar lids. Instead of being cut into materials, knuckle threads are typically rolled into sheet metal or formed into cast parts.

Thread Series

A *thread series* is a group of diameter-pitch combinations for certain thread form. Thread series are distinguished from one another by the number of threads per inch for a series of specific diameters. For example, the standard series is a screw thread series of coarse (UNC/UNRC), fine (UNF/UNRF), and extra-fine (UNEF/UNREF) graded pitches and eight series with constant pitches. **See Figure 9-5.**

STANDARD SERIES THREADS — GRADED PITCHES						
NOMINAL DIAMETER*	UNC/UNRC		UNF/UNRF		UNEF/UNREF	
	TPI	TAP DRILL	TPI	TAP DRILL	TPI	TAP DRILL
0 (.0600)			80	3/64		
1 (.0730)	64	No. 53	72	No. 53		
2 (.0860)	56	No. 50	64	No. 50		
3 (.0990)	48	No. 47	56	No. 45		
4 (.1120)	40	No. 43	48	No. 42		
5 (.1250)	40	No. 38	44	No. 37		
6 (.1380)	32	No. 36	40	No. 33		
8 (.1640)	32	No. 29	36	No. 29		
10 (.1900)	24	No. 25	32	No. 21		
12 (.2160)	24	No. 16	28	No. 14	32	No. 13
1/4 (.2500)	20	No. 7	28	No. 3	32	7/32
5/16 (.3125)	18	F	24	I	32	9/32
3/8 (.3750)	16	5/16	24	Q	32	11/32
7/16 (.4375)	14	U	20	25/64	28	13/32
1/2 (.5000)	13	27/64	20	29/64	28	15/32
9/16 (.5625)	12	31/64	18	33/64	24	33/64
5/8 (.6250)	11	17/32	18	37/64	24	37/64
11/16 (.6875)					24	41/64
3/4 (.7500)	10	21/32	16	11/18	20	45/64
13/16 (.8125)					20	49/64
7/8 (.8750)	9	49/64	14	13/16	20	53/64
15/16 (.9375)					20	57/64
1 (1.000)	8	7/8	12	59/64	20	61/64

* in in.

Figure 9-5. Thread series are groups of diameter-pitch combinations.

A *graded pitch series* is a standard thread series with a different number of threads per inch for most diameters. Generally, the smaller diameters of a graded pitch series have more threads per inch. For example, a number 2 (0.0860″) UNC thread has 56 threads per inch while a 3/4″ (0.7500″) UNC thread has 10 threads per inch. Graded pitch series are the most common series of threads.

A *constant pitch series* is a standard thread series with a set number of threads per inch for a range of diameters. Common constant pitch series have 4, 6, 8, 12, 16, 20, 28, or 32 threads per inch. The series is designated by the pitch and thread form, such as 8-UN for a UN series of diameters that all have a pitch of 8 threads per inch.

A given diameter may be included in several series and, therefore, could have several possible pitches. For larger diameters, there are fewer threads per inch in the range of possible series. For example, a 2½″ diameter bolt may have 4, 6, 8, 12, 16, or 20 threads per inch, while a 1/4″ diameter bolt may have 20, 28, or 32 threads per inch.

A *special series* is a screw thread series with combinations of diameter and pitch not in the standard screw thread series. Preference is given to standard series coarse- and fine-graded pitch threads.

Pitch is the distance between corresponding points on adjacent thread forms. Pitch is always measured parallel to the axis. To calculate pitch, the following formula is used:

$P = 1/N$

where

P = pitch (in in.)

N = threads per inch

For example, what is the pitch of a thread form having 16 threads per inch?

$P = 1/N$

$P = 1/16$

$P =$ **0.0625 in.**

Metric threads are measured in millimeters. The same formula can be used to determine the approximate pitch for metric threads. Metric thread series may also be either graded pitch or constant pitch.

Thread Classes

The class of a thread indicates its tolerance and allowance. Together, tolerance and allowance determine the fit of mating threads, loose or tight. The thread classes have been developed for the thread forms based on desired fit.

Unified Threads. Threaded classes in the Unified Standard are designated by a numeral followed by the letter A for external threads and B for internal threads. The three classes of external threads are 1A, 2A, and 3A. The three classes of internal threads are 1B, 2B, and 3B.

Classes 1A and 1B have the greatest amount of allowance. They are intended for applications that require frequent and rapid assembly and disassembly with minimum binding, even with slightly damaged or dirty threads.

Classes 2A and 2B are considered standard for general-purpose threads on bolts, nuts, and screws. They provide standard allowances to ensure minimum clearance between external and internal threads, which minimizes galling and seizing in high-cycle wrench assembly. Classes 2A and 2B are widely used for mass-production purposes.

Classes 3A and 3B are suitable for applications requiring closer tolerances than those provided by classes 2A and 2B. They are designated for set screws, socket head cap screws, aircraft bolts, and higher strength materials where it is necessary to limit the variations of the thread elements.

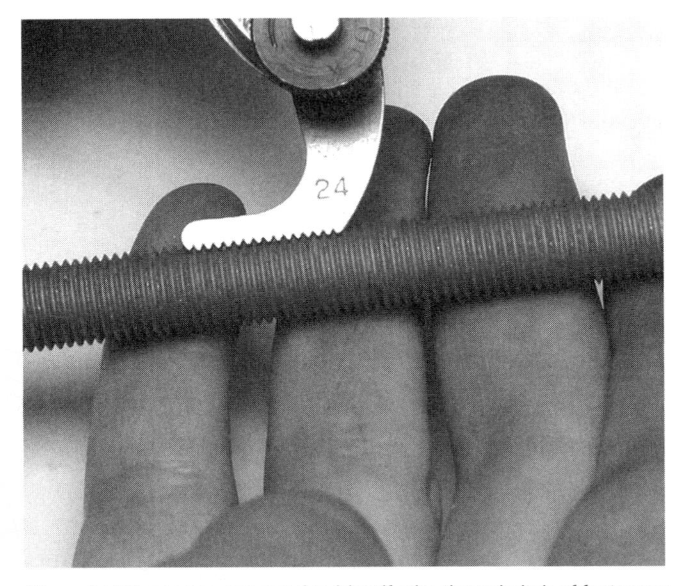

Thread pitch gauges are used to identify the thread pitch of fasteners and threaded parts.

The requirements for thread fits depend upon their end use. Combination of thread classes for fasteners is possible. For example, a Class 2A external thread may be used with a Class 1B, 2B, or 3B internal thread. Cost generally increases proportionately to the accuracy required. For economy, no closer thread fit should be used than is needed for the proper functioning of the components.

Metric Threads. Metric threads follow a combination number and letter system for classifying thread tolerances and fit. The number specifies the tolerance grade and ranges from 3 to 9, depending on the application. Smaller numbers indicate tighter tolerances, which is needed for critical applications. The most common grades are 4, 5, and 6, for general-purpose applications. The letter specifies the fundamental deviation, which is the distance of the tolerance from the basic size of the thread profile. Capital letters G and H are used for internal threads, and lowercase letters e, f, g, and h are used for external threads. A metric thread note may include one or more number and letter combinations.

Acme Threads. ANSI lists two types of Acme threads, the general purpose (G) and the centralizing (C). The general-purpose Acme threads have three classes: 2G, 3G, and 4G. These classes provide a relatively loose fit for the frequent rotation of the threaded parts. Centralizing threads have five classes: 2C, 3C, 4C, 5C, and 6C. The centralizing threads have limited clearance for the major diameter, which helps maintain proper alignment of the thread axes. It is not recommended to use combinations of classes for internal and external Acme threads.

Buttress Threads. Three classes have been standardized for buttress threads: Class 1 (free), Class 2 (standard), and Class 3 (precision).

Thread Designations

Threads are designated on prints by thread notes. **See Figure 9-6.** The thread note specifies in sequence the nominal size, number of threads per inch or pitch, thread form and series, and thread class. For example, the thread note ½-13 UNC-2A specifies ½″ nominal diameter, 13 threads per inch, Unified National Coarse thread form, Class 2 fit, and external thread. For a metric example, the thread note M6×1–6g specifies M profile, 6 mm nominal diameter, 1 mm pitch, external thread, and 6g tolerance class.

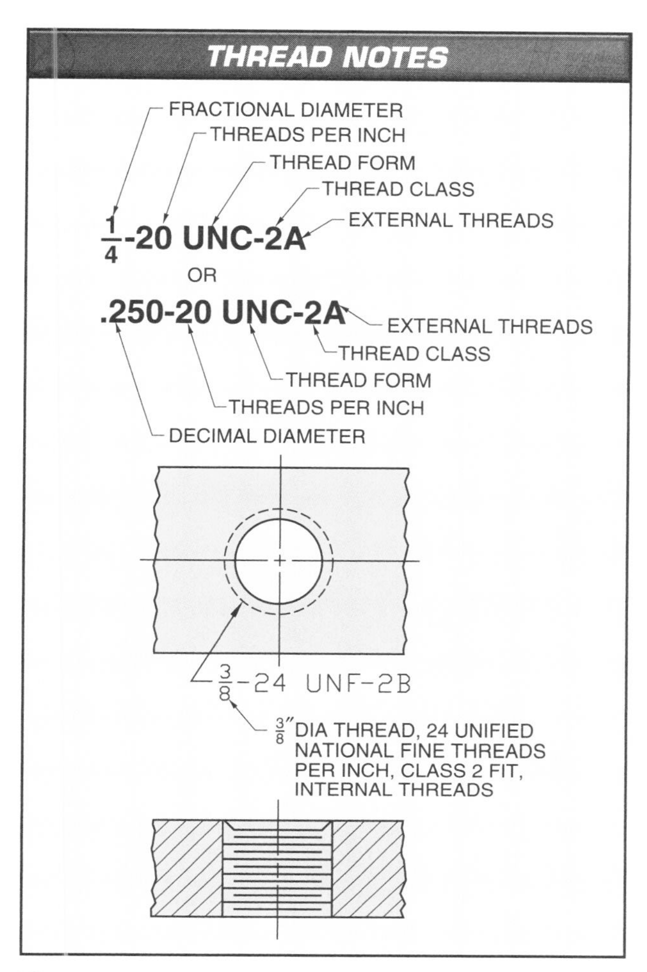

Figure 9-6. Threads are designated by thread notes.

The nominal size of the diameter may be stated in fractional or decimal dimensions or by number. Diameters smaller than ¼″ are numbered from 12 to 0, with smaller numbers indicating smaller diameters. The letters LH follow the thread note for left-hand threads. If not specified, the thread is a right-hand thread. The thread form designations begin with UN (for Unified), M (for metric), ACME, or BUTT (for buttress threads). The length of a threaded fastener may be included at the end of the thread note.

Thread Representation

The representation of threads in drawings is based on the conventions in ANSI/ASME Y14.6, *Screw Thread Representation*. Threads are represented on drawings by three methods: simplified representation, schematic representation, and detailed representation. **See Figure 9-7.**

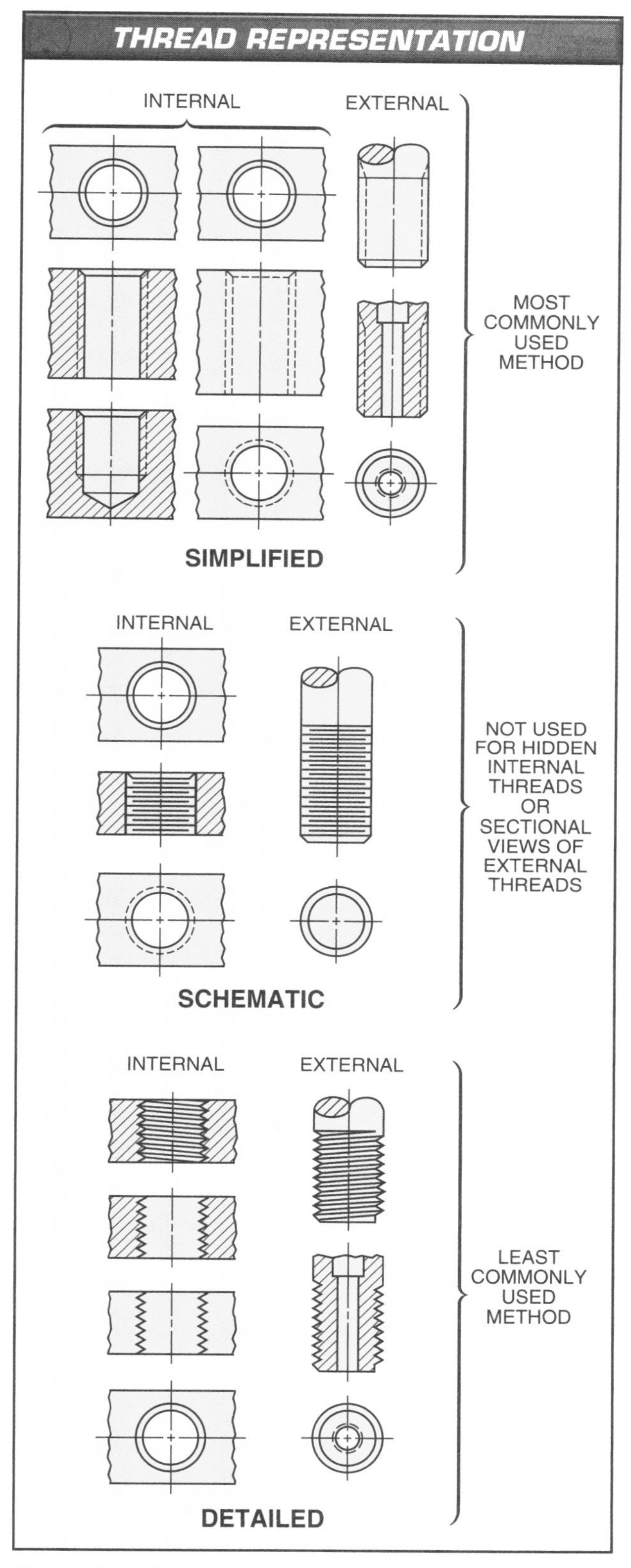

Figure 9-7. Thread representation is a method of drawing that shows a threaded part.

Typically, only one method of thread representation is used throughout a drawing, although more than one method may be used on the same drawing if this improves clarity. **See Figure 9-8.** End purpose and use of drawings, drafting time, and other factors influence the selection and use of the conventions.

Simplified Representation. *Simplified representation* is a method of thread representation in which hidden lines are drawn parallel to the axis at the approximate depth (minor diameter) of the thread. Simplified representation is the most commonly used method of thread representation. Various combinations of internal, external, and sectional views of threads are shown with this method.

Schematic Representation. *Schematic representation* is a method of thread representation in which solid lines perpendicular to the axis represent roots and crests. This method is not used for hidden internal threads or sectional views of external threads.

Threaded fasteners are available in a wide variety of sizes, threads, head styles, and materials.

Detailed Representation. *Detailed representation* is a method of thread representation in which the thread profiles are drawn realistically. Detailed representation is the least commonly used method of thread representation because it is time-consuming to draw.

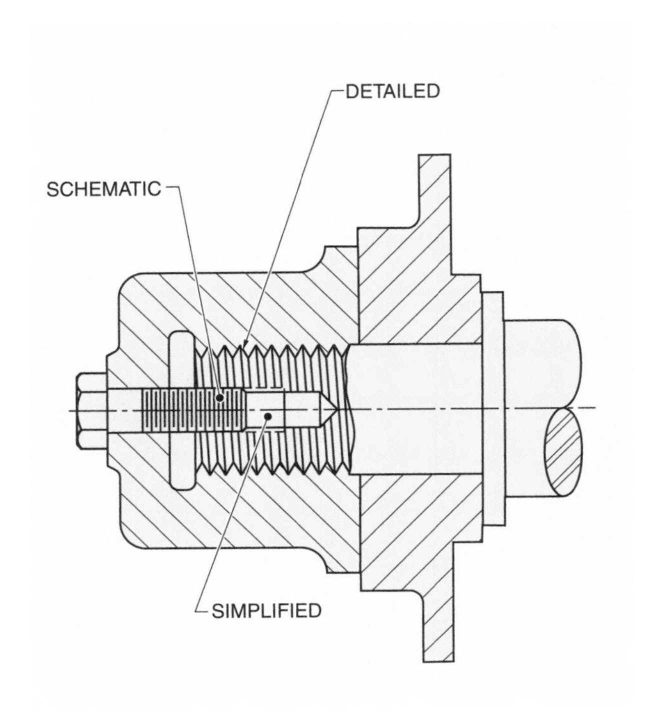

Figure 9-8. Simplified, schematic, and detailed thread representation conventions may be used on a single drawing.

Bolts, Screws, and Nuts

Bolts, screws, and nuts are purchased parts. Consequently, they are generally shown on prints as simplified thread representations with specification notes. Bolts and screws are available in a wide range of sizes (diameters and lengths), strengths, head styles, materials, and drives. **See Figure 9-9.** The *drive* is the shape of the recess on the fastener head that fits the tool used to rotate the fastener.

The length of a bolt or screw is the distance from the bearing surface of the head to the tip, measured parallel to the axis. Usually only part of the length is threaded. The thread length for standard bolts is generally twice the diameter plus ¼″ for bolts up to 6″. For bolts over 6″, the thread length is generally twice the diameter plus ½″.

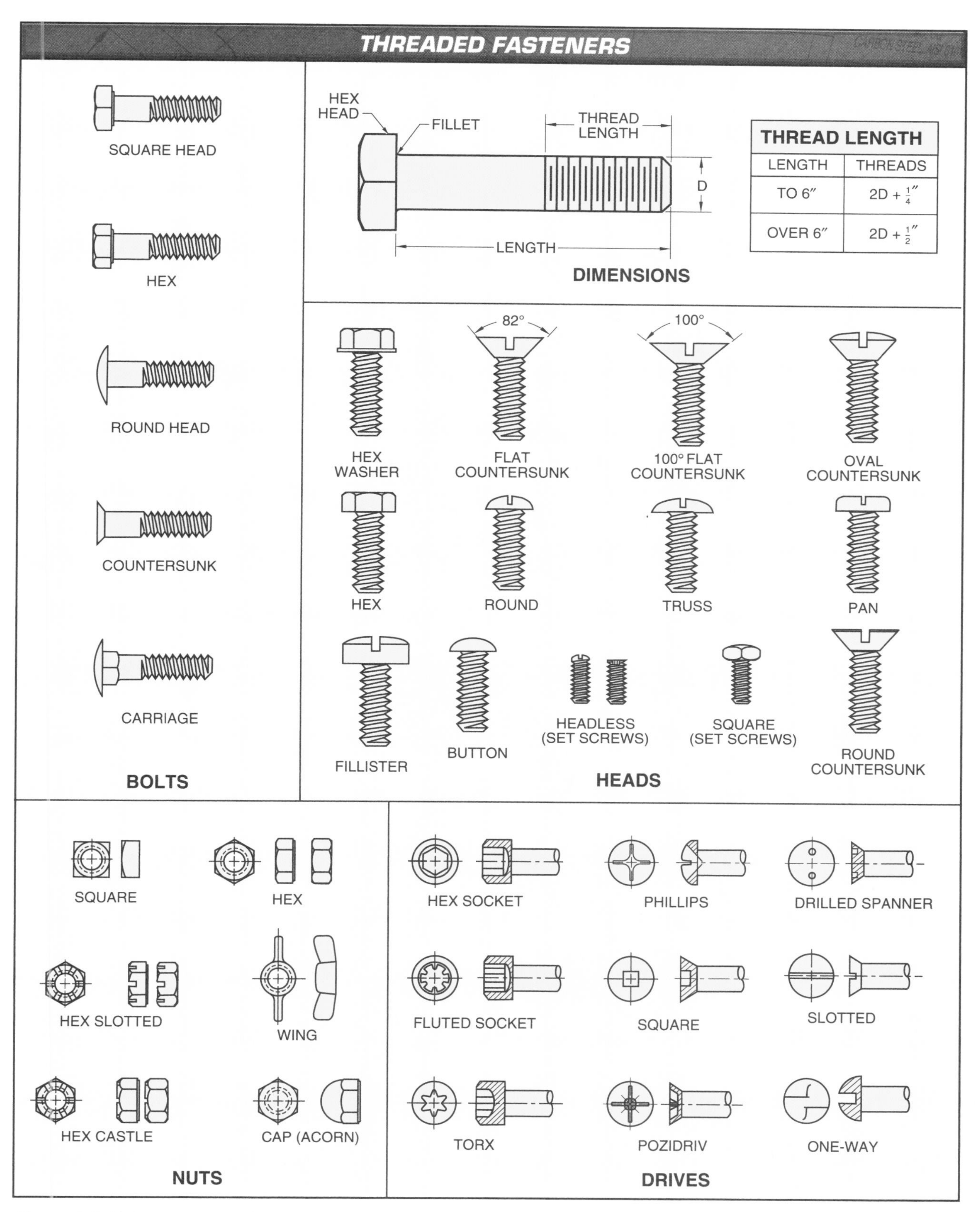

Figure 9-9. Bolts, screws, and nuts are threaded fasteners.

Nuts are either square (having four sides) or hexagonal (having six sides). The distance across the flats corresponds to standard dimensions to facilitate driving with wrenches and ratchet sockets.

The Society of Automotive Engineers (SAE) and ASTM International have established standards for classifying bolts and screws into grades based on their tensile strength and yield strength. Markings of radial lines on the bolt head indicate the grade. **See Figure 9-10.** This information must be referenced when the fastener is initially used and during service and repair of an assembled component. Threaded fasteners that are replaced must match the characteristics of the original threaded fastener.

Washers

The three basic types of washers are plain, spring lock, and tooth lock. All three types are available in standard sizes for use with standard bolts and screws.

Plain Washers. Plain washers are round and flat. They are used under the head of a screw or bolt or under a nut to spread a load over a greater area. They are also used to prevent the marring of the parts during assembly as a result of turning the screw, bolt, or nut. **See Figure 9-11.**

Spring Lock Washers. Spring lock washers are split on one side and helical in shape. These washers have several functions. As springing devices, they provide good bolt tension and protect against looseness developed from vibration or corrosion. **See Figure 9-12.** Spring lock washers also act as hardened bearing surfaces and provide uniform load distribution.

Lock wire is another method of preventing fasteners from loosening. It is also called safety wire because it prevents critical components from unfastening. Lock wire is used with nuts or bolt heads that have been drilled with a small hole. The wire is threaded through the hole, twisted tightly to remove any slack, threaded through or around a stationary component, and twisted together at the ends. Fasteners can also be lock-wired together in chains. Lock-wired fasteners cannot be removed, purposefully or accidentally, without cutting the wire.

BOLT GRADE MARKINGS

BOLT HEAD MARKING	SAE/ASTM DEFINITIONS	MATERIAL	MINIMUM TENSILE STRENGTH*
NO MARKS	SAE Grade 1 SAE Grade 2 Indeterminate quality	Low carbon steel	65,000
3 MARKS	SAE Grade 5 ASTM A 449 Common commercial quality	Medium carbon steel, quenched and tempered	120,000
5 MARKS	SAE Grade 7	Medium carbon alloy steel, quenched and tempered, roll threaded after heat treatment	133,000
6 MARKS	SAE Grade 8 ASTM A 354 Best commercial quality	Medium carbon alloy steel, quenched and tempered	150,000

* in psi

Figure 9-10. Bolt head markings indicate grade.

PLAIN WASHERS

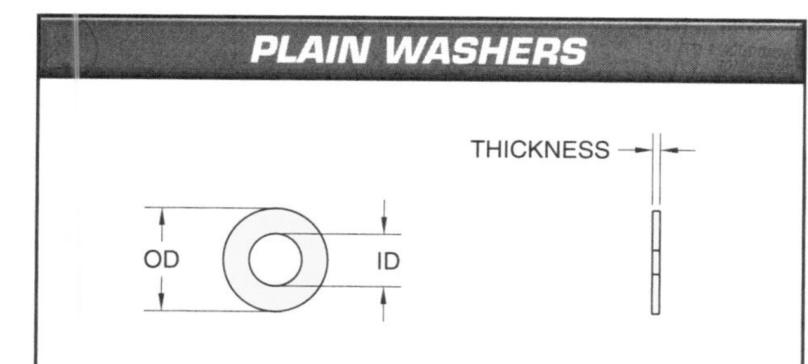

TYPE	BOLT SIZE*	ID*	OD*	THICKNESS*
ZINC	1/4	5/16	3/4	1/16
	5/16	3/8	7/8	5/64
	3/8	7/16	1	5/64
	7/16	1/2	1 1/4	5/64
	1/2	9/16	1 3/8	7/64
	5/8	11/16	1 3/4	9/64
	3/4	13/16	2	5/32
	7/8	15/16	2 1/4	11/64
	1	1 1/16	2 1/2	11/64
STAINLESS STEEL	#4	0.125	0.312	0.031
	#6	0.149	0.375	0.031
	#8	0.174	0.375	0.031
	#10	0.203	0.437	0.031
	1/4	9/32	5/8	0.050
	5/16	11/32	3/4	0.050
	3/8	13/32	7/8	0.050
	1/2	17/32	1 1/4	0.062
	5/8	11/16	1 1/2	0.078
SAE ZINC	#8	3/16	1/2	3/64
	#10	7/32	1/2	3/64
	1/4	9/32	5/8	1/16
	5/16	11/32	11/16	1/16
	3/8	13/32	13/16	1/16
	7/16	15/32	59/64	1/16
	1/2	17/32	1 1/16	3/32
	9/16	19/32	1 5/32	3/32
	5/8	21/32	1 5/16	3/32
	3/4	13/16	1 1/2	9/64
	7/8	15/16	1 3/4	9/64
	1	1 1/16	2	9/64
FENDER	3/16	17/64	1	3/64
	1/4	17/64	1	3/64
	1/4	9/32	1 1/4	3/64
	1/4	9/32	1 1/2	3/64
	5/16	11/32	1 1/4	3/64
	5/16	11/32	1 1/2	3/64
	3/8	13/32	1 1/4	3/64
	3/8	13/32	1 1/2	3/64
	1/2	17/32	2	3/64
SAE GRADE ZINC	1/4	9/32	5/8	1/16
	5/16	11/32	11/16	1/16
	3/8	13/32	13/16	1/16
	7/16	15/32	59/64	1/16
	1/2	17/32	1 1/16	3/32
	5/8	21/32	1 5/16	3/32
	3/4	13/16	1 1/2	9/64
	7/8	15/16	1 3/4	9/64
	1	1 1/16	2	9/64

* in in.

Figure 9-11. Plain washers are round and flat.

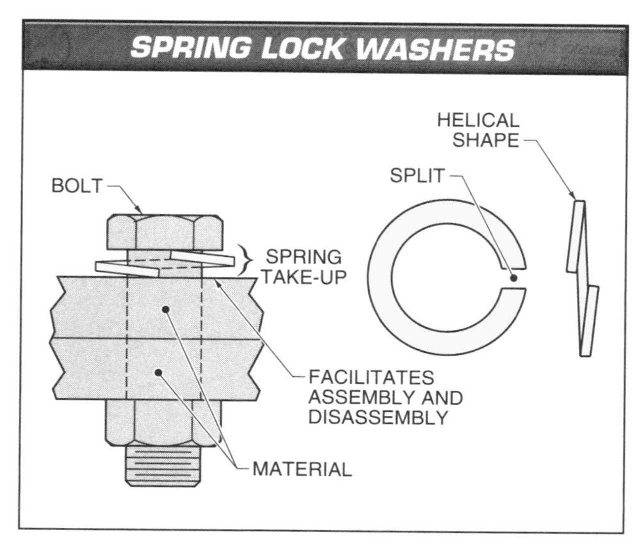

Figure 9-12. Spring lock washers are split and helical in shape.

Tooth Lock Washers. Tooth lock washers have hardened teeth along an edge that are twisted offset to bite or grip both the work surface and the bolt head or nut. This locks fasteners to the assembly or increases the friction between the fastener and the assembly. The teeth can be external, internal, internal-external, or countersunk external. **See Figure 9-13.** They also make good electrical contacts. Unlike spring lock washers, they do not provide spring action to counteract wear or stretch in the parts of an assembly.

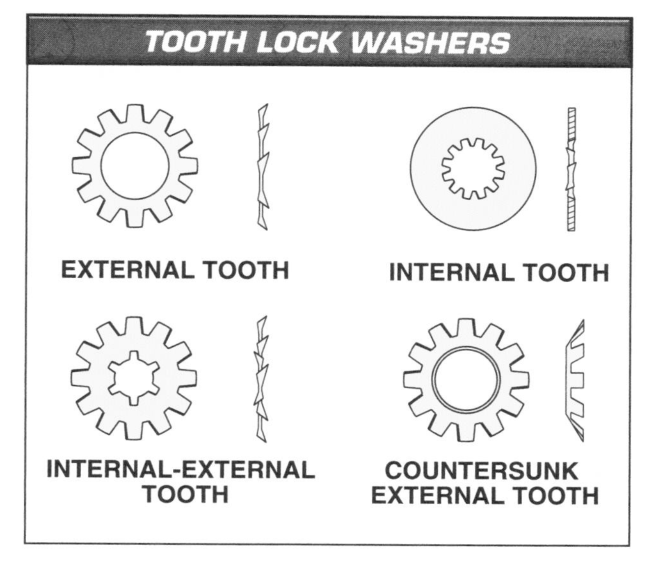

Figure 9-13. Tooth lock washers have teeth that are offset to bite or grip the bolt or nut and the work surface.

The external tooth lock washer is the most commonly used of the tooth lock washers. The internal tooth lock washer is used where it is necessary to consider appearance and to ensure engagement of teeth with the bearing surface of the fastener.

Where additional locking ability is required or there is need for a large bearing surface, such as over a clearance hole, the internal-external tooth lock washer may be used. Countersunk external tooth lock washers are used with flat head and oval head machine screws.

Thread Lock Coatings. A *thread lock coating* is a liquid coating applied to a threaded fastener to prevent the loosening of assembled parts from vibration, shock, and/or chemical leakage. **See Figure 9-14.** Thread lock coatings can be used in place of lock washers. The thread lock coating selected is determined by the fastener and application. Special high-temperature and chemical-resistant thread lock coatings are available. Fasteners can also be purchased with the threads precoated.

I.T.W. Devcon

Figure 9-14. Thread lock coating is applied to threads to prevent loosening of fastened parts.

Pipe Threads

The two standard forms of pipe threads are regular and Dryseal. Regular pipe thread is the standard for the plumbing trade. Dryseal pipe thread is the standard for automotive fittings, refrigeration and hydraulic tube and pipe fittings, lubrication fittings, and drain cocks.

Regular pipe thread forms allow crest and root clearance when the flanks contact. Leakage occurs if this clearance is not filled. With Dryseal pipe thread forms, there is no crest and root clearance and sealer is not needed.

Regular and Dryseal threads come in two forms, straight and tapered. The tapered thread ensures a tighter joint. Regular threads are designated as NPS (straight) and NPT (tapered). Dryseal threads are identified as NPSF (straight) and NPTF (tapered).

Pipe threads should be shown on a drawing by means of the simplified method. **See Figure 9-15.** The taper threads are drawn the same as the straight threads except that the thread lines should form an angle of approximately 3° with the axis. The designation of pipe threads on a drawing should include the nominal size, number of threads per inch, thread form, and thread series symbols. For example, the thread note ⅛-27 Dryseal NPTF specifies ⅛″ diameter pipe, 27 threads per inch, Dryseal thread form, and American National Standard Taper pipe thread series.

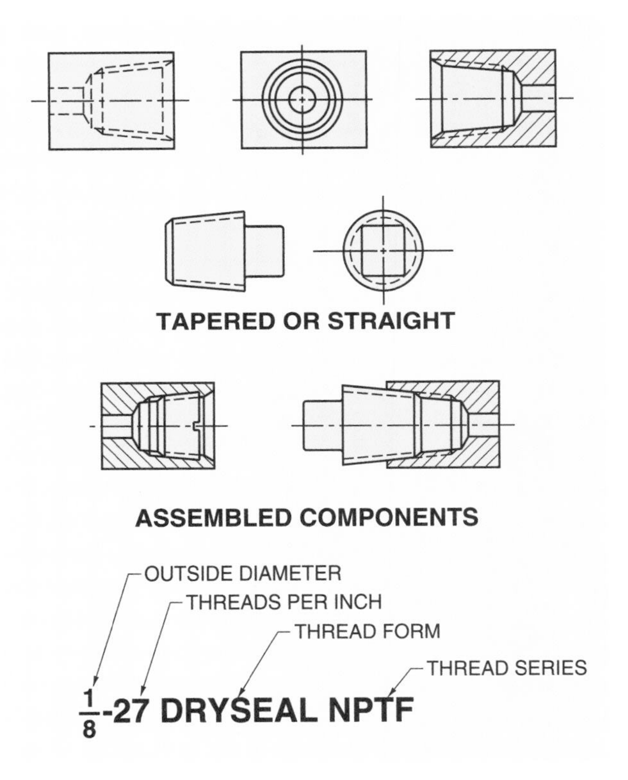

Figure 9-15. Pipe threads are shown on a drawing by the conventional simplified representation.

NONTHREADED FASTENERS

A *nonthreaded fastener* is a device that joins parts together without threads. Nonthreaded fasteners may be either temporary (easily removed) or permanent (removable only with special tools, great force, or destruction of the fastener). The most common types of nonthreaded fasteners are rivets, pins, keys, and adhesives.

Rivets

The most common nonthreaded fastener is the rivet. A *rivet* is a cylindrical metal pin that is deformed in order to hold parts together. The rivet shank is inserted through holes and pressed or beaten into a second head. The *shank* is the cylindrical body of a rivet. The shape of the preformed head and the length and diameter of the shank distinguish one rivet from another. **See Figure 9-16.**

Two parts are joined together by the grip of a rivet, which fits through predrilled holes slightly larger than the shank of the rivet. The length of the shank must exceed the thickness of the two parts to be joined by enough material to allow the shank to be shaped into the final form. The *grip* is the effective holding length of a rivet. The size of the rivet required is determined by the thickness of the parts being joined.

Rivets are relatively inexpensive and are generally manufactured from ductile metals such as steel, aluminum, copper, brass, or bronze. Riveting can also be used to join materials that cannot be welded, such as dissimilar metals, plastics, or materials that could be damaged by heat.

A riveted joint is not easily disassembled and is considered permanent. However, rivets can loosen under stress and become ineffective. Rivets are also subject to corrosion by liquids and generally cannot hold pressure because of the possibility of leaks.

Large rivets are rivets with a shank of ½″ or greater in diameter. The second head of large rivets can only be formed by applying force to the rivet after it has been heated red-hot. Small rivets are rivets with a shank of 7⁄16″ or less in diameter. The second head of some small rivets can be formed by force alone, without heating. A *blind rivet* is a type of rivet that can be used when there is access from one side only.

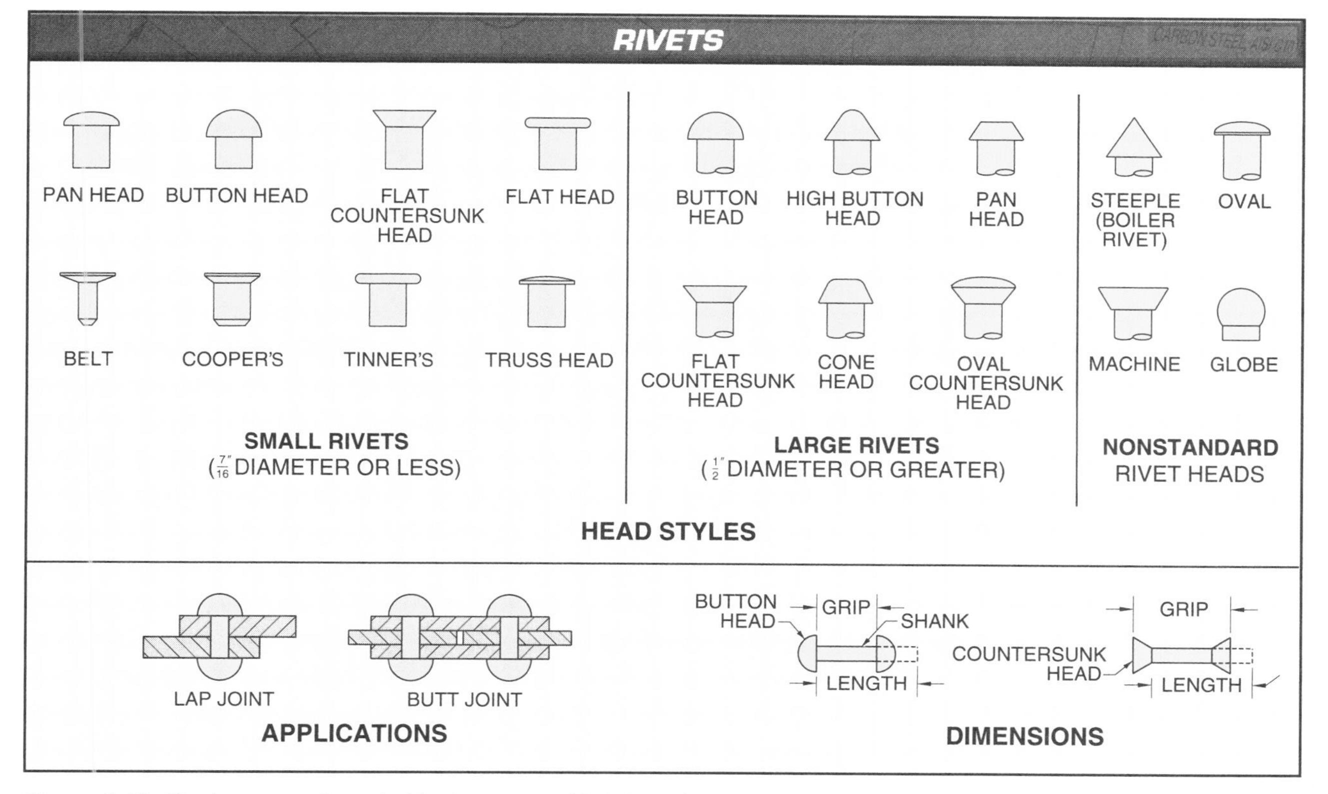

Figure 9-16. Rivets are nonthreaded fasteners used to join or fasten parts.

Rivets are shown on prints with conventional representation symbols. **See Figure 9-17.** Shop rivets are shown as clear circles with slash marks indicating countersinking, flattening, near side, far side, and both sides. A *shop rivet* is a rivet placed in a shop. Field rivets are shown as darkened circles with slash marks indicating countersinking, flattening, near side, far side, and both sides. A *field rivet* is a rivet placed in the field.

Rivet placement is controlled by the thickness of the material being riveted, pitch, and margin. *Rivet pitch* is the distance from the center of one rivet to the center of the next rivet in the same row. *Back (transverse) pitch* is the distance from the center of one row of rivets to the center of the adjacent row of rivets. *Diagonal pitch* is the distance between the centers of rivets nearest each other in adjacent rows. *Margin* is the distance from the edge of a plate to the centerline of the nearest row of rivets.

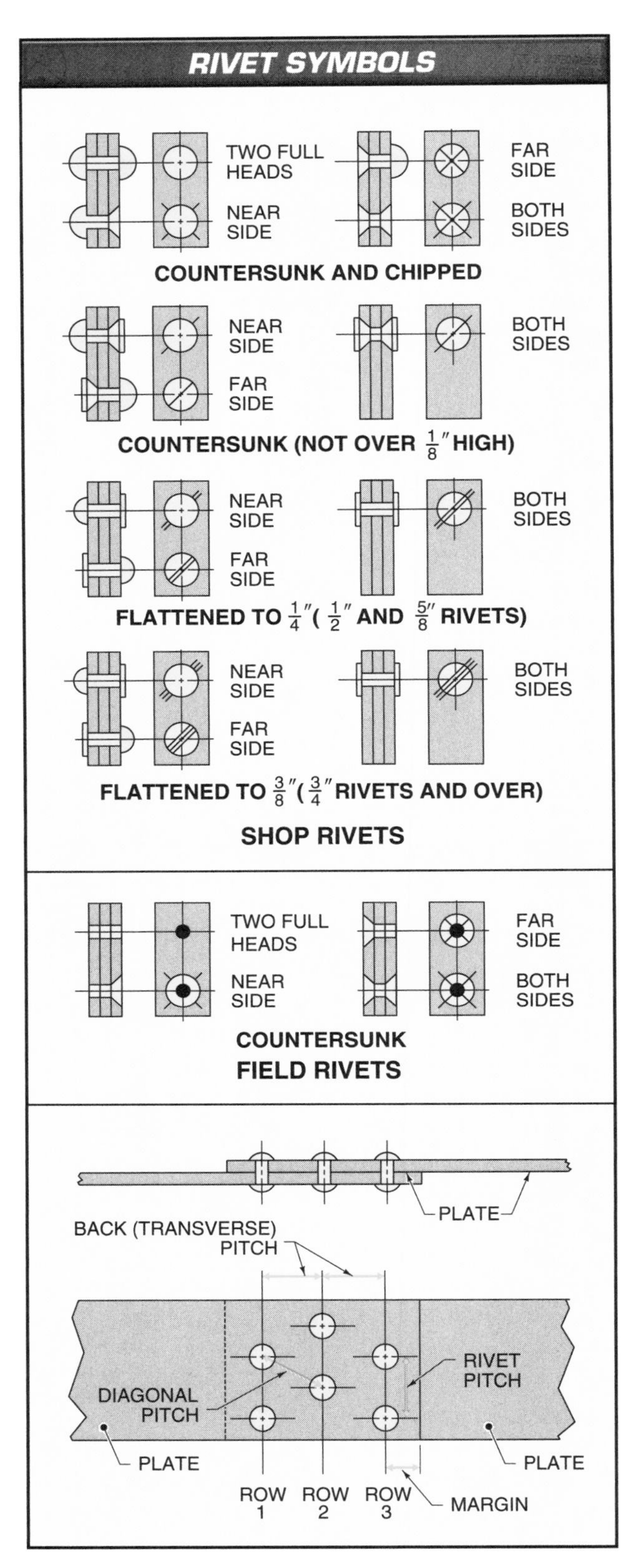

Figure 9-17. Conventional rivet symbols are used to show rivets on prints.

Pins

A *pin* is a cylindrical, nonthreaded fastener that is placed into a hole to secure the position of two or more parts. A wide variety of pin types, sizes, and materials are commercially available. Standard pins include straight, dowel, taper, clevis, cotter, slotted spring, spirally coiled, and grooved. Pins are designated on prints by their name, type, diameter, length, material, and protective finish (if required). Examples of pin designations on a print are pin, chamfered straight, 1/8 × 1.500, steel; pin, taper (precision class), .219 × 1.750, steel, zinc plated; and pin, Type A grooved, 3/32 × 3/4, steel. **See Appendix.**

Straight Pins. Straight pins are usually fabricated from bar stock. The ends are either square or chamfered. They are often used to transmit torque in round shafts. **See Figure 9-18.**

Many of our most important structures are constructed with rivets. There are approximately six million rivets in the Sydney Harbour Bridge, the world's largest steel-span bridge. The Eiffel Tower contains 2,550,000 rivets. There are three million fasteners on a Boeing 747 commercial aircraft, and half of those are rivets. The battleship USS Missouri was constructed with 1,185,000 rivets. The original denim blue jeans made in the 1880s contained 11 copper rivets.

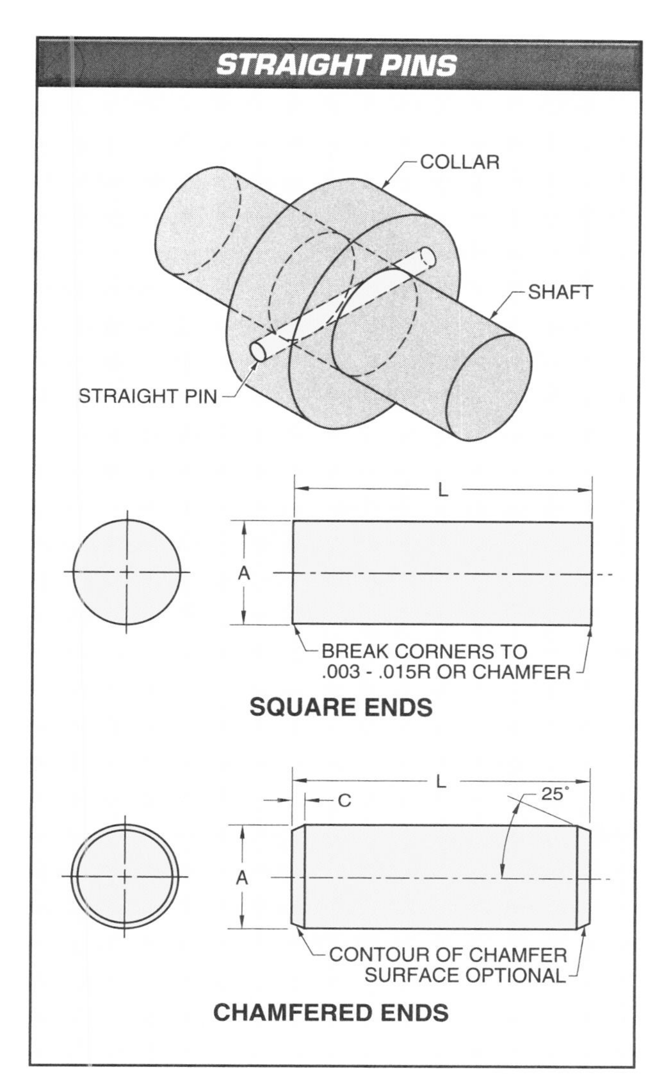

Figure 9-18. Straight pins are often used to secure collars to shafts.

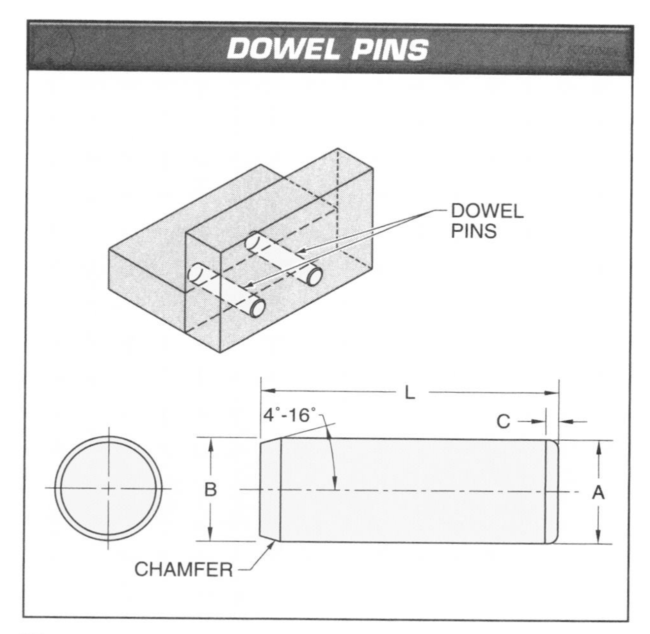

Figure 9-19. Dowel pins are often used to retain parts in a fixed position.

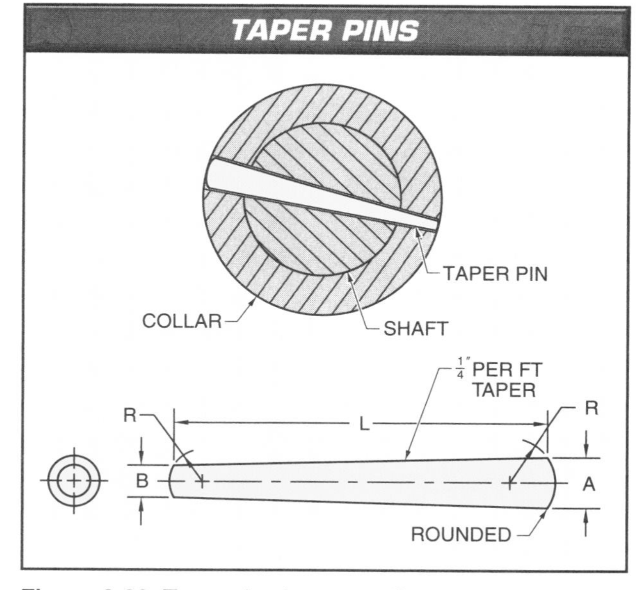

Figure 9-20. Taper pins have a uniform taper that allows easy assembly.

Dowel Pins. Dowel pins are fabricated from bar stock. Hardened dowel pins are bullet-nosed on the entry end. Soft dowel pins are chamfered on both ends. Dowel pins are used in machine and tool fabrication and to retain parts in a fixed position or to preserve alignment. **See Figure 9-19.**

Taper Pins. Taper pins are fabricated from bar stock and have rounded ends and a uniform taper of ¼″ per foot of length. Taper pins are available in commercial and precision classes, the latter having tighter tolerances. Taper pins are used to transmit small torques or to position parts. **See Figure 9-20.**

Clevis Pins. Clevis pins are fabricated from bar stock. The heads are radiused, and the entry ends have broken corners. They are used to attach clevises to rod ends and rigging and to serve as bearings. Clevis pins are held in place by cotter pins. **See Figure 9-21.**

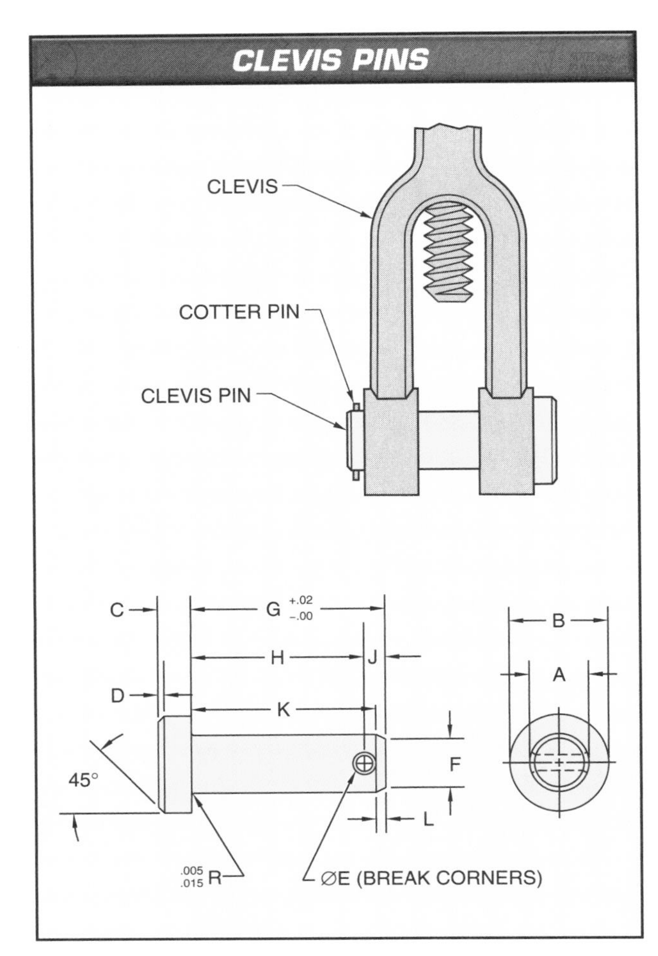

Figure 9-21. Clevis pins are large pins that are often used to attach rigging.

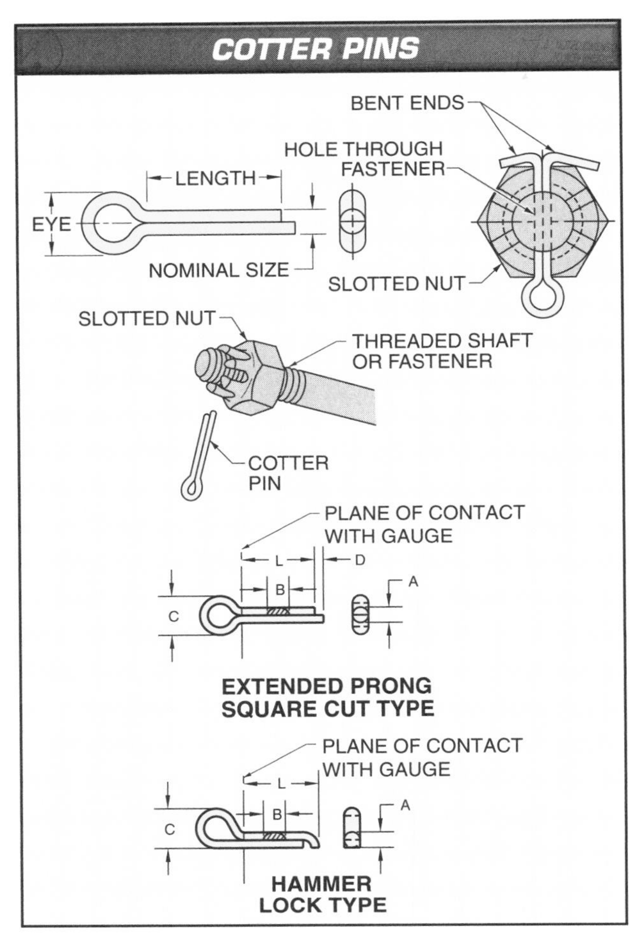

Figure 9-22. Cotter pins are used to secure clevis pins.

Cotter Pins. Cotter pins are used with clevis pins or castellated nuts to prevent the assembly from disengaging while in normal use. The entry ends are bent open after insertion to keep the cotter pin in place. **See Figure 9-22.** Cotter pins are also known as split pins, particularly in countries where "cotter pin" refers to a different type of fastener.

Spring Pins. Spring pins are rolled from spring stock and rely on spring tension to stay in place. The pins are squeezed into slightly smaller diameter holes in order to compress the spring. These pins seat firmly, so they can withstand greater torque than many other pins. Spring pins include slotted and coiled spring pins.

Slotted spring pins are tubular with one longitudinal slot. **See Figure 9-23.** Spirally coiled spring pins are rolled from a longer and thinner piece of metal. **See Figure 9-24.** Both types have at least one chamfered end to facilitate insertion.

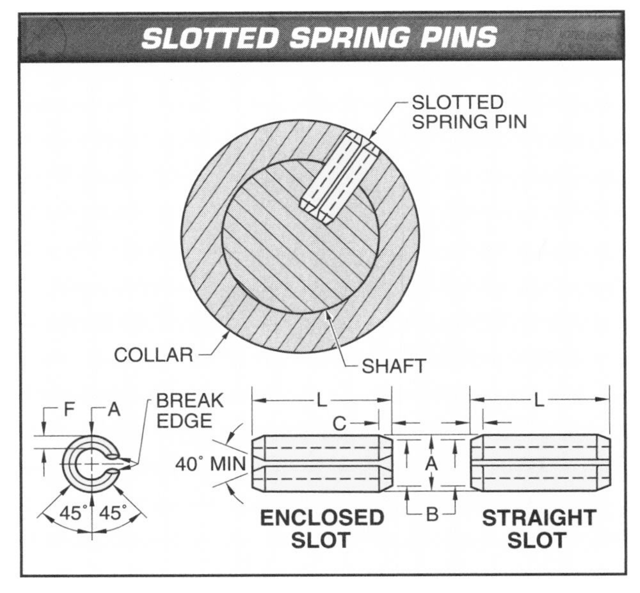

Figure 9-23. Slotted spring pins are tubular with one elongated slot.

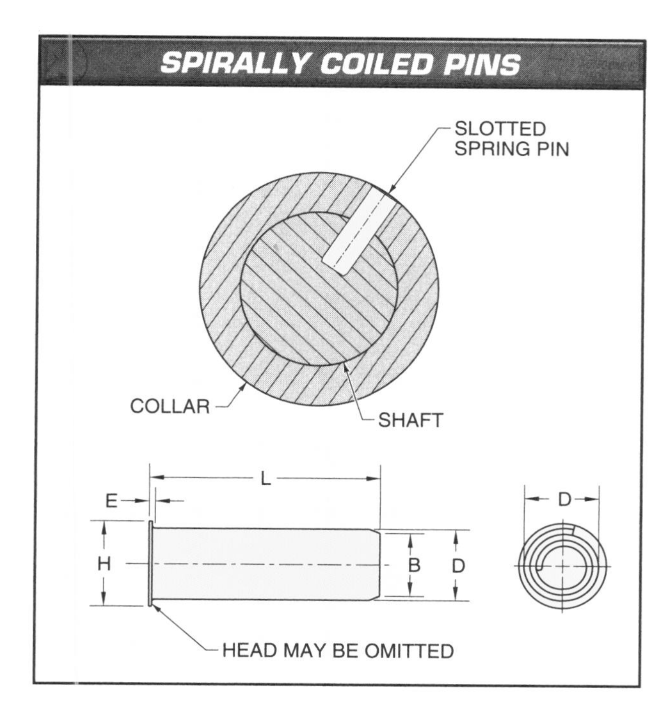

Figure 9-24. Spirally coiled pins are rolled from spring stock.

Grooved Pins. Grooved pins are solid with three parallel, equally spaced grooves. The grooves provide a tight fit and a locking feature. They are used for semipermanent fastening of levers, collars, gears, cams, and other parts to shafts. **See Figure 9-25.**

Keys

A *key* is a removable fastener that provides a positive means of transmitting torque between a shaft and a hub when mounted in a keyseat. A *keyseat* is a rectangular groove along the axis of a shaft or hub that mates with a key. The basic shapes of keys include parallel, taper, and Woodruff. All are available in standard stock sizes. **See Figure 9-26. See Appendix.**

> Standard sizes and shapes for parallel and taper keys and keyseats are given in ANSI/ASME B17.1, *Keys and Keyseats*. Standard Woodruff key specifications are given in ANSI/ASME B17.2, *Woodruff Keys and Keyseats*.

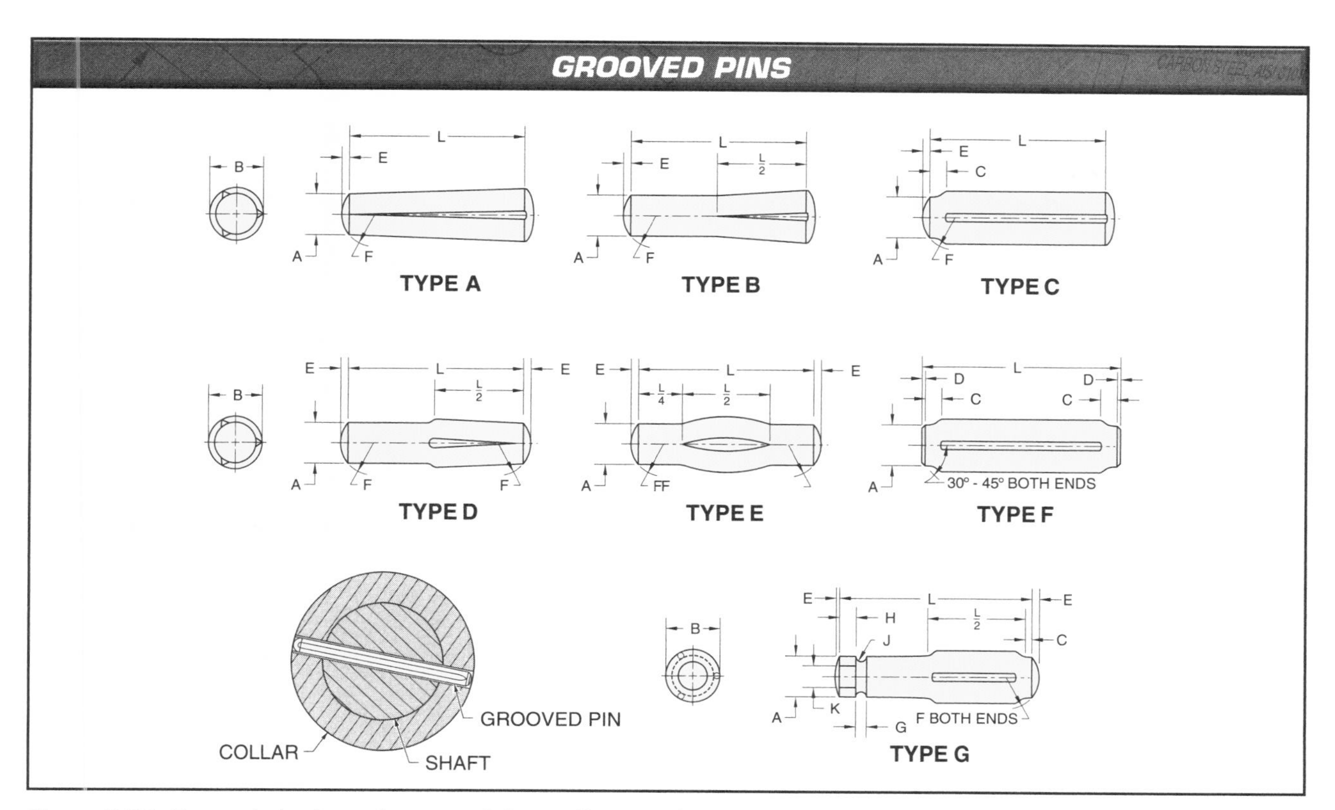

Figure 9-25. Grooved pins have three parallel, equally spaced grooves that provide a tight fit.

Parallel Keys. Parallel keys are square or rectangular in shape. They are used for transmitting unidirectional torques in shafts and hubs that do not have heavy starting loads. Parallel keys may be easily withdrawn. Parallel keys are designated on prints by their name, type, and dimensions.

Taper Keys. Taper keys may be either plain taper, alternate plain taper, or gib head taper. They are used for transmitting heavy unidirectional torques in shafts and hubs that are reversed frequently and subject to vibration. Taper keys may be easily withdrawn. Taper keys are designated on prints by their name, type, and dimensions.

Woodruff Keys. Woodruff keys are half-moon in shape. They may have a full radius or flat bottom. Woodruff keys are used for transmitting light torques or locating parts on tapered shafts. Woodruff keys are designated by key numbers, which indicate the nominal dimensions. The last two digits give the nominal diameter in eighths of an inch. The digits in front of the last two digits give the nominal width in thirty-seconds of an inch. For example, a #204 Woodruff key is ½″ in nominal diameter and 1⁄16″ in width.

Keys and keyseats provide a strong connection between shafts and other components.

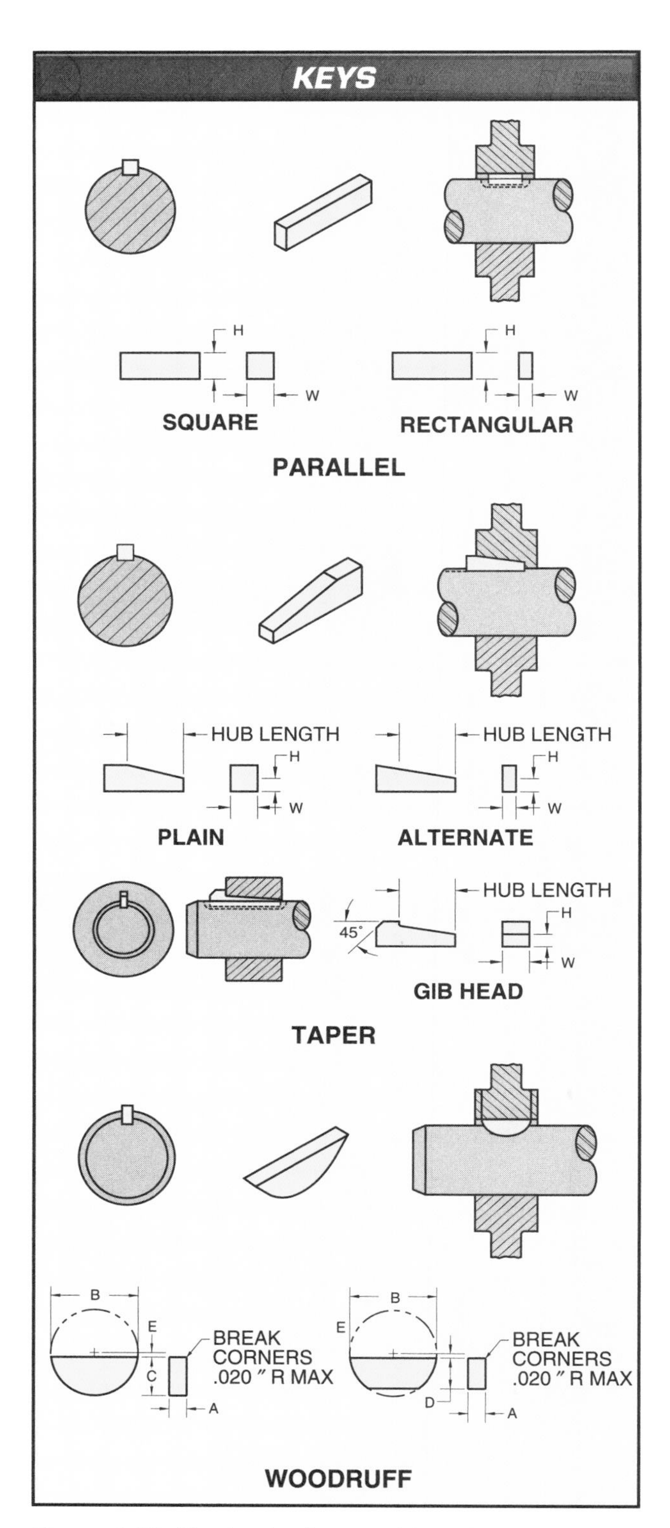

Figure 9-26. The basic shapes of keys include parallel, taper, and Woodruff.

REVIEW QUESTIONS 9

Name ______________________________ Date ____________________

Completion

______________ 1. The length of a bolt or screw is the distance from the bearing surface of the head to the tip, measured parallel to the ___.

______________ 2. The ___ is the effective holding length of a rivet.

______________ 3. The ___ is the distance a threaded fastener advances per revolution

______________ 4. A(n) ___ rivet can be used when there is access from only one side.

______________ 5. ___ washers are round and flat.

______________ 6. ___ is the distance between corresponding points on adjacent raised projections.

______________ 7. ___ is the distance from the edge of a plate to the centerline of the nearest row of rivets.

______________ 8. A thread ___ is a group of diameter-pitch combinations for a certain thread form.

______________ 9. A(n) ___ pitch series is a standard thread series with a different number of threads per inch for most diameters.

______________ 10. Threaded fasteners are based upon an inclined plane wrapped around a(n) ___ cylinder.

______________ 11. ___ keys are used for transmitting unidirectional torque in shafts and hubs that do not have heavy starting loads.

______________ 12. UN threads are based on a(n) ___° standard thread angle.

______________ 13. The thread note specifies in sequence the nominal size, number of threads per inch or pitch, thread form and series, and thread ___.

______________ 14. The letter ___ designates an external thread.

______________ 15. The size of the rivet required is determined by the ___ of the parts being joined.

______________ 16. Hardened dowel pins are ___ on the entry end.

______________ 17. A(n) ___ is a cylindrical, nonthreaded fastener that is placed into a hole to secure the position of two or more parts.

______________ 18. ___ pins are used with clevis pins or castellated nuts to prevent the assembly from disengaging while in normal use.

______________ 19. ___ keys are half-moon in shape.

______________ 20. A(n) ___-hand thread always slopes up to the right.

True-False

T F **1.** Together, tolerance and allowance determine the fit of mating threads.

T F **2.** Acme threads were developed from the Unified thread form.

T F **3.** Metric threads share the same basic profile as UN threads.

T F **4.** Thread series are distinguished from one another by the number of threads per inch for a series of specific diameters.

T F **5.** Combination of thread classes for fasteners is not possible.

T F **6.** A keyseat is a rectangular groove along the axis of a shaft or hub.

T F **7.** The nominal size of the diameter of a thread may be stated in fractional or decimal or by a number.

T F **8.** The letters LH follow the thread note for left-hand threads.

T F **9.** The length of a threaded fastener may be included at the end of the thread note.

T F **10.** The length of the shank of a rivet must be the same dimension as the two parts to be joined.

T F **11.** Riveted joints are not easily disassembled.

T F **12.** A large rivet has a shank of ½″ or greater in diameter.

T F **13.** A taper thread is a thread formed on a cone.

T F **14.** A complete thread has full form at both crest and root.

T F **15.** Spring lock washers are split on two sides.

T F **16.** Regular pipe thread is the standard for the plumbing trade.

T F **17.** Soft dowel pins are square on both ends.

T F **18.** Spirally coiled spring pins are tubular with one longitudinal slot.

T F **19.** Taper keys may be easily withdrawn.

T F **20.** Sealer is required with Dryseal pipe threads.

Identification — Rivets

_______________ 1. Shop rivet; countersunk and chipped; far side

_______________ 2. Field rivet; countersunk; far side

_______________ 3. Shop rivet; flattened to ⅜″; near side

_______________ 4. Shop rivet; countersunk; not over ⅛″ high; near side

_______________ 5. Field rivet; countersunk; both sides

_______________ 6. Shop rivet; countersunk and chipped; both sides

_______________ 7. Shop rivet; flattened to ¼″; far side

_______________ 8. Shop rivet; countersunk and chipped; near side

_______________ 9. Shop rivet; countersunk; not over ⅛″ high; far side

_______________ 10. Field rivet; countersunk; near side

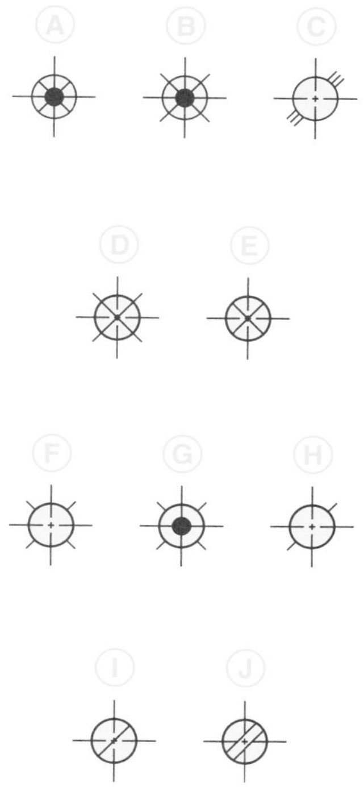

A B C D E F G H I J

Identification — Thread Representation

_______________ 1. Simplified; external; section

_______________ 2. Schematic; external

_______________ 3. Schematic; internal; section

_______________ 4. Detailed; external

_______________ 5. Simplified; internal; section

_______________ 6. Simplified; internal

_______________ 7. Detailed; external; section

_______________ 8. Detailed; internal; section

_______________ 9. Detailed; internal

_______________ 10. Simplified; external

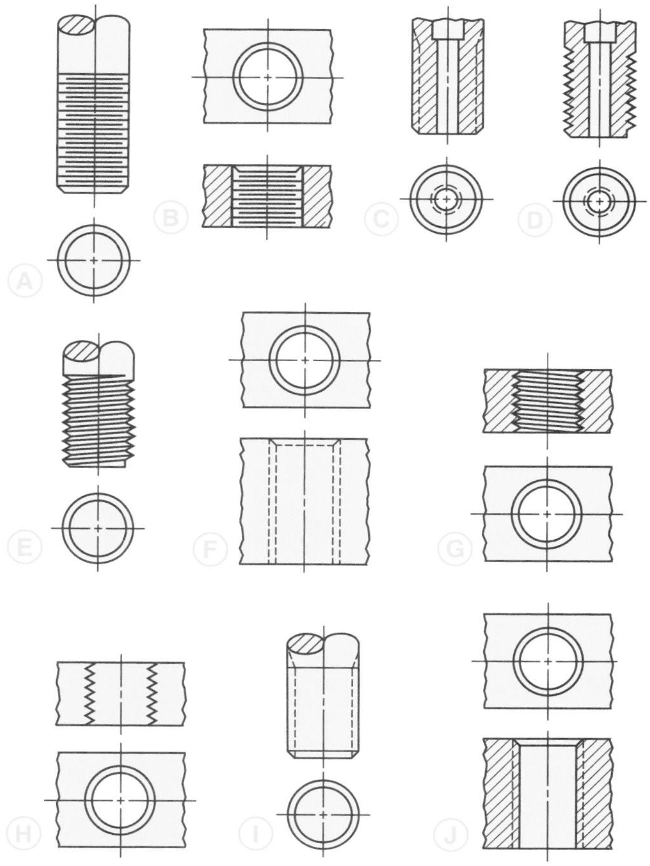

A B C D E F G H I J

Identification — Threaded Fasteners

_____ 1. Hex bolt
_____ 2. Cap (acorn) nut
_____ 3. Hex castle nut
_____ 4. Square set screw
_____ 5. Square nut
_____ 6. Round head bolt
_____ 7. Hex nut
_____ 8. Hex slotted nut
_____ 9. Square head bolt
_____ 10. Countersunk bolt

A B C D E F G H I J

Identification — Screw Thread

_____ 1. Thread angle
_____ 2. Internal thread
_____ 3. Root (internal thread)
_____ 4. Root (external thread)
_____ 5. Crest (internal thread)
_____ 6. Crest (external thread)
_____ 7. Pitch
_____ 8. Major diameter
_____ 9. Minor diameter
_____ 10. External thread

A B C D E F G H I J

Identification — Thread Forms

_____ 1. Unified
_____ 2. Buttress
_____ 3. Knuckle
_____ 4. Acme

A B C D

TRADE COMPETENCY TEST 9

Name ______________________ **Date** ______________

Punch

Refer to print on page 233.

______________	**1.**	A No. ___ drill is used to drill the two holes.
______________	**2.**	Decimal tolerances for the Punch are ±___″.
______________	**3.**	The overall width of the part is ___″.
______________	**4.**	Fractional tolerances for the Punch are ±___″.
______________	**5.**	Two holes are drilled, countersunk, and tapped for ___ cap screws.
T F	**6.**	Two holes are tapped for 10-32 threads.
______________	**7.**	The center-to-center dimension for the two holes is ___″.
______________	**8.**	The maximum decimal center-to-center dimension for the two holes is ___″.
______________	**9.**	The minimum overall decimal height of the Punch is ___″.
T F	**10.**	The minimum fractional radius is $\frac{3}{32}$″.
T F	**11.**	The Punch drawing was drawn by TM.
T F	**12.**	The overall depth of the Punch is .946″.
T F	**13.**	The Punch shaft is represented by hidden lines in the side view.
T F	**14.**	The front view and the side view are identical.
T F	**15.**	Through holes are drilled into the Punch.
______________	**16.**	The print was originally drawn to the scale of ___″ = 1′-0″.
______________	**17.**	The manufacturer of the Punch is located in ___, IL.
______________	**18.**	The Punch shaft has a diameter of ___″.
______________	**19.**	The .250″ dimension in the front view is a(n)___ dimension.
______________	**20.**	Approval of the print was completed on ___.

Machine Screw

Refer to print on page 233.

____________________	**1.**	A(n) ___ provides the self-locking feature to the Machine Screw.
____________________	**2.**	The maximum angle of the head is ___.
____________________	**3.**	The maximum overall length of the Machine Screw is ___″.
____________________	**4.**	The Machine Screw has a cross recess type ___ drive.
T F	**5.**	All sharp edges are to be deburred.
T F	**6.**	The drawing may be scaled.
____________________	**7.**	The material for the Machine Screw is ___ series stainless steel.
____________________	**8.**	The Machine Screw conforms to ANSI ___.
____________________	**9.**	The depth of the head is ___″.
____________________	**10.**	The class of thread specified is ___.
____________________	**11.**	The nominal diameter of the Machine Screw is ___″.
____________________	**12.**	There are ___ threads per inch on the Machine Screw.
____________________	**13.**	The thread form on the Machine Screw is ___.
T F	**14.**	The nylon patch may be started on the second thread of the Machine Screw.
T F	**15.**	The print was checked on 12-19.
T F	**16.**	The Machine Screw is drawn twice its actual size.
T F	**17.**	ANSI Y14.5 applies to the drawing methods used.
T F	**18.**	The Machine Screw is to be plated.
____________________	**19.**	The overall length of the nylon patch is ___″.
____________________	**20.**	The sheet size is ___.

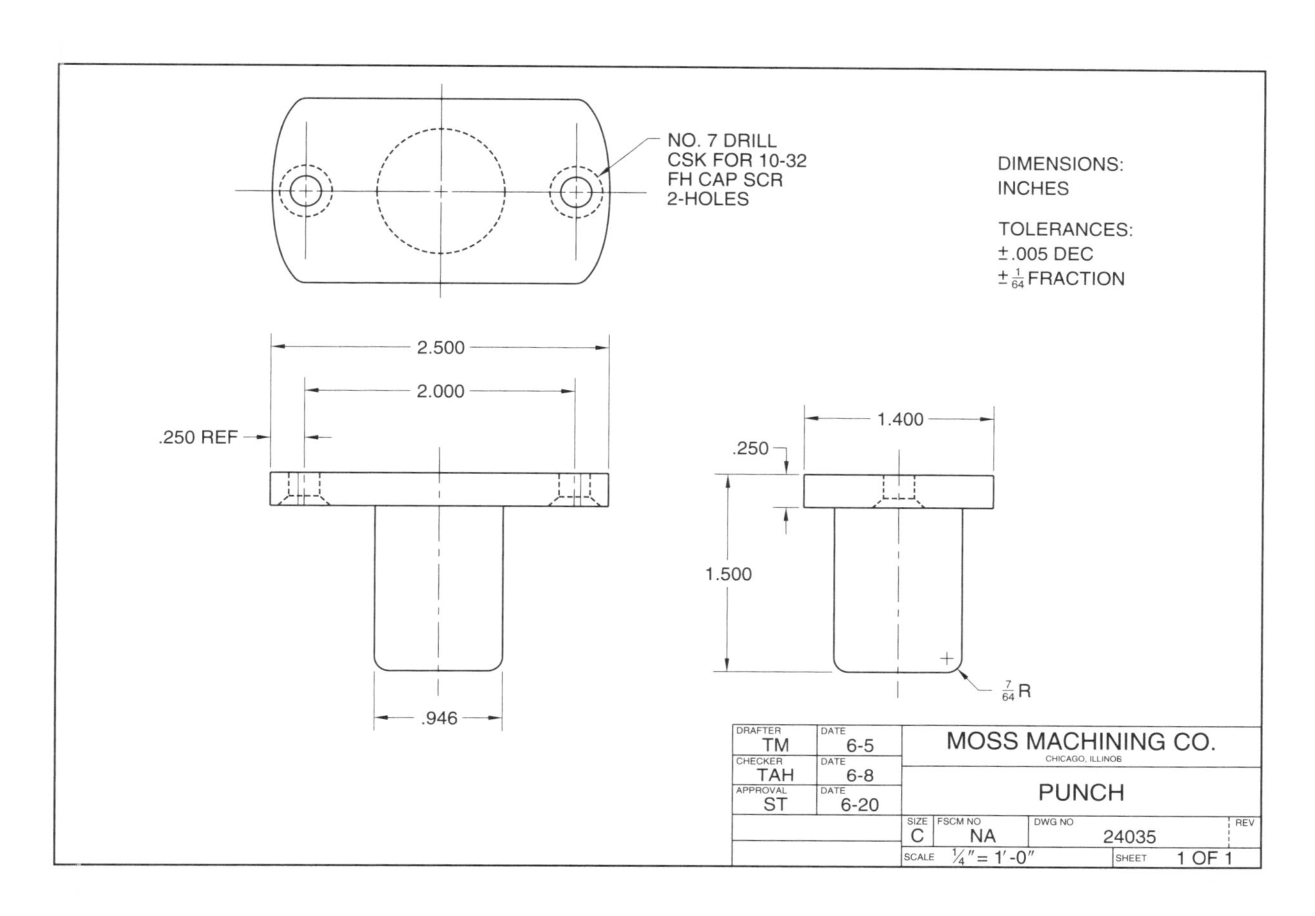

PUNCH

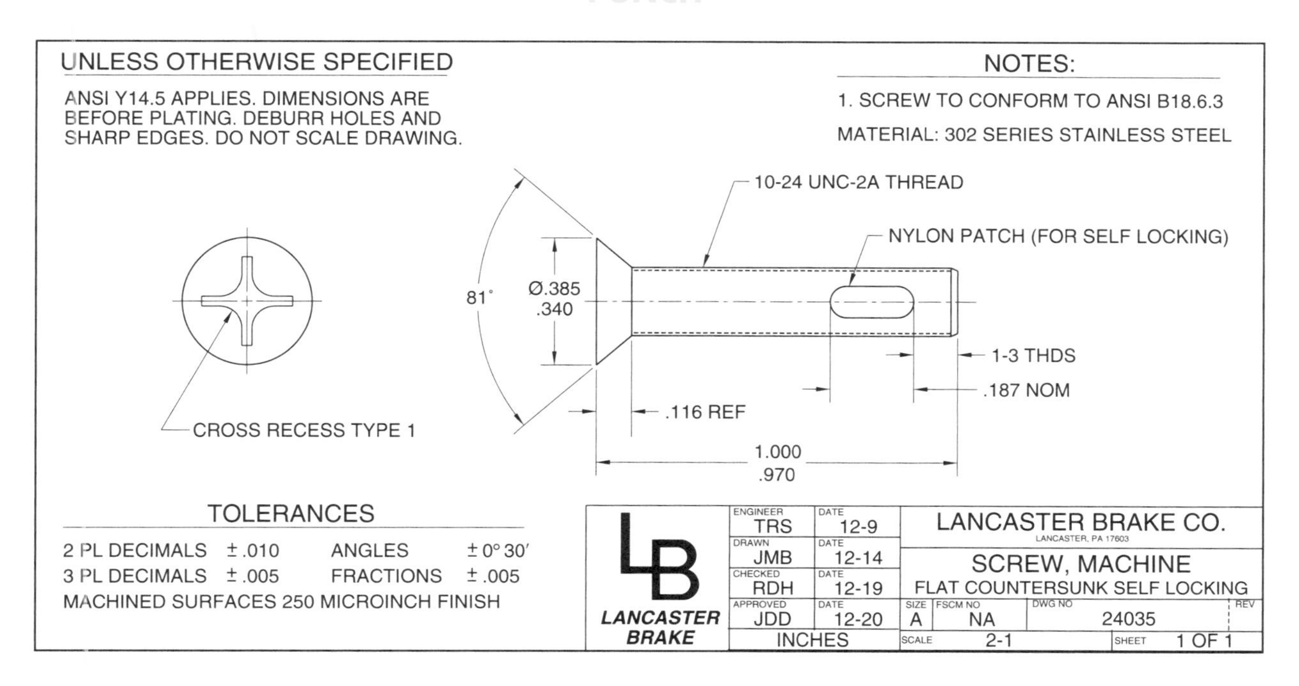

MACHINE SCREW

Plunger

Refer to print below.

______ **1.** Both ends of the Plunger are chamfered at ___°.

______ **2.** The diameter of the external thread is ___″.

______ **3.** The internal thread has ___ threads per inch.

______ **4.** The hole for the internal thread is ___″ deep.

______ **5.** The slot must be square to the flat within ___°.

______ **6.** The U-cut is made to the ___ of the external thread.

T F **7.** The external thread has 13 threads per inch.

T F **8.** The minimum width for the U-cut is 7/64″.

______ **9.** The overall length of the Plunger is ___.

______ **10.** The thread form for the external thread is ___.

T F **11.** The Plunger is drawn full-scale.

T F **12.** Decimal tolerances are ±.005.

______ **13.** The center of the two holes is ___″ from the right end of the Plunger.

______ **14.** The two holes are drilled to a maximum diameter of ___″.

______ **15.** The slot is ___″ deep.

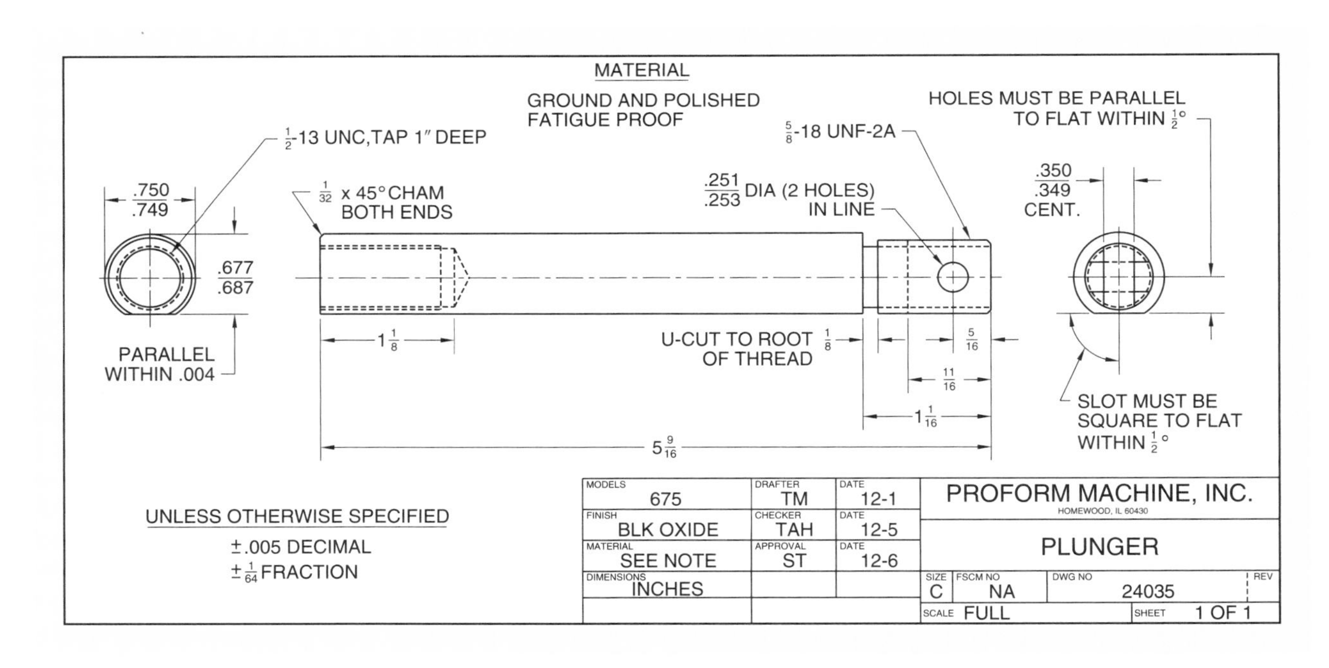

PLUNGER

10 Gears and Cams

Gears are toothed wheels used in pairs to transmit power or motion from one shaft to another. The American Gear Manufacturers Association (AGMA) provides specifications and tolerances for gear manufacture and applications. Cams are used with cam followers to transfer loads, change direction, or change the speed of a machine. Assemblies using cams must include cam prints because cams must be custom made for their particular applications.

GEARS

A *gear* is a toothed wheel that is used with other gears in order to transmit rotational power from one shaft to another. The meshing of gear teeth prevents slippage, which can otherwise occur with belt and pulley systems. Slippage is lost motion, which also creates friction heat, reducing the efficiency of the power transmission. However, gear shafts must be in close proximity. Longer distances require the use of a chain and sprockets.

Meshing gears rotate in opposite directions. A *pinion gear* is the smallest of meshing gears. With two meshing gears, one gear rotating counterclockwise causes the other gear to rotate clockwise. **See Figure 10-1.** An added third gear rotates in the same direction as the first gear.

Gear Ratio

Meshed gears of equal diameter rotate at the same rate. Since the circumferences are also equal, both return to the same position after each rotation. One complete revolution of one gear produces one complete revolution of the other gear.

If gears have different diameters, they turn at different rates. *Gear ratio* is the relationship between the sizes of two meshing gears. The gear ratio can be calculated by dividing the diameter of the driving gear by the diameter of the driven gear. **See Figure 10-2.** Another method of calculating gear ratio is to divide the number of teeth on the respective gears. Gear ratio is important to determining how rotational speed and forces change as they are transmitted through gears.

Boston Gear

Gears are made in many different sizes, shapes, and materials.

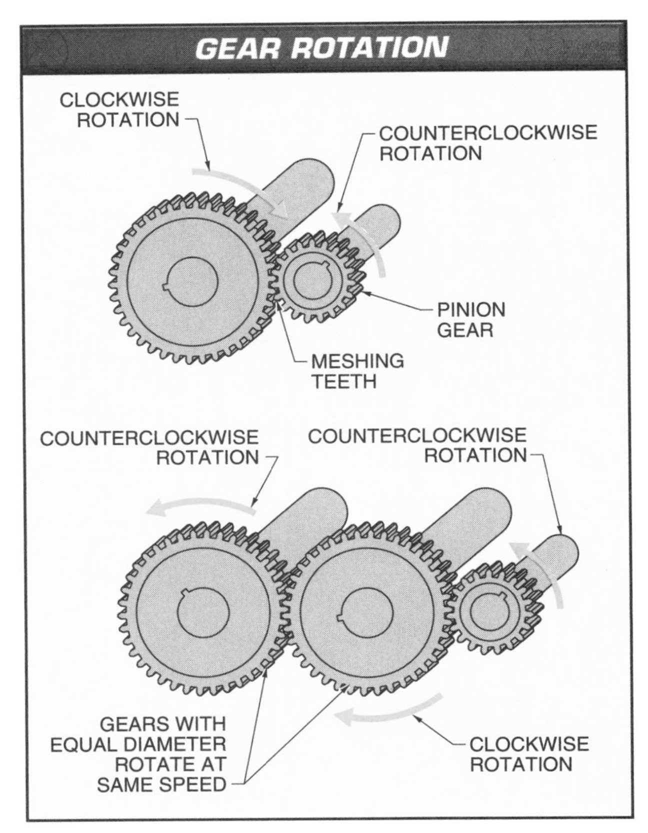

Figure 10-1. Adjacent gears turn in opposite directions.

For example, if Gear A is twice the diameter of Gear B, they have a 2:1 ratio. The smaller Gear B must turn two revolutions for each revolution of larger Gear A. Since Gear B revolves twice as much as Gear A in a period, it rotates twice as fast. The speed ratio is the opposite of the size ratio, or 1:2 in this case. The smaller gear always turns faster than the larger gear.

Torque also changes through gear transmission. *Torque* is rotational force. Speed multiplied by torque equals power, which remains constant (or very nearly constant) from gear to gear. Therefore, if speed increases, torque decreases; and if speed decreases, torque increases. Torque between gears follows the same ratio as the size ratio.

Power equals rotational speed multiplied by torque. In gear trains, the output power is equal to the input power, neglecting a small loss from inefficiency. Since gear trains are designed to change speed or torque, then the opposite quantity must also change if the power is to remain the same. For example, a gear train that reduces speed increases torque; and one that increases speed decreases torque.

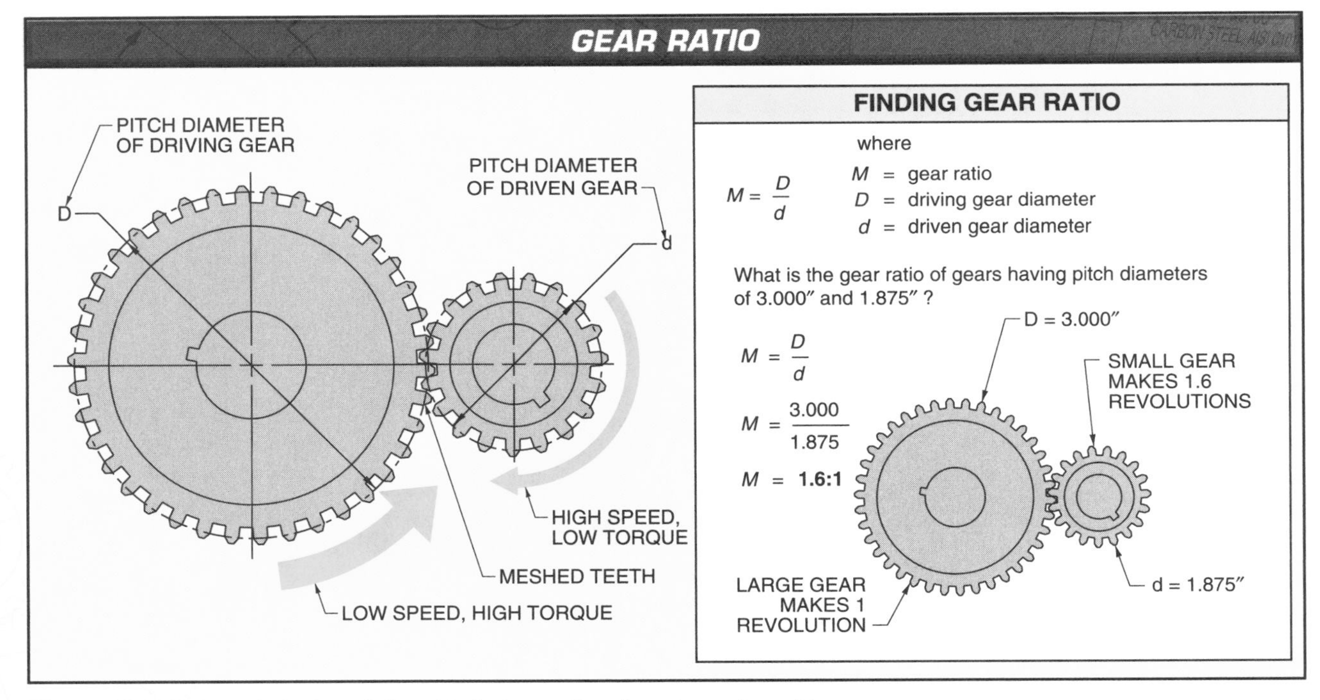

Figure 10-2. Gear ratio is the relationship between the diameters of meshing gears.

Compound Gearing

Compound gearing is the use of several gears to achieve high gear ratios. Ratios are calculated between pairs of gears and multiplied to determine the overall gear ratio from the first to last gear. For example, if the ratio between Gear A and Gear B is 1.5:1 and the ratio between Gear B and Gear C is 2:1, then the overall ratio between Gear A and Gear C is 3:1. **See Figure 10-3.**

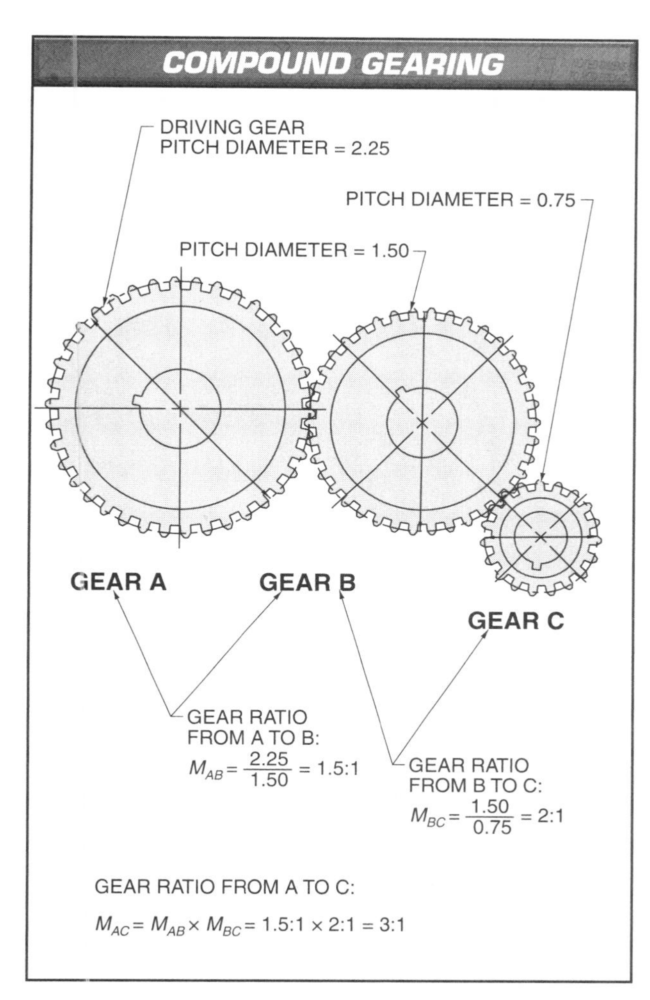

Figure 10-3. Gear pairs can be chained together to achieve higher overall gear ratios.

An automobile transmission uses many gears to reduce the fast rotational speed from the engine to a slower speed for the wheels. This results in a high overall gear ratio. However, when designing gear drive systems, lost motion, strains, and space limitations required by compound gearing must be considered.

Backlash

Gears are designed to have a limited amount of backlash between meshing teeth for maximum life and efficiency. *Backlash* is the small amount that one gear can rotate before a meshing gear begins to move. Backlash results when the tooth space exceeds the size of the meshing tooth. Backlash is required between meshing gears to prevent full contact on both sides of the teeth. The space allows lubricant between the contacting surfaces.

Too little backlash can cause undesirable resistance, resulting in the overheating or jamming of the meshing gears. Lubricant can become trapped at the base of the teeth. Excessive backlash can be noisy and can cause problems if the load is reversed frequently.

GEAR SPECIFICATION

Gears are designed by engineers and specified on prints using ANSI/ASME Y14.7.1, *Gear and Spline Drawing Standards – Part 1 for Spur, Helical, Double Helical and Rack,* and Y14.7.2, *Gear and Spline Drawing Standards – Part 2 for Bevel and Hypoid Gears.* The American Gear Manufacturers Association (AGMA) also contributes to the development of these standards, which include common gear terminology and drawing practices. **See Figure 10-4.**

Boston Gear

Bevel gears transmit power between shafts at right angles.

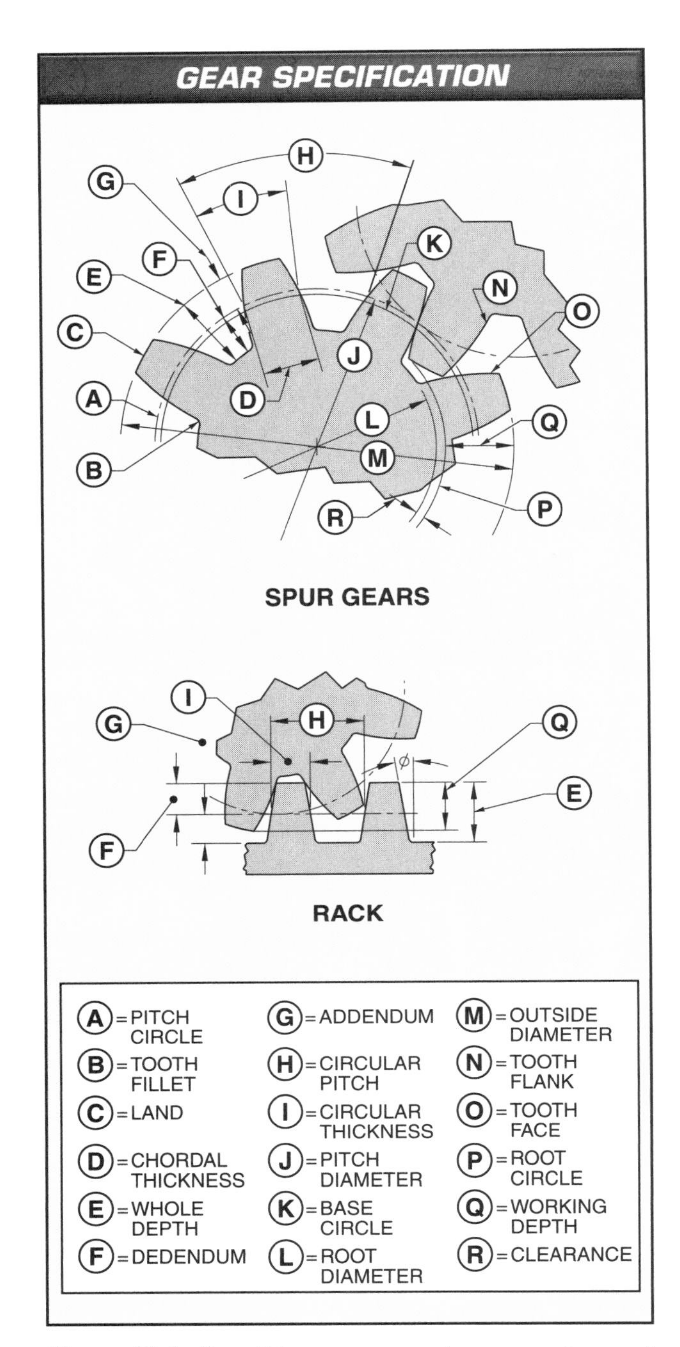

Figure 10-4. Specifying gears requires many types of dimensions.

The involute curve that determines tooth shape on most gears is designed to produce a constant rotational speed throughout the time of contact between opposing teeth. The point of contact follows a line that determines the pitch circle.

Gear Diameters

The *pitch circle* is the imaginary circle located approximately halfway between the tops and the roots of raised projections. The pitch circle is the line of contact between two circles having the same ratios as the gears. It cannot be directly measured, but is important to calculating gear ratios and other gear characteristics. The *pitch diameter* is the diameter of the pitch circle.

The *outside diameter* is the diameter of the circle formed by the tops of the gear teeth. The *addendum* is the portion of the gear teeth between the pitch circle and the outside diameter circle. The outside diameter circle can also be called the addendum circle. The *root circle* is a circle formed by the bottom of the tooth spaces. The *root diameter* is the diameter of the root circle.

Diametral Pitch

Diametral pitch is the number of raised projections per unit of pitch diameter. A 16-pitch gear has 16 teeth for each inch of pitch diameter. Diametral pitch is usually specified on a print with the number followed by the letter P. **See Figure 10-5.**

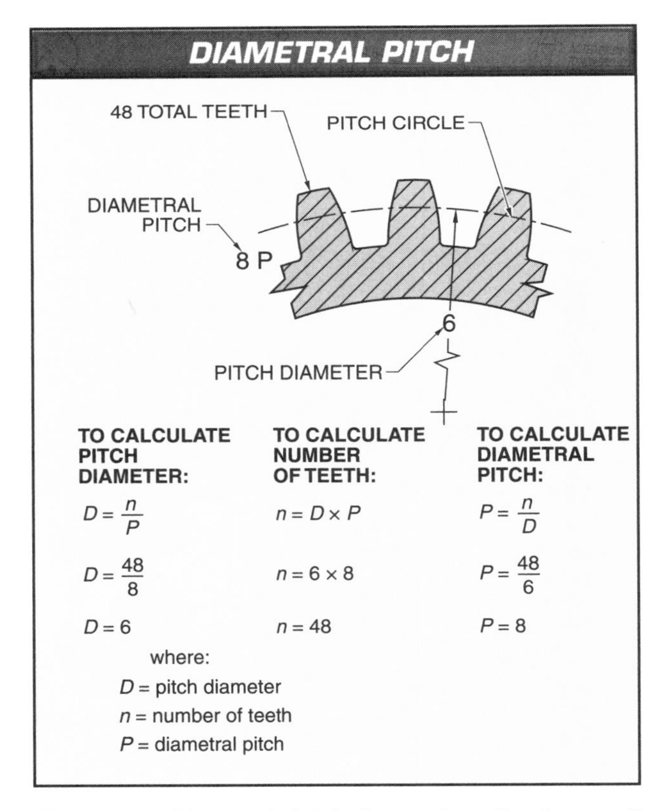

Figure 10-5. Diametral pitch is the number of teeth per unit of pitch diameter.

Gears must have the same diametral pitch in order to mesh correctly. For example, a 2″ diameter gear with 24 teeth can be paired with a 3″ diameter gear with 36 teeth because they are both 12-pitch gears. The opposite of diametral pitch is circular pitch. *Circular pitch* is the distance between corresponding points of two adjacent gear teeth on the pitch circle. Regardless of diameter, a 12-pitch gear has teeth spaced 0.262″ apart.

A print may include any two of the three most important specifications for gears: the pitch diameter, number of teeth, and diametral pitch. The third may be calculated from a simple formula. For instance, to calculate pitch diameter, apply the following formula:

$D = n / P$

where

D = pitch diameter

n = number of teeth

P = diametral pitch

For example, what is the pitch diameter of a gear having 18 teeth and a diametral pitch of 2?

$D = n / P$

$D = 18 / 2$

$D = \mathbf{9}$

The formula may be rearranged to solve for number of teeth or diametral pitch instead, depending upon the given information.

Gear Teeth

Gear teeth can be made in different forms, or shapes. The intended application determines optimal tooth form because the shape can affect efficiency, noise, contact area, and tooth strength. The most common type of form is calculated geometrically from the involute. The *involute* is the curve formed by the path of a point on a straight line as it unwinds from a round surface. **See Figure 10-6.** This is used to determine the geometric profile of gear teeth.

> Metal gears can be manufactured in a large number of ways, such as casting, extrusion, cold forming, and machining. Casting and extrusion produce relatively low-quality gears. Cold forming rolls gear dies against metal blanks to form the teeth. This process generally improves the properties of the metal and produces good-quality teeth. Machining produces the best gear teeth but is the most expensive process.

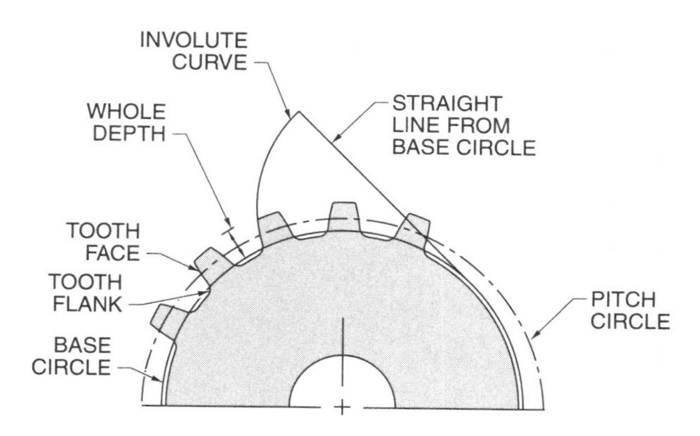

Figure 10-6. The involute curve forms the shape of the sides of the teeth.

The *tooth face* is the curved surface of a tooth located above the pitch circle. The *tooth flank* is the curved surface of a gear tooth located below the pitch circle. The *whole depth* is the total height of the gear tooth from the root circle to the outside diameter circle.

Clearance is the space between the bottom of a gear tooth space and the tip of a tooth fully meshed into that tooth space. The *working depth* is the depth a tooth extends into the tooth space when in full mesh with proper clearance.

GEAR TYPES

Gears most commonly used in industry include spur, helical, herringbone, bevel, and worm gears. Gears are selected based on their specific application in the machine. **See Figure 10-7.**

Gears can be grouped into categories based on shaft orientation of meshing gears. Parallel shaft gears include spur, helical, and herringbone gears. Intersecting shaft gears are gears with shafts that intersect each other in the same plane, usually at 90°, such as bevel gears. Nonintersecting shaft gears are gears with shafts that do not intersect in the same plane, such as worm gears and some helical gears.

Spur Gears

A *spur gear* is a gear with straight teeth that are parallel to the shaft axis. **See Figure 10-8.** Spur gears are the most common gear type and are used where the meshing gears are mounted on parallel shafts. They are suitable for gear ratios ranging from 1:1 to 1:6 and surface speeds of up to 1000 ft/min.

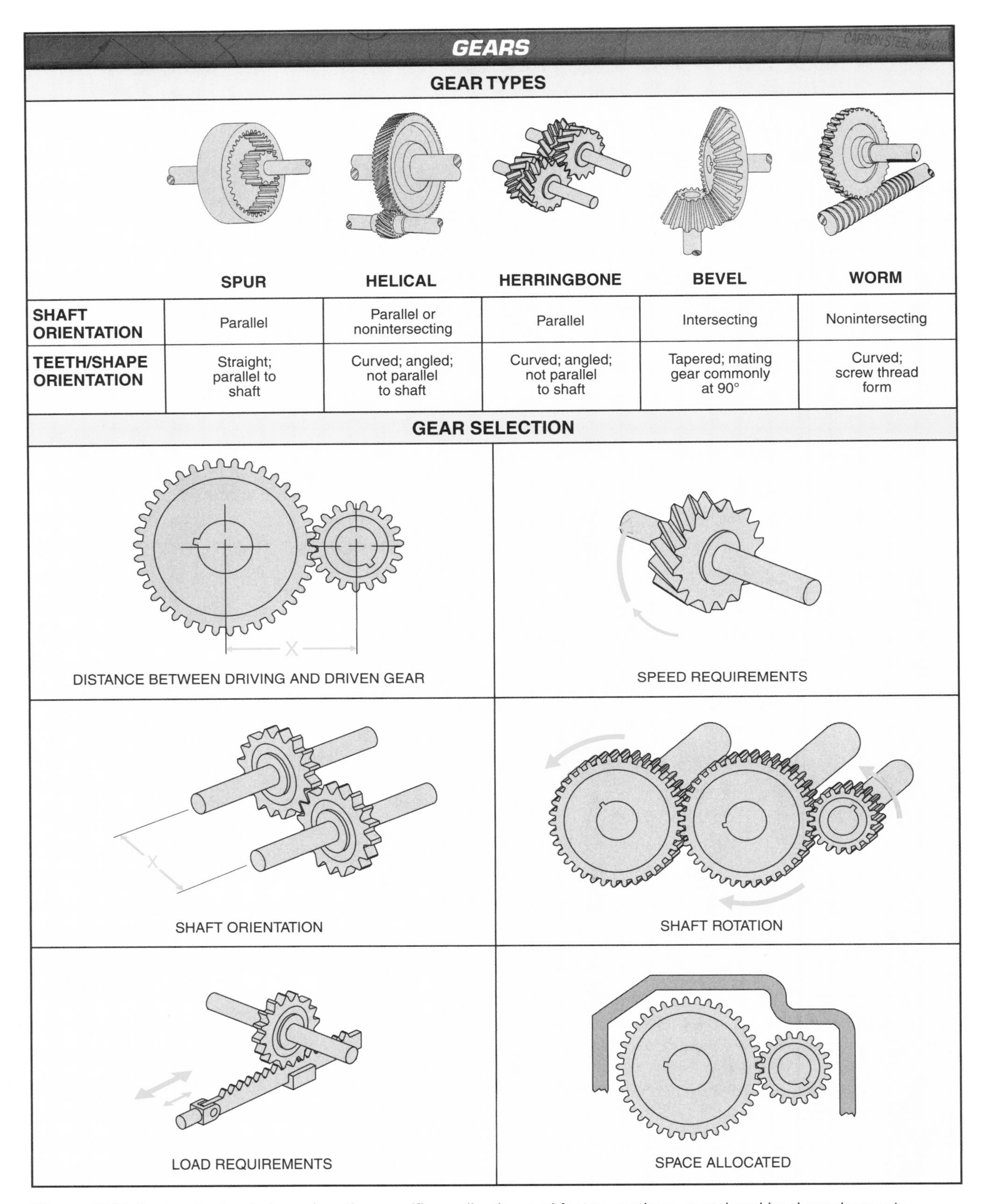

	SPUR	HELICAL	HERRINGBONE	BEVEL	WORM
SHAFT ORIENTATION	Parallel	Parallel or nonintersecting	Parallel	Intersecting	Nonintersecting
TEETH/SHAPE ORIENTATION	Straight; parallel to shaft	Curved; angled; not parallel to shaft	Curved; angled; not parallel to shaft	Tapered; mating gear commonly at 90°	Curved; screw thread form

Figure 10-7. Gear selection is based on the specific application and factors such as speed and load requirements.

An *internal gear* is a spur gear that meshes on its inside circumference. This permits a larger ratio of reduction in a small space. A *rack* is a spur gear that is flat rather than concentric. A pinion is used with a rack to convert rotary motion into linear motion.

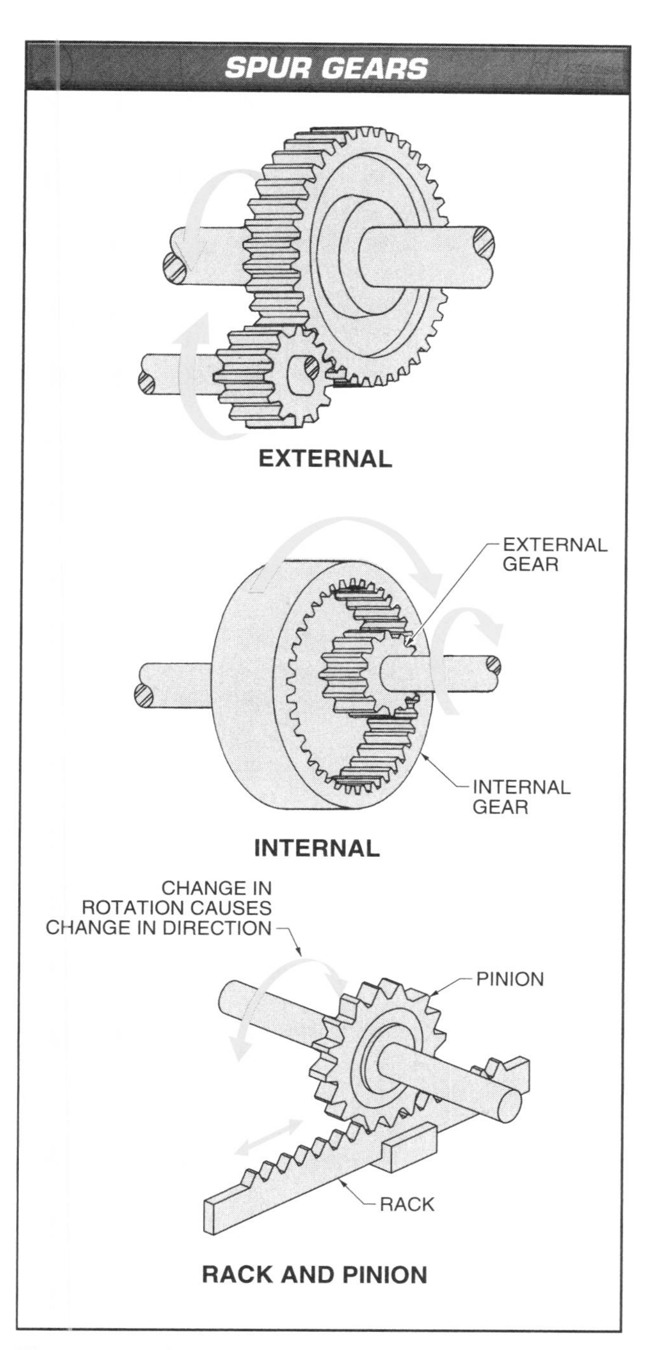

Figure 10-8. Spur gear types include internal, external, and rack.

A *spline* is a series of teeth or parallel surfaces machined into a shaft (external splines) or hub (internal splines). **See Figure 10-9.** Spline tooth design specifications are similar to spur gears but are commonly machined to half the typical tooth depth. Splines provide positive transmission of power and permit assembly and disassembly as required. Splines transmit heavy loads without slippage where single keys would not provide sufficient strength. Splines are straight-sided or involute in form. The involute design is more common and is machined to match mating teeth on the shaft or hub.

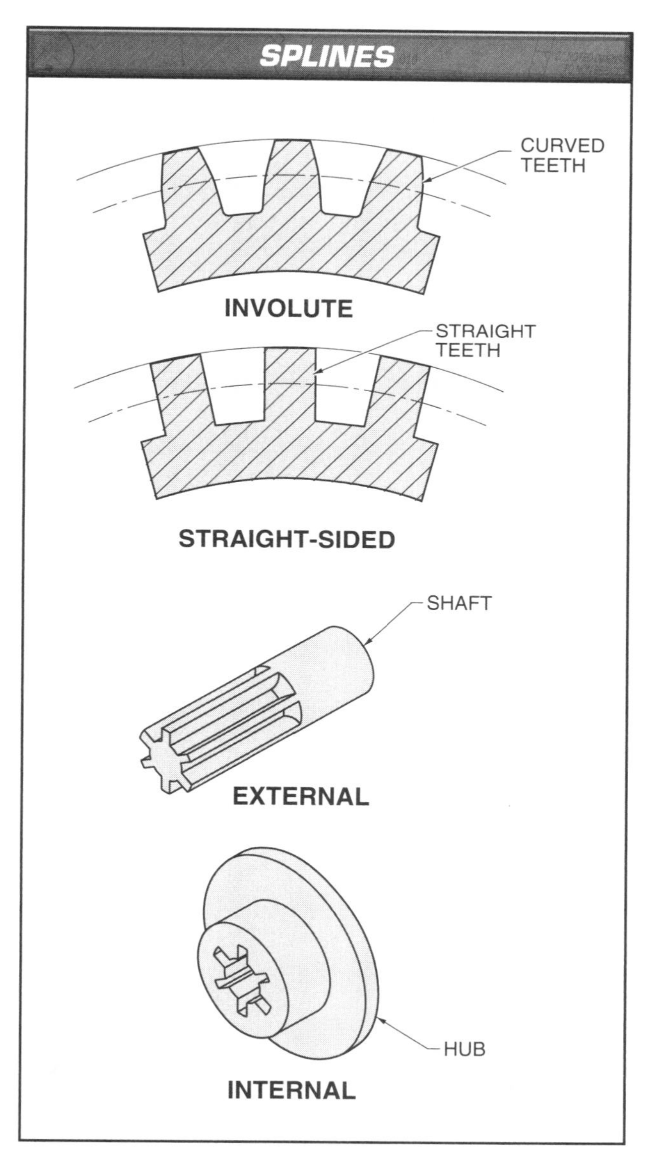

Figure 10-9. Splines prevent the rotation of a hub on a shaft.

Spur gears are drawn as cylinder shapes on prints and can be represented in any type of view. **See Figure 10-10.** It is typically not necessary to actually draw the gear teeth. Usually a gear is drawn with the outside and root circles as object or phantom lines and the pitch circle in between as a centerline. Adjacent notes specify the diametral pitch and any other relevant information to completely describe the gear.

Bevel Gears

A *bevel gear* is a gear with tapered teeth that is used in applications where shaft axes intersect. **See Figure 10-11.** Bevel gears resemble a cone shape rather than the cylinder shape of spur gears. Shafts of meshing bevel gears are most commonly positioned at 90°. In addition to speed reduction, bevel gears also change the axis and direction of rotation. A *miter gear* is one of a pair of bevel gears having the same number of teeth. Miter gears change the axis and direction of rotation, but not the speed.

Gears cause forces and vibrations to their shafts, which are typically supported with ball or roller bearings. These bearings are composed of inner and outer sleeves with balls or rollers in between. The balls or rollers allow the shaft, which contacts the inside sleeve, to rotate smoothly under high forces.

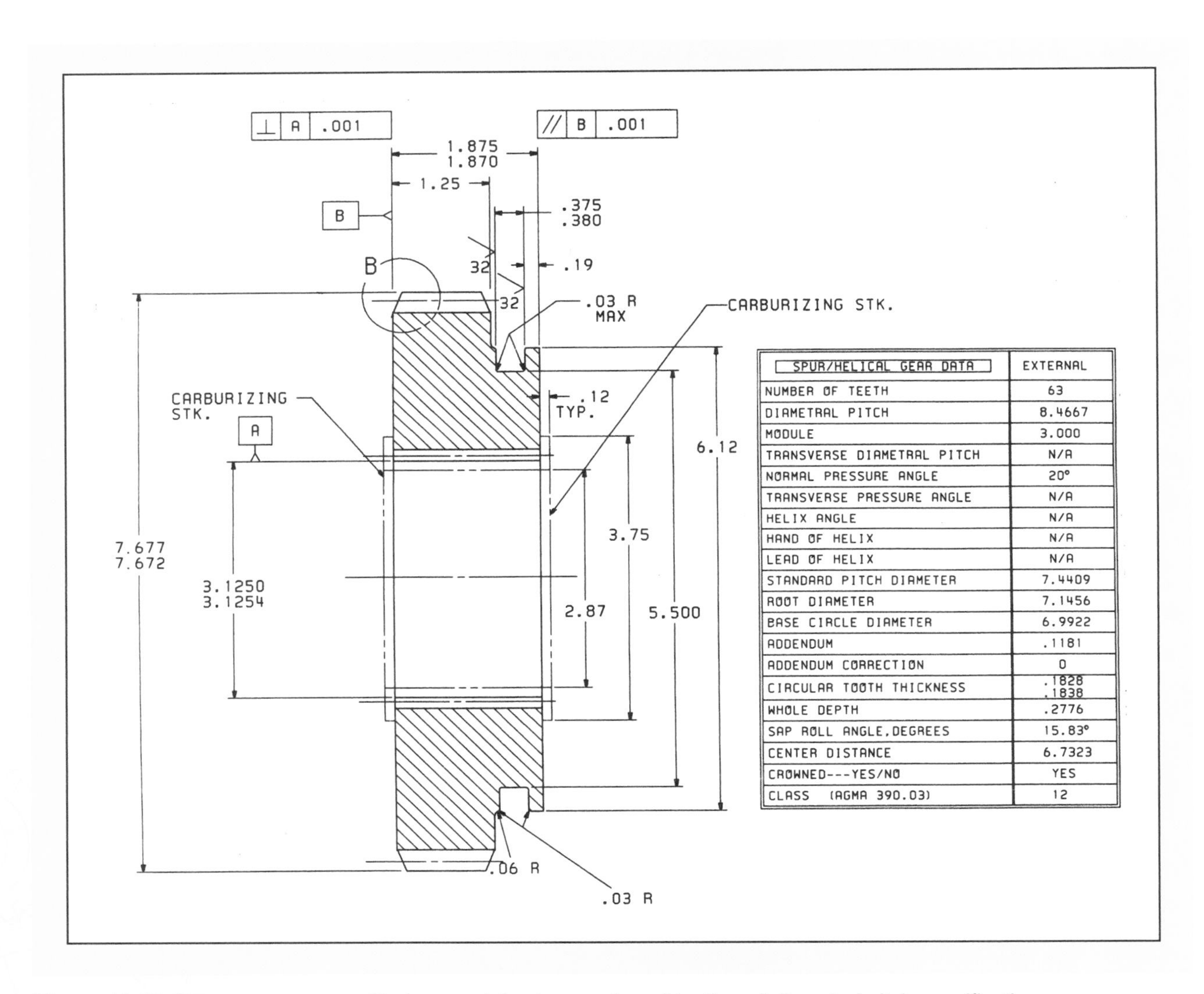

SPUR/HELICAL GEAR DATA	EXTERNAL
NUMBER OF TEETH	63
DIAMETRAL PITCH	8.4667
MODULE	3.000
TRANSVERSE DIAMETRAL PITCH	N/A
NORMAL PRESSURE ANGLE	20°
TRANSVERSE PRESSURE ANGLE	N/A
HELIX ANGLE	N/A
HAND OF HELIX	N/A
LEAD OF HELIX	N/A
STANDARD PITCH DIAMETER	7.4409
ROOT DIAMETER	7.1456
BASE CIRCLE DIAMETER	6.9922
ADDENDUM	.1181
ADDENDUM CORRECTION	0
CIRCULAR TOOTH THICKNESS	.1828 .1838
WHOLE DEPTH	.2776
SAP ROLL ANGLE, DEGREES	15.83°
CENTER DISTANCE	6.7323
CROWNED---YES/NO	YES
CLASS (AGMA 390.03)	12

Figure 10-10. Spur gears are specified on a print using number of teeth and diametral pitch specifications.

Bevel gears are manufactured in pairs to assure matching tapers. They are suitable for speed ratios ranging from 1:1 to 1:4. Bevel gears are primarily used in drive trains requiring shafts at right angles to each other.

The teeth of bevel gears may be straight or spiral. Straight bevel gears have straight teeth that would intersect at a point if extended outward. Straight bevel gears are most common and are economical to produce, but may be impractical for some applications. Spiral bevel gears have curved teeth. The curved teeth provide an overlapping contact action that provides smoother and quieter operation at high speeds.

Bevel gears are typically represented on a print in a section view. **See Figure 10-12.** In addition to the basic gear specifications, bevel gears will also include the pitch angle, root angle, axis of mating gear, and extension lines from the root, pitch, and outside surfaces that meet at points outside the gear.

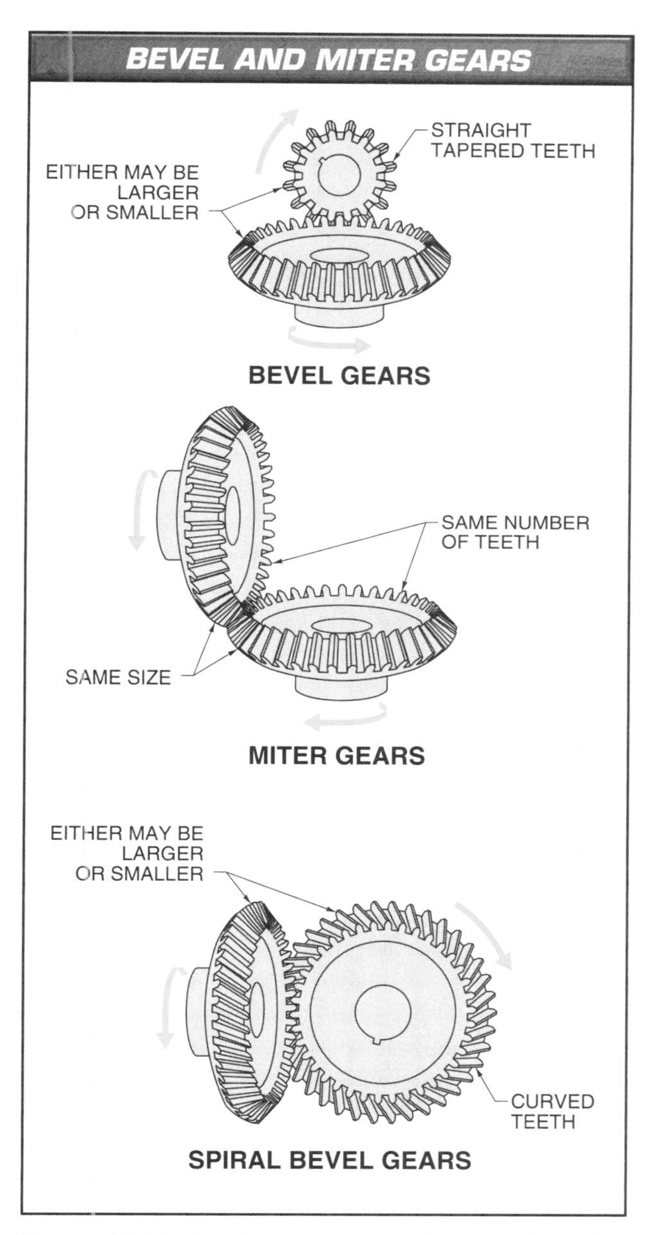

Figure 10-11. Bevel gears commonly transmit motion between shafts intersecting at 90°.

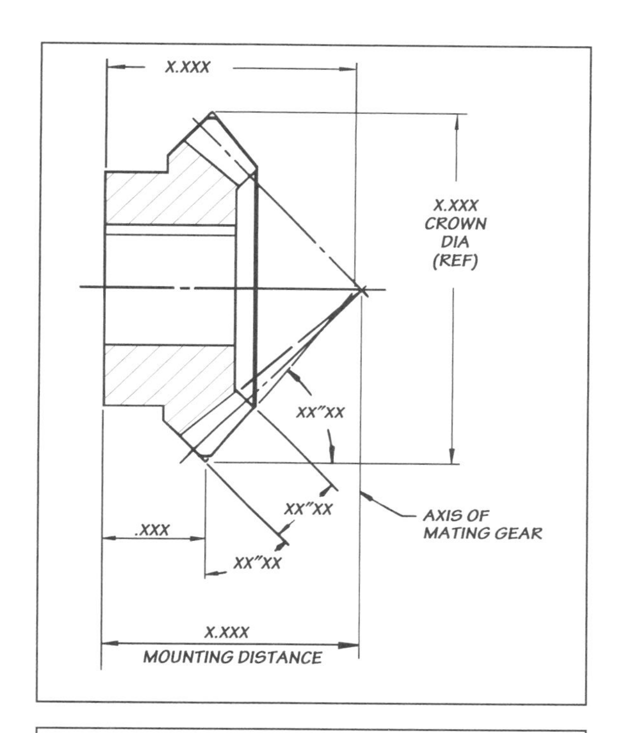

TOOTH CUTTING DATA

NUMBER OF TEETH	XX
DIAMETRAL PITCH	XX
PRESSURE ANGLE	XX°XX′
CONE DISTANCE	X.XXX
PITCH DIAMETER	X.XXX
CIRCULAR THICKNESS (REF)	.XXXX
PITCH ANGLE	X°XX′
ROOT ANGLE	XX°XX′
ADDENDUM	.XXX
WHOLE DEPTH (APPROX)	.XXX
CHORDAL ADDENDUM	.XXX
CHORDAL THICKNESS	.XXX
PART NUMBER OF MATING GEAR	XXXXX
TEETH IN MATING GEAR	XX
SHAFT ANGLE	XX°XX′
BACKLASH (ASSEMBLED)	.XXXX
TOOTH ANGLE (APPROX)	X°XX′
LIMIT POINT WIDTH	.XXX
TOOL EDGE RADIUS	XXX

Figure 10-12. Bevel gear specifications are shown in a section view on a print.

Helical Gears

A *helical gear* is a gear with teeth that follow a helix (spiral) shape and are not parallel to the shaft axis. The helix angles of a pair of helical gears must match in order to mesh. Depending upon their tooth orientation, shafts with helical gears may be parallel or nonintersecting. **See Figure 10-13.** Helical gears are smoother in operation than spur gears and are commonly used in automotive drive trains.

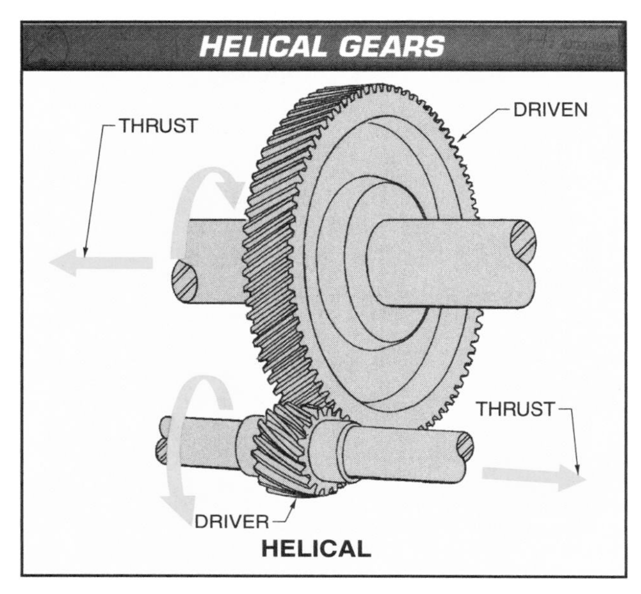

Figure 10-13. Helical gears have angled teeth that produce end thrust under load.

The teeth of helical gears are in greater contact with each other, but their angle causes end thrust, so the shafts require thrust bearings to absorb the load. The amount of end thrust produced is proportional to the helix angle of the teeth.

On prints, helical gears are drawn similarly to spur gears. Phantom lines indicate the root and outside diameters, and a centerline indicates the pitch diameter. Other information specific to helical gears, such as helix angle, may be included in notes.

Herringbone Gears

A *herringbone gear* is a gear that is composed of two rows of helical teeth. **See Figure 10-14.** The two sets of teeth have opposite helical angles, so they form a chevron, or herringbone, shape. There may or may not be a groove between them. Herringbone gears have parallel shaft alignment.

The pair of angled gear teeth keeps the gears aligned, retaining the advantage of helical gears' smooth operation. It also distributes the transmitted load evenly, eliminating the problem of end thrust. Herringbone gears provide quiet and efficient operation.

Herringbone gears are represented on drawings identically to helical gears, but as a pair. Some type of delineation, such as a line or gap, is shown between the sets of opposing teeth.

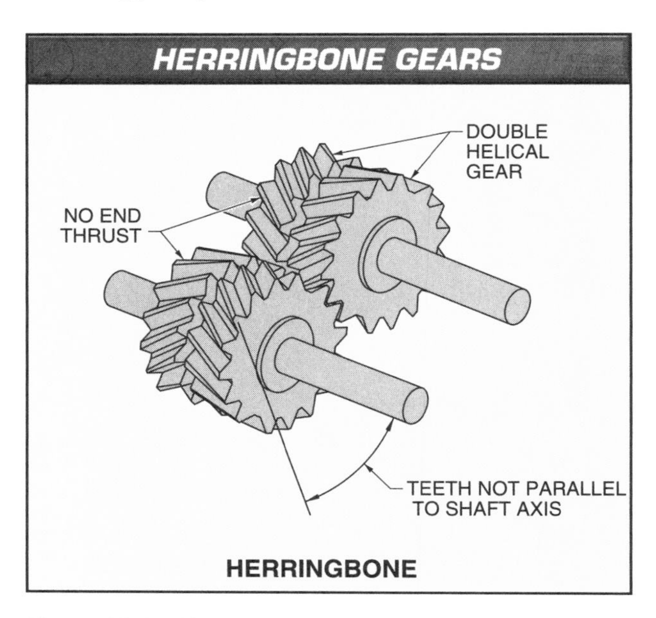

Figure 10-14. Herringbone gears are double-helical gears that cancel out the end thrust.

> It is important to note the shaft forces induced by gear meshing because it affects the design of the entire gear train and housing. The magnitudes and directions of the forces affect the necessary bearing types, gear materials, and gear arrangements.

Worms and Worm Gears

Pairs of gears consisting of a worm and worm gear are used for large speed reductions. **See Figure 10-15.** A *worm* is a threaded rod that rotates a worm gear. A *worm gear* is a special type of spur gear that is driven by a worm. The axes of the worm and worm gear do not intersect and are not parallel. Worm gear teeth are cut on an angle and in a concave shape to mate securely to the worm.

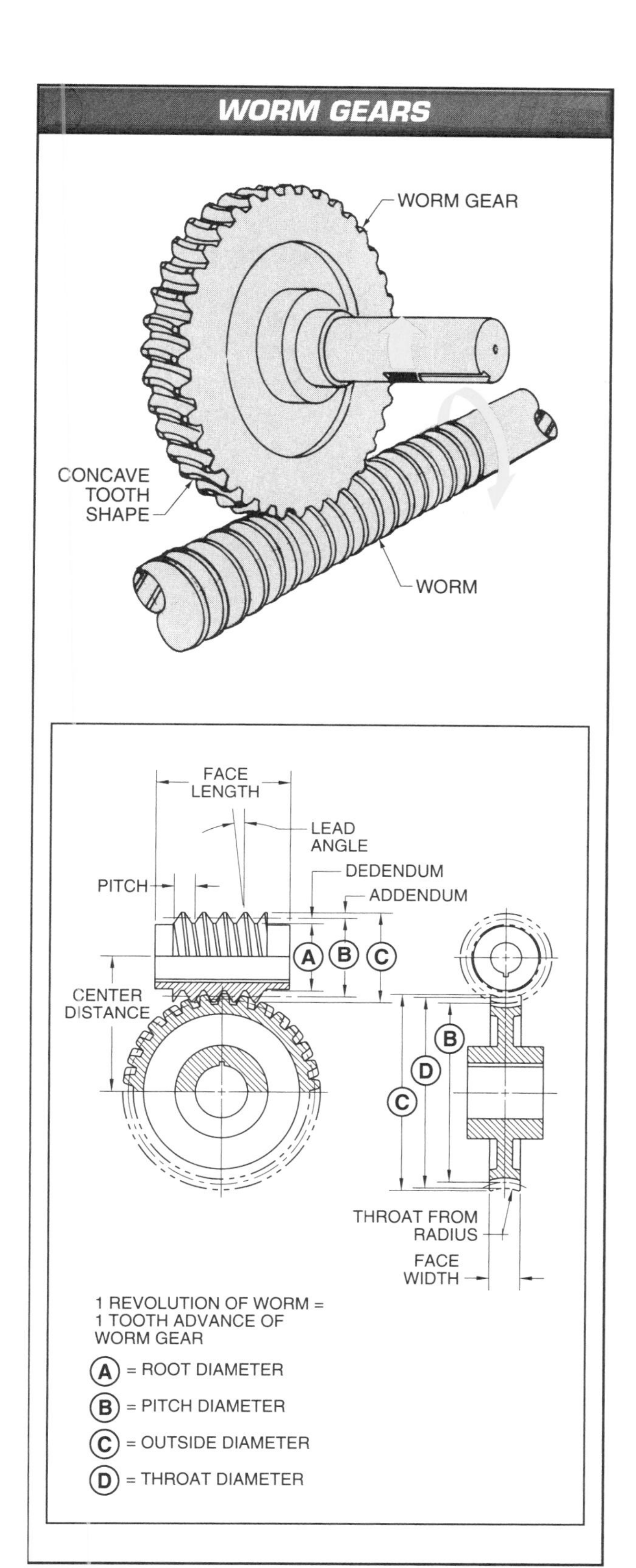

Figure 10-15. Worm gears are specified on the print with a section view detailing the gear teeth profile.

The worm gear is commonly designed with a single thread to advance one tooth for every revolution of the worm. Worms with multiple threads can drive more than one tooth per revolution. This provides excellent speed reduction capability and smooth, quiet service. Only the worm can be driven, however, never the worm gear.

Worm and worm gear dimensions are commonly represented in a full or half section. Specifications are included on the sectional views or in accompanying tables on the print.

Boston Gear

Worm and worm gear combinations produce very high gear ratios.

CAMS

A *cam* is a machine part that transmits a pattern of motion using an irregular external or internal surface. Cams are useful in creating complex motions that would be difficult to reproduce using other mechanical parts.

Cams require cam followers. A *cam follower* is a machine part in contact with a cam path that moves as the cam rotates. Cam followers can be arranged to derive either liner or rotary motion.

Cam Types

The shape of the cam determines the direction and motion of the cam follower. Cams can be used with other cams to produce a combination of motions. Common cam designs include the plate cam, face cam, barrel cam, and yoke cam. **See Figure 10-16.**

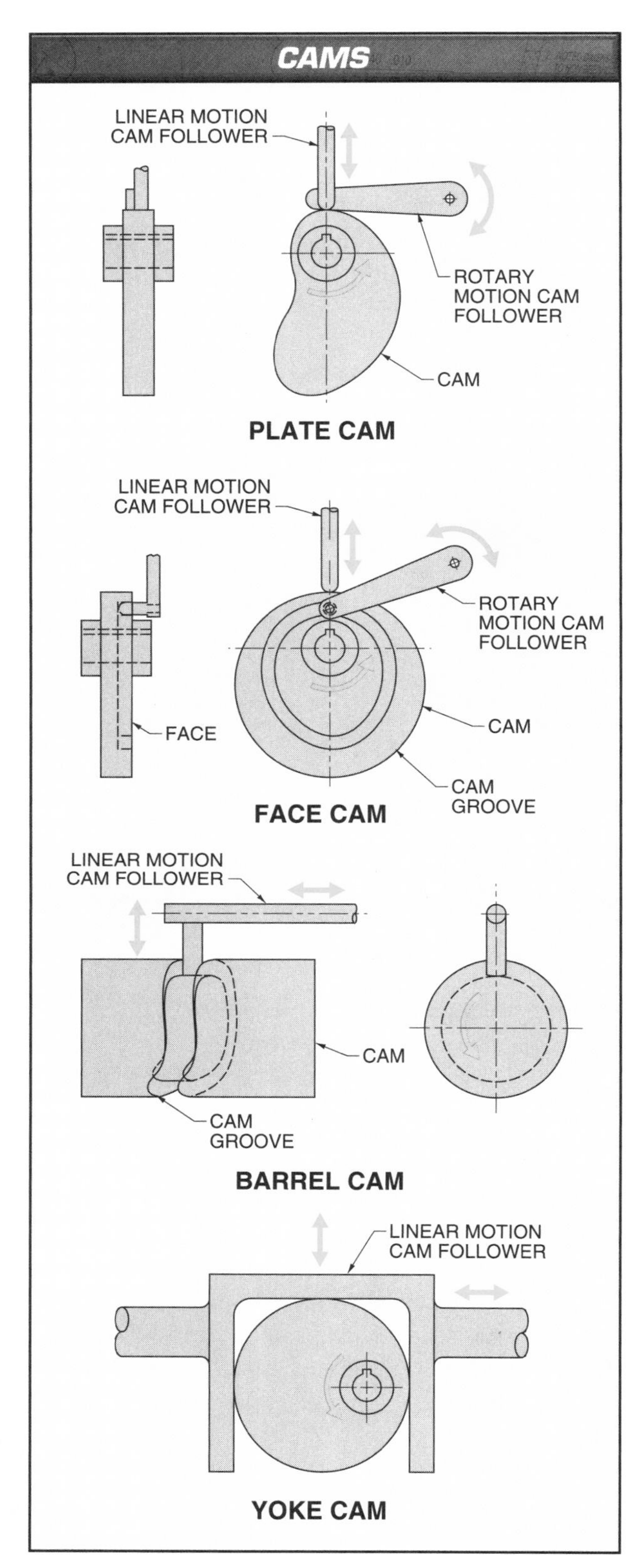

Figure 10-16. Complex machine motions can be accomplished with cams and cam followers.

A *plate cam* is a cam made from a flat plate that contacts the follower at its perimeter. The cam's radius changes as it rotates, changing the distance between the follower and cam surface. A follower on a plate cam only moves up and down. A *face cam* is a cam that uses a groove cut into its face, instead of or in addition to the perimeter, to contact a follower.

A *barrel cam* is a cylindrical cam with a groove in its surface that guides the follower. Followers on barrel cams can move both vertically and laterally. A *yoke cam* is a cam that rotates within an enclosure in order to translate the rotary motion into both vertical and lateral movement. Yoke cams will typically be round, but with an offset axis of rotation.

Cam Followers

Cam followers are designed for specific applications requiring certain loads and speeds. Common cam followers include flat, pointed, edge, spherical, and roller. Each of the cam followers must have a spring to maintain contact with the cam. **See Figure 10-17.** Because of the surface contact, most cam followers must have hardened bearing surfaces.

A *flat cam follower* is a follower that contacts the cam with a flat surface. Flat cam followers are used primarily with cams that rotate slowly and have gently curving profiles.

A *pointed cam follower* is a follower that ends in a conical tip. An *edge cam follower* is a follower shaped like a chisel that contacts the cam with an edge. Pointed and edge cam followers are required when cam profiles have more abrupt feature changes.

A *spherical cam follower* is a follower with a rounded end. These followers produce smooth motion in all directions. Round grooves can be machined in the cam to mate with spherical cam followers. A *roller cam follower* is a follower tipped with a rotating wheel that significantly reduces friction.

Unlike gears, which are stocked by vendors in standard sizes and pitches, cams are usually custom made for a particular application. This is why assemblies using cams must include cam prints. The cam print is a plot of displacement versus degrees of rotation. A cam print shows all dimensions and/or formulas needed to describe a unique pattern of displacement for each revolution.

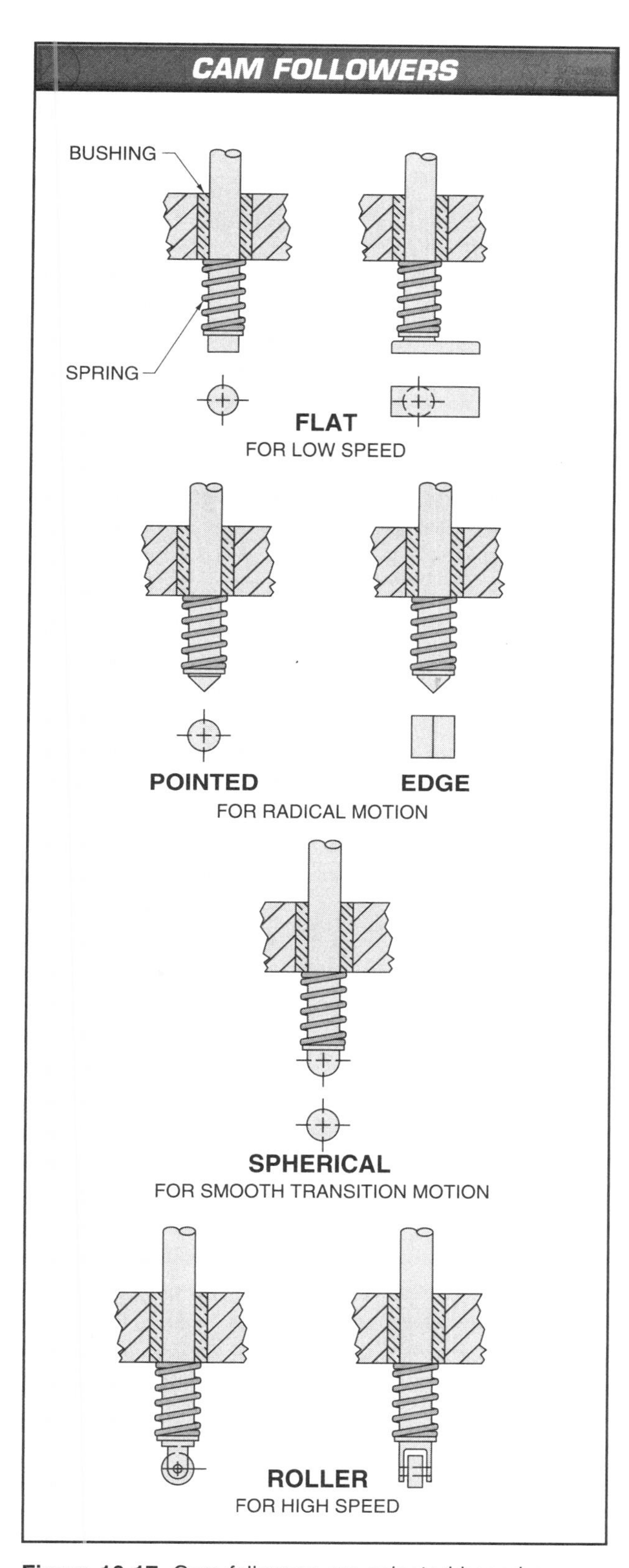

Figure 10-17. Cam followers are selected based on cam design and operating speed.

CAM SPECIFICATIONS

Cams are custom designed for each unique application, so there are no standard sizes or shapes to be designated. In most instances where a cam is specified on a print, the cam will require a detail drawing of its own to completely describe its shape. Certain features are given standard names, but each must be separately defined.

Cam Profiles

A *cam profile* is the shape of the cam perimeter or another path that actuates the cam follower. **See Figure 10-18.** A *lobe* is a projecting part of a cam that causes the cam follower to be displaced. Depending upon the required motion, the cam may have several lobes to obtain the desired motions.

A *trace point* is the point of contact between the cam and cam follower. The *cam displacement* is the maximum distance from the lowest to the highest point of the cam. The *cam rise* is the highest point on a cam. The *cam drop* is the lowest point on a cam. The *base circle* is the circle formed at the radius of a cam drop. *Dwell* is the time during the cam revolution in which there is no motion of the cam follower.

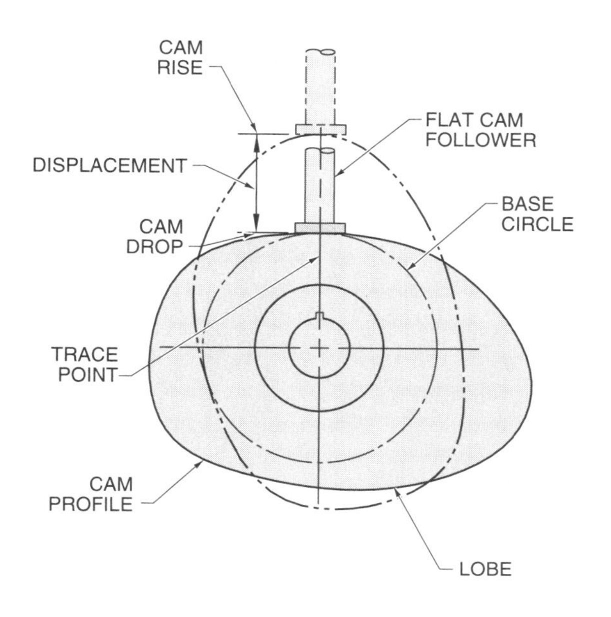

Figure 10-18. The cam profile shows the motion of the cam follower during machine operation.

Automotive engine cams control the movement and timing of the valves.

Cam Diagrams

Producing a cam to meet design specifications requires interpretation of the print information. Cam design is specified using a diagram that plots displacement against degrees of rotation. A complete revolution (360°) of the cam must be represented on the cam diagram. **See Figure 10-19.**

Plate and yoke cams are specified on a print showing the amount of motion or rise using radii against travel. The radii may be measured from the cam centerpoint or from an established base circle. Face cams can use a set of two such diagrams, one for the groove and one for the perimeter. Barrel cams are specified on a print with a pattern that shows the motion of the cam as it would appear unrolled.

> Cam motions may follow mathematically generated functions, such as parabolic motion, or be designed as irregular motions. Depending upon the manufacturing process, cam shape may be specified as mathematical formulas or as a group of points that are dimensioned from datums.

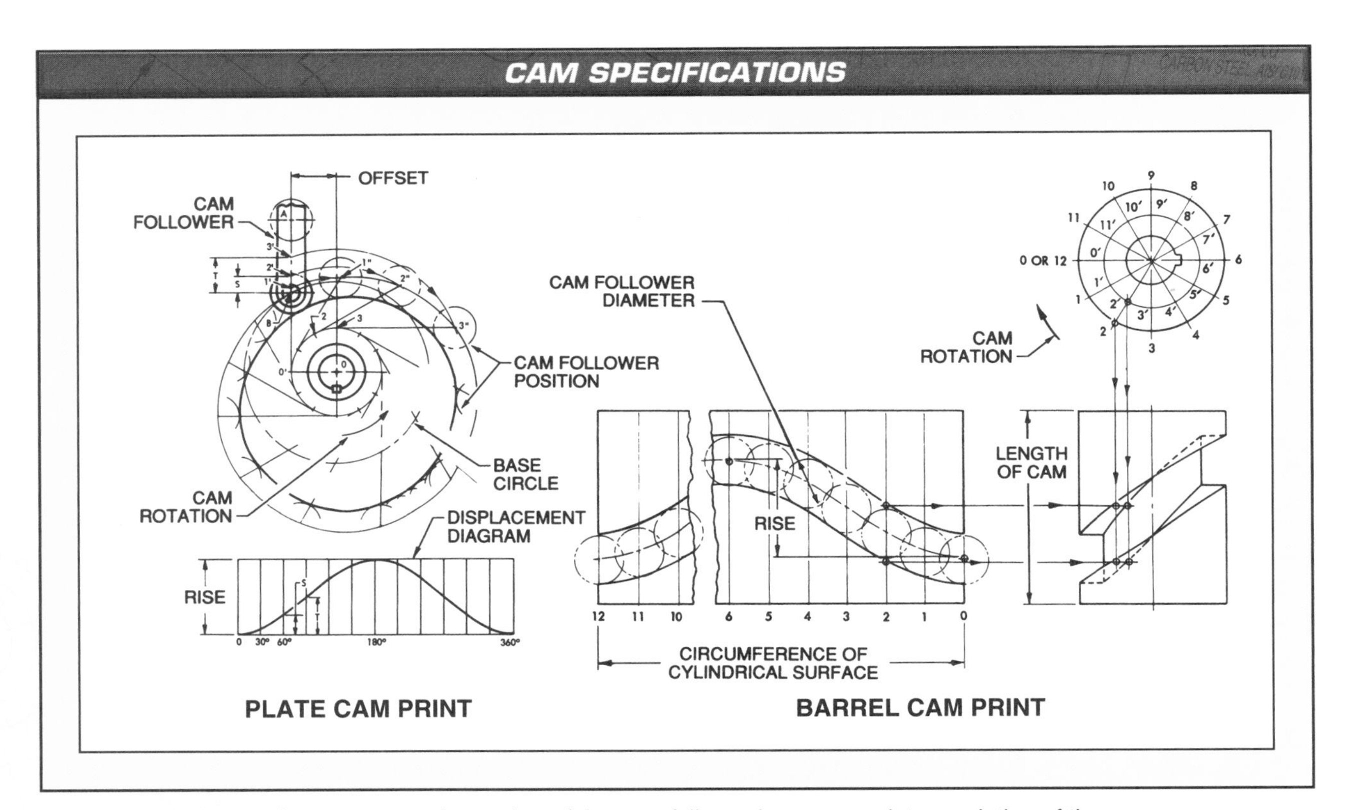

Figure 10-19. A cam diagram traces the motion of the cam follower in one complete revolution of the cam.

Cam Motion

Cam designs often use the same lobe shapes for portions of their profiles involving rising or falling motions. These common lobe shapes produce constant-velocity, parabolic, or harmonic motion. **See Figure 10-20.**

Constant-velocity motion is motion at a constant rate of speed. This is the simplest type of motion, but the disadvantage is the abrupt change at the beginning and end with virtually no transition to the constant speed.

Parabolic motion is motion in which displacement follows a parabolic curve and the acceleration and deceleration is constant in absolute value. The resulting motion is not at a constant speed as with constant-velocity motion. The speed rises linearly until the midpoint and then falls linearly back to zero.

Harmonic motion is motion in which acceleration changes smoothly, causing displacement to vary according to a sinusoidal curve. The speed changes gradually in the midpoint and the motion begins and ends in smooth arcs. If the displacement pattern is acceptable for the application, these smooth transitions are desirable because they cause less wear and shock to the follower.

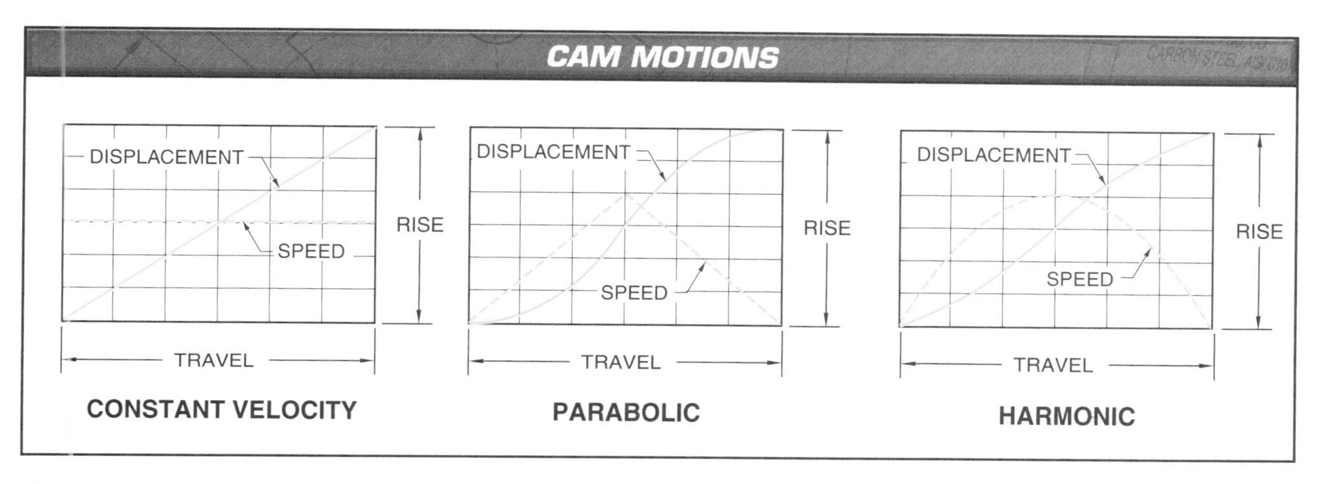

Figure 10-20. Motions of cams and cam followers are designed to provide efficient operation and maximum life expectancy.

REVIEW QUESTIONS 10

Name ______________________________ Date ______________

Completion

_______________ **1.** A(n) ___ gear is the smallest of meshing gears.

_______________ **2.** Meshed gears of equal ___ rotate at the same rate.

_______________ **3.** A(n) ___ gear has straight teeth that are parallel to the shaft axis.

_______________ **4.** ___ gears change the axis and direction of rotation, but not the speed.

_______________ **5.** A cam ___ is a machine part in contact with a cam path that moves as the cam rotates.

_______________ **6.** A(n) ___ is a spur gear that is flat rather than concentric.

_______________ **7.** The ___ is the portion of the gear teeth between the pitch circle and the outside diameter circle.

_______________ **8.** ___ bevel gears have curved teeth.

_______________ **9.** ___ are useful in creating complex motions that would be difficult to reproduce using other mechanical parts.

_______________ **10.** A(n) ___ cam follower is tipped with a rotating wheel that significantly reduces friction.

True-False

T F **1.** The axes of the worm and worm gear are parallel.

T F **2.** Helical gear teeth are not parallel to the shaft axis.

T F **3.** A lobe is a projecting part of the cam.

T F **4.** Parabolic cam motion follows a parabolic curve and the acceleration and deceleration is constant in absolute value.

T F **5.** Flat cam followers are designed for high-speed applications.

T F **6.** Cam design is specified using a diagram that plots displacement against degrees of rotation.

T F **7.** Backlash should always be eliminated for maximum gear efficiency.

T F **8.** Meshing gears all rotate in the same direction.

T F **9.** The involute is the curve used to determine the geometric profile of gear teeth.

T F **10.** Bevel gears are manufactured in pairs to assure matching tapers.

Multiple Choice

_______________ **1.** ___ is the small amount that one gear can rotate before a meshing gear begins to move.
A. Circular pitch
B. Tooth flank
C. Backlash
D. Lost mesh

_______________ **2.** ___ contributes to standards on the specification and drawing practices of gears.
A. AGMA
B. AISI
C. UNS
D. ASTM International

_______________ **3.** Gear ___ is the relationship between the sizes of two meshing gears.
A. ratio
B. fraction
C. compounding
D. pitch

_______________ **4.** All bevel gears have ___ teeth.
A. screw thread
B. tapered
C. curved
D. helical

_______________ **5.** Cam ___ is the maximum distance from the lowest to the highest point of the cam.
A. profile
B. lobe
C. circle
D. displacement

_______________ **6.** The ___ is the distance between corresponding points of two adjacent gear teeth on the pitch circle.
A. involute
B. addendum
C. circular pitch
D. profile

_______________ **7.** A ___ is a series of teeth or parallel surfaces machined into a shaft or hub.
A. cam
B. spline
C. lobe
D. key

______ **8.** Shafts of meshing ___ gears are commonly positioned at 90°.

A. spur
B. internal
C. herringbone
D. bevel

______ **9.** ___ gears have two rows of helical teeth.

A. Worm
B. Spur
C. Miter
D. Herringbone

______ **10.** ___ is the number of teeth per unit of pitch diameter.

A. Diametral pitch
B. Pitch circle
C. Circular pitch
D. Root diameter

Identification — Gear Types

______ **1.** Helical

______ **2.** Bevel

______ **3.** Herringbone

______ **4.** Worm

______ **5.** Spur

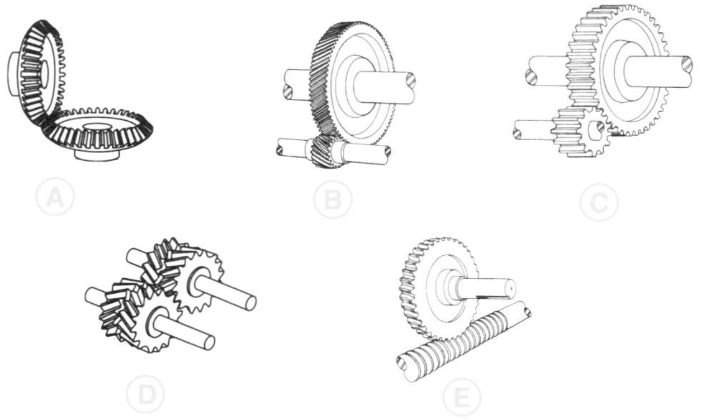

Identification — Gear Specification

______ **1.** Tooth fillet

______ **2.** Working depth

______ **3.** Tooth flank

______ **4.** Root diameter

______ **5.** Pitch circle

______ **6.** Outside diameter

______ **7.** Addendum

______ **8.** Pitch diameter

______ **9.** Clearance

______ **10.** Circular pitch

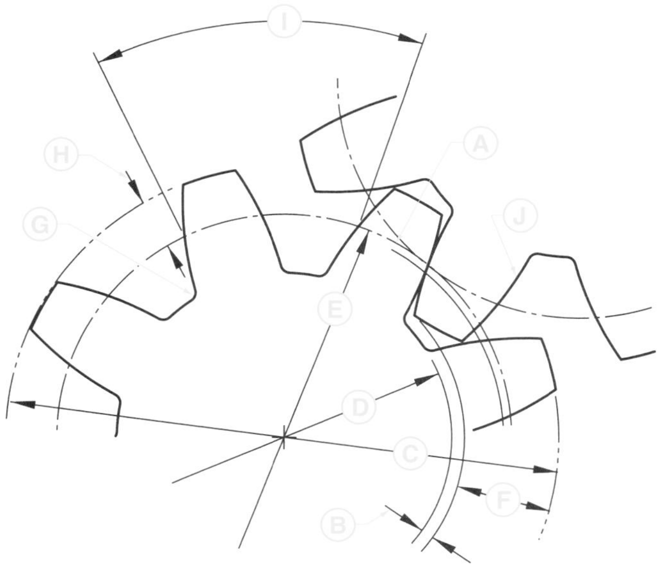

Identification — Cam Types

______________________ **1.** Plate

______________________ **2.** Barrel

______________________ **3.** Face

______________________ **4.** Yoke

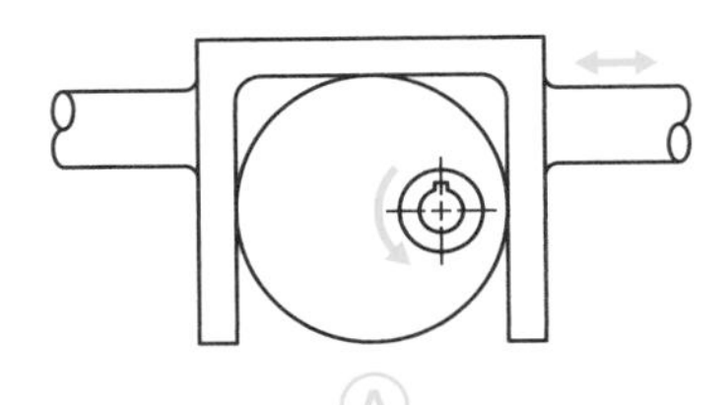

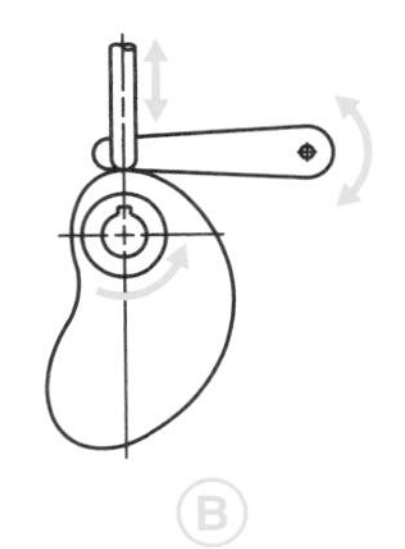

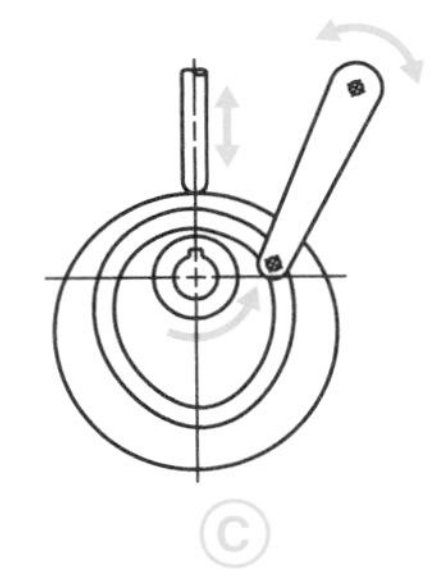

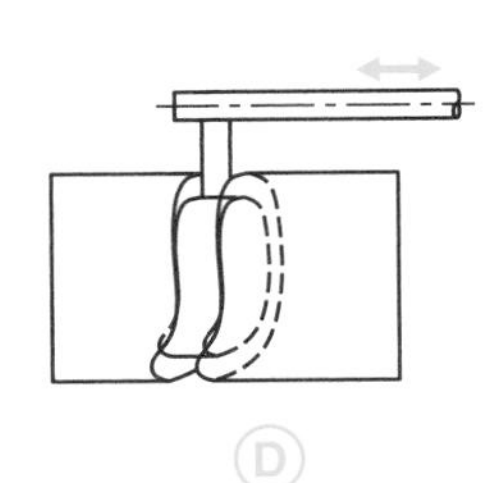

Identification — Cam Specification

______________________ **1.** Base circle

______________________ **2.** Cam rise

______________________ **3.** Cam follower

______________________ **4.** Displacement

______________________ **5.** Cam drop

______________________ **6.** Lobe

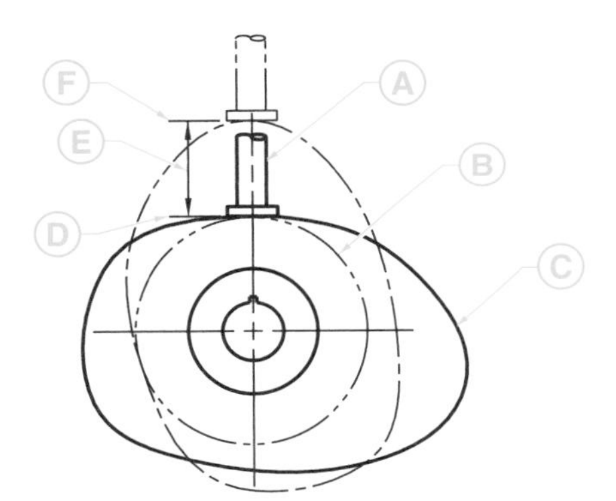

TRADE COMPETENCY TEST 10

Gear 1070

Refer to print on page 256.

______________ **1.** The specified width of the gear at the face is ___″.

______________ **2.** A chamfer having a(n) ___″ depth is specified.

______________ **3.** The keyseat is specified for a(n) ___″ maximum depth.

______________ **4.** The ratio of the gears referred to on the print is ___.

______________ **5.** The maximum outside diameter is ___″.

______________ **6.** The gear has ___ (number of) teeth.

______________ **7.** The gear shown on the print is designed to mesh with gear number ___.

______________ **8.** The keyseat is specified for a(n) ___″ minimum width.

______________ **9.** The total tolerance for the hole diameter is ___″.

______________ **10.** The maximum hub width specified is ___″.

______________ **11.** The gear has a diametral pitch of ___.

______________ **12.** The maximum hole size specified is ___″.

______________ **13.** Tooth form for the gear is ___.

______________ **14.** The drawing number is ___.

______________ **15.** The root diameter specified is ___″.

T F **16.** The part number for the gear is 1215287.

T F **17.** The 5″ dimension specified is a manufacturer's option.

T F **18.** The maximum allowable width of the gear at the face is 3.730″.

T F **19.** The minimum outside diameter specified is 9.368″.

T F **20.** The gear is case hardened to a depth of .025″ to .035″ after grinding.

______________ **21.** Symbol ___ specifies circular runout tolerance on the chamfer side.

______________ **22.** The gear shown on the print meshes with a gear that has ___ (number of) teeth.

______________ **23.** The keyseat is specified for a(n) ___″ maximum width.

______________ **24.** The helix angle specified is ___.

______________________ **25.** Symbol A specifies a tolerance zone of ___″.

______________________ **26.** The minimum hub width is ___″.

______________________ **27.** Finished weight of the gear is ___ lb.

______________________ **28.** The pitch circle diameter specified is ___″.

______________________ **29.** Symbol ___ specifies cylindricity tolerance.

______________________ **30.** The minimum hole size is ___″.

______________________ **31.** The gear is specified to be carburized and hardened per specification number ___.

______________________ **32.** Unless specified on the print, tolerances to one decimal place are ±___″.

______________________ **33.** The material specified for the gear is ___ steel.

______________________ **34.** The keyseat is specified for a(n) ___″ minimum depth.

______________________ **35.** A total tolerance of ___″ is specified for the outside diameter of the gear.

______________________ **36.** Symbol A references datum ___.

______________________ **37.** The ___ is indicated by a hidden line on the print.

______________________ **38.** The gear specified is a(n) ___ helical gear.

______________________ **39.** Unless otherwise specified, angular tolerance is ±___°.

______________________ **40.** The gear is specified for a(n) ___ helix.

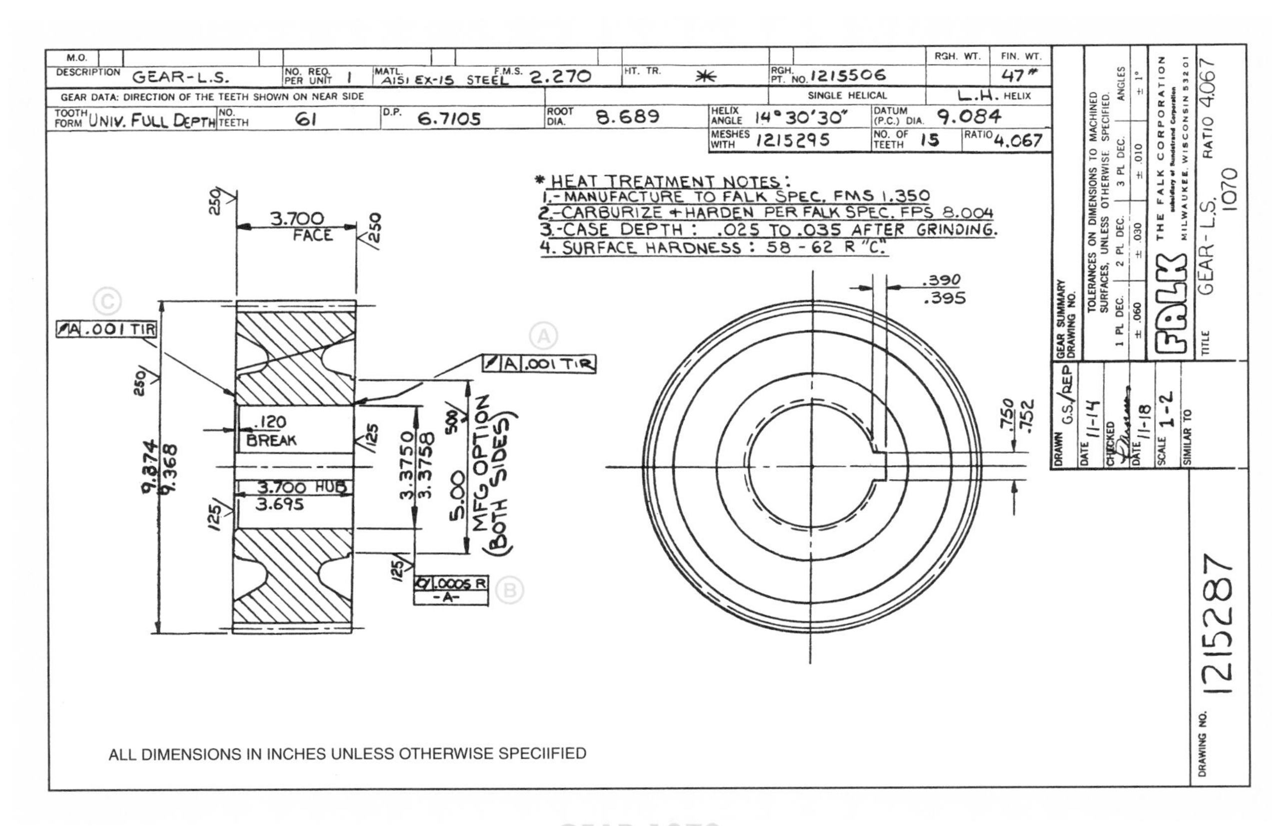

11 Numerical Control

Numerical control is widely used in modern machine shops. Numerical control documents define the machining processes used to produce a part. The advantages of this form of tool control include repeatability, productivity, and efficiency. The prints are dimensioned and annotated in a way that makes the programming easier. Numerical control programs are composed of commands and codes that define each step in the manufacturing process.

NUMERICAL CONTROL PROCESSES

Numerical control (NC) is a process of automatically controlling the motion of machine tools using a set of programmed commands. The process is repeatable because the same set of commands and operations is completed in the same manner every time. This allows for control of operations and close tolerances to be maintained. All common machine operations, such as drilling, turning, milling, grinding, boring, tapping, and cutting, can be numerically controlled.

A related type of machining is actually only an aid to manual machining and does not directly control the machine. A *digital readout (DRO)* is a device on a machine tool that provides a digital display of the cutting tool's position in space. **See Figure 11-1.** This helps the machine operator make precise movements and cuts while still controlling the machine manually.

NC Machining

The first generations of NC machines used punch tape to transfer control codes to the machine. Using prints of the desired part, programmers wrote code to control the movements of the machine and make the cuts necessary to produce the part. The code was translated into a series of holes in a punch tape that the machine could read and interpret. This type of machine control is rarely used now.

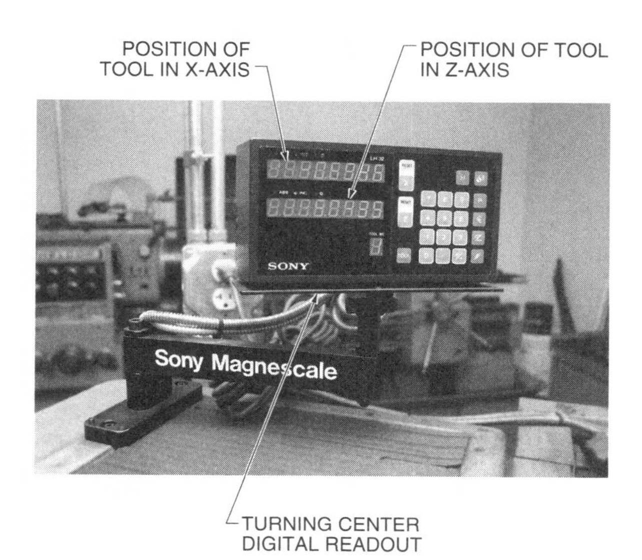

Figure 11-1. Digital readouts are NC aids for manual machining.

Computer Numerical Control Machining

Most modern machines use a computer to interpret geometric part information and control the machining process. These are called computer numerical control (CNC) machines. The computers may be connected to the machines directly or over a network. Information on part geometry can be entered manually from a paper print or can be converted from a CAD drawing. The geometric information is translated into code that controls the movement of the machine to produce the part with little or no assistance from the operator. The most common CNC machines are machining centers and turning centers. **See Figure 11-2.**

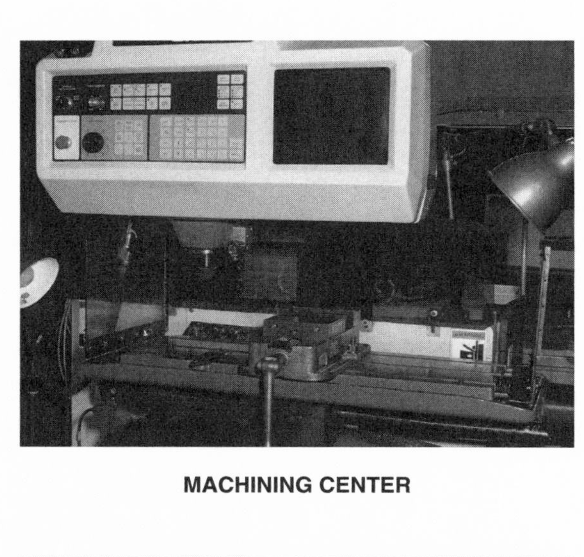

MACHINING CENTER

TURNING CENTER

Figure 11-2. Two NC machines are a machining center and a turning center.

Rapid Prototyping

Rapid prototyping is an NC process that builds parts in an additive process rather than the traditional subtractive (material removing) processes. This means that these special machines build parts from the bottom up in successive layers, usually from some type of thermoplastic. **See Figure 11-3.** Rapid prototyping was originally only used to quickly build concept prototypes for feasibility testing. The development of more sophisticated processes using a variety of materials has allowed some machines to build even production-quality parts ready for final assembly or sale.

RAPID PROTOTYPING

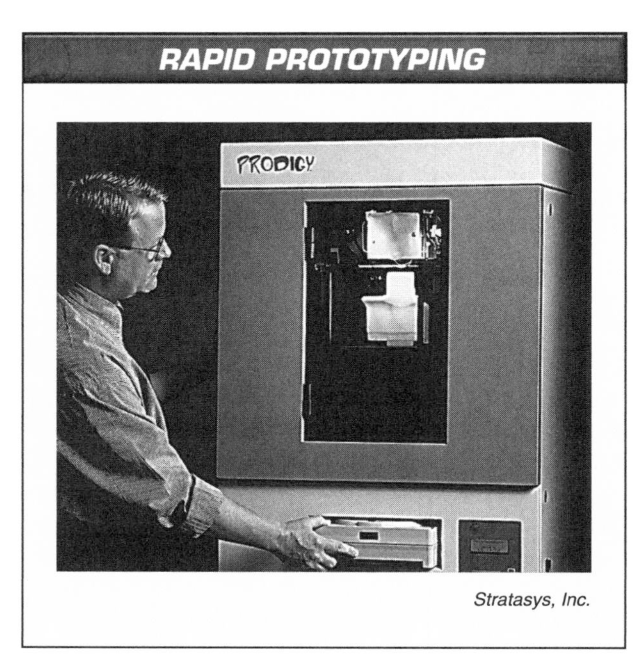

Stratasys, Inc.

Figure 11-3. Rapid prototyping machines produce parts quickly and without tooling.

NUMERICAL CONTROL FEATURES

NC offers many advantages. Productivity is enhanced because operator error is virtually eliminated and fixed production times can be set and followed. Production typically stops only for maintenance shutdowns. In addition, there is a reduced startup time on the production floor since all programs can be tested on the computer prior to sending the commands to the machines.

NC programs generally operate more efficiently because they permit rapid tool changes and reduced tooling requirements. Tool carousels contain multiple tools that can be automatically changed during machining operations. This allows many types of machine processes to be completed without machine changes. Most machining and turning centers have a means of holding the workpiece that is more flexible than manual machines. This eliminates the need for costly tooling of jigs and fixtures.

Cartesian Coordinate System

All NC programs are based on the Cartesian coordinate system. The *Cartesian coordinate system* is a system of defining a position or motion in space by its distance in three perpendicular directions from an origin point. The three directions (axes) are designated x, y, and z. **See Figure 11-4.** All programming begins at the origin. For this reason, many drawings are dimensioned from an origin to allow for easier programming.

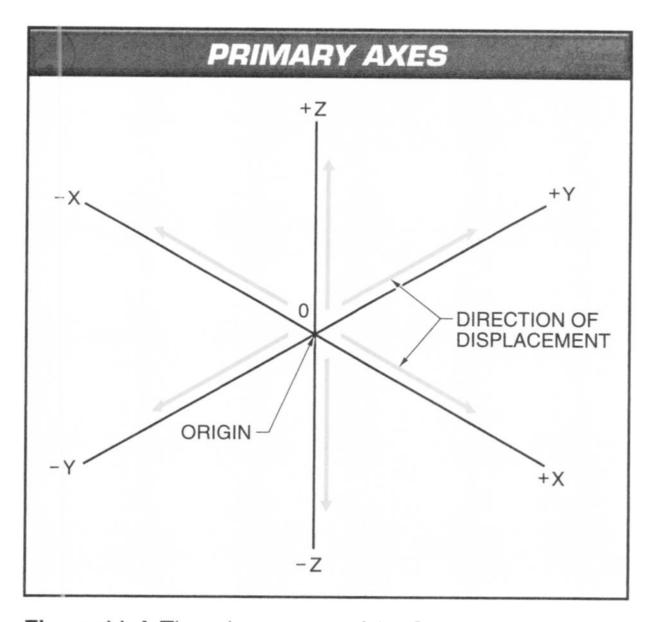

Figure 11-4. The primary axes of the Cartesian coordinate system are the x-, y-, and z-axes.

In addition to the three primary axes, machining centers may have up to three additional axes of operation based on rotation around the primary axes. These are designated with A for rotation around the x-axis, B for rotation around the y-axis, and C for rotation around the z-axis. **See Figure 11-5.**

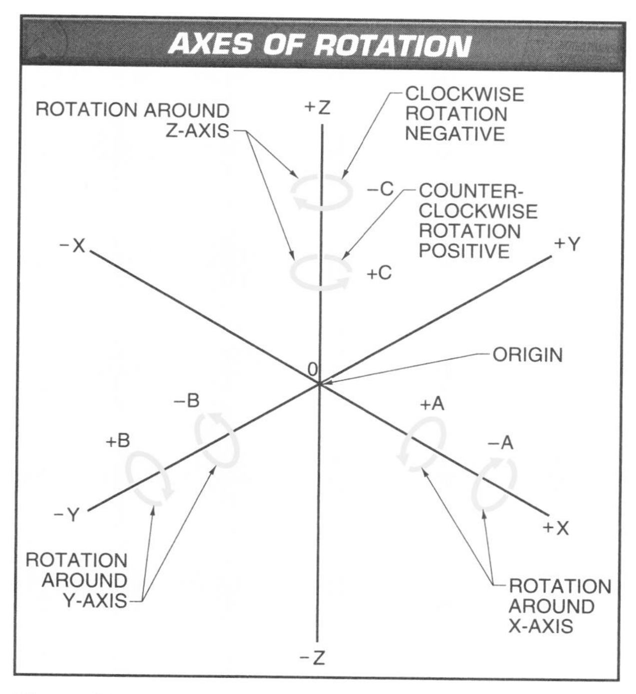

Figure 11-5. Axes of rotation provide three additional directions to control motion.

Travel along or around these axes allows numerically controlled machines to produce parts with complex shapes. Machines have the ability to control movement by referencing two or more axes. A flame cutter controls the x- and y-axes, while a turning center controls the x- and z-axes. Machining centers control at least the three primary axes, though many control axes of rotation as well. **See Figure 11-6.**

A lathe controls the x- and z-axes.

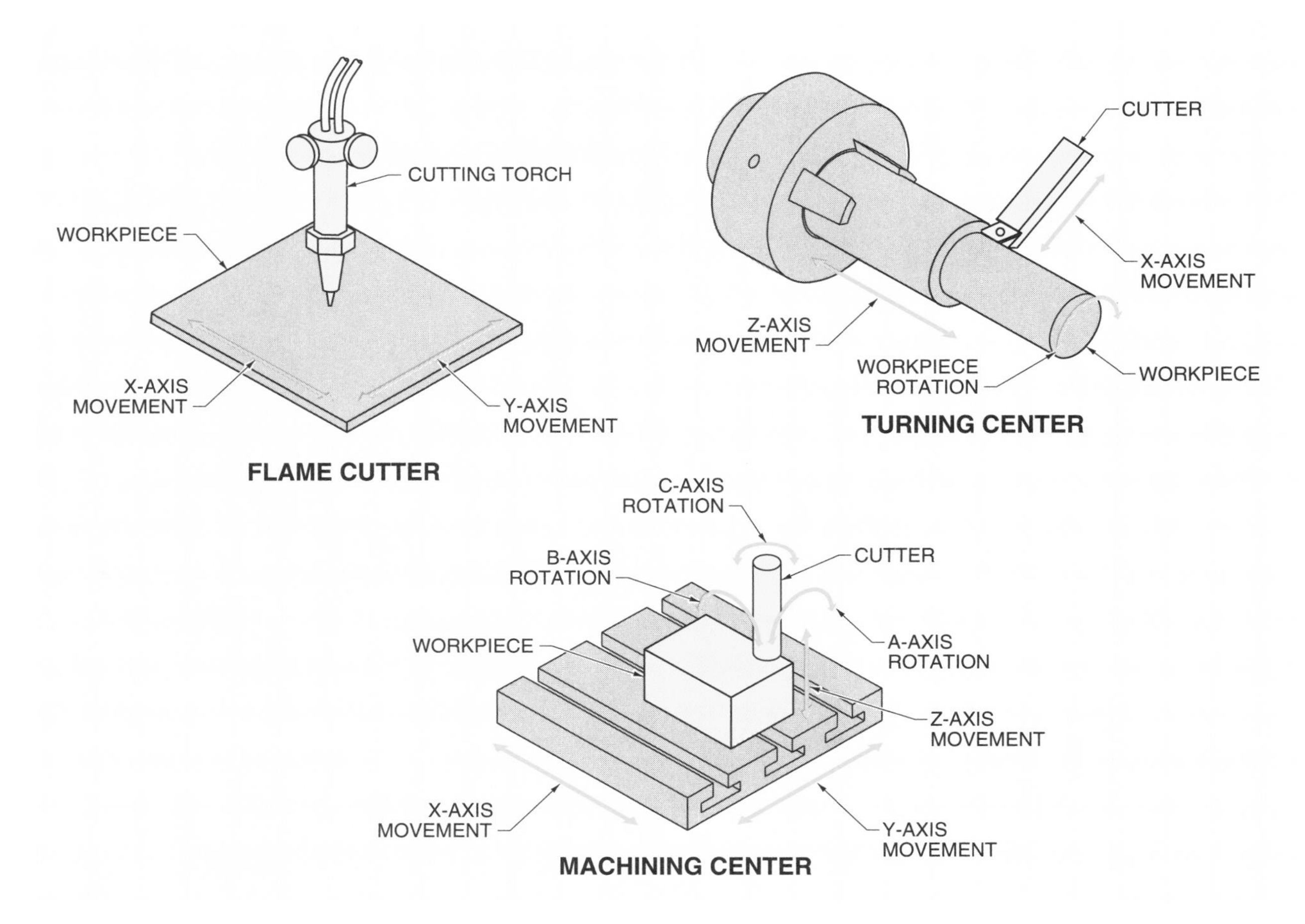

Figure 11-6. Different manufacturing equipment requires control of specific axes of movement.

NC Functions

NC systems provide a powerful and flexible way to manufacture a variety of complex parts quickly. Once set up, the systems control every aspect of the machining operation, including the following:

- sequence of operations
- time intervals between operations
- positions and motions of the cutting tool
- positions and motions of the workpiece
- spindle speeds of the cutting tool
- automatic changing of cutting tools
- application of coolant
- status of the machine and cutting tool
- machine shutdown procedures

NUMERICAL CONTROL COMMANDS

An *NC program* is a collection of machining commands used to produce a specific feature or part. A *command* is a line of the necessary codes to complete one step in an operation. Each line begins with an N number, which identifies that command and puts all the commands in the proper sequence. The commands are stored on the appropriate media, usually an electronic file.

> It is often unnecessary to write G-code commands manually. Most CAD software packages can automatically generate the G-codes necessary to machine a design once certain workpiece size and set-up information is provided.

Programming Methods

For each part or feature, there are many different ways to program the machining process. The method and sequence used may depend upon the machine, tooling, workpiece material, and tolerances. The two primary methods are incremental and absolute programming. **See Figure 11-7.**

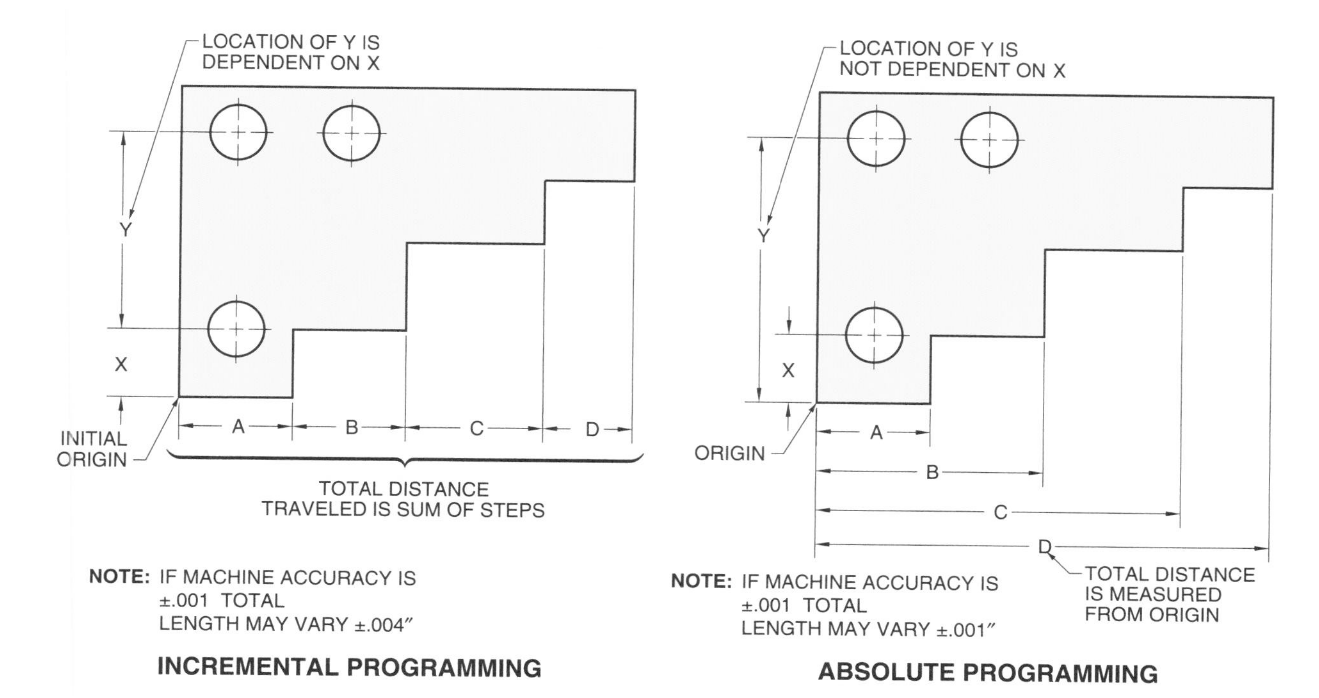

Figure 11-7. NC programming may be incremental or absolute.

Incremental Programming. *Incremental programming* is a method of generating NC commands where each point is specified in relation to the last point where the machine stopped. This concept is similar to the idea of point-to-point dimensioning. Likewise, this method of programming is less accurate because of the addition of tolerances for each consecutive movement.

Absolute Programming. *Absolute programming* is a method of generating NC commands where each point is specified in relation to the origin. This is similar to rectangular coordinate dimensioning. With this method, there is no accumulation of tolerances as occurs in incremental programming.

NC Codes

Numerically controlled machines are programmed to perform a series of operations using a system of codes. A *code* is an instruction that consists of a letter to designate the type of action required and a number to specify the command code, distance, or speed. Each line of instruction, or command, includes one or more codes.

G-Codes. NC machine tools may use a proprietary code, but most use or incorporate G-code. *G-code* is a programming language for NC machining that consists of a series of standardized letter and number codes for machining operations. Each code consists of a letter prefix and a number. Almost every letter in the alphabet is assigned a meaning. **See Figure 11-8.** G-code is named for one of the most common letter prefixes. Depending on the letter prefix, the number may identify a standardized instruction or quantify a distance, speed, or other value.

Since they require no manual actions after being set up, CNC machining centers are often set in enclosures for safety.

STANDARD G-CODE DESIGNATIONS	
LETTER PREFIX	DESCRIPTION
A	Absolute or incremental position of A axis (rotational around x-axis)
B	Absolute or incremental position of B axis (rotational around y-axis)
C	Absolute or incremental position of C axis (rotational around z-axis)
D	Diameter or radial offset (for cutter size compensation)
E	Precision feed rate for threading on lathes
F	Feed rate
G	Initialization commands
H	Tool length offset
I	Arc size in x-axis
J	Arc size in y-axis
K	Arc size in z-axis
L	Fixed cycle loop count
M	Miscellaneous functions
N	Program line number
O	Program name
P	Parameter address for various G and M codes
Q	Peck increment in canned cycles
R	Size of arc radius or retract height
S	Speed, either spindle speed or surface speed
T	Tool selection
U	Incremental axis parallel to x-axis
V	Incremental axis parallel to y-axis
W	Incremental axis parallel to z-axis
X	Absolute or incremental position of x-axis
Y	Absolute or incremental position of y-axis
Z	Absolute or incremental position of z-axis

Figure 11-8. The G-code programming language includes many types of letter codes.

Computer-aided part programming (CAPP) is a process that uses a computer to generate an NC program from a computer model of a desired part. CAPP programming is a capability of most current CAD software. After the program is generated, it must be processed with a postprocessor. This software customizes the program to meet the specific requirements of individual machines.

G-codes are initialization commands used to set up the machine for a specific operation. There are dozens of commands, such as G00, which moves the tool to a point, and G04, which holds the machine for a specified number of milliseconds. **See Figure 11-9.** For example, the code G04 P1500 results in the machine pausing at a point for 1.5 sec. In addition, G commands specify the type of programming used. G91 specifies incremental programming and G90 specifies absolute programming.

SELECTED G-CODES	
CODE	DESCRIPTION
G00	Rapid positioning
G01	Linear interpolation
G02	Circular interpolation, clockwise
G03	Circular interpolation, counterclockwise
G04	Dwell
G07	Imaginary axis designation
G10	Programmable data input
G11	Data write cancel
G12	Full-circle interpolation, clockwise
G13	Full-circle interpolation, counterclockwise
G17	XY plane selection
G18	ZX plane selection
G19	YZ plane selection
G20	Programming in inches
G21	Programming in millimeters
G28	Return to home position
G30	Return to secondary home position
G31	Skip function
G33	Constant pitch threading
G34	Variable pitch threading
G40	Tool radius compensation off
G41	Tool radius compensation left
G42	Tool radius compensation right
G43	Tool height offset compensation negative
G44	Tool height offset compensation positive

Figure 11-9. G-codes specify various machine tool setup or initialization functions.

X-, Y-, and Z-Codes. X-, Y-, and Z-codes are used to specify movement along one of the primary axes. The letter is followed by a positive or negative coordinate or distance to travel. The code G01 X3.0 Y4.0, in absolute programming, commands the machine to move to a point

3.0 units from the origin on the x-axis and 4.0 units from the origin on the y-axis. If the programming method were incremental instead, the code would command the machine to move from its starting point 3.0 units along the x-axis and 4.0 units along the y-axis.

I-, J-, and K-Codes. I-, J-, and K-codes are used to specify the centerpoint of tool paths that are arcs and circles. I gives a dimension along the x-axis, J along the y-axis, and K along the z-axis. G codes G02 and G03 give the movement direction in clockwise and counterclockwise directions, respectively. For instance, the absolute programming code G02 X4.6 Y0.8 I2.2 J1.6 instructs the machine to move from its starting point to a new point (X4.6 Y0.8) in a clockwise arc (G02) centered at a point (I2.2 J1.6).

F-Codes. F-codes are used to specify feed rate. Tools that rotate, such as drill bits and taps, can be specified to turn at the optimal speed for each operation. G-codes specify the feed rate units, such as G94 for in./min and G95 for mm/min.

S-Codes. S-codes are used to specify spindle speed. This is the speed of either the cutter of a machining center or the workpiece of a turning center. G-codes specify the spindle speed units, such as G97 for revolutions per minute.

T-Codes. T-codes are used to specify the desired tool number. Machines with tool holders can be programmed to automatically select among several tools and perform several machining operations in one program.

M-Codes. M-codes are used to control miscellaneous actions. **See Figure 11-10.** Some M-codes may vary by manufacturer and type of machine. They are used to start and stop the machine, turn coolant on or off, specify spindle rotation direction, and many other functions.

SELECTED M-CODES	
CODE	**DESCRIPTION**
M00	Compulsory stop
M01	Optional stop
M02	End of program
M03	Spindle on (clockwise rotation)
M04	Spindle on (counterclockwise rotation)
M05	Spindle stop
M06	Automatic tool change (ATC)
M07	Coolant on (mist)
M08	Coolant on (flood)
M09	Coolant off
M10	Pallet clamp on
M11	Pallet clamp off
M13	Spindle on (clockwise rotation) and coolant on (flood)
M19	Spindle orientation

Figure 11-10. M-codes are a collection of miscellaneous commands related to CNC machine tools.

> Manufacturers may add proprietary codes to the standard G-code programming language to accommodate the special features or functions of their machine tools.

NC Programs

A sequence of multiple commands forms a program. The program contains all the instruction a CNC system needs to complete the machining operations. **See Figure 11-11.** Programs may consist of only several commands up to thousands of commands.

This system of commands allows program changes to be made easily. Once the program is written, it can be easily modified to a new sequence by inserting new commands or moving commands from one location in the program to another. Certain common sequences may even be saved separately and copied into multiple programs for similar machining operations.

NUMERICAL CONTROL PRINTS

In addition to the normal information included on detail prints, prints created for NC parts usually have additional information unique to NC machining. Much of this information, such as tooling requirements, relates to the piece of equipment to be used to produce the part. These special features may be specified as part of the drawing or as notes to the drawing.

An NC print also specifies the origin of the part. The origin is designated at a corner of the part or with a specified offset from the part. **See Figure 11-12.**

The drawing is generally oriented on the page so the plan view appears as the part appears in the machine. Often the cutter size and offset are specified. The operator's copy of the print may include cutter path information.

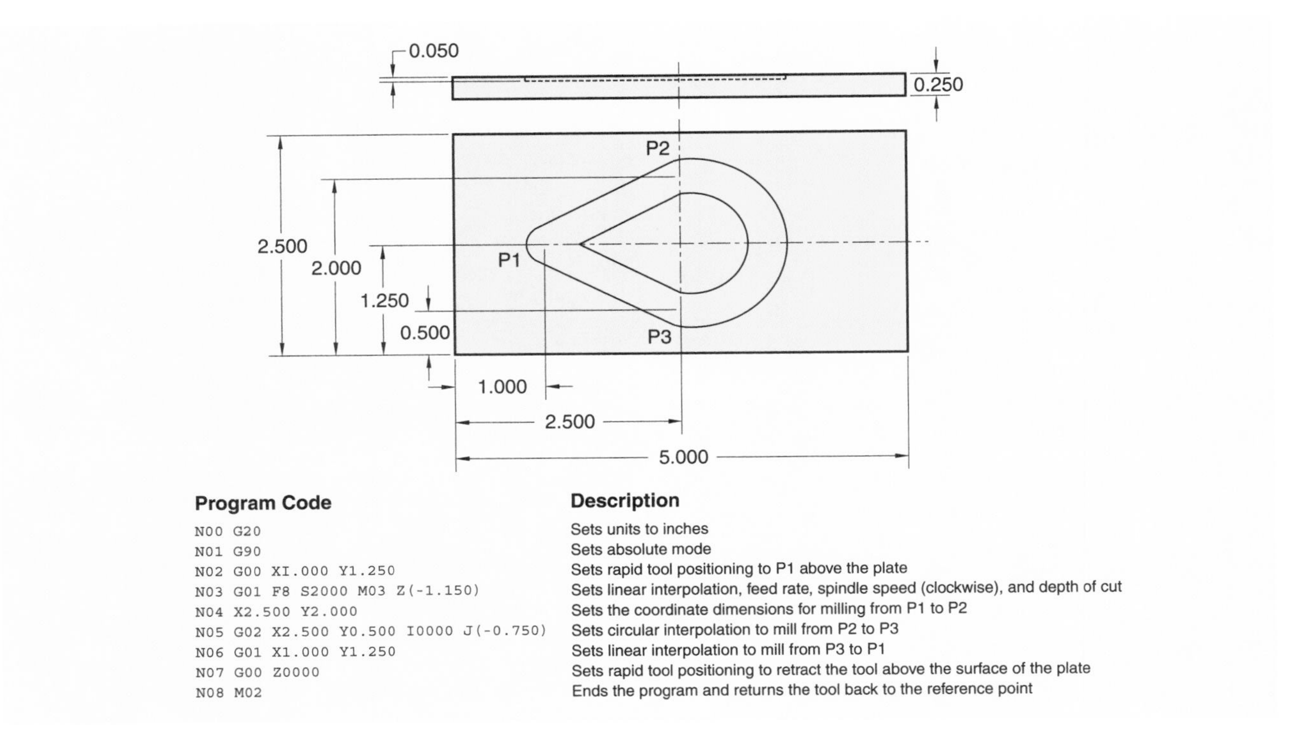

Program Code	Description
N00 G20	Sets units to inches
N01 G90	Sets absolute mode
N02 G00 XI.000 Y1.250	Sets rapid tool positioning to P1 above the plate
N03 G01 F8 S2000 M03 Z(-1.150)	Sets linear interpolation, feed rate, spindle speed (clockwise), and depth of cut
N04 X2.500 Y2.000	Sets the coordinate dimensions for milling from P1 to P2
N05 G02 X2.500 Y0.500 I0000 J(-0.750)	Sets circular interpolation to mill from P2 to P3
N06 G01 X1.000 Y1.250	Sets linear interpolation to mill from P3 to P1
N07 G00 Z0000	Sets rapid tool positioning to retract the tool above the surface of the plate
N08 M02	Ends the program and returns the tool back to the reference point

Figure 11-11. An NC program includes all the commands necessary to initialize the machine tool, perform the machining operations, and end the sequence.

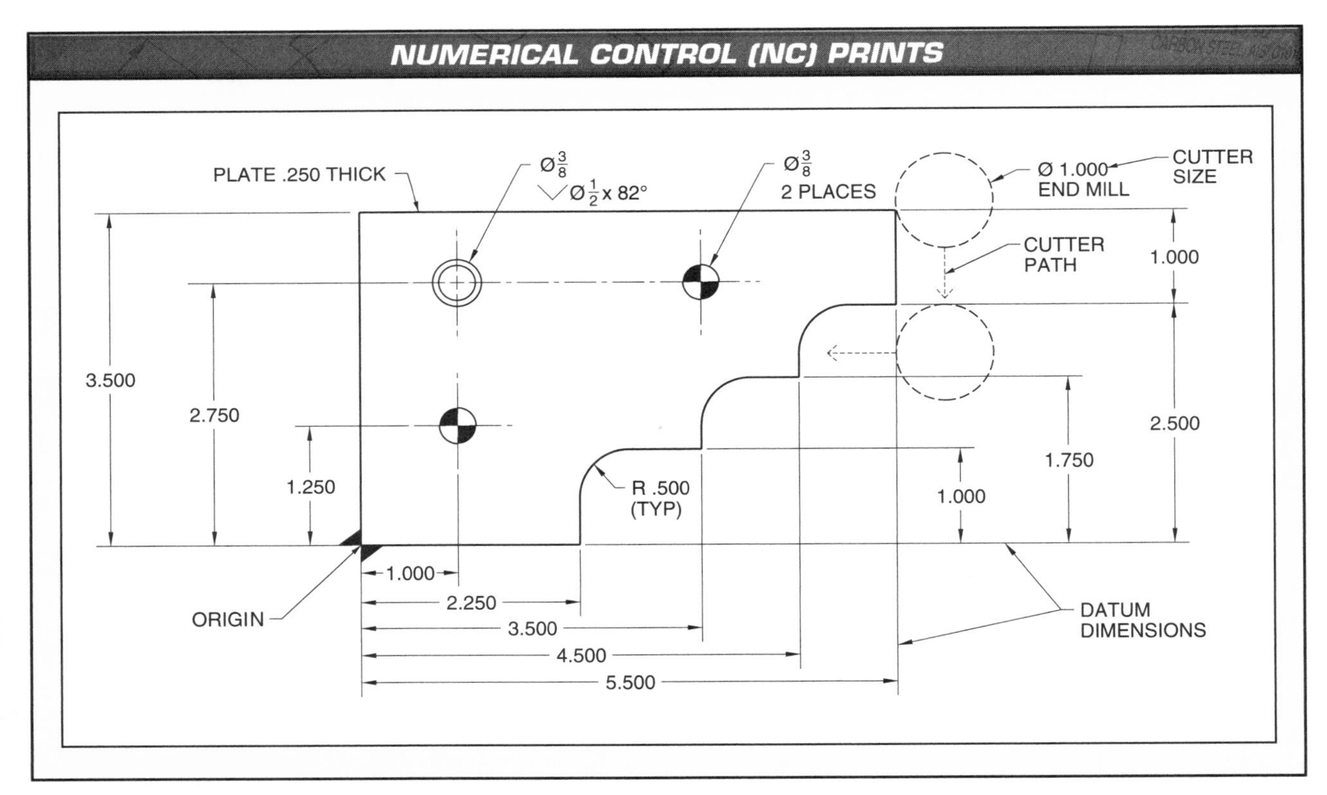

Figure 11-12. NC prints may specify the origin and cutter path and size information.

REVIEW QUESTIONS 11

Name ______________________________ Date ______________

Multiple Choice

_______________ 1. A turning center controls the ___-axes.
- A. x- and y
- B. x- and z
- C. x-, y-, and z
- D. y- and z

_______________ 2. ___-codes are initialization commands used to set up a machine for a specific operation.
- A. G
- B. M
- C. S
- D. X, Y, and Z

_______________ 3. Rapid prototyping builds parts in a(n) ___ process.
- A. subtractive
- B. additive
- C. manual
- D. none of the above

_______________ 4. A flame cutter controls the ___-axes.
- A. x- and y
- B. x- and z
- C. x-, y-, and z
- D. y- and z

_______________ 5. ___ processes can be numerically controlled.
- A. Milling
- B. Turning
- C. Drilling
- D. all of the above

_______________ 6. NC systems can control ___.
- A. workpiece positions and motions
- B. tool positions and motions
- C. spindle speeds
- D. all of the above

_____________________ **7.** A machining center controls the ___-axes.
A. x- and y
B. x- and z
C. x-, y-, and z
D. y- and z

_____________________ **8.** The three directions (axes) in the Cartesian coordinate system are designated ___.
A. A, B, and C
B. I, J, and K
C. s, t, and u
D. x, y, and z

_____________________ **9.** Each command line begins with a(n) ___ number.
A. G
B. M
C. N
D. X

_____________________ **10.** ___-codes are used to specify spindle speed.
A. G
B. M
C. S
D. T

Completion

_____________________ **1.** ___ machines build parts from the bottom up in successive layers.

_____________________ **2.** All NC programs are based on the ___ coordinate system.

_____________________ **3.** Rotation around the primary axes are designated with letters ___.

_____________________ **4.** I, J, and K codes specify the ___ of tool paths that are arcs and circles.

_____________________ **5.** ___ programming is a method of generating NC commands where each point is specified in relation to the origin.

_____________________ **6.** Tool ___ carry multiple tools that can be automatically changed during machining operations.

_____________________ **7.** ___-codes are used to specify the desired tool number.

_____________________ **8.** The primary axes in the Cartesian coordinate system are designated with letters ___.

_____________________ **9.** ___ programming is a method of generating NC commands where each point is specified in relation to the last point where the machine stopped.

_____________________ **10.** A digital ___ is a device on a machine tool that provides a digital display of the cutting tool's position in space.

True-False

T F **1.** NC machining is repeatable because the same set of commands and operations is completed in the same manner every time.

T F **2.** A digital readout directly controls a machine.

T F **3.** The most common CNC machines are machining centers and turning centers.

T F **4.** All NC programming begins at the origin.

T F **5.** A sequence of multiple programs forms a command.

T F **6.** An NC print specifies the origin of the part.

T F **7.** M-codes are used only for manufacturer-specific commands.

T F **8.** Each code consists of a letter prefix and a number.

T F **9.** NC machines have slower tool changes than conventional machine tools.

T F **10.** I-, J-, and K-codes are used to specify movement along one of the primary axes.

TRADE COMPETENCY TEST 11

Name ______________________________ **Date** ________________

Bar

Refer to print below.

____________ **1.** The drilled holes are ___ to datum C.

____________ **2.** The maximum length for the Bar is ___″.

T F **3.** The origin is at the intersection of datums A, B, and C.

T F **4.** Incremental dimensioning is used.

T F **5.** There is 1.500″ between centers of the drilled holes.

____________ **6.** Depth A is ___″ at its maximum depth.

____________ **7.** The drilled holes are positioned with reference to datum(s) ___.

____________ **8.** At a maximum material condition, the diameter of the drilled holes is ___″.

T F **9.** The drilled holes are .875″ from datum B.

____________ **10.** The center for radius B is ___″ from datum B.

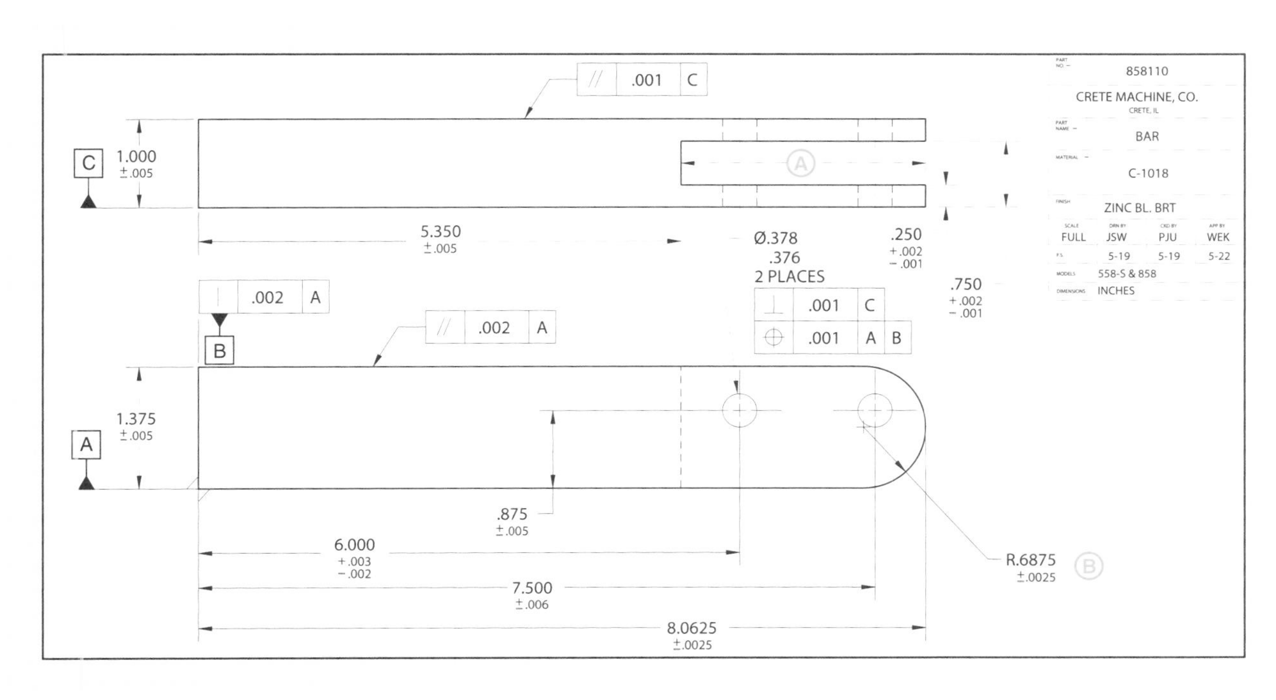

BAR

Bearing

Refer to print below.

_______________ **1.** The origin is located at ___.

_______________ **2.** The Bearing is oil impregnated with Skil Std. ___ or approved equivalent.

_______________ **3.** The inside diameter is ___ with diameter B.

_______________ **4.** Angle C may be a maximum of ___°.

_______________ **5.** The outside diameter has a tolerance of ±___″.

T F **6.** All chamfers are 1/64″ deep.

T F **7.** All decimal dimensions have a tolerance of ±.010″.

T F **8.** The diameter of datum A is from .8290″ to .8300″.

_______________ **9.** Before revision, the diameter at A1 was ___″.

_______________ **10.** The collar at D is ___″ thick.

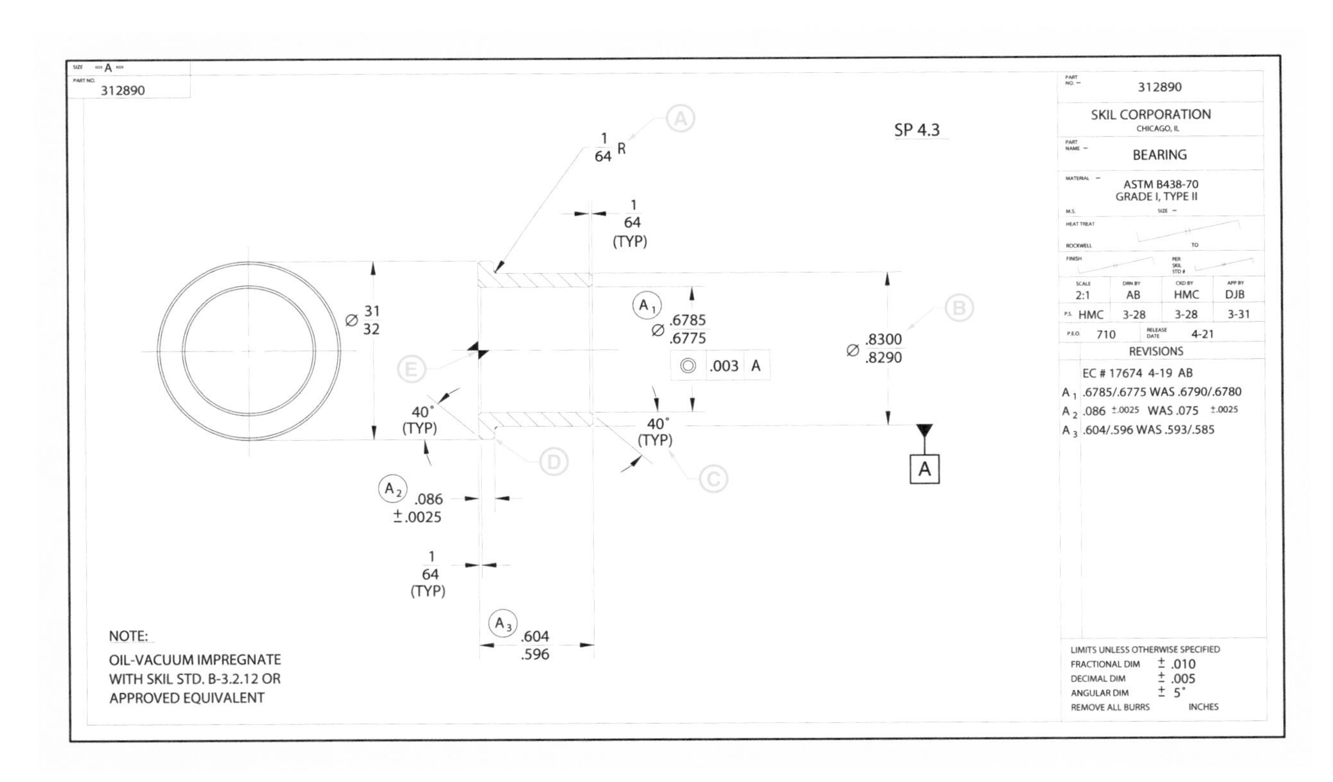

BEARING

Trade Test 1

TRADE COMPETENCY TEST 12

Name ______________________________ Date ______________

Parallel Clamp

Refer to print in foldout print section.

__________ **1.** The final assembly contains ___ different parts.
A. three
B. four
C. five
D. none of the above

__________ **2.** The ___ is a purchased part.
A. loose jaw
B. e-clip
C. threaded jaw
D. clamping screw

__________ **3.** The hole in the clamping screw is ___″ diameter.
A. 1/4
B. 5/16
C. 3/8
D. 9/16

__________ **4.** The longest part is ___″ long.
A. 4
B. 5 3/4
C. 6 1/8
D. 6 3/4

__________ **5.** The maximum allowable angle specified for the clamping screw end is ___°.
A. 43
B. 47
C. 61
D. 72

__________ **6.** On the threaded jaw, there is ___″ of material between the end with the radius and the edge of the nearest threaded hole, assuming nominal hole size.
A. 1/8
B. 3/16
C. 1/4
D. none of the above

_____________________ **7.** The plans specify ___ e-clip(s) required.

A. 1
B. 2
C. ± 3
D. 4

_____________________ **8.** There are ___ threads per inch specified for the holes in the threaded jaw.

A. ⅜
B. ¾
C. 24
D. 48

_____________________ **9.** The hole with a specified depth in the loose jaw is ___″ in diameter.

A. ¼
B. 5⁄16
C. 13⁄32
D. ⅜

_____________________ **10.** The drawing is ___ scale.

A. ½
B. ¾
C. full
D. not to

_____________________ **11.** The spotface on the threaded jaw is ___″ deep.

A. .003
B. 1⁄16
C. 1⁄32
D. ⅜

_____________________ **12.** The threaded portion of the clamping screw is ___″ long.

A. ½
B. 1⅛
C. 3½
D. 4⅜

_____________________ **13.** The maximum allowable width of the threaded jaw is ___″.

A. ¾
B. .751
C. 49⁄64
D. 51⁄64

_____________________ **14.** In the threaded fastener designation of ⅜-24 UNF-2B, the "2B" represents ___.

A. external threads
B. fractional diameter
C. threads per inch
D. internal threads

______________ **15.** After assembly of the parallel clamp, the e-clip is located next to the ___.

A. threaded jaw
B. clamping screw
C. loose jaw
D. none of the above

T F **16.** A medium diamond pattern is specified for the knurls.

T F **17.** The angled surfaces on the jaws are dimensioned in degrees.

______________ **18.** The hole through the loose jaw is ___″ diameter.

A. 5⁄16
B. 3⁄8
C. 13⁄32
D. 3⁄4

______________ **19.** The threaded portion of the adjusting screw is ___″ long.

A. 1 1⁄8
B. 3 1⁄2
C. 4 3⁄8
D. 6 1⁄8

______________ **20.** All parts that are not purchased are made of ___.

______________ **21.** The clamping screw is inserted into the loose jaw ___″.

______________ **22.** The length of the spotface on the threaded jaw is ___″.

______________ **23.** The note is a direction to ___ all sharp corners.

______________ **24.** The threaded jaw is heat treated by ___.

______________ **25.** In the threaded fastener designation of 3⁄8-24 UNF-2A, the "2A" represents ___.

A. threads per inch
B. internal threads
C. external threads
D. fractional diameter

______________ **26.** The true centers of the threaded jaw are specified to be within ±___″ of the centerline.

A. .0005
B. .002
C. 1⁄64
D. .003

______________ **27.** The knurled surface is specified for a ___″ diameter before knurling.

A. 9⁄16
B. .725
C. 3⁄4
D. 1 1⁄8

_____________________ **28.** The spotface surface of the loose jaw is ___″ wide.
A. 1⁄16
B. 3⁄16
C. 5⁄16
D. 3⁄4

_____________________ **29.** The centers of the holes on the threaded jaw are ___″ apart.
A. 3⁄4
B. 1 5⁄8
C. 1 3⁄4
D. 2 3⁄8

_____________________ **30.** In the threaded fastener designation of 3⁄8-24 UNF-2A, the "UNF" represents ___.
A. unified national factor 2A thread
B. unified national fine thread
C. universal N class fit
D. unified national fractional tolerance

Trade Test 2

TRADE COMPETENCY TEST 12

Name ______________________ Date ______________

Support, UV 1″–18″ Arm

Refer to print on page 295.

______________ **1.** The thread form for the two studs is ___.
A. UNC
B. UNF
C. UNEF
D. NPT

______________ **2.** A ___ auxiliary view shows details of the obround.
A. full
B. partial
C. double
D. revolved

______________ **3.** The four corners are joined with ___.

______________ **4.** The inside bend radius is ___″.

______________ **5.** ___ lines are drawn to indicate sections AA and BB.

______________ **6.** The diameter of the threaded stud is ___″.

______________ **7.** The tolerance for all holes is ±___″.

______________ **8.** The bottom piece is bent at two ___° angles to form the sides.

______________ **9.** The overall length of the Support, not including the studs, is ___″.

______________ **10.** Side A measures ___″.

______________ **11.** The maximum center-to-center dimension of B is ___″.

______________ **12.** The minimum overall height of the Support is ___″.

______________ **13.** The dimension of the bottom piece at C is ___″.

______________ **14.** The view shown at D is the ___ view.

______________ **15.** The dimensions of the sheet for the original print are ___.

T F **16.** The studs may be zinc plated or stainless steel.

T F **17.** Section BB is taken from the top view.

T F **18.** The Support is drawn to full scale.

_______________ **19.** The center-to-center distance between E and F is ___″.

_______________ **20.** The two cutting planes are ___″ apart.

T F **21.** All obrounds are the same size.

T F **22.** Every dimension has a tolerance of .030″.

T F **23.** Each corner of the Support is welded.

T F **24.** Section AA gives location dimensions for obrounds.

_______________ **25.** Section BB is taken ___″ from the left end of the Support.

_______________ **26.** Centerlines G and H are ___″ apart.

_______________ **27.** The drawing number for the Support is ___.

T F **28.** All external threads are covered before painting.

T F **29.** The material for the Support is .598″ thick.

T F **30.** All dimensioning on the drawing is unidirectional.

_______________ **31.** The maximum distance between centers of the threaded studs is ___″.

_______________ **32.** There are ___ obrounds on the Support.

_______________ **33.** The drawing was completed on ___.

_______________ **34.** The Support is to be finished in ___ wrinkle powder coat.

_______________ **35.** The two ∅.281″ holes have a minimum diameter of___″.

_______________ **36.** The drawing was approved by ___.

T F **37.** The ∅.312″ holes are located .375″ from the end of the Support.

T F **38.** The centers of the studs are located .387″ from the bottom of the Support.

T F **39.** The obrounds at E and F are equally spaced in from the ends of the Support.

T F **40.** The studs have 20 threads per inch.

Trade Test 3

TRADE COMPETENCY TEST 12

Name ______________________ Date ______________

A1-ACC Adaptor

Refer to print on page 296.

__________ **1.** All linear dimensions are given in ___.

__________ **2.** The overall thickness of the Adaptor is ___ mm.

T F **3.** The Adaptor is manufactured from cast iron.

T F **4.** The cutting plane line is offset to pass through more internal features.

__________ **5.** The Adaptor must be cylindrical within a(n) ___ mm tolerance zone.

__________ **6.** Hole C must be perpendicular to datum ___ within 0.001 mm.

__________ **7.** Datum B has a maximum diameter of ___ mm.

T F **8.** The angular tolerance, unless otherwise specified, is ±0′15″.

T F **9.** The project engineer approved the print on the same day it was checked.

__________ **10.** The counterbored hole at A is ___° from the horizontal.

__________ **11.** Circle B is referred to as a(n) ___ circle.

__________ **12.** The blind hole must be at a true ___ within a 0.002 tolerance zone with regard to datum B.

__________ **13.** Surface F has a total runout within a(n) ___ tolerance zone around datum B.

__________ **14.** All true position tolerances are toleranced around datum ___.

__________ **15.** The counterbored hole at C has a counterbore depth of ___ mm.

T F **16.** Section A-A is a removed section.

T F **17.** The maximum decimal tolerance, unless otherwise specified, is ±0.005.

T F **18.** The Adaptor is to be polished to a brilliant finish.

T F **19.** The overall diameter of the Adaptor is 133 mm.

T F **20.** Section lines show where the cutting plane was passed through the object.

__________ **21.** The minimum diameter of D is ___ mm.

__________ **22.** An M5 × 0.8 thread, 8 mm deep, is used in ___ places on the Adaptor.

__________________ **23.** The drawing number of the Adaptor is ___.

__________________ **24.** The print is drawn to ___ scale.

__________________ **25.** The back edge contains a 2 mm × ___° chamfer.

__________________ **26.** The maximum diameter at E is ___ mm.

T F **27.** The bottom recess slopes 7°.

T F **28.** Datum A is parallel within a tolerance zone of 0.01 mm.

__________________ **29.** The maximum distance hole C may be from datum axis B is ___ mm.

__________________ **30.** The minimum distance the center of threaded hole G may be from datum axis B is ___ mm.

__________________ **31.** Piedmont Tool Co. is located in ___.

__________________ **32.** The six tapped holes are located on a(n) ___ mm bolt circle.

__________________ **33.** The blind hole on the 105 BC is ___ mm deep.

T F **34.** The centers of the ∅5.5 holes are located 6 mm from datum A.

T F **35.** The back view of the object is represented by a simplified view.

T F **36.** There are four ∅12 holes drilled on the 105 BC.

T F **37.** The drawing was checked by TRE.

T F **38.** The drawing may be scaled to determine the overall size.

__________________ **39.** Hole H may be a maximum of ___° offset from the vertical centerline.

__________________ **40.** The project engineer was ___.

Trade Test 4

TRADE COMPETENCY TEST 12

Name ______________________ **Date** ______________

Punch Assembly

Refer to print in foldout print section.

______________ **1.** The Punch Holder is made from ___.
- A. aluminum
- B. spring steel
- C. cold rolled steel
- D. heat treated RC58

______________ **2.** The 3⁄16″ punch has a ___″ outside diameter.
- A. .250
- B. .375
- C. .500
- D. .625

______________ **3.** The maximum allowable diameter of the largest hole on the pivot arm is ___″.

______________ **4.** All punches require a(n) ___° chamfer.

______________ **5.** The bushing is made from ___.

______________ **6.** The small diameter of the 5⁄16″ punch is ___″.

______________ **7.** The center of the smallest hole on the Pivot Arm is located ___″ from the large end.

______________ **8.** The UNC threads specified for the Pivot Screw have ___ threads per inch.

______________ **9.** The length of the index pin is ___″.

______________ **10.** The set screw joins the Knob with the ___.
- A. Pivot Screw
- B. Index Pin
- C. Bushing
- D. 1⁄8″ Punch

______________ **11.** The five holes of the same size on the Die are centered on a ___″ radius.

______________ **12.** The maximum allowable height of the Punch Holder is ___″.

______________ **13.** The largest unthreaded hole specified for the Die is ___″ diameter.
- A. .261
- B. .323
- C. .386
- D. .500

____________________ **14.** The knurl on the Die is located ___″ from the edge.

____________________ **15.** The maximum diameter allowable for the flange of the Index Pin is ___″.

T F **16.** The scale for the drawing is 2:1.

____________________ **17.** The largest hole in the Knob is ___″.

____________________ **18.** The smallest allowable diameter of the ⅛″ Punch is ___″.

____________________ **19.** The thickness of the Pivot Arm is ___″.

____________________ **20.** The Pivot Screw can be tightened to the Die with a(n) ___″ wrench.

____________________ **21.** In the threaded fastener designation of ½-20 UNF-2A on the Pivot Screw, the "20" represents ___.

A. threads per inch
B. internal threads
C. external threads
D. fractional diameter

____________________ **22.** The ⅜″ Punch specifies a ___″ chamfer.

____________________ **23.** A counterbored hole with no threads is specified on the ___.

A. Bushing
B. Knob
C. Pivot Arm
D. Punch Holder

____________________ **24.** The overall length of the Pivot Arm is ___″.

____________________ **25.** Without allowance for the chamfer, the length of the large diameter of the ¼″ Punch is ___″.

A. .250
B. .375
C. 2.250
D. 2.500

T F **26.** There are 16 parts required for the Punch Assembly.

T F **27.** The spring is made from .003″ diameter spring steel.

T F **28.** The length of the Bushing is .750″.

T F **29.** The Pivot Screw is the only part with external threads.

T F **30.** The larger hole in the Knob is .625″.

____________________ **31.** The holes in the Punch Holder are ___″ in diameter.

____________________ **32.** The uncompressed length of the Spring is ___″.

____________________ **33.** The smallest hole diameter on the Pivot Arm is ___″.

____________________ **34.** The specified depth for the holes in the Punch Holder is ___″.

____________________ **35.** The outside diameter of the Knob is ___″.

TRADE COMPETENCY TEST 12

Name ______________________________ **Date** ____________________

Bearing Retainer Assembly

Refer to prints on pages 297–300.

T F **1.** A half section is shown of the Left Retainer.

__________ **2.** Sheet 1 of 4 is a(n) ___ drawing of the Bearing Retainer Assembly.

__________ **3.** The bearing has a(n) ___″ × 45° chamfer on both ends.

__________ **4.** The maximum spherical diameter of the Right Retainer is ___″.

T F **5.** Rivets are purchased items for the Bearing Retainer Assembly.

T F **6.** The bearing should show movement within a weight range of 5 to 30 oz only.

__________ **7.** A(n) ___ section of the Right Retainer is shown.

__________ **8.** The Bearing Retainer Assembly fits a(n) ___″ diameter pin.

__________ **9.** The Bearing is made of Grade I, Type II ___.

__________ **10.** When assembled, the Right Retainer must be flat within a(n) ___″ tolerance zone.

T F **11.** The release date of the Bearing print is 7-20.

T F **12.** No burrs are permitted on the Right Retainer.

T F **13.** No burrs are permitted on the Left Retainer.

__________ **14.** The overall height of the Left Retainer is ___″.

__________ **15.** The overall length of the Right Retainer is ___″.

T F **16.** The three ∅.116 holes in the Right Retainer align with the three .096 holes in the Left Retainer.

__________ **17.** The typical radius of Retainer corners is ___″.

__________ **18.** The thickness of the Right Retainer is ___″.

__________ **19.** The Left Retainer is made of hardened ___ steel.

T F **20.** All drawings of the Bearing Retainer Assembly are drawn at full scale.

T F **21.** Part No. 318714 was formerly Part No. 341456.

T F **22.** The outside diameter of the Bearing is .502″.

T F **23.** RSW drew Part No. 318714 on 7-14.

T F **24.** All bearings are tumbled.

____________________ **25.** Rivet heads are placed on the side of the assembly containing the ___ Retainer.

____________________ **26.** Part No. ___ is the only part in which alternate material may be used.

____________________ **27.** Part No. 318717 is to be finished to Skil standard ___.

____________________ **28.** Omnilube ___ is used to oil vacuum impregnate the Bearing.

____________________ **29.** The radius at B on the Right Retainer is ___″.

____________________ **30.** The stock size at A on Part No. 318717 is ___″.

T F **31.** Fractional dimensions are toleranced to ±.010″ on all manufacturing drawings unless otherwise specified.

____________________ **32.** The hole at A on Part No. 318716 has a diameter of ___″.

____________________ **33.** Datum A is ___″ in diameter at maximum material condition.

____________________ **34.** The Right Retainer has a thickness tolerance of ±___″.

____________________ **35.** When all parts are at maximum material condition, there is ___″ clearance between the Bearing slot and the tab on the Left Retainer.

____________________ **36.** The three horizontally aligned holes on Part No. 318717 are located a maximum of ___″ from the centerpoint of the Left Retainer.

____________________ **37.** All decimal dimensions have a tolerance of ±___″ unless otherwise specified.

____________________ **38.** The maximum stock thickness of Part No. 318716 is ___″.

____________________ **39.** The Right Retainer is made of cold rolled strip ___.

____________________ **40.** Three clips on the Retainer are each ___″ wide.

T F **41.** The flatness of the Right Retainer can vary .001″ less when assembled than when manufactured.

T F **42.** Surface A on the Bearing is cylindrical with datum A within a .010″ tolerance zone.

____________________ **43.** The Left Retainer has a(n) ___ finish.

T F **44.** The approval date is the same for all the drawings.

____________________ **45.** The Left Retainer is ___ within a .010″ tolerance zone.

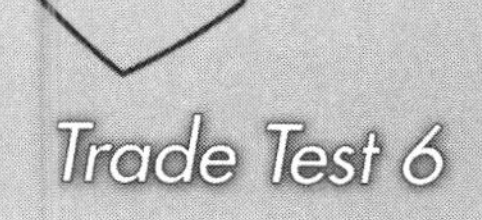

Trade Test 6

TRADE COMPETENCY TEST 12

Name ______________________________ **Date** ____________________

Holder–Punch

Refer to print in foldout print section.

______________ **1.** Section A-A is a(n) ___ section.

______________ **2.** The section lining symbol indicates ___.
- A. finishing type
- B. surface tolerance
- C. sectioned views
- D. all of the above

______________ **3.** The maximum overall diameter of the Holder-Punch is ___″.

______________ **4.** All chamfers are cut at a(n) ___° angle.

______________ **5.** The minimum overall depth of the Holder-Punch is ___″.

______________ **6.** The air hole is ___.
- A. ∅3/16
- B. drilled thru
- C. 30° from vertical
- D. all of the above

______________ **7.** The maximum diameter at A is ___″.

______________ **8.** Fillets are specified to have a(n) ___″ radius.

T F **9.** Datum A is perpendicular to datum B within a .00025″ tolerance zone.

T F **10.** The width of the chamfer at B is .06″.

______________ **11.** Drawing No. 217203 supersedes and replaces Drawing No. ___.

______________ **12.** The distance across flats on the center shaft is ___″.

______________ **13.** Eight ∅.62 holes are ___″ deep.

______________ **14.** Dimensions of 1.104″ and 1.562″ show the location of ___ of eight drilled holes.

______________ **15.** The diameter at C is ___″.

______________ **16.** Four counterbored holes are located ___″ from the centerpoint.
- A. .67
- B. .875
- C. 1.104
- D. 1.562

_______________ **17.** The depth at D is ___″.

_______________ **18.** The depth at F is ___″.

A. 1.795
B. 1.985
C. 2.175
D. 2.72

_______________ **19.** The 2.2010 diameter at A is ___ to datum B.

_______________ **20.** The depth at E is ___″.

_______________ **21.** The four counterbored holes have a counterbore diameter of ___″.

T F **22.** The holes are dimensioned with polar coordinates.

_______________ **23.** The cutting plane line for Section A-A is represented by a(n) ___ .

_______________ **24.** Surface G is parallel to datum A within a(n) ___″ tolerance zone.

_______________ **25.** Datum B has a(n) ___″ diameter under a least material condition.

_______________ **26.** The drawing was completed on ___.

_______________ **27.** The Holder-Punch is to be marked ___.

_______________ **28.** All flat-bottom holes are ___″ in diameter.

_______________ **29.** The scale of the drawing is ___.

_______________ **30.** All three-place dimensions have a tolerance of ±___″ unless otherwise specified.

_______________ **31.** The four counterbored holes have a counterbore depth of ___″.

T F **32.** The Holder-Punch is finished to a 125 microinch finish unless otherwise specified.

T F **33.** The part is marked with a part number.

T F **34.** The air hole is ∅.20.

T F **35.** The part is hardened to RC 60-62.

Trade Test 7

TRADE COMPETENCY TEST 12

Name ______________________________ **Date** ________________

Detail, Bottom Die Holder

Refer to print in foldout print section.

T F **1.** All decimal dimensions are ±.005″ unless otherwise specified.

T F **2.** Item 3 can be machined from a piece of plate steel 4⅞″ long.

T F **3.** All tapped holes are drilled through.

______________ **4.** The drilled hole at A is taper-reamed to a depth of ___″.

______________ **5.** The maximum overall length of the Bottom Die Holder is ___″.

A. 12 13/32
B. 12 27/64
C. 12 7/16
D. none of the above

______________ **6.** The dimension at M is ___″.

______________ **7.** The overall height of the Bottom Die Holder is ___.

T F **8.** Items 2 and 3 are welded together.

T F **9.** Section A-A is taken from the front orthographic view.

T F **10.** The dimension at I is a location dimension.

______________ **11.** A total of ___ bronze bushings are required for the assembly.

______________ **12.** A(n) ___″ radius is typical for all rounded corners.

T F **13.** The Bottom Die Holder is symmetrical in the top view.

______________ **14.** The centerpoints for drilled and tapped holes at D are located ___″ above the base.

A. 3.656
B. 4.406
C. 5.156
D. 5¼

______________ **15.** The drilled hole at E is ___.

A. located 3.656″ above the base
B. the centerpoint for the radiused profile
C. located 1½″ below the top
D. none of the above

____________________ **16.** The section lining symbol at Section A-A indicates ___.
- A. surface roughness
- B. brass material
- C. heat treatment
- D. sectioned material

____________________ **17.** The centerline for the valve is ___″ above the base.

____________________ **18.** The dimension at F is ___″.

____________________ **19.** The thread note at H specifies a(n) ___ thread form.
- A. UNC
- B. UNF
- C. UNEF
- D. NPT

____________________ **20.** The standard tolerance for fractions, unless otherwise specified, is ±___″.

____________________ **21.** The fractional distance at J is ___″.

T F **22.** The surface at K is finished.

T F **23.** The bronze bushings are 7/16″ in length.

T F **24.** The maximum radius of the semicircular valve seat area is .5625″.

T F **25.** The Bottom Die Holder is drawn at half size.

____________________ **26.** The fractional length of L is ___″.

____________________ **27.** The overall width of the Bottom Die Holder is ___″.

T F **28.** All counterbores are bored to the same depth.

T F **29.** The plans for the Bottom Die Holder have been revised.

T F **30.** Section A-A is a revolved section.

____________________ **31.** The diameter of the counterbore at C is ___″.

____________________ **32.** The depth at B is ___″.

____________________ **33.** The overall height of Item 2 is ___″.

____________________ **34.** The inside diameter of the bronze bushings is ___″.

____________________ **35.** The diameter of the hole at G is ___″.

TRADE COMPETENCY TEST 12

Name ______________________________ Date ____________________

O.D. Grinder

Refer to print on pages 301–303.

______________ **1.** A total of___ separate component types are required for the assembly.

T F **2.** Item No. 6 is a purchased item.

T F **3.** Item No. 9 is a manufactured item.

______________ **4.** Fractional tolerances, unless otherwise specified, are ±___″.

______________ **5.** The drawing number of the assembly is ___.

______________ **6.** The bolt at A is a ___.
- A. #10-24 UNC
- B. ¼-20 UNC
- C. 5⁄16-18 UNC
- D. ½-20 UNF

______________ **7.** The maximum length of the Guide Bar Support is ___″.

______________ **8.** The length of the Bearing Plate is ___″.

______________ **9.** The part at B is manufactured by ___.

______________ **10.** The Adjusting Wheel is made from a purchased ___.

______________ **11.** Datum surface X on the Guide Bar must be ___ within a .001″ tolerance zone.

______________ **12.** The slot at C is ___″ long.

______________ **13.** The bronze bushings are placed in the Bearing Plate so they are ___ to datum X.

T F **14.** Datum X on the Bearing Plate is straight within a .001″ tolerance zone.

T F **15.** All slots are .25 cm wide.

T F **16.** The minimum thickness of the Guide Bar is .22″.

T F **17.** Holes in the Guide Bar are tapped with a ¼-20 UNF 2B THD.

T F **18.** The maximum radius allowed at the end of the Guide Bar is .125″.

______________ **19.** Bronze bushings are pressed into holes in the ___.
- A. Guide Bar Support
- B. Bearing Plate
- C. Guide Bar
- D. none of the above

_______________ **20.** The 3″ threaded rod is tightened to the Guide Bar Support with a ___ set screw.
A. slotted headless
B. square head
C. socket head
D. none of the above

_______________ **21.** The minimum dimension at D is ___″.
A. 1.35
B. 1.38
C. 1.385
D. 1.41

_______________ **22.** The overall height of the Guide Bar Support is ___″.

_______________ **23.** The overall depth of the Bearing Plate is ___″.

_______________ **24.** The Guide Bar is hardened to ___ Rc.

_______________ **25.** The flat surfaces on the Guide Bar Support are parallel within a(n) ___″ tolerance zone.

T F **26.** The Bearing Plate is symmetrical in the front view.

T F **27.** All sharp edges are broken on the Guide Bar Support.

T F **28.** The Bearing Plate is manufactured from cold-rolled steel (CRS).

T F **29.** The shoulder bolt is 3″ long.

_______________ **30.** The hole in the Adjusting Wheel is ∅___″ before threading.

_______________ **31.** Surface E on the Guide Bar Support is ___ to datum X.

_______________ **32.** The .62 DIA hole provides ___″ clearance for the threaded rod.

_______________ **33.** All geometric tolerances have a(n) ___″ tolerance zone.

_______________ **34.** The overall height of the Bearing Plate is ___″.
A. .75
B. .87
C. 1.25
D. 2.25

_______________ **35.** Tapped holes at F are located to within ___″.

Trade Test 9

TRADE COMPETENCY TEST 12

Name ______________________________ Date ____________________

Sine Vise

Refer to print in foldout print section.

_______________ **1.** The maximum allowable radius on the Base is ___″.
- A. .0002
- B. .1870
- C. .2505
- D. .2550

_______________ **2.** The angled holes on the Vise Body have ___ threads per inch.
- A. .313
- B. 10
- C. 16
- D. 32

_______________ **3.** The Pivot material is ___″ in diameter.
- A. .2656
- B. .375
- C. .750
- D. 1.9375

_______________ **4.** The Pivot Cap has a ___″ inside radius.
- A. .0938
- B. .2500
- C. .5000
- D. 1.2500

_______________ **5.** The Rods are counterbored to a ___″ depth.
- A. .193
- B. .300
- C. .375
- D. .500

_______________ **6.** The Pivot Cap is made from ___.

_______________ **7.** The length of the Jaw Screw is ___″.

_______________ **8.** The Pivot Cap is secured to the Base with a ___ screw.

_______________ **9.** The width of the Base is ___″.

_______________ **10.** The nameplate section has a depth of ___″.

__________________ **11.** The minimum allowable length of the Base is ___″.
A. 1.9998
B. 6.6248
C. 6.624
D. 6.625

__________________ **12.** The nameplate section is ___″ long.
A. .125
B. .438
C. 1.000
D. 1.250

__________________ **13.** The centers of the holes in the Rods are ___″ apart.
A. .219
B. 1.500
C. 1.719
D. 1.9375

__________________ **14.** The hole in the Base is ___″ diameter.
A. .1870
B. .193
C. .250
D. none of the above

__________________ **15.** Four holes in the Vise Body are drilled at a ___° angle to the vertical surface.
A. 30
B. 45
C. 50
D. 75

__________________ **16.** The width of the nameplate is ___″.
A. .0094
B. 1/32
C. 1/4
D. 9/32

__________________ **17.** The maximum allowable length of the Rod is ___″.
A. 1.719
B. 1.938
C. 1.939
D. 2.001

__________________ **18.** The height of the Base is ___″.
A. .0938
B. 1.2500
C. 2.0000
D. none of the above

__________________ **19.** A 1/16″ radius undercut is specified on the ___.

A. Pivot Cap
B. Jaw
C. Base
D. Vise Body

__________________ **20.** The maximum dimension allowed when manufacturing the Pivot Cap is ___″.

A. .7501
B. 1.000
C. 1.252
D. none of the above

__________________ **21.** When assembled, the Jaw Screw is resting on the ___.

__________________ **22.** The seven holes specified on the Vise Body are reamed to a ___″ diameter.

__________________ **23.** The total number of parts to be machined to specified dimensions is ___.

__________________ **24.** The maximum allowable length of the Vise Body is ___″.

__________________ **25.** The large hole in the Jaw Screw is to be reamed to a ___″ diameter.

T F **26.** The drawing is not to scale.

T F **27.** The Jaw Screw requires a 5/32″ diameter hole.

T F **28.** The seven holes on a line on the Vise Body are countersunk.

T F **29.** There are two Jaws required for the Sine Vise assembly.

T F **30.** The height of the Jaw is 1.6870″.

Final Exam

TRADE COMPETENCY TEST 12

Name ______________________________ **Date** ______________

3.5 Vise

Refer to prints in foldout print section.

______ **1.** The Vise Screw has a(n) ___ thread.

______ **2.** The Handle Cap has a(n) ___″ chamfer on both edges.

______ **3.** The Vise Screw is ___ to within a .001″ tolerance zone.
- A. concentric
- B. runout
- C. cylindrical
- D. round

______ **4.** The maximum width of the Pressure Plate is ___″.
- A. 1.875
- B. 1.8775
- C. 1.880
- D. 1.885

______ **5.** The shoulder cut in the bottom of the Sliding Jaw is ___″ deep.

T F **6.** The Base has an overall width of 3.500″.

T F **7.** Datum A is parallel within a .002″ tolerance zone.

T F **8.** The Vise Screw is lubricated with oil prior to assembly.

______ **9.** The 3.5 Vise is painted shades of ___ prior to assembly.

______ **10.** All machined surfaces are finished to a(n) ___ finish.

______ **11.** The Vise Screw extends ___″ beyond the Base when fully opened.

______ **12.** To fully use the 3.5 Vise, a space of ___ is required.
- A. 7.55″ × 3.50″
- B. 8.50″ × 4.50″
- C. 10.915″ × 3.50″
- D. 11.965″ × 3.50″

______ **13.** The Vise Screw is held in the Sliding Jaw by a(n) ___.
- A. slot head set screw
- B. Phillips head set screw
- C. round head machine screw
- D. hex head set screw

_______________ **14.** The Base is ___ to its finished length.

A. machined
B. cast
C. forged
D. stamped

_______________ **15.** The maximum length of the assembled Handle is ___″.

A. 3.355
B. 3.560
C. 3.575
D. 3.585

T F **16.** Datum Z is flat within a .001″ tolerance zone.

T F **17.** The minimum distance between centers on the Pressure Plate is .620″.

T F **18.** All threaded holes have a ¼-28 UNF thread.

T F **19.** The nominal clearance between the Base and Sliding Jaw is .020″.

_______________ **20.** The bottom of the Base is ___ to datum A.

_______________ **21.** The Base Jaw Plate is made with ___ steel.

_______________ **22.** The handle has a(n) ___ of .004″ with reference to datum axis Y.

_______________ **23.** The 3.5 Vise assembly drawing was checked on ___.

T F **24.** The Sliding Jaw Plate is parallel to the Base Jaw Plate within a .006″ tolerance zone.

T F **25.** All parts are drawn to the same scale.

T F **26.** All fractional dimensions are accurate within a 1/32″ wide tolerance zone.

_______________ **27.** Drawings were approved by ___.

_______________ **28.** The 3.5 Vise has a maximum opening of ___″.

A. 3.365
B. 3.500
C. 4.415
D. 7.550

T F **29.** The Handle is made of cold-rolled steel.

_______________ **30.** All sizes on the cast parts that are not machined later are given as ___ dimensions.

A. reference
B. basic
C. nominal
D. missing

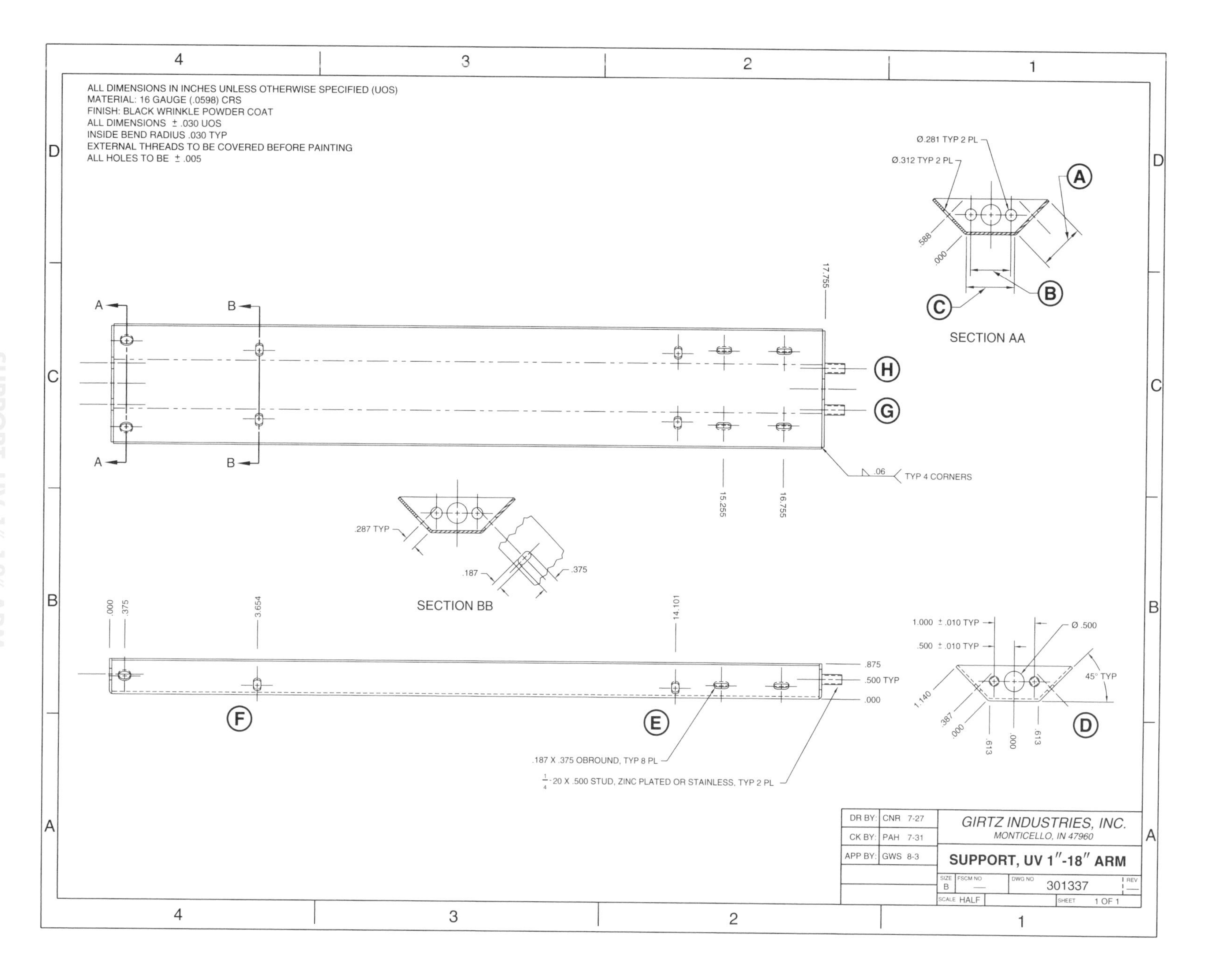

ALL DIMENSIONS IN INCHES UNLESS OTHERWISE SPECIFIED (UOS)
MATERIAL: 16 GAUGE (.0598) CRS
FINISH: BLACK WRINKLE POWDER COAT
ALL DIMENSIONS ± .030 UOS
INSIDE BEND RADIUS .030 TYP
EXTERNAL THREADS TO BE COVERED BEFORE PAINTING
ALL HOLES TO BE ± .005
Ø.281 TYP 2 PL
Ø.312 TYP 2 PL
.588
.000
SECTION AA
17.755
16.755
15.255
.06
TYP 4 CORNERS
.287 TYP
.187
.375
SECTION BB
.000
.375
3.654
14.101
.875
.500 TYP
.000
.187 X .375 OBROUND, TYP 8 PL
1/4-20 X .500 STUD, ZINC PLATED OR STAINLESS, TYP 2 PL
1.000 ± .010 TYP
.500 ± .010 TYP
Ø.500
45° TYP
1.140
.387
.000
.613
.000
.613
DR BY: CNR 7-27
CK BY: PAH 7-31
APP BY: GWS 8-3
GIRTZ INDUSTRIES, INC.
MONTICELLO, IN 47960
SUPPORT, UV 1"-18" ARM
SIZE B
DWG NO 301337
SCALE HALF
SHEET 1 OF 1

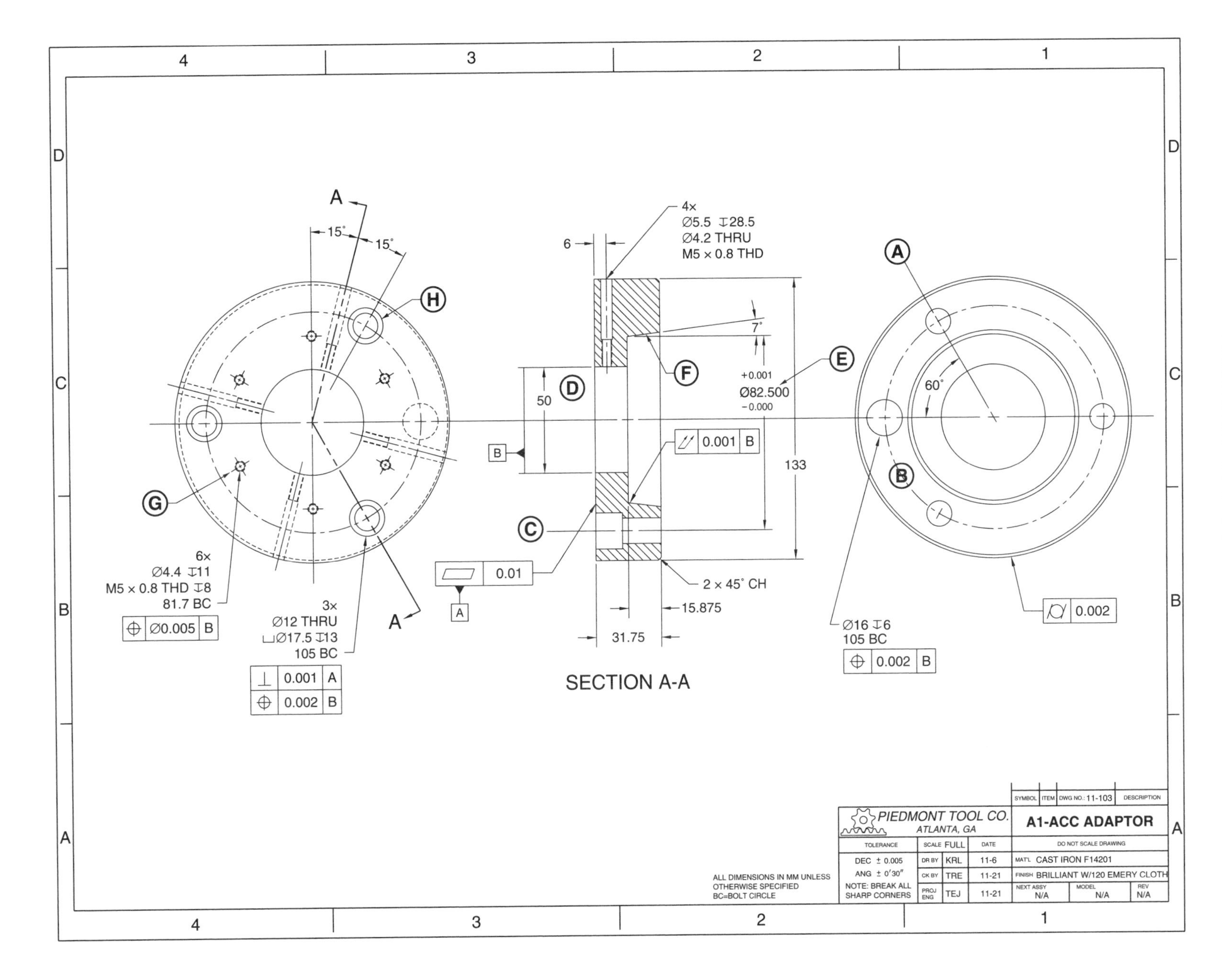

A1-ACC ADAPTOR

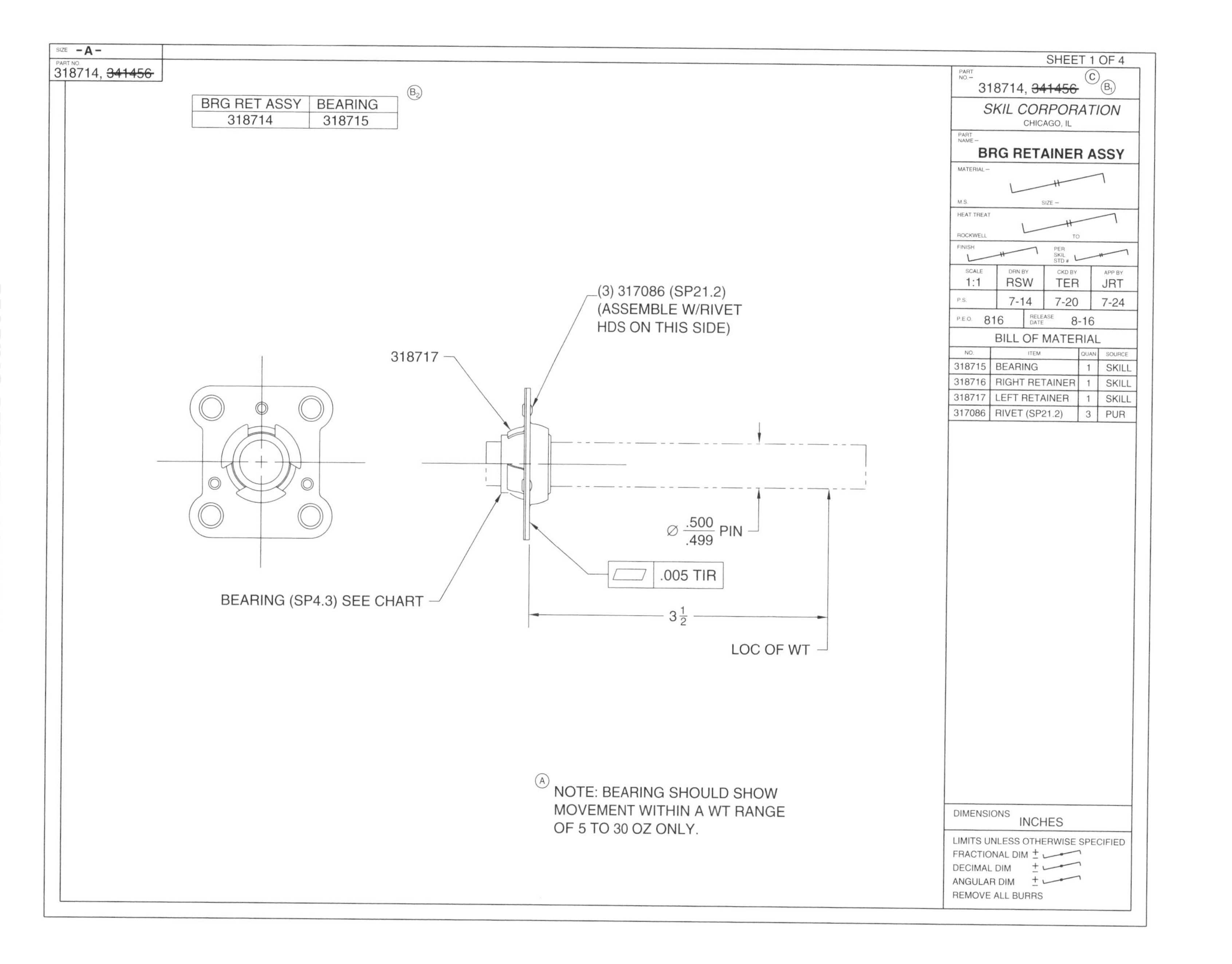

BEARING RETAINER ASSEMBLY

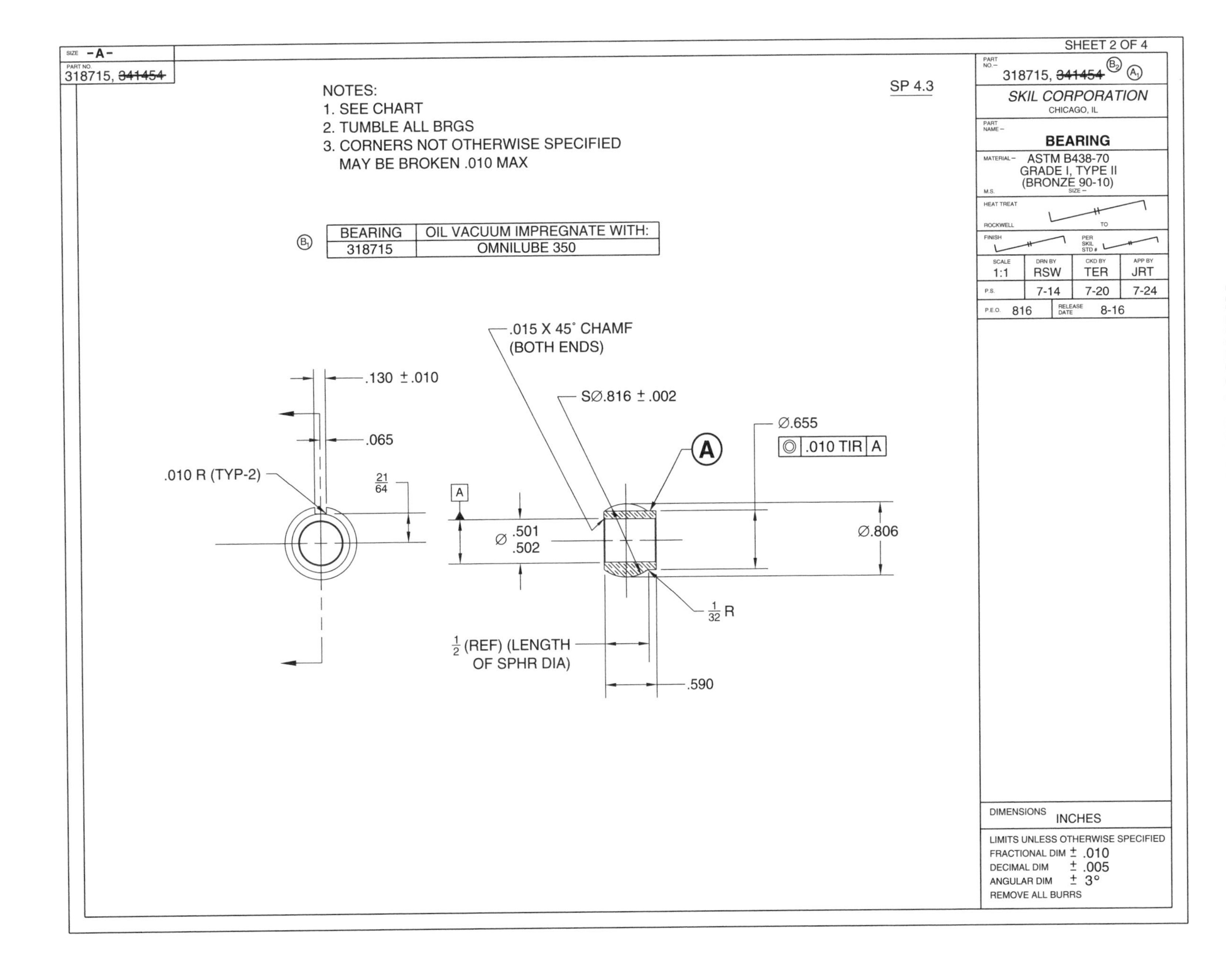

SIZE -A-
PART NO. 318715, 341454
NOTES:
1. SEE CHART
2. TUMBLE ALL BRGS
3. CORNERS NOT OTHERWISE SPECIFIED
MAY BE BROKEN .010 MAX
SP 4.3
BEARING | OIL VACUUM IMPREGNATE WITH:
318715 | OMNILUBE 350
.015 X 45° CHAMF (BOTH ENDS)
.130 ± .010
S∅.816 ± .002
∅.655
.065
A
.010 TIR A
.010 R (TYP-2)
21/64
A
∅ .501 .502
∅.806
1/32 R
1/2 (REF) (LENGTH OF SPHR DIA)
.590
SHEET 2 OF 4
PART NO. – 318715, 341454
SKIL CORPORATION
CHICAGO, IL
PART NAME – BEARING
MATERIAL – ASTM B438-70 GRADE I, TYPE II (BRONZE 90-10)
M.S. SIZE –
HEAT TREAT
ROCKWELL TO
FINISH
PER SKIL STD #
SCALE 1:1
DRN BY RSW
CKD BY TER
APP BY JRT
P.S. 7-14 7-20 7-24
P.E.O. 816
RELEASE DATE 8-16
DIMENSIONS INCHES
LIMITS UNLESS OTHERWISE SPECIFIED
FRACTIONAL DIM ± .010
DECIMAL DIM ± .005
ANGULAR DIM ± 3°
REMOVE ALL BURRS

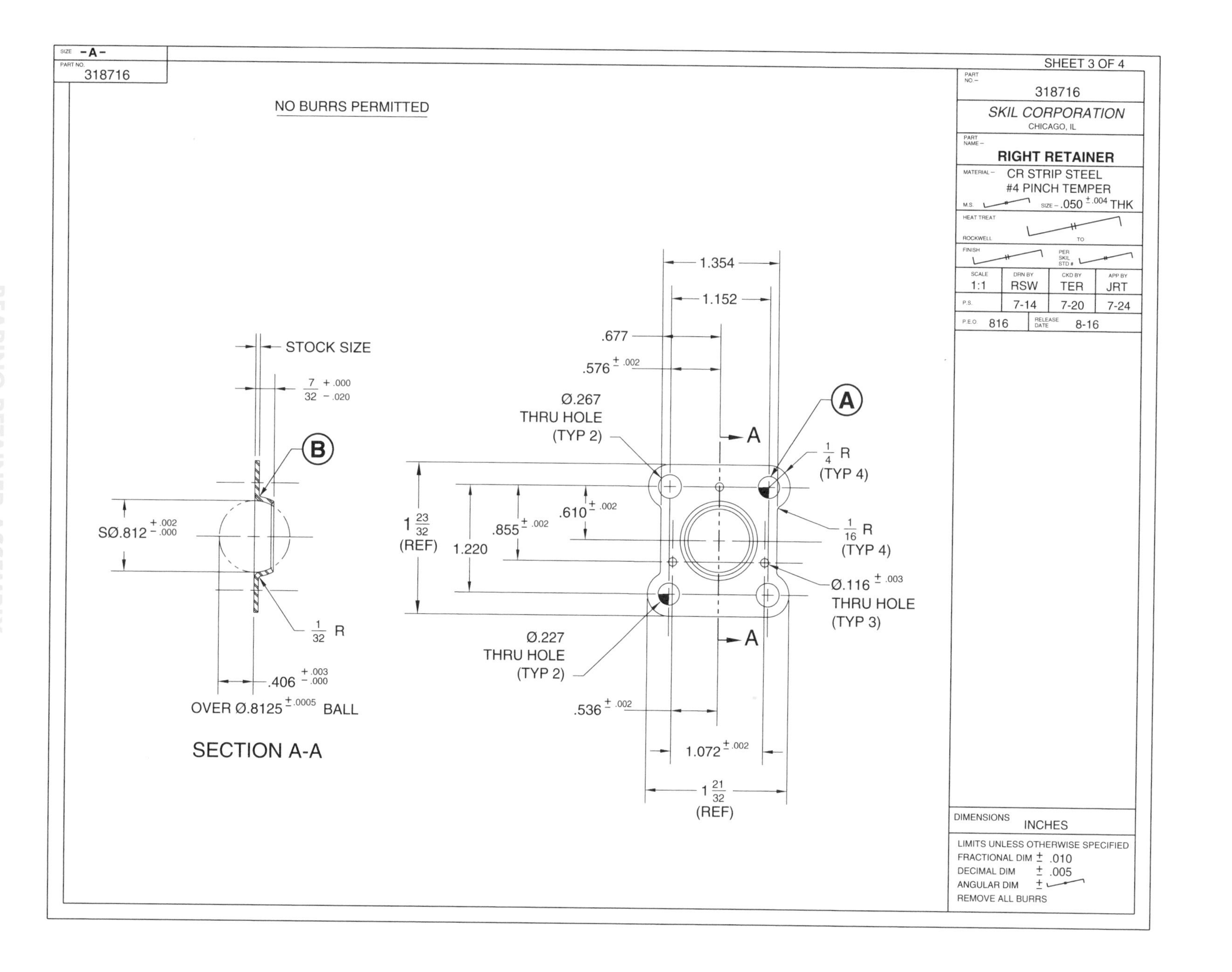

SIZE -A-
PART NO. 318716
SHEET 3 OF 4
PART NO.- 318716
SKIL CORPORATION
CHICAGO, IL
PART NAME- RIGHT RETAINER
MATERIAL- CR STRIP STEEL
#4 PINCH TEMPER
M.S.
SIZE - .050 ± .004 THK
HEAT TREAT
ROCKWELL
TO
FINISH
PER SKIL STD #
SCALE 1:1
DRN BY RSW
CKD BY TER
APP BY JRT
P.S.
7-14
7-20
7-24
P.E.O. 816
RELEASE DATE 8-16
DIMENSIONS INCHES
LIMITS UNLESS OTHERWISE SPECIFIED
FRACTIONAL DIM ± .010
DECIMAL DIM ± .005
ANGULAR DIM ±
REMOVE ALL BURRS
NO BURRS PERMITTED
STOCK SIZE
7/32 +.000 −.020
B
SØ.812 +.002 −.000
1/32 R
.406 +.003 −.000
OVER Ø.8125 ±.0005 BALL
SECTION A-A
1.354
1.152
.677
.576 ± .002
Ø.267
THRU HOLE
(TYP 2)
A
1 23/32
(REF)
1.220
.855 ± .002
.610 ± .002
1/4 R
(TYP 4)
1/16 R
(TYP 4)
Ø.116 ± .003
THRU HOLE
(TYP 3)
Ø.227
THRU HOLE
(TYP 2)
.536 ± .002
1.072 ± .002
1 21/32
(REF)

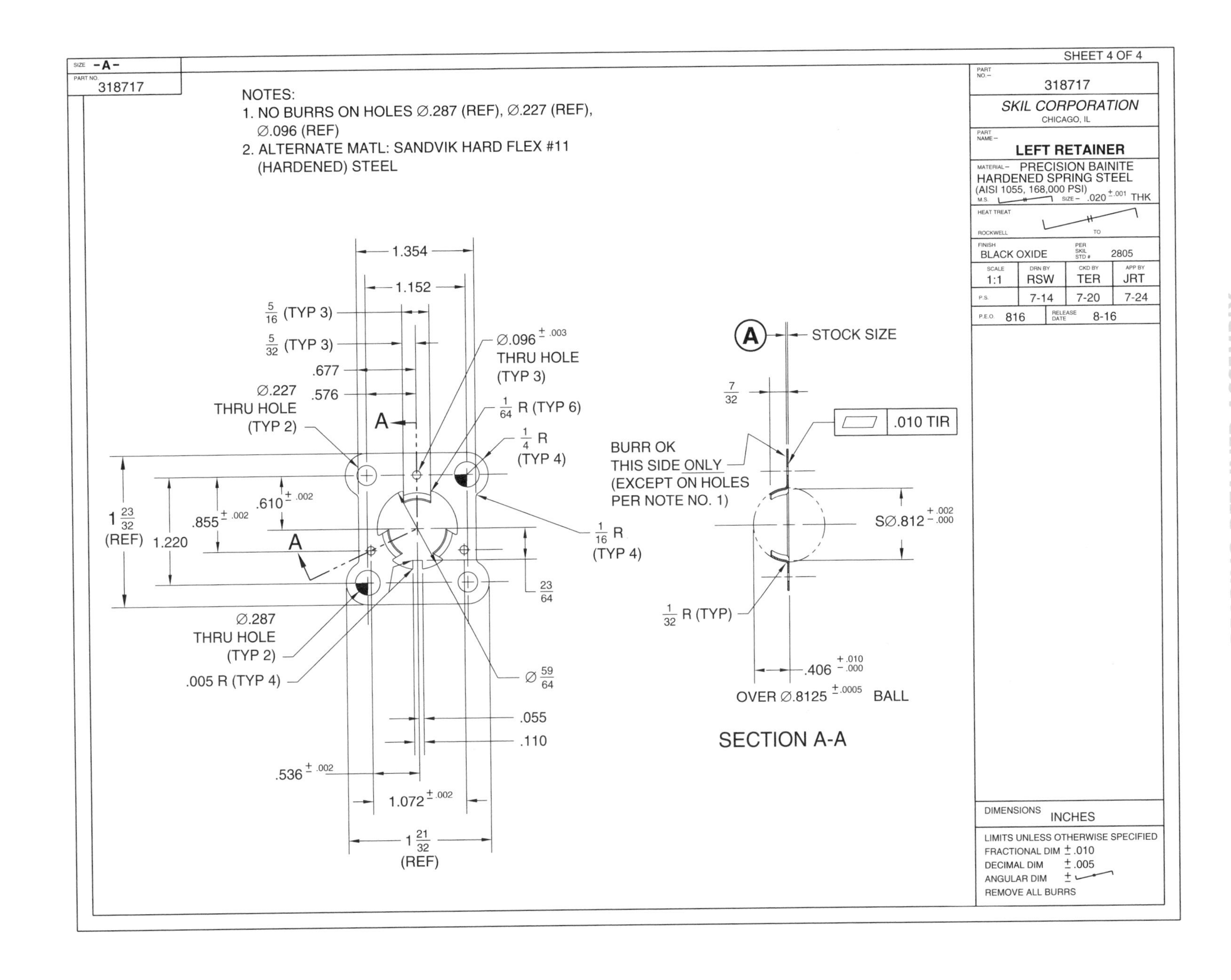
SIZE -A-
PART NO. 318717
NOTES:
1. NO BURRS ON HOLES Ø.287 (REF), Ø.227 (REF), Ø.096 (REF)
2. ALTERNATE MATL: SANDVIK HARD FLEX #11 (HARDENED) STEEL
1.354
1.152
5/16 (TYP 3)
5/32 (TYP 3)
.677
.576
Ø.227 THRU HOLE (TYP 2)
Ø.096 ± .003 THRU HOLE (TYP 3)
1/64 R (TYP 6)
1/4 R (TYP 4)
A
.610 ± .002
.855 ± .002
1 23/32 (REF)
1.220
1/16 R (TYP 4)
23/64
Ø.287 THRU HOLE (TYP 2)
.005 R (TYP 4)
Ø 59/64
.055
.110
.536 ± .002
1.072 ± .002
1 21/32 (REF)
STOCK SIZE
7/32
.010 TIR
BURR OK THIS SIDE ONLY (EXCEPT ON HOLES PER NOTE NO. 1)
SØ.812 +.002 -.000
1/32 R (TYP)
.406 +.010 -.000
OVER Ø.8125 ±.0005 BALL
SECTION A-A
SHEET 4 OF 4
PART NO. 318717
SKIL CORPORATION
CHICAGO, IL
PART NAME - LEFT RETAINER
MATERIAL - PRECISION BAINITE HARDENED SPRING STEEL (AISI 1055, 168,000 PSI)
M.S. SIZE - .020 ± .001 THK
HEAT TREAT
ROCKWELL TO
FINISH BLACK OXIDE
PER SKIL STD # 2805
SCALE 1:1
DRN BY RSW
CKD BY TER
APP BY JRT
P.S. 7-14 7-20 7-24
P.E.O. 816
RELEASE DATE 8-16
DIMENSIONS INCHES
LIMITS UNLESS OTHERWISE SPECIFIED
FRACTIONAL DIM ± .010
DECIMAL DIM ± .005
ANGULAR DIM ±
REMOVE ALL BURRS

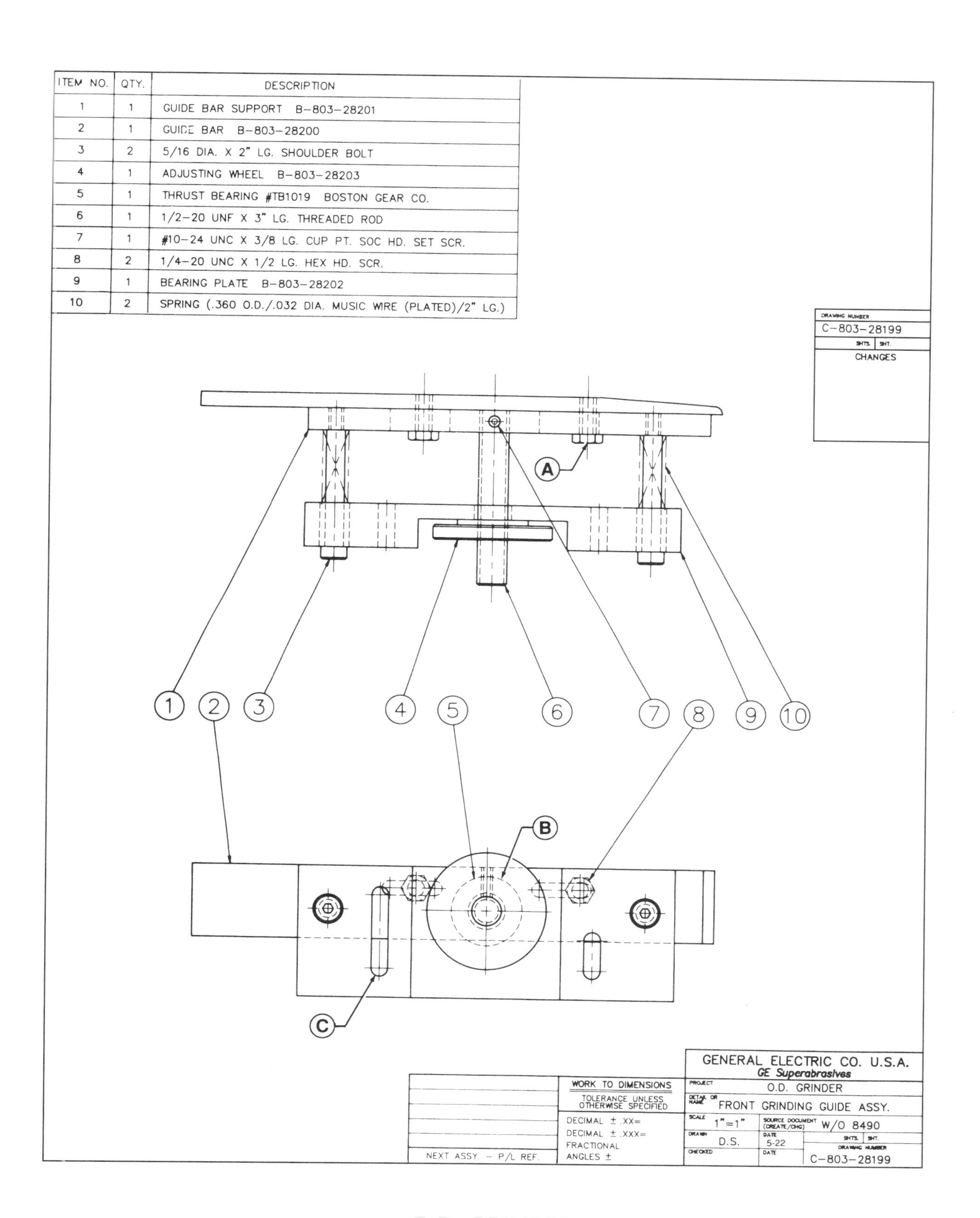

ITEM NO.	QTY.	DESCRIPTION
1	1	GUIDE BAR SUPPORT B-803-28201
2	1	GUIDE BAR B-803-28200
3	2	5/16 DIA. X 2" LG. SHOULDER BOLT
4	1	ADJUSTING WHEEL B-803-28203
5	1	THRUST BEARING #TB1019 BOSTON GEAR CO.
6	1	1/2-20 UNF X 3" LG. THREADED ROD
7	1	#10-24 UNC X 3/8 LG. CUP PT. SOC HD. SET SCR.
8	2	1/4-20 UNC X 1/2 LG. HEX HD. SCR.
9	1	BEARING PLATE B-803-28202
10	2	SPRING (.360 O.D./.032 DIA. MUSIC WIRE (PLATED)/2" LG.)

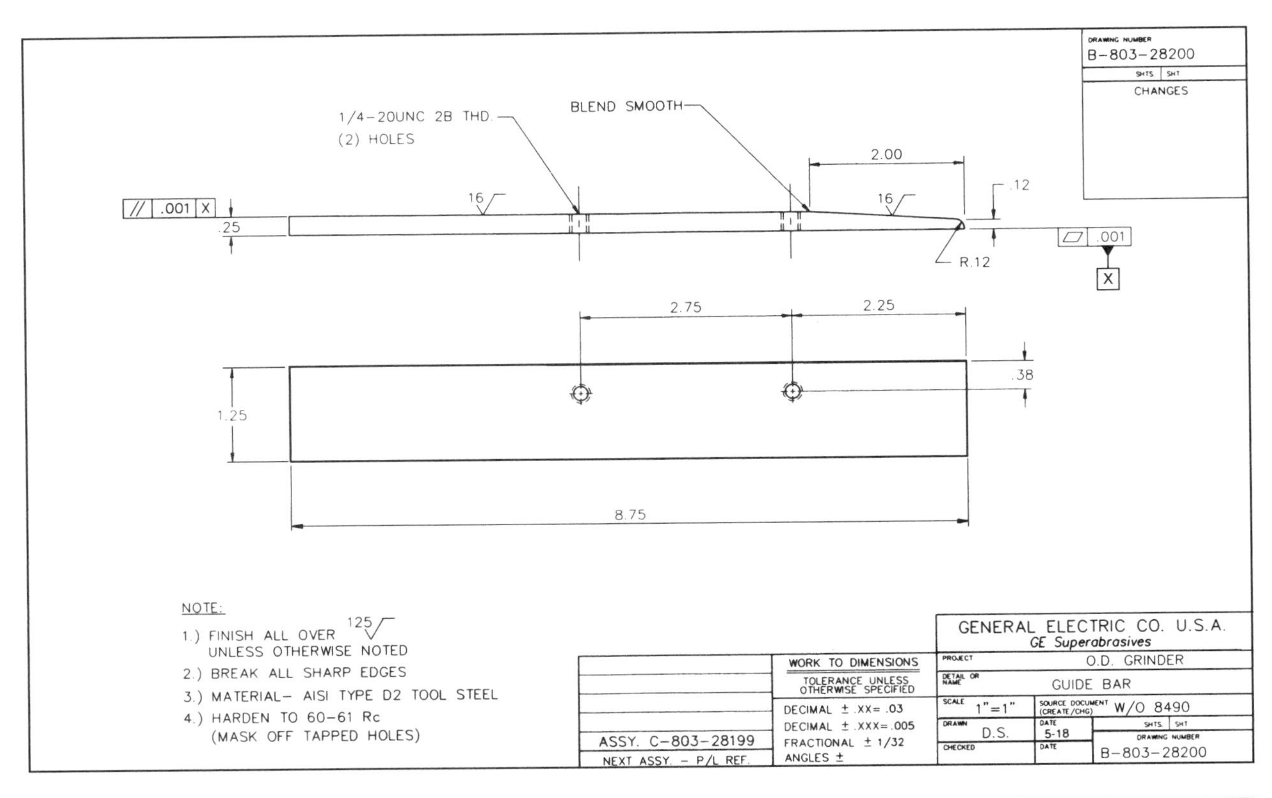

B-803-28200
CHANGES
1/4-20UNC 2B THD.
(2) HOLES
BLEND SMOOTH
2.00
.12
16
.001 X
.25
.001
R.12
X
2.75
2.25
.38
1.25
8.75
NOTE:
1.) FINISH ALL OVER 125 UNLESS OTHERWISE NOTED
2.) BREAK ALL SHARP EDGES
3.) MATERIAL- AISI TYPE D2 TOOL STEEL
4.) HARDEN TO 60-61 Rc (MASK OFF TAPPED HOLES)
GENERAL ELECTRIC CO. U.S.A.
GE Superabrasives
WORK TO DIMENSIONS
TOLERANCE UNLESS OTHERWISE SPECIFIED
DECIMAL ± .XX= .03
DECIMAL ± .XXX= .005
FRACTIONAL ± 1/32
ANGLES ±
O.D. GRINDER
GUIDE BAR
1"=1"
W/O 8490
D.S.
5-18
ASSY. C-803-28199
NEXT ASSY. - P/L REF.
B-803-28200

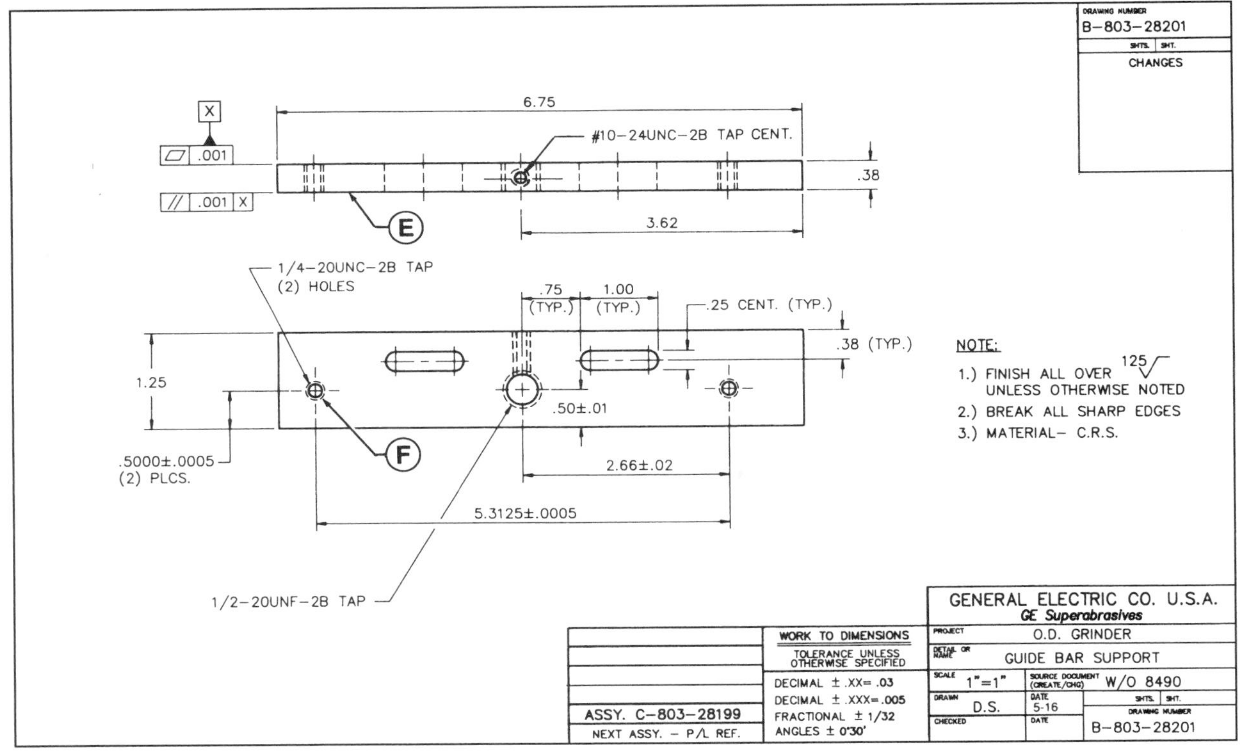

B-803-28201
CHANGES
X
.001
.001 X
6.75
#10-24UNC-2B TAP CENT.
.38
E
3.62
1/4-20UNC-2B TAP
(2) HOLES
.75 (TYP.)
1.00 (TYP.)
.25 CENT. (TYP.)
.38 (TYP.)
1.25
.50±.01
F
.5000±.0005
(2) PLCS.
2.66±.02
5.3125±.0005
1/2-20UNF-2B TAP
NOTE:
1.) FINISH ALL OVER 125 UNLESS OTHERWISE NOTED
2.) BREAK ALL SHARP EDGES
3.) MATERIAL- C.R.S.
GENERAL ELECTRIC CO. U.S.A.
GE Superabrasives
WORK TO DIMENSIONS
TOLERANCE UNLESS OTHERWISE SPECIFIED
DECIMAL ± .XX= .03
DECIMAL ± .XXX= .005
FRACTIONAL ± 1/32
ANGLES ± 0°30'
O.D. GRINDER
GUIDE BAR SUPPORT
1"=1"
W/O 8490
D.S.
5-16
ASSY. C-803-28199
NEXT ASSY. - P/L REF.
B-803-28201

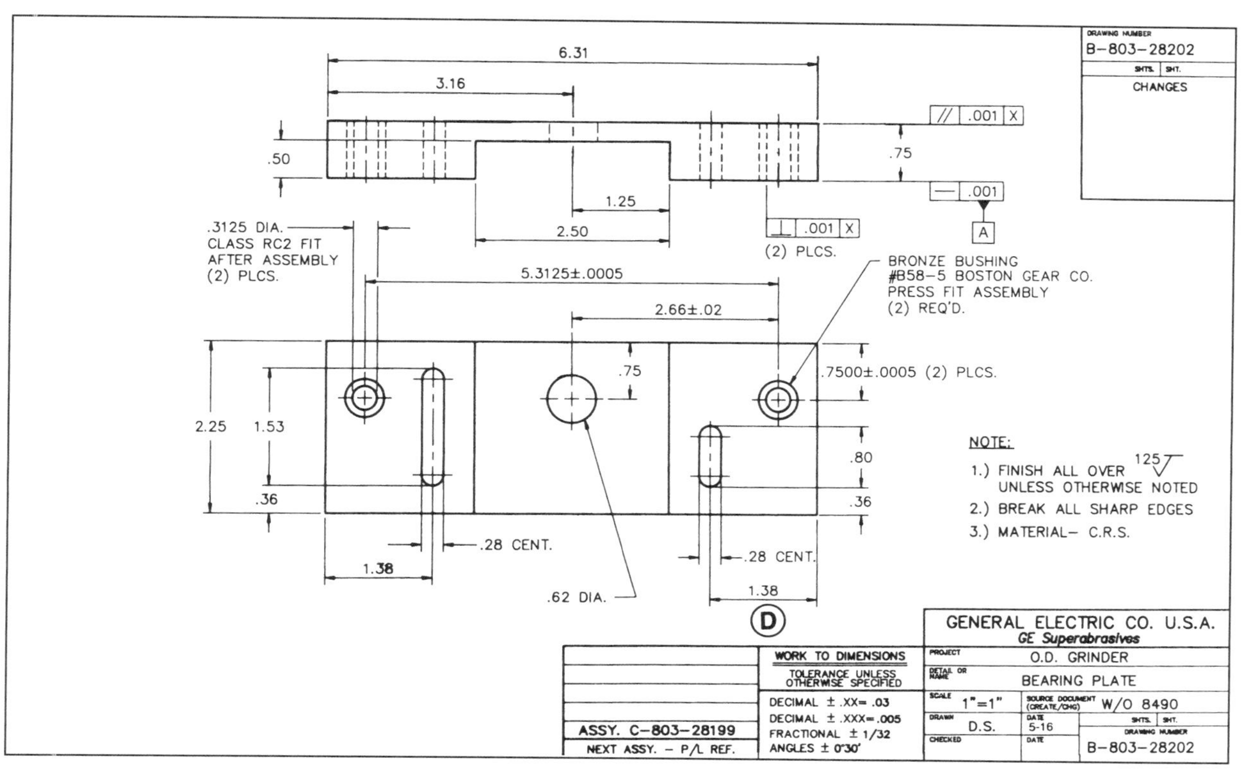
6.31
3.16
.50
.75
// .001 X
— .001
A
1.25
2.50
⊥ .001 X
(2) PLCS.
.3125 DIA.
CLASS RC2 FIT
AFTER ASSEMBLY
(2) PLCS.
5.3125±.0005
2.66±.02
BRONZE BUSHING
#B58-5 BOSTON GEAR CO.
PRESS FIT ASSEMBLY
(2) REQ'D.
.75
.7500±.0005 (2) PLCS.
2.25
1.53
.36
.80
.36
.28 CENT.
1.38
.62 DIA.
.28 CENT.
1.38
D
NOTE:
1.) FINISH ALL OVER 125
UNLESS OTHERWISE NOTED
2.) BREAK ALL SHARP EDGES
3.) MATERIAL- C.R.S.
DRAWING NUMBER
B-803-28202
CHANGES
GENERAL ELECTRIC CO. U.S.A.
GE Superabrasives
WORK TO DIMENSIONS
TOLERANCE UNLESS OTHERWISE SPECIFIED
DECIMAL ± .XX= .03
DECIMAL ± .XXX=.005
FRACTIONAL ± 1/32
ANGLES ± 0°30'
PROJECT O.D. GRINDER
DETAIL OR NAME BEARING PLATE
SCALE 1"=1"
SOURCE DOCUMENT (CREATE/CHG) W/O 8490
DRAWN D.S.
DATE 5-16
CHECKED
DATE
DRAWING NUMBER B-803-28202
ASSY. C-803-28199
NEXT ASSY. - P/L REF.

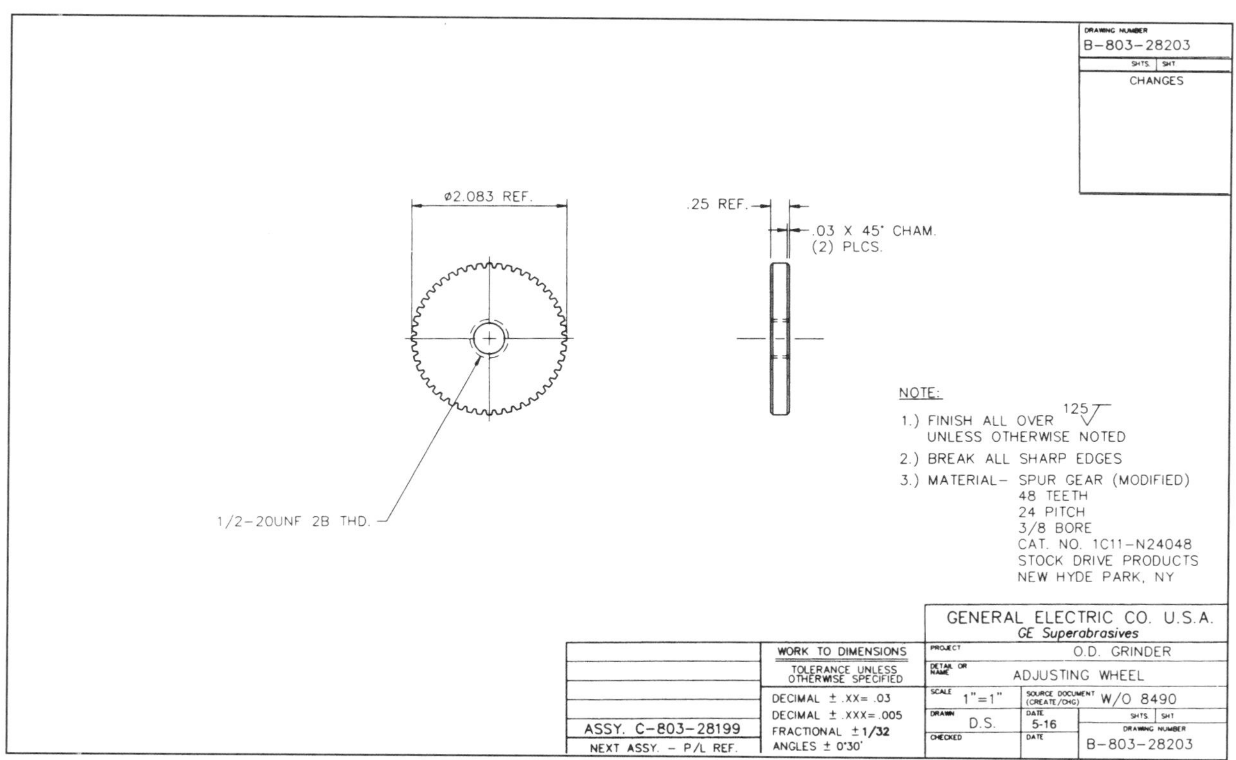
DRAWING NUMBER
B-803-28203
CHANGES
⌀2.083 REF.
.25 REF.
.03 X 45° CHAM.
(2) PLCS.
1/2-20UNF 2B THD.
NOTE:
1.) FINISH ALL OVER 125
UNLESS OTHERWISE NOTED
2.) BREAK ALL SHARP EDGES
3.) MATERIAL- SPUR GEAR (MODIFIED)
48 TEETH
24 PITCH
3/8 BORE
CAT. NO. 1C11-N24048
STOCK DRIVE PRODUCTS
NEW HYDE PARK, NY
GENERAL ELECTRIC CO. U.S.A.
GE Superabrasives
WORK TO DIMENSIONS
TOLERANCE UNLESS OTHERWISE SPECIFIED
DECIMAL ± .XX= .03
DECIMAL ± .XXX=.005
FRACTIONAL ±1/32
ANGLES ± 0°30'
PROJECT O.D. GRINDER
DETAIL OR NAME ADJUSTING WHEEL
SCALE 1"=1"
SOURCE DOCUMENT (CREATE/CHG) W/O 8490
DRAWN D.S.
DATE 5-16
CHECKED
DATE
DRAWING NUMBER B-803-28203
ASSY. C-803-28199
NEXT ASSY. - P/L REF.

APPENDIX

ABBREVIATIONS . . .

A

Term	Abbreviation
Absolute	ABS
Actual	ACT
Adapter	ADPT
Addendum	ADD
Adjust	ADJ
Advance	ADV
Allowance	ALLOW
Alloy	ALY
Altitude	ALT
Aluminum	AL
American Standard	AMER STD
American Wire Gauge	AWG
Amount	AMT
Anneal	ANL
Apparatus	APPAR
Approved	APP
Approximate	APPROX
Arc Weld	ARC/W
Area	A
Arrangement	ARR
Assemble	ASSEM
Assembly	ASSY
Authorized	AUTH
Auxiliary	AUX

B

Term	Abbreviation
Babbitt	BAB
Back-feed	BF
Back Pressure	BP
Ball Bearing	BB
Base Line	BL
Base Plate	BP
Bearing	BRG
Benchmark	BM
Bending Moment	M
Between	BET
Bevel	BEV
Bill of Material	B/M
Bolt Circle	BC
Both Faces	BF
Both Sides	BS
Both Ways	BW
Bottom	BOT
Bottom Chord	BC
Bracket	BRKT
Brass	BRS
Brazing	BRZG
Break	BRK
Brinell Hardness	BH
British Standard	BR STD
Broach	BRO
Bronze	BRZ
Bushing	BUSH

C

Term	Abbreviation
Cabinet	CAB
Cadmium Plate	CD PL
Capacity CAP	
Cap Screw	CAP SCR
Carbon	C
Carburize	CARB
Carriage	CRG
Case Harden	CH
Cast	C
Cast Iron	CI
Cast Iron Pipe	CIP
Cast Steel	CS
Casting	CSTG
Castle Nut	CAS NUT
Center	CTR
Centerline	CL
Center of Gravity	CG
Center Punch	CP
Ceramic	CER
Chamfer	CH or CHAM
Channel	CHAN
Chrome Molybdenum	CR MOLY
Chromium Plate	CR PL
Chrome Vanadium	CR VAN
Circle	CIR
Circular Pitch	CP
Circumference	CIRC
Clearance	CL
Clockwise	CW
Coated	CTD
Cold Drawn	CD
Cold Drawn Steel	CDS
Cold Finish	CF
Cold Punched	CP
Cold Rolled Steel	CRS
Concentric	CONC
Copper Plate	COP PL
Corrosion Resistant	CRE
Corrosion Resistant Steel	CRES
Cotter	COT
Counterclockwise	CCW
Counterbore	CB or CBORE
Counterdrill	CD or CDRILL
Countersink	CSK or CSINK
Countersink Other Side	CSKO
Coupling	CPLG
Cross Section	XSECT
Cubic	CU
Cubic Foot	CU FT
Cubic Inch	CU IN
Cylinder	CYL

D

Term	Abbreviation
Decimal	DEC
Dedendum	DED
Depth	DP or DEEP
Deep Drawn	DD
Degree	DEG
Density	D
Design	DSGN
Detail	DET
Diagonal	DIAG
Diagram	DIAG
Diameter	DIA
Diametral Pitch	DP
Dimension	DIM
Dovetail	DVTL
Dowel	DWL
Drafting	DFTG
Draftsman	DFTSMN
Drawing	DWG
Drill	DR
Drive	DR
Drop Forge	DF
Duplicate	DUP

E

Term	Abbreviation
Each	EA
Eccentric	ECC
Electric	ELEC
Elongation	ELONG
Enclose	ENCL
Engineer	ENGR
Envelope	ENV
Equipment	EQUIP
Equivalent	EQUIV
Existing	EXIST
Extension	EXT
Extrude	EXTR

F

Term	Abbreviation
Fabricate	FAB
Far Side	FS
Feet	FT
Feet Per Minute	FPM
Feet Per Second	FPS
Figure	FIG
Fillet	FIL
Finish	FIN
Finish All Over	FAO
Fitting	FTG
Fixture	FIX
Flange	FLG
Flashing	FL
Flat	F
Flat Head	FH
Flexible	FLEX
Forged Steel	FST
Forging	FORG
Forward	FWD
Foundry	FDRY
Fractional	FRAC
Furnish	FURN

G

Term	Abbreviation
Gauge	GA
Galvanize	GALV
Galvanized	Iron GI
Galvanized	Steel GS
Gasket	GSKT
General	GEN
Glass	GL
Grade	GR
Grind	GRD or GND
Groove	GRV

H

Term	Abbreviation
Half-Round	½RD
Hard	H
Hard Drawn	HD
Harden	HDN
Hardware	HDW
Head	HD
Headless	HDLS

. . . ABBREVIATIONS . . .

Term	Abbreviation
Heat	HT
Heat Treat	HT TR
Heavy	HVY
Height	HGT
Hexagon	HEX
High-Speed	HS
High-Speed Steel	HSS
High-Tensile Cast Iron	HTCI
High-Tensile Steel	HTS
Horizontal	HORIZ
Hot Rolled	HR
Hot Rolled Steel	HRS
I	
Impregnate	IMPG
Inch	IN.
Inches Per Minute	IPM
Indicate	IND
Inside Diameter	ID
Install	INSTL
Internal	INT
International Pipe Standard	IPS
Intersect	INT
Iron	I
Irregular	IRREG
J	
Joint	JT
Junction	JCT
K	
Key	K
Keyseat	KST
Keyway	KWY
Knockout	KO
L	
Laboratory	LAB
Laminate	LAM
Lateral	LAT
Left Hand	LH
Length	LG
Limit	LIM
Linear	LIN
Locate	LOC
Low-Speed	LS
Lubricate	LUB
M	
Machine	MACH
Machine Steel	MS
Malleable	MALL
Malleable	Iron MI
Manual	MAN
Manufacture	MFR
Manufactured	MFD
Manufacturing	MFG
Material	MAT or MATL
Material List	ML
Maximum	MAX
Mechanical	MECH
Metal	MET
Micrometer	MIC
Military	MIL
Millimeter	MM
Minimum	MIN
Minute	MIN
Miscellaneous	MISC
Mold Line	ML
Molded	MLD
Molding	MLDG
Morse Taper	MOR T
Mounted	MTD
Mounting	MTG
Multiple	MULT
N	
National	NATL
Near Face	NF
Near Side	NS
New British Standard (Imperial Wire Gauge)	NBS
Nipple	NIP
Nominal	NOM
Normal	NORM
Not To Scale	NTS
Number	NO.
O	
Octagon	OCT
On Center	OC
Opening	OPNG
Opposite	OPP
Original	ORIG
Outside Diameter	OD
Overall	OA
P	
Pair	PR
Parallel	PAR
Part	PT
Patent	PAT
Pattern	PATT
Permanent	PERM
Perpendicular	PERP
Phenolic	PHEN
Pitch	P
Pitch Circle	PC
Pitch Diameter	PD
Plate	PL
Point	PT
Position	POS
Pound	LB
Pounds Per Square Inch	PSI
Precast	PRCST
Prefabricated	PREFAB
Preferred	PFD
Primary	PRIM
Production	PROD
Profile	PF
Project	PROJ
Proposed	PROP
Punch	PCH
Q	
Quadrant	QUAD
Quality	QUAL
Quantity	QTY
Quarter Round	¼RD
R	
Radial	RAD
Radians	RAD
Radius	R
Ream	RM
Reassemble	REASM
Received	RECD
Rectangle	RECT
Reference	REF
Reference Line	REF L
Reinforce	REINF
Relief	REL
Remove	REM
Require	REQ
Required	REQD
Return	RET
Reverse	REV
Revolutions Per Minute	RPM
Right Hand	RH
Rivet	RIV
Rockwell Hardness	RH
Roller Bearing	RB
Root Diameter	RD
Root Mean Square	RMS
Round	RD
S	
Schedule	SCH
Schematic	SCHEM
Screw	SCR
Secondary	SEC
Section	SECT
Semi-Finished	SF
Set Screw	SS
Shaft	SFT
Sheet	SH
Shop Order	SO
Shoulder	SHLD
Side	S
Similar	SIM
Sketch	SK
Sleeve	SLV
Sleeve Bearing	SB
Slotted	SLOT
Socket	SOC
Space	SP
Special Treatment Steel	STS
Specification	SPEC
Speed	SP
Spherical	SPHER
Spotface	SF or SFACE
Spring	SPG
Square	SQ
Stainless	STN
Stainless Steel	SST or SS
Standard	STD

. . . ABBREVIATIONS

Steel	STL
Stock	STK
Straight	STR
Stress Anneal	SA
Structural	STR
Supplement	SUPP
Supply	SUP
Surface	SURF
Symbol	SYM
Symmetrical	SYM
Synthetic	SYN
T	
Tachometer	TACH
Tangent	TAN
Taper	TPR
Technical	TECH
Tee	T
Teeth	T
Teeth Per Inch	TPI
Temperature	TEMP
Template	TEMP
Tensile Strength	TS
Tension	TENS
Thick	THK
Thread	THD
Threads Per Inch	TPI
Through	THRU
Tolerance	TOL
Tool Steel	TS
Total	TOT
Total Indicator Reading	TIR
Tubing	TUB
Typical	TYP
U	
United States Gauge	USG
United States Standard	USS
V	
Velocity	V
Vertical	VERT
Vibrate	VIB
Volume	VOL
W	
Washer	WASH
Weight	WT
Wheel Base	WB
Width	W
Wire	W
With	W/
Without	W/O
Woodruff	WDF
Wrought	WRT
Wrought Iron	WI

PRINTREADING SYMBOLS

MEANING	SYMBOL	MEANING	SYMBOL
Straightness	—	Projected Tolerance Zone	Ⓟ
Flatness	▱	Diameter	⌀
Circularity	○	Basic dimension	[50]
Cylindricity	⌭	Reference dimension	(50)
Profile of a line	⌒	Conical taper	⌲
Profile of a surface	⌓	Taper	⌳
All around	↙○	Counterbore/Spotface	⌴
Angularity	∠	Countersink	⌵
Perpendicularity	⊥	Depth/deep	↧
Parallelism	//	Square	□
Position	⌖	Dimension not to scale	$\underline{15}$
Concentricity	◎	Arc length	$\overset{\frown}{105}$
Symmetry	⌯	Radius	R
Circular runout	*↗	Spherical radius	SR
Total runout	*⌰	Spherical diameter	SØ
Maximum Material Condition	Ⓜ	Between	*↔
Least Material Condition	Ⓛ	Statistical tolerance	⟨ST⟩

* may be filled or not filled

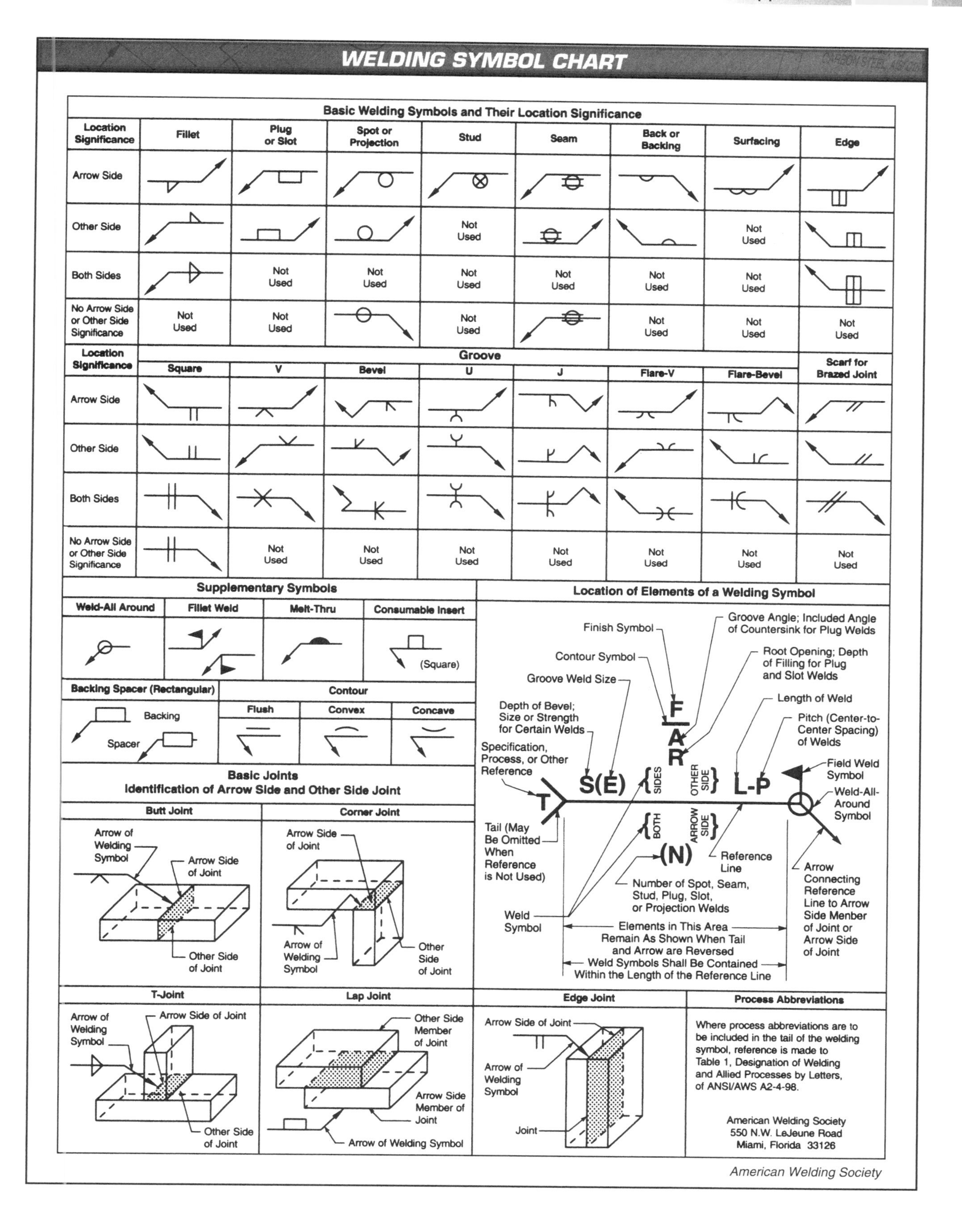
WELDING SYMBOL CHART
Basic Welding Symbols and Their Location Significance
Location Significance
Fillet
Plug or Slot
Spot or Projection
Stud
Seam
Back or Backing
Surfacing
Edge
Arrow Side
Other Side
Both Sides
No Arrow Side or Other Side Significance
Not Used
Groove
Square
V
Bevel
U
J
Flare-V
Flare-Bevel
Scarf for Brazed Joint
Supplementary Symbols
Weld-All Around
Fillet Weld
Melt-Thru
Consumable Insert
(Square)
Backing Spacer (Rectangular)
Backing
Spacer
Contour
Flush
Convex
Concave
Basic Joints
Identification of Arrow Side and Other Side Joint
Butt Joint
Arrow of Welding Symbol
Arrow Side of Joint
Other Side of Joint
Corner Joint
Arrow Side of Joint
Arrow of Welding Symbol
Other Side of Joint
T-Joint
Arrow of Welding Symbol
Arrow Side of Joint
Other Side of Joint
Lap Joint
Other Side Member of Joint
Arrow Side Member of Joint
Arrow of Welding Symbol
Location of Elements of a Welding Symbol
Finish Symbol
Groove Angle; Included Angle of Countersink for Plug Welds
Contour Symbol
Root Opening; Depth of Filling for Plug and Slot Welds
Groove Weld Size
Length of Weld
Depth of Bevel; Size or Strength for Certain Welds
Pitch (Center-to-Center Spacing) of Welds
Specification, Process, or Other Reference
Field Weld Symbol
Weld-All-Around Symbol
F
A
R
S(E)
SIDES
OTHER SIDE
L-P
T
BOTH
ARROW SIDE
(N)
Tail (May Be Omitted When Reference is Not Used)
Reference Line
Number of Spot, Seam, Stud, Plug, Slot, or Projection Welds
Weld Symbol
Elements in This Area Remain As Shown When Tail and Arrow are Reversed
Weld Symbols Shall Be Contained Within the Length of the Reference Line
Arrow Connecting Reference Line to Arrow Side Member of Joint or Arrow Side of Joint
Edge Joint
Arrow Side of Joint
Arrow of Welding Symbol
Joint
Process Abbreviations
Where process abbreviations are to be included in the tail of the welding symbol, reference is made to Table 1, Designation of Welding and Allied Processes by Letters, of ANSI/AWS A2-4-98.
American Welding Society
550 N.W. LeJeune Road
Miami, Florida 33126
American Welding Society

ROUGHNESS PER MANUFACTURING PROCESS

PROCESS	2000* (50)†	1000 (25)	500 (12.5)	250 (6.3)	125 (3.2)	63 (1.6)	32 (0.80)	16 (0.40)	8 (0.20)	4 (0.10)	2 (0.05)	1 (0.025)	0.5 (0.012)
FLAME CUTTING													
PLANING, SHAPING													
DRILLING													
MILLING													
BROACHING													
REAMING													
BORING, TURNING													
GRINDING													
POLISHING													
LAPPING													
SUPERFINISHING													
SAND CASTING													
HOT ROLLING													
FORGING													
EXTRUDING													
COLD ROLLING, DRAWING													

* in microinches (μin.)
† in micrometers (μm)

STANDARD SERIES THREADS — GRADED PITCHES

NOMINAL DIAMETER	UNC		UNF		UNEF	
	TPI	TAP DRILL	TPI	TAP DRILL	TPI	TAP DRILL
0 (.0600)			80	3/64		
1 (.0730)	64	No. 53	72	No. 53		
2 (.0860)	56	No. 50	64	No. 50		
3 (.0990)	48	No. 47	56	No. 45		
4 (.1120)	40	No. 43	48	No. 42		
5 (.1250)	40	No. 38	44	No. 37		
6 (.1380)	32	No. 36	40	No. 33		
8 (.1640)	32	No. 29	36	No. 29		
10 (.1900)	24	No. 25	32	No. 21		
12 (.2160)	24	No. 16	28	No. 14	32	No.13
1/4 (.2500)	20	No. 7	28	No. 3	32	7/32
5/16 (.3125)	18	F	24	I	32	9/32
3/8 (.3750)	16	5/16	24	Q	32	11/32
7/16 (.4375)	14	U	20	25/64	28	13/32
1/2 (.5000)	13	27/64	20	29/64	28	15/32
9/16 (.5625)	12	31/64	18	33/64	24	33/64
5/8 (.6250)	11	17/32	18	37/64	24	37/64
11/16 (.6875)					24	41/64
3/4 (.7500)	10	21/32	16	11/16	20	45/64
13/16 (.8125)					20	49/64
7/8 (.8750)	9	49/64	14	13/16	20	53/64
15/16 (.9375)					20	57/64

TWIST DRILL FRACTIONAL, NUMBER, AND LETTER SIZES

Drill No.	Frac	Deci	Drill No.	Frac	Deci	Drill No.	Frac	Deci	Drill No.	Frac	Deci
80	—	.0135	42	—	.0935	7	—	.201	X	—	.397
79	—	.0145	—	3/32	.0938	—	13/64	.203	Y	—	.404
—	1/64	.0156				6	—	.204			
78	—	.0160	41	—	.0960	5	—	.206	—	13/32	.406
77	—	.0180	40	—	.0980	4	—	.209	Z	—	.413
			39	—	.0995				—	27/64	.422
76	—	.0200	38	—	.1015	3	—	.213	—	7/16	.438
75	—	.0210	37	—	.1040	—	7/32	.219	—	29/64	.453
74	—	.0225				2	—	.221			
73	—	.0240	36	—	.1065	1	—	.228	—	15/32	.469
72	—	.0250	—	7/64	.1094	A	—	.234	—	31/64	.484
			35	—	.1100				—	1/2	.500
71	—	.0260	34	—	.1110	—	15/64	.234	—	33/64	.516
70	—	.0280	33	—	.1130	B	—	.238	—	17/32	.531
69	—	.0292				C	—	.242			
68	—	.0310	32	—	.116	D	—	.246	—	35/64	.547
—	1/32	.0313	31	—	.120	—	1/4	.250	—	9/16	.562
			—	1/8	.125				—	37/64	.578
67	—	.0320	30	—	.129	E	—	.250	—	19/32	.594
66	—	.0330	29	—	.136	F	—	.257	—	39/64	.609
65	—	.0350				G	—	.261			
64	—	.0360	—	9/64	.140	—	17/64	.266	—	5/8	.625
63	—	.0370	28	—	.141	H	—	.266	—	41/64	.641
			27	—	.144				—	21/32	.656
62	—	.0380	26	—	.147	I	—	.272	—	43/64	.672
61	—	.0390	25	—	.150	J	—	.277	—	11/16	.688
60	—	.0400				—	9/32	.281			
59	—	.0410	24	—	.152	K	—	.281	—	45/64	.703
58	—	.0420	23	—	.154	L	—	.290	—	23/32	.719
			—	5/32	.156				—	47/64	.734
57	—	.0430	22	—	.157	M	—	.295	—	3/4	.750
56	—	.0465	21	—	.159	—	19/64	.2297	—	49/64	.766
—	3/64	.0469				N	—	.302			
55	—	.0520	20	—	.161	—	5/16	.313	—	25/32	.781
54	—	.0550	19	—	.166	O	—	.316	—	51/64	.797
			18	—	.170				—	13/16	.813
53	—	.0595	—	11/64	.172	P	—	.323	—	53/64	.828
—	1/16	.0625	17	—	.173	—	21/64	.328	—	27/32	.844
52	—	.0635				Q	—	.332			
51	—	.0670				R	—	.339			
50	—	.0700	16	—	.177	—	11/32	.344	—	55/64	.859
			15	—	.180				—	7/8	.875
49	—	.0730	14	—	.182	S	—	.348	—	57/64	.891
48	—	.0760	13	—	.185	T	—	.358	—	29/32	.906
—	5/64	.0781	—	3/16	.188	—	23/64	.359	—	59/64	.922
47	—	.0785				U	—	.368			
46	—	.0810	12	—	.189	—	3/8	.375	—	15/16	.938
			11	—	.191				—	61/64	.953
45	—	.0820	10	—	.194	V	—	.377	—	31/32	.969

METRIC SCREW THREADS

Coarse (general purpose)		Fine	
Nom Size & Thd Pitch	Tap Drill Dia (mm)	Nom Size & Thd Pitch	Tap Drill Dia (mm)
M1.6 × 0.35	1.25	—	—
M1.8 × 0.35	1.45	—	—
M2 × 0.4	1.6	—	—
M2.2 × 0.45	1.75	—	—
M2.5 × 0.45	2.05	—	—
M3 × 0.5	2.50	—	—
M3.5 × 0.6	2.90	—	—
M4 × 0.7	3.30	—	—
M4.5 × 0.75	3.75	—	—
M5 × 0.8	4.20	—	—
M6.3 × 1	5.30	—	—
M7 × 1	6.00	—	—
M8 × 1.25	6.80	M8 × 1	7.00
M9 × 1.25	7.75		
M10 × 1.5	8.50	M10 × 1.25	8.75
M11 × 1.5	9.50		
M12 × 1.75	10.30	M12 × 1.25	10.50
M14 × 2	12.00	M14 × 1.5	12.50
M16 × 2	14.00	M16 × 1.5	14.50
M18 × 2.5	15.50	M18 × 1.5	16.50
M20 × 2.5	17.50	M20 × 1.5	18.50
M22 × 2.5	19.50	M22 × 1.5	20.50
M24 × 3	21.00	M24 × 2	22.00
M27 × 3	24.00	M27 × 2	25.00
M30 × 3.5	26.50	M30 × 2	28.00
M33 × 3.5	29.50	M30 × 2	31.00
M36 × 4	32.00	M36 × 3	33.00
M39 × 4	35.00	M39 × 3	36.00
M42 × 4.5	37.50	M42 × 3	39.00
M45 × 4.5	40.50	M45 × 3	42.00
M48 × 5	43.00	M48 × 3	45.00
M52 × 5	47.00	M52 × 3	49.00
M56 × 5.5	50.50	M56 × 4	52.00
M60 × 5.5	54.50	M60 × 4	56.00
M64 × 6	58.00	M64 × 4	60.00
M68 × 6	62.00	M68 × 4	64.00
M72 × 6	66.00	—	—
M80 × 6	74.00	—	—
M90 × 6	84.00	—	—

SAE STEEL ALLOY DESIGNATION SYSTEM

Type	Designation	Nominal Alloy Content
Carbon	10xx	Plain carbon (1% Mn max)
	11xx	Resulfurized
	12xx	Resulfurized and rephosphorized
	15xx	Plain carbon (1% Mn – 1.65% Mn max)
Manganese	13xx	1.75% Mn
Nickel	23xx	3.5 Ni
	25xx	5% Ni
Nickel-Chromium	31xx	1.25% Ni; .65% Cr and .80% Cr
	32xx	1.75% Ni; 1.07% Cr
	33xx	3.5% Ni; 1.50% Cr and 1.57% Cr
	34xx	3% Ni; .77% Cr
Molybdenum	40xx	.20% Mo and .25% Mo
	44xx	.40% Mo and .52% Mo
Chromium-molybdenum	41xx	.50% Cr, .80% Cr and .95% Cr; .12% Mo; .20% Mo; .25% Mo; and .30% Mo
Nickel-Chromium-Molybdenum	43xx	1.82% Ni .50% Cr and .80% Cr; .25 % Mo
	43BVxx	1.82% Ni .50% Cr; 12% Mo and .25 % Mo; .03% Vmin
	47xx	1.05% Ni; .45% Cr; .20% Mo and .35 % Mo
	81xx	.30% Ni; .40% Cr; .12% Mo
	86xx	.55% Ni; .50% Cr; .20% Mo
Nickel-Chromium-Molybdenum (cont.)	87xx	.55% Ni; .50% Cr; .25% Mo
	88xx	.55% Ni; .50% Cr; .35% Mo
	93xx	3.25% Ni; 1.20% Cr; .12% Mo
	94xx	0.45% Ni; .40% Cr; .12% Mo
	97xx	.55% Ni; .20% Cr; .20% Mo
	98xx	1% Ni; .80% Cr; .25% Mo
Nickel-Molybdenum	46xx	.85% Ni and 1.82% Ni; .20% Mo and .25% Mo
	48xx	3.50% Ni; .25% Mo;
Chromium	50xx	.27% Cr; 40% Cr; .50% Cr; and .65% Cr
	51xx	.80% Cr; .87% Cr; .92% Cr; .95% Cr; 1% Cr and 1.05% Cr
Chromium	50xx	.50% Cr (C 1% min)
	51xx	1.02% Cr (C 1% min)
	52xx	1.45% Cr (C 1% min)
Chromium-Vanadium	61xx	.60% Cr, .80% Cr; and .95% Cr, .10% V and .15%V min
Tungsten-Chromium	72xx	1.75% W ; .75% Cr
Silicon-Manganese	92xx	1.40% Si; and 2.00% Si; .65% Mn; .82% Mn; and .85% Mn; 0% Cr and .65% Cr
High-Strength Low-Alloy	9xx	Various SAE grades
Boron	xxBxx	B denotes Boron steel
Leaded	xxLxx	L denotes Leaded steel

UNIFIED NUMBERING SYSTEM (UNS) FOR METALS AND ALLOYS

Nonferrous Metals		Ferrous Metals	
Designation	**Metal**	**Designation**	**Metal**
A00001 to A99999	Aluminum and aluminum alloys	D00001 to D99999	Specified mechanical property steels
C00001 to C99999	Copper and copper alloys	F00001 to F99999	Cast irons
E00001 to E99999	Rare earth and rare earth-like metals and alloys	G00001 to G99999	AISI and SAE carbon and alloy steels (except tool steels)
L00001 to L99999	Low melting metals and alloys	H00001 to H99999	AISI H-steels
M00001 to M99999	Miscellaneous nonferrous metals and alloys	J00001 to J99999	Cast steels (except tool steels)
P00001 to P99999	Precious metals and alloys	K00001 to K99999	Miscellaneous steels and ferrous alloys
R00001 to R99999	Reactive and refractory metals and alloys	S00001 to S99999	Heat and corrosion resistant (stainless) steels
Z00001 to Z99999	Zinc and zinc alloys		

ROUND HEAD MACHINE SCREWS

Nom Size	D	A		H		J		T	
	Screw Max Dia	Head Dia		Head Height		Slot Width		Slot Depth	
		Max	Min	Max	Min	Max	Min	Max	Min
0	0.060	0.113	0.099	0.053	0.043	0.023	0.016	0.039	0.029
1	0.073	0.138	0.122	0.061	0.051	0.026	0.019	0.044	0.033
2	0.086	0.162	0.146	0.069	0.059	0.031	0.023	0.048	0.037
3	0.099	0.187	0.169	0.078	0.067	0.035	0.027	0.053	0.040
4	0.112	0.211	0.193	0.086	0.075	0.039	0.031	0.058	0.044
5	0.125	0.236	0.217	0.095	0.083	0.043	0.035	0.063	0.047
6	0.138	0.260	0.240	0.103	0.091	0.048	0.039	0.068	0.051
8	0.164	0.309	0.287	0.120	0.107	0.054	0.045	0.077	0.058
10	0.190	0.359	0.334	0.137	0.123	0.060	0.050	0.087	0.065
12	0.216	0.408	0.382	0.153	0.139	0.067	0.056	0.096	0.072
1/4	0.250	0.472	0.443	0.175	0.160	0.075	0.064	0.109	0.082
5/16	0.3125	0.590	0.557	0.216	0.198	0.084	0.072	0.132	0.099
3/8	0.375	0.708	0.670	0.256	0.237	0.094	0.081	0.155	0.117
7/16	0.4375	0.750	0.707	0.328	0.307	0.094	0.081	0.196	0.148
1/2	0.500	0.813	0.766	0.355	0.332	0.106	0.091	0.211	0.159
9/16	0.5625	0.938	0.887	0.410	0.385	0.118	0.102	0.242	0.183
5/8	0.625	1.000	0.944	0.438	0.411	0.133	0.116	0.258	0.195
3/4	0.750	1.250	1.185	0.547	0.516	0.149	0.131	0.320	0.242

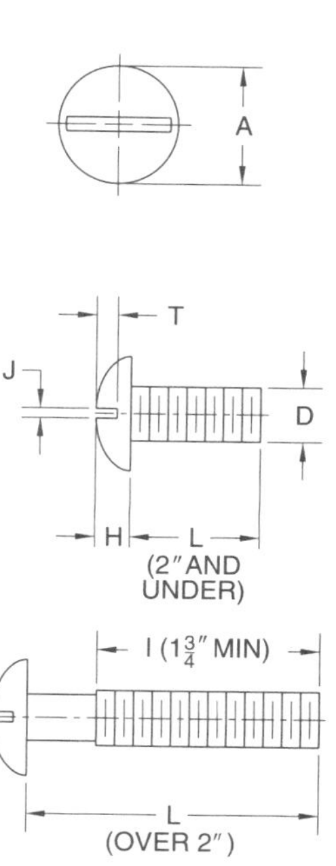

FLAT HEAD MACHINE SCREWS

Nom Size	D	A			H		J		T	
	Screw Max Dia	Head Dia			Head Height		Slot Width		Slot Depth	
		Max Sharp	Min Sharp	Abs Min*	Max	Min	Max	Min	Max	Min
0	0.060	0.119	0.105	0.101	0.035	0.026	0.023	0.016	0.015	0.010
1	0.073	0.146	0.130	0.126	0.043	0.033	0.026	0.019	0.019	0.012
2	0.086	0.172	0.156	0.150	0.051	0.040	0.031	0.023	0.023	0.015
3	0.099	0.199	0.181	0.175	0.059	0.048	0.035	0.027	0.027	0.017
4	0.112	0.225	0.207	0.200	0.067	0.055	0.039	0.031	0.030	0.020
5	0.125	0.252	0.232	0.225	0.075	0.062	0.043	0.035	0.034	0.022
6	0.138	0.279	0.257	0.249	0.083	0.069	0.048	0.039	0.038	0.024
8	0.164	0.332	0.308	0.300	0.100	0.084	0.054	0.045	0.045	0.029
10	0.190	0.385	0.359	0.348	0.116	0.098	0.060	0.050	0.053	0.034
12	0.216	0.438	0.410	0.397	0.132	0.112	0.067	0.056	0.060	0.039
1/4	0.250	0.507	0.477	0.462	0.153	0.131	0.075	0.064	0.070	0.046
5/16	0.3125	0.635	0.600	0.581	0.191	0.165	0.084	0.072	0.088	0.058
3/8	0.375	0.762	0.722	0.700	0.230	0.200	0.094	0.081	0.106	0.070
7/16	0.4375	0.812	0.771	0.743	0.223	0.190	0.094	0.081	0.103	0.066
1/2	0.500	0.875	0.831	0.802	0.223	0.186	0.106	0.091	0.103	0.065
9/16	0.5625	1.000	0.950	0.919	0.260	0.220	0.118	0.102	0.120	0.077
5/8	0.625	1.125	1.069	1.035	0.298	0.253	0.133	0.116	0.137	0.088
3/4	0.750	1.375	1.306	1.267	0.372	0.319	0.149	0.131	0.171	0.111

* with maximum sharpness

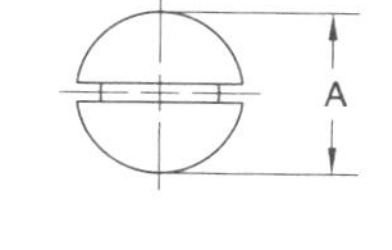

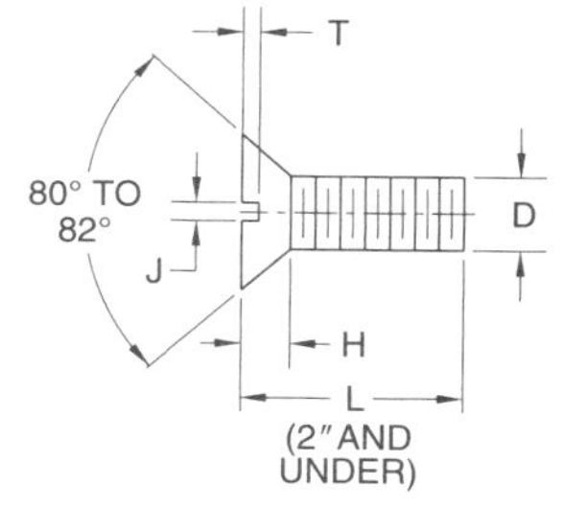

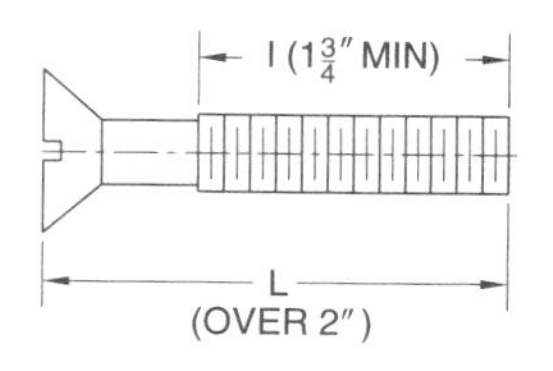

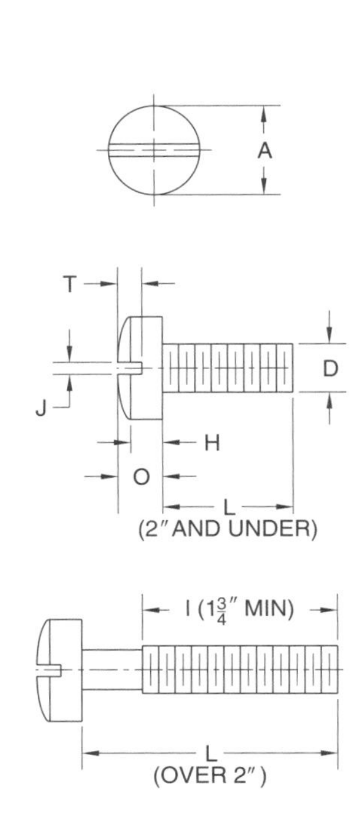

FILLISTER HEAD MACHINE SCREWS

Nom Size	D	A		H		O		J		T	
	Screw Max Dia	Head Dia		Head Height		Total Head Height		Slot Width		Slot Depth	
		Max	Min	Max	Min	Max	Min	Max	Min	Max	Min
0	0.060	0.096	0.083	0.045	0.037	0.059	0.043	0.023	0.016	0.025	0.015
1	0.073	0.118	0.104	0.053	0.045	0.071	0.055	0.026	0.019	0.031	0.020
2	0.086	0.140	0.124	0.062	0.053	0.083	0.066	0.031	0.023	0.037	0.025
3	0.099	0.161	0.145	0.070	0.061	0.095	0.077	0.035	0.027	0.043	0.030
4	0.112	0.183	0.166	0.079	0.069	0.107	0.088	0.039	0.031	0.048	0.035
5	0.125	0.205	0.187	0.088	0.078	0.120	0.100	0.043	0.035	0.054	0.040
6	0.138	0.226	0.208	0.096	0.086	0.132	0.111	0.048	0.039	0.060	0.045
8	0.164	0.270	0.250	0.113	0.102	0.156	0.133	0.054	0.045	0.071	0.054
10	0.190	0.313	0.292	0.130	0.118	0.180	0.156	0.060	0.050	0.083	0.064
12	0.216	0.357	0.334	0.148	0.134	0.205	0.178	0.067	0.056	0.094	0.074
1/4	0.250	0.414	0.389	0.170	0.155	0.237	0.207	0.075	0.064	0.109	0.087
5/16	0.3125	0.518	0.490	0.211	0.194	0.295	0.262	0.084	0.072	0.137	0.110
3/8	0.375	0.622	0.590	0.253	0.233	0.355	0.315	0.094	0.081	0.164	0.133
7/16	0.4375	0.625	0.589	0.265	0.242	0.368	0.321	0.094	0.081	0.170	0.135
1/2	0.500	0.750	0.710	0.297	0.273	0.412	0.362	0.106	0.091	0.190	0.151
9/16	0.5625	0.812	0.768	0.336	0.308	0.466	0.410	0.118	0.102	0.214	0.172
5/8	0.625	0.875	0.827	0.375	0.345	0.521	0.461	0.133	0.116	0.240	0.193
3/4	0.750	1.000	0.945	0.441	0.406	0.612	0.542	0.149	0.131	0.281	0.226

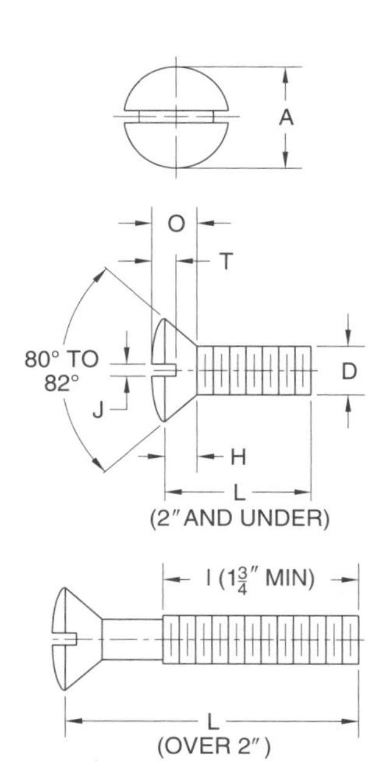

OVAL HEAD MACHINE SCREWS

Nom Size	D	A			H		O		J		T	
	Screw Max Dia	Head Dia			Head Height		Total Head Height		Slot Width		Slot Depth	
		Max Sharp	Min Sharp	Abs Min*	Max	Min	Max	Min	Max	Min	Max	Min
0	0.060	0.119	0.105	0.101	0.035	0.026	0.056	0.041	0.023	0.016	0.030	0.025
1	0.073	0.146	0.130	0.126	0.043	0.033	0.068	0.052	0.026	0.019	0.038	0.031
2	0.086	0.172	0.156	0.150	0.051	0.040	0.080	0.063	0.031	0.023	0.045	0.037
3	0.099	0.199	0.181	0.175	0.059	0.048	0.092	0.073	0.035	0.027	0.052	0.043
4	0.112	0.225	0.207	0.200	0.067	0.055	0.104	0.084	0.039	0.031	0.059	0.049
5	0.125	0.252	0.232	0.225	0.075	0.062	0.116	0.095	0.043	0.035	0.067	0.055
6	0.138	0.279	0.257	0.249	0.083	0.069	0.128	0.105	0.048	0.039	0.074	0.060
8	0.164	0.332	0.308	0.300	0.100	0.084	0.152	0.126	0.054	0.045	0.088	0.072
10	0.190	0.385	0.359	0.348	0.116	0.098	0.176	0.148	0.060	0.050	0.103	0.084
12	0.216	0.438	0.410	0.397	0.132	0.112	0.200	0.169	0.067	0.056	0.117	0.096
1/4	0.250	0.507	0.477	0.462	0.153	0.131	0.232	0.197	0.075	0.064	0.136	0.112
5/16	0.3125	0.635	0.600	0.581	0.191	0.165	0.290	0.249	0.084	0.072	0.171	0.141
3/8	0.375	0.762	0.722	0.700	0.230	0.200	0.347	0.300	0.094	0.081	0.206	0.170
7/16	0.4375	0.812	0.771	0.743	0.223	0.190	0.345	0.295	0.094	0.081	0.210	0.174
1/2	0.500	0.875	0.831	0.802	0.223	0.186	0.354	0.299	0.106	0.091	0.216	0.176
9/16	0.5625	1.000	0.950	0.919	0.260	0.220	0.410	0.350	0.118	0.102	0.250	0.207
5/8	0.625	1.125	1.069	1.035	0.298	0.253	0.467	0.399	0.133	0.116	0.285	0.235
3/4	0.750	1.375	1.306	1.267	0.372	0.319	0.578	0.497	0.149	0.131	0.353	0.293

* with maximum sharpness

HEXAGONAL SOCKET HEAD CAP SCREWS

D			A		H		S			J		T
Body Dia			Head Dia		Head Height		Head Side-Height			Socket Width Across Flats		Key Depth
Nom	Max	Min	Max	Min	Max	Min	Nom	Max	Min	Max	Min	Min
0	0.060	0.0583	0.0960	0.0926	0.0600	0.0574	0.055	0.056	0.054	0.051	0.050	0.025
1	0.0730	0.0711	0.1180	0.1142	0.0730	0.0702	0.067	0.068	0.066	0.051	0.050	0.031
2	0.0860	0.0840	0.140	0.136	0.086	0.083	0.079	0.081	0.078	0.0635	1/16	0.038
3	0.0990	0.0968	0.161	0.157	0.099	0.096	0.091	0.093	0.089	0.0791	5/64	0.044
4	0.1120	0.1096	0.183	0.178	0.112	0.109	0.103	0.105	0.101	0.0791	5/64	0.051
5	0.1250	0.1226	0.205	0.200	0.125	0.122	0.115	0.117	0.113	0.0947	3/32	0.057
6	0.1380	0.1353	0.226	0.221	0.138	0.134	0.127	0.129	0.125	0.0947	3/32	0.064
8	0.1640	0.1613	0.270	0.265	0.164	0.160	0.150	0.152	0.148	0.1270	1/8	0.077
10	0.1900	0.1867	5/16	0.306	0.190	0.185	0.174	0.176	0.172	0.1582	5/32	0.090
12	0.2160	0.2127	11/32	0.337	0.216	0.211	0.198	0.200	0.196	0.1582	5/32	0.103
1/4	0.2500	0.2464	3/8	0.367	1/4	0.244	0.229	0.232	0.226	0.1895	3/16	0.120
5/16	0.3125	0.3084	7/16	0.429	5/16	0.306	0.286	0.289	0.283	0.2207	7/32	0.151
3/8	0.3750	0.3705	9/16	0.553	3/8	0.368	0.344	0.347	0.341	0.3155	5/16	0.182
7/16	0.4375	0.4326	5/8	0.615	7/16	0.430	0.401	0.405	0.397	0.3155	5/16	0.213
1/2	0.5000	0.4948	3/4	0.739	1/2	0.492	0.458	0.462	0.454	0.3780	3/8	0.245
9/16	0.5625	0.5569	13/16	0.801	9/16	0.554	0.516	0.520	0.512	0.3780	3/8	0.276
5/8	0.6250	0.6191	7/8	0.863	5/8	0.616	0.573	0.577	0.569	0.5030	1/2	0.307
3/4	0.7500	0.7436	1	0.987	3/4	0.741	0.688	0.693	0.684	0.5655	9/16	0.370
7/8	0.8750	0.8680	1 1/8	1.111	7/8	0.865	0.802	0.807	0.797	0.5655	9/16	0.432
1	1.0000	0.9924	1 5/16	1.297	1	0.989	0.917	0.922	0.912	0.6290	5/8	0.495
1 1/8	1.1250	1.1165	1 1/2	1.483	1 1/8	1.113	1.031	1.037	1.025	0.7540	3/4	0.557
1 1/4	1.2500	1.2415	1 3/4	1.733	1 1/4	1.238	1.146	1.152	1.140	0.7540	3/4	0.620
1 3/8	1.3750	1.3649	1 7/8	1.855	1 3/8	1.361	1.260	1.267	1.253	0.7540	3/4	0.682

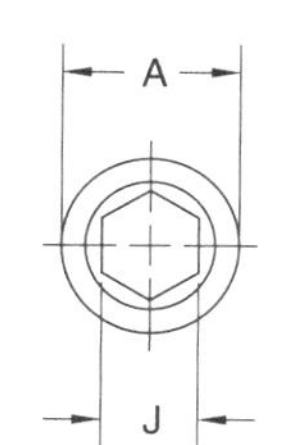

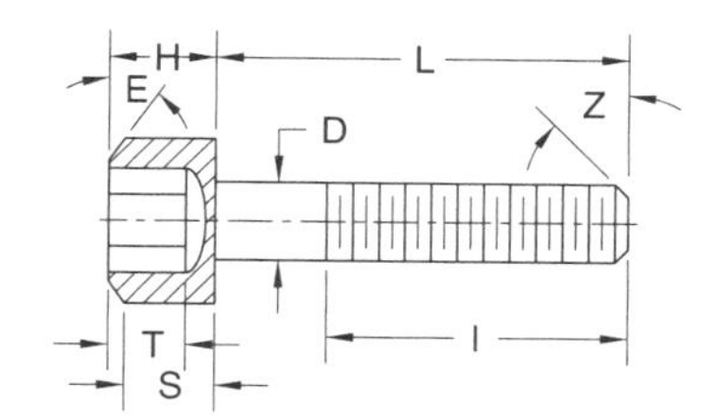

HEXAGONAL SOCKET SET SCREWS

D	C		R	Y		P		Q	q	J		T
Nom Dia	Cup and Flat Point Dia		Oval Point Radius	Cone Point Angle		Full and Half Dog Point				Socket Width Across Flats		Key Depth
				118° ± 2°*	90° ± 2°†	Dia		Full	Half			
	Max	Min				Max	Min			Max	Min	Min
0	0.033	0.027	3/64	1/16	5/64	0.040	0.037	0.030	0.015	0.0285	0.028	0.022
1	0.040	0.033	0.055	5/64	3/32	0.049	0.045	0.037	0.019	0.0355	0.035	0.028
2	0.047	0.039	1/16	3/32	7/64	0.057	0.053	0.043	0.022	0.0355	0.035	0.028
3	0.054	0.045	5/64	7/64	1/8	0.066	0.062	0.050	0.025	0.051	0.050	0.040
4	0.061	0.051	0.084	1/8	5/32	0.075	0.070	0.056	0.028	0.051	0.050	0.040
5	0.067	0.057	3/32	1/8	3/16	0.083	0.078	0.06	0.03	0.0635	1/16	0.050
6	0.074	0.064	7/64	1/8	3/16	0.092	0.087	0.07	0.035	0.0635	1/16	0.050
8	0.087	0.076	1/8	3/16	1/4	0.109	0.103	0.08	0.04	0.0791	5/64	0.062
10	0.102	0.088	9/64	3/16	1/4	0.127	0.120	0.09	0.045	0.0947	3/32	0.075
12	0.115	0.101	5/32	3/16	1/4	0.144	0.137	0.11	0.055	0.0947	3/32	0.075
1/4	0.132	0.118	3/16	1/4	5/16	5/32	0.149	1/8	1/16	0.1270	1/8	0.100
5/16	0.172	0.156	15/64	5/16	3/8	13/64	0.195	5/32	5/64	0.1582	5/32	0.125
3/8	0.212	0.194	9/32	3/8	7/16	1/4	0.241	3/16	3/32	0.1895	3/16	0.150
7/16	0.252	0.232	21/64	7/16	1/2	19/64	0.287	7/32	7/64	0.2207	7/32	0.175
1/2	0.291	0.270	3/8	1/2	9/16	11/32	0.334	1/4	1/8	0.2520	1/4	0.200
9/16	0.332	0.309	27/64	9/16	5/8	25/64	0.379	9/32	9/64	0.2520	1/4	0.200
5/8	0.371	0.347	15/32	5/8	3/4	15/32	0.456	5/16	5/32	0.3155	5/16	0.250
3/4	0.450	0.425	9/16	3/4	7/8	9/16	0.549	3/8	3/16	0.3780	3/8	0.300
7/8	0.530	0.502	21/32	7/8	1	21/32	0.642	7/16	7/32	0.5030	1/2	0.400
1	0.609	0.579	3/4	1	1 1/8	3/4	0.734	1/2	1/4	0.5655	9/16	0.450
1 1/8	0.689	0.655	27/32	1 1/8	1 1/4	27/32	0.826	9/16	9/32	0.5655	9/16	0.450
1 1/4	0.767	0.733	15/16	1 1/4	1 1/2	15/16	0.920	5/8	5/16	0.6290	5/8	0.500
1 3/8	0.848	0.808	1 1/32	1 3/8	1 5/8	1 1/32	1.011	11/16	11/32	0.6290	5/8	0.500
1 1/2	0.926	0.886	1 1/8	1 1/2	1 3/4	1 1/8	1.105	3/4	3/8	0.7540	3/4	0.600
1 3/4	1.086	1.039	1 5/16	1 3/4	2	1 5/16	1.289	7/8	7/16	1.0040	1	0.800

* for these lengths and under
† for these lengths and over

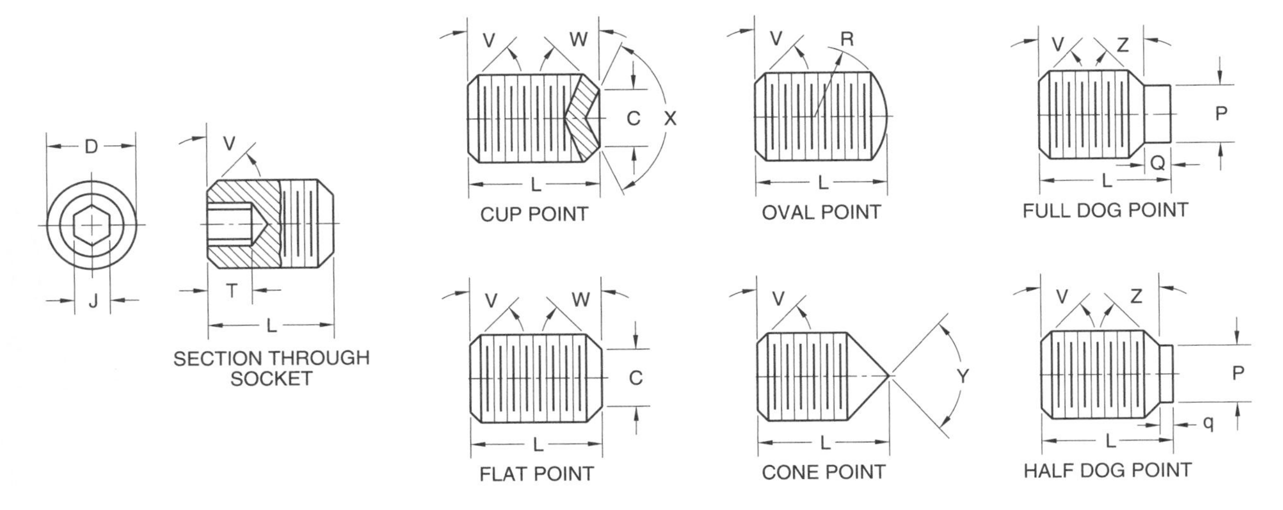

COTTER PINS

Nom Size	Dia A & Width B Max	Wire Width B Min	Head Dia C Min	Prong Length D Min	Hole Size
1/32	.032	.022	0.06	.01	.047
3/64	.048	.035	0.09	.02	.062
1/16	.060	.044	0.12	.03	.078
5/64	.076	.057	0.16	.04	.094
3/32	.090	.069	0.19	.04	.109
7/64	.104	.080	0.22	.05	.125
1/8	.120	.093	0.25	.06	.141
9/64	.134	.104	0.28	.06	.156
5/32	.150	.116	0.31	.07	.172
3/16	.176	.137	0.38	.09	.203
7/32	.207	.161	0.44	.10	.234
1/4	.225	.176	0.50	.11	.266
5/16	.280	.220	0.62	.14	.312
3/8	.335	.263	0.75	.16	.375
7/16	.406	.320	0.88	.20	.438
1/2	.473	.373	1.00	.23	.500
5/8	.598	.472	1.25	.30	.625
3/4	.723	.572	1.50	.36	.750

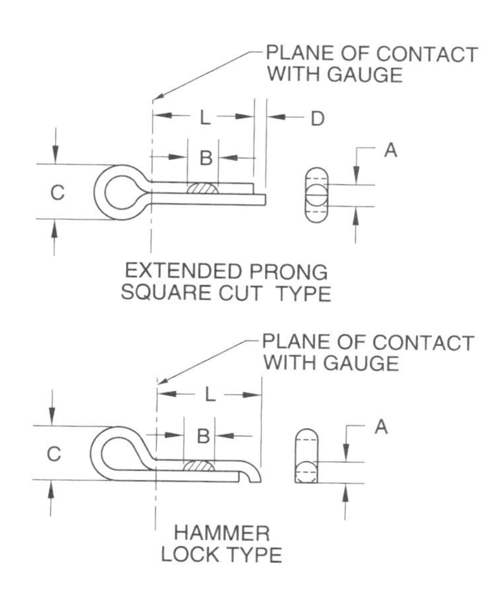

DOWEL PINS

Nom Size Pin Dia		Pin Dia A						Point Dia B		Crown Height or Radius C		Range of Preferred Lengths L	Double Shear Load*	Suggested Hole Dia	
		Standard Series Pins			Oversize Series Pins										
Frac	Deci	Basic	Max	Min	Basic	Max	Min	Max	Min	Max	Min			Max	Min
1/16	0.0625	0.0627	0.0628	0.0626	0.0635	0.0636	0.0634	0.058	0.048	0.020	0.008	3/16 – 3/4	800	0.0625	0.0620
5/64	0.0781	0.0783	0.0784	0.0782	0.0791	0.0792	0.0790	0.074	0.064	0.026	0.010	—	1240	0.0781	0.0776
3/32	0.0938	0.0940	0.0941	0.0939	0.0948	0.0949	0.0947	0.089	0.079	0.031	0.012	5/16 – 1	1800	0.0937	0.0932
1/8	0.1250	0.1252	01253	0.1251	0.1260	0.1261	0.1259	0.120	0.110	0.041	0.016	3/8 – 2	3200	0.1250	0.1245
5/32	0.1562	0.1564	0.1565	0.1563	0.1572	0.1573	0.1571	0.150	0.140	0.052	0.020	—	5000	0.1562	0.1557
3/16	0.1875	0.1877	0.1878	0.1876	0.1885	0.1886	0.1884	0.180	0.170	0.062	0.023	1/2 – 2	7200	0.1875	0.1870
1/4	0.2500	0.2502	0.2503	0.2501	0.2510	0.2511	0.2509	0.240	0.230	0.083	0.031	1/2 – 2 1/2	12,800	0.2500	0.2495
5/16	0.3125	0.3127	0.3128	0.3126	0.3135	0.3136	0.3134	0.302	0.290	0.104	0.039	1/2 – 2 1/2	20,000	0.3125	0.3120
3/8	0.3750	0.3752	0.3753	0.3751	0.3760	0.3761	0.3759	0.365	0.350	0.125	0.047	1/2 – 3	28,700	0.3750	0.3745
7/16	0.4375	0.4377	0.4378	0.4376	0.4385	0.4386	0.4384	0.424	0.409	0.146	0.055	7/8 – 3	39,100	0.4375	0.4370
1/2	0.5000	0.5002	0.5003	0.5001	0.5010	0.5011	0.5009	0.486	0.471	0.167	0.063	3/4, 1 – 4	51,000	0.5000	0.4995
5/8	0.6250	0.6252	0.6253	0.6251	0.6260	0.6261	0.6259	0.611	0.595	0.208	0.078	1 1/4 – 5	79,800	0.6250	0.6245
3/4	0.7500	0.7502	0.7503	0.7501	0.7510	0.7511	0.7509	0.735	0.715	0.250	0.094	1 1/2 – 6	114,000	0.7500	0.7495
7/8	0.8750	0.8752	0.8753	0.8751	0.8760	0.8761	0.8759	0.860	0.840	0.293	0.109	2, 2 1/2 – 6	156,000	0.8750	0.8745
1	1.0000	1.0002	1.0003	1.0001	1.0010	1.0011	1.0009	0.980	0.960	0.333	0.125	2, 2 1/2 – 5, 6	204,000	1.0000	0.9995

* *minimum lb*

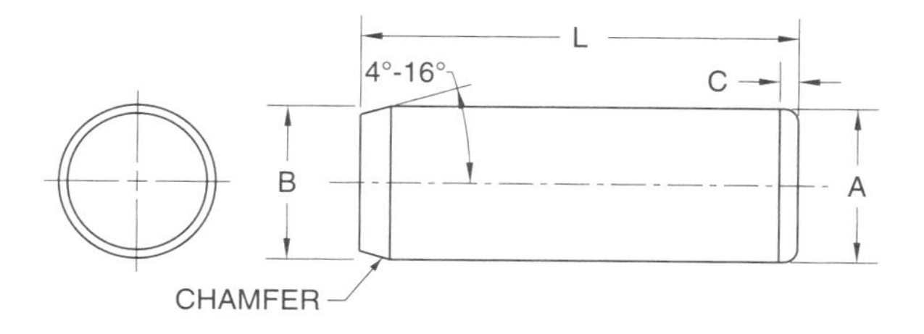

STRAIGHT PINS

Nom Size or Basic Pin Dia		Pin Dia *A*		Chamfer Length *C*	
		Max	Min	Max	Min
1/16	0.062	0.0625	0.0605	0.025	0.005
3/32	0.094	0.0937	0.0917	0.025	0.005
7/64	0.109	0.1094	0.1074	0.025	0.005
1/8	0.125	0.1250	0.1230	0.025	0.005
5/32	0.156	0.1562	0.1542	0.025	0.005
3/16	0.188	0.1875	0.1855	0.025	0.005
7/32	0.219	0.2187	0.2167	0.025	0.005
1/4	0.250	0.2500	0.2480	0.025	0.005
5/16	0.312	0.3125	0.3105	0.040	0.020
3/8	0.375	0.3750	0.3730	0.040	0.020
7/16	0.438	0.4375	0.4355	0.040	0.020
1/2	0.500	0.5000	0.4980	0.040	0.020
5/8	0.625	0.6250	0.6230	0.055	0.035
3/4	0.750	0.7500	0.7480	0.055	0.035
7/8	0.875	0.8750	0.8730	0.055	0.035
1	1.000	1.0000	0.9980	0.055	0.035

** All dimensions are in inches.*

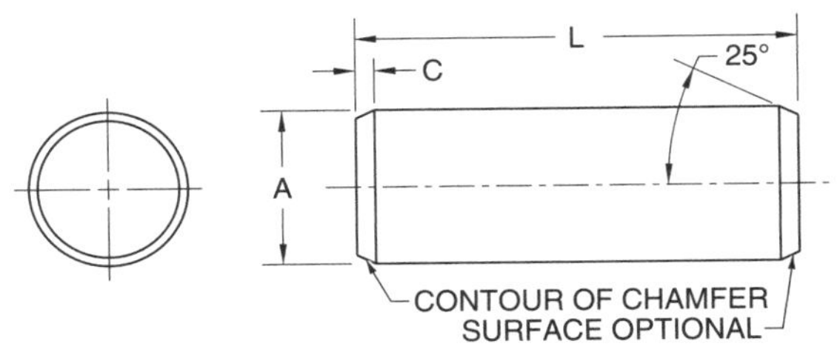

CHAMFERED ENDS

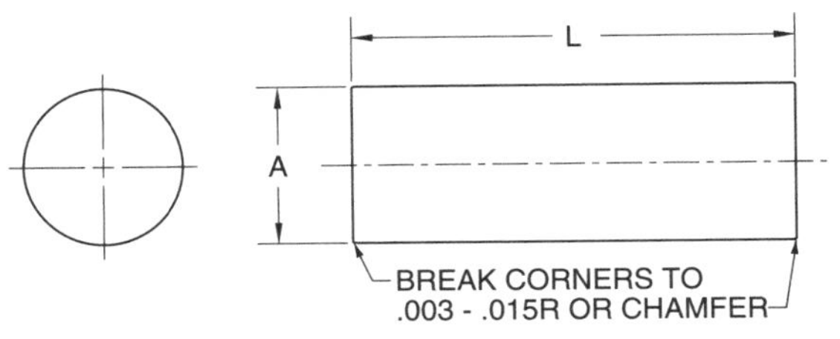

SQUARE ENDS

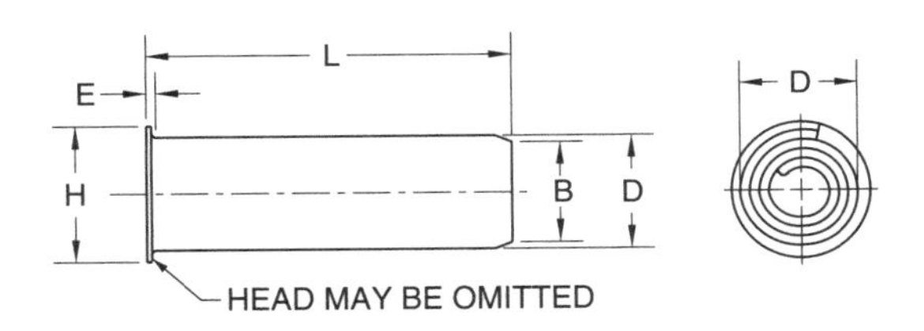

SPIRALLY COILED PINS

Nom Dia *D*	Decimal Dimension	Recommended Hole Tolerance	*H*		*B*	*D*		Min Double Shear	*E*†
		+	Min	Max		Min	Max	1070	
1/16	0.062	0.003	0.084	0.099	0.059	0.067	0.072	330	1/64
5/64	0.078	0.003	0.105	0.122	0.075	0.083	0.088	550	1/64
3/32	0.094	0.003	0.125	0.144	0.091	0.099	0.105	775	1/32
7/64	0.109	0.003	0.146	0.167	0.106	0.114	0.120	1050	1/32
1/8	0.125	0.004	0.166	0.189	0.121	0.131	0.138	1400	1/32
5/32	0.156	0.004	0.207	0.234	0.152	0.163	0.171	2200	3/64
3/16	0.187	0.005	0.248	0.279	0.182	0.196	0.205	3150	1/16

L Lengths Available from Stock*

Nom Dia *D**	3/16	1/4	5/16	3/8	7/16	1/2	9/16	5/8	11/16	3/4	13/16	7/8	15/16	1	1 1/8	1 1/4	1 3/8	1 1/2	1 5/8	1 3/4	1 7/8	2
1/16																						
5/64																						
3/32																						
7/64																						
1/8																						
5/32																						
3/16																						

* *Diameters 1/16″ through 1/8″, L ±0.015. Diameters 5/32″ and 3/16″, L ±0.020.*

† *reference only*

GROOVED PINS

Size	Pin Dia *A*		Pilot Length *C*	Pilot Length *D*	Crown Height *E*		Crown Radius *F*		Neck Width *G*		Shoulder Length *H*		Neck Rad *J*	Neck Dia *K*	
Frac	Max	Min	Ref	Min	Max	Min	Max	Min	Max	Min	Max	Min	Ref	Max	Min
1/32	0.0312	0.0302	0.015	—	—	—	—	—	—	—	—	—	—	—	—
3/64	0.0469	0.0459	0.031	—	—	—	—	—	—	—	—	—	—	—	—
1/16	0.0625	0.0615	0.031	0.016	0.0115	0.0015	0.088	0.098	—	—	—	—	—	—	—
5/64	0.0781	0.0771	0.031	0.016	0.0137	0.0037	0.104	0.084	—	—	—	—	—	—	—
3/32	0.0938	0.928	0.031	0.016	0.0141	0.0041	0.135	0.115	0.038	0.028	0.041	0.031	0.016	0.067	0.057
7/64	0.1094	0.1074	0.031	0.016	0.0160	0.0060	0.150	0.130	0.038	0.028	0.041	0.031	0.016	0.082	0.072
1/8	0.1250	0.1230	0.031	0.016	0.0180	0.0080	0.166	0.146	0.069	0.059	0.041	0.031	0.031	0.088	0.078
5/32	0.1563	0.1543	0.062	0.031	0.0220	0.0120	0.198	0.178	0.069	0.059	0.057	0.047	0.031	0.109	0.099
3/16	0.1875	0.1855	0.062	0.031	0.0230	0.0130	0.260	0.240	0.069	0.059	0.057	0.047	0.031	0.130	0.120
7/32	0.2188	0.2168	0.062	0.031	0.0270	0.0170	0.291	0.271	0.101	0.091	0.072	0.062	0.047	0.151	0.141
1/4	0.2500	0.2480	0.062	0.031	0.0310	0.0210	0.322	0.302	0.101	0.091	0.072	0.062	0.047	0.172	0.162
5/16	0.3125	0.3105	0.094	0.047	0.0390	0.0290	0.385	0.365	0.132	0.122	0.104	0.094	0.062	0.214	0.204
3/8	0.3750	0.3730	0.094	0.047	0.0440	0.0340	0.479	0.459	0.132	0.122	0.135	0.125	0.062	0.255	0.245
7/16	0.4375	0.4355	0.094	0.047	0.0520	0.0420	0.541	0.521	0.195	0.185	0.135	0.125	0.094	0.298	0.288
1/2	0.5000	0.4980	0.094	0.047	0.0570	0.0470	0.635	0.615	0.195	0.185	0.135	0.125	0.094	0.317	0.307

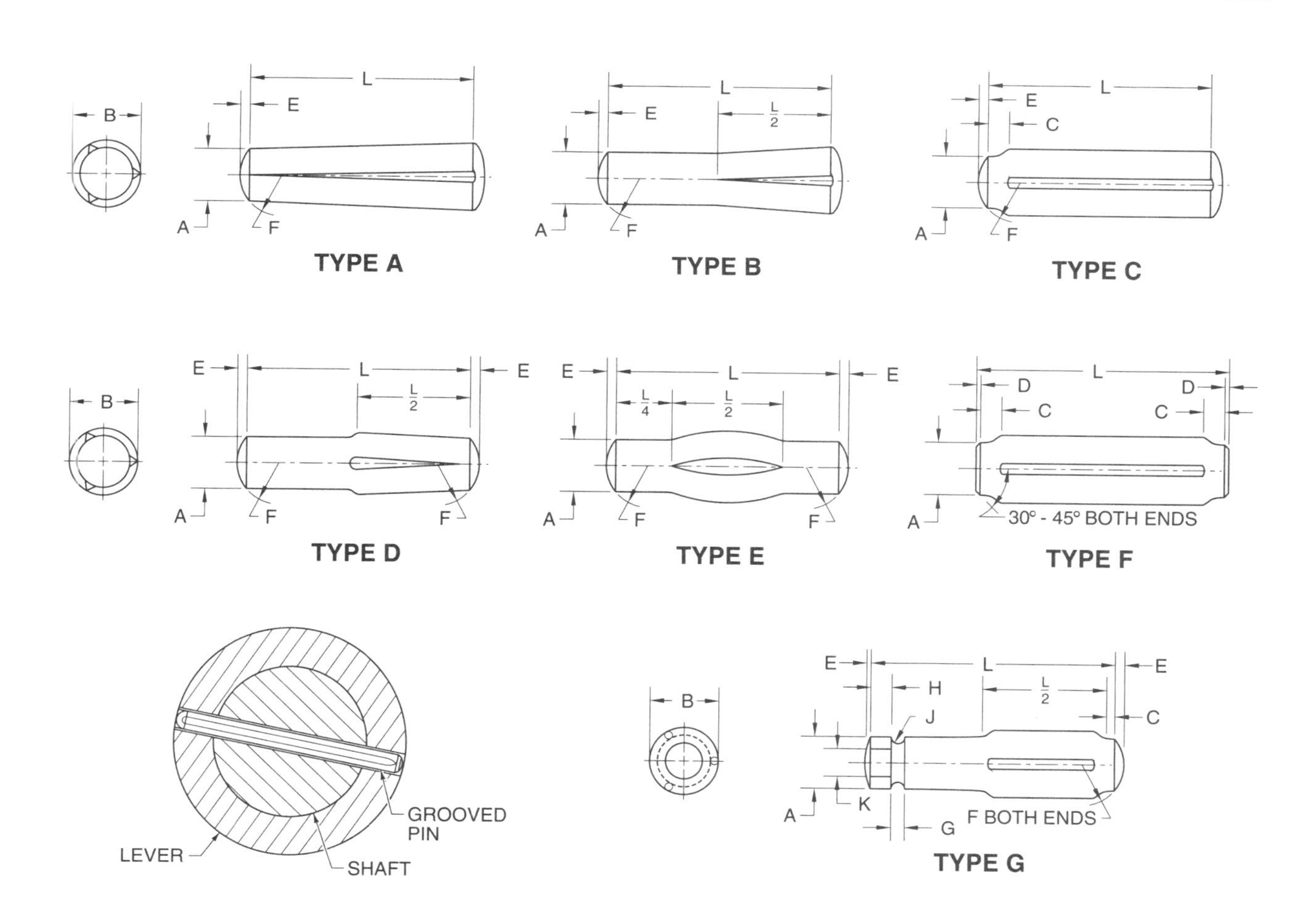

SLOTTED SPRING PINS

Pin Dia		Pin Dia A		Chamfer Dia B	Chamfer Length C		Stock Thickness F	Recommended Hole Size		Material AISI 1070-1095 and 420	Material AISI 302	Material Beryllium Copper
Frac	Deci	Max	Min	Max	Max	Min	Basic	Max	Min	Min Double Shear Load (lb)		
1/16	0.062	0.069	0.066	0.059	0.028	0.007	0.012	0.065	0.062	425	350	270
5/64	0.078	0.086	0.083	0.075	0.032	0.008	0.018	0.081	0.078	650	550	400
3/32	0.094	0.103	0.099	0.091	0.038	0.008	0.022	0.097	0.094	1000	800	660
1/8	0.125	0.135	0.131	0.122	0.044	0.008	0.028	0.129	0.125	2100	1500	1200
9/64	0.141	0.149	0.145	0.137	0.044	0.008	0.028	0.144	0.140	2200	1600	1400
5/32	0.156	0.167	0.162	0.151	0.048	0.010	0.032	0.160	0.156	3000	2000	1800
3/16	0.188	0.199	0.194	0.182	0.055	0.011	0.040	0.192	0.187	4400	2800	2600
7/32	0.219	0.232	0.226	0.214	0.065	0.011	0.048	0.224	0.219	5700	3550	3700
1/4	0.250	0.264	0.258	0.245	0.065	0.012	0.048	0.256	0.250	7700	4600	4500
5/16	0.312	0.328	0.321	0.306	0.080	0.014	0.062	0.318	0.312	11,500	7095	6800
3/8	0.375	0.392	0.385	0.368	0.095	0.016	0.077	0.382	0.375	17,600	10,000	10,100
7/16	0.438	0.456	0.448	0.430	0.095	0.017	0.077	0.445	0.437	20,000	12,000	12,200
1/2	0.500	0.521	0.513	0.485	0.110	0.025	0.094	0.510	0.500	25,800	15,500	16,800
5/8	0.625	0.650	0.640	0.608	0.125	0.030	0.125	0.636	0.625	46,000	18,800	—
3/4	0.750	0.780	0.769	0.730	0.150	0.030	0.150	0.764	0.750	66,000	23,200	—

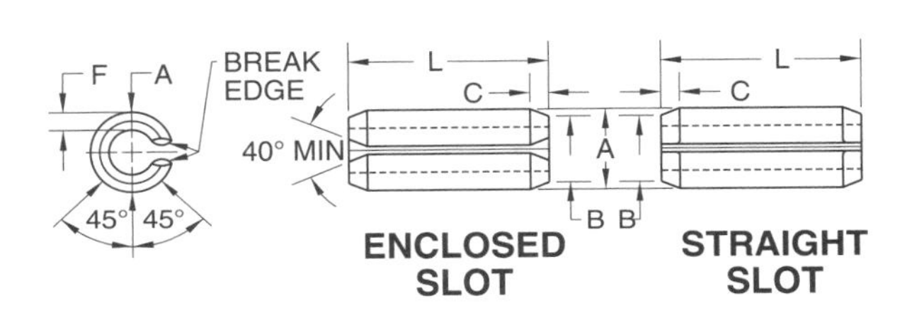

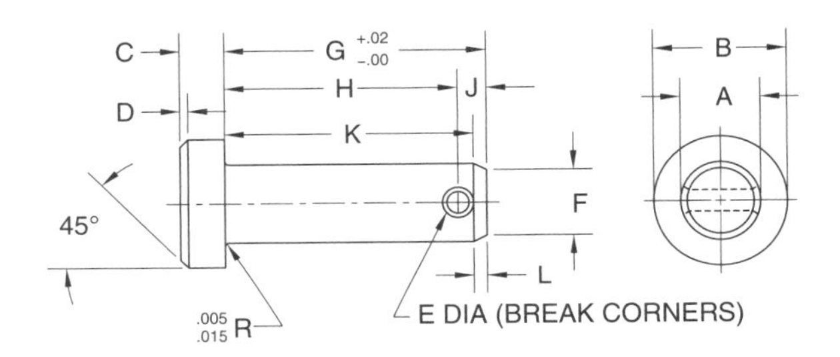

CLEVIS PINS

Pin		A Shank Dia		B Head Dia		C Head Height		D Head Chamfer	E Hole Dia		F Point Dia		G* Pin Length	H Head to Center of Hole		J† End to Center Ref	K‡ Head to Edge of Hole Ref		L Point Length		Rec Cotter Pin Nom Size	
Frac	Deci	Max	Min	Max	Min	Max	Min	±0.01	Max	Min	Max	Min	Basic	Max	Min	Basic	Max	Min	Max	Min		
3/16	0.188	0.186	0.181	0.32	0.30	0.07	0.05	0.02	0.088	0.073	0.15	0.14	0.58	0.504	0.484	0.09	0.548	0.520	0.055	0.035	1/16	0.062
1/4	0.250	0.248	0.243	0.38	0.36	0.10	0.08	0.03	0.088	0.073	0.21	0.20	0.77	0.692	0.672	0.09	0.736	0.708	0.055	0.035	1/16	0.062
5/16	0.312	0.311	0.306	0.44	0.42	0.10	0.08	0.03	0.119	0.104	0.26	0.25	0.94	0.832	0.812	0.12	0.892	0.864	0.071	0.049	3/32	0.093
3/8	0.375	0.373	0.368	0.51	0.49	0.13	0.11	0.03	0.119	0.104	0.33	0.32	1.06	0.958	0.938	0.12	1.018	0.990	0.071	0.049	3/32	0.093
7/16	0.438	0.436	0.431	0.57	0.55	0.16	0.14	0.04	0.119	0.104	0.39	0.38	1.19	1.082	1.062	0.12	1.142	1.114	0.071	0.049	3/32	0.093
1/2	0.500	0.496	0.491	0.63	0.61	0.16	0.14	0.04	0.151	0.136	0.44	0.43	1.36	1.223	1.203	0.15	1.298	1.271	0.089	0.063	1/8	0.125
5/8	0.625	0.621	0.616	0.82	0.80	0.21	0.19	0.06	0.151	0.136	0.56	0.55	1.61	1.473	1.453	0.15	1.548	1.521	0.089	0.063	1/8	0.125
3/4	0.750	0.746	0.741	0.94	0.92	0.26	0.24	0.07	0.182	0.167	0.68	0.67	1.91	1.739	1.719	0.18	1.830	1.802	0.110	0.076	5/32	0.156
7/8	0.875	0.871	0.866	1.04	1.02	0.32	0.30	0.09	0.182	0.167	0.80	0.79	2.16	1.989	1.969	0.18	2.080	2.052	0.110	0.076	5/32	0.156
1	1.000	0.996	0.991	1.19	1.17	0.35	0.33	0.10	0.182	0.167	0.93	0.92	2.41	2.239	2.219	0.18	2.330	2.302	0.110	0.076	5/32	0.156

** Lengths tabulated are intended for use with standard clevises without spacers.*

† J dimension is for calculating hole location from underside of head on pins of lengths not tabulated.

‡ reference dimension

TAPER PINS

Pin		Major Dia (Large End) A				End Crown Radius R		Range of Lengths L	
		Commercial Class		Precision Class					
Size #	Dia B	Max	Min	Max	Min	Max	Min	Reamer	Other
7/0	0.0625	0.0638	0.0618	0.0635	0.0625	0.072	0.052	—	¼ – 1
6/0	0.0780	0.0793	0.0773	0.0790	0.0780	0.088	0.068	—	¼ – 1½
5/0	0.0940	0.0953	0.0933	0.0950	0.0940	0.104	0.084	¼ – 1	1¼, 1½
4/0	0.1090	0.1103	0.1083	0.1100	0.1090	0.119	0.099	¼ – 1	1¼ – 2
3/0	0.1250	0.1263	0.1243	0.1260	0.1250	0.135	0.115	¼ – 1	1¼ – 2
2/0	0.1410	0.1423	0.1403	0.1420	0.1410	0.151	0.131	½ – 1¼	1½ – 2½
0	0.1560	0.1573	0.1553	0.1570	0.1560	0.166	0.146	½ – 1¼	1½ – 3
1	0.1720	0.1733	0.1713	0.1730	0.1720	0.182	0.162	¾ – 1¼	1½ – 3
2	0.1930	0.1943	0.1923	0.1940	0.1930	0.203	0.183	¾ – 1½	1¾ – 3
3	0.2190	0.2203	0.2183	0.2200	0.2190	0.229	0.209	¾ – 1¾	2 – 4
4	0.2500	0.2513	0.2493	0.2510	0.2500	0.260	0.240	¾ – 2	2¼ – 4
5	0.2890	0.2903	0.2883	0.2900	0.2890	0.299	0.279	1 – 2½	2¾ – 6
6	0.3410	0.3423	0.3403	0.3420	0.3410	0.351	0.331	1¼ – 3	3¼ – 6
7	0.4090	0.4103	0.4083	0.4100	0.4090	0.419	0.399	1¼ – 3¾	4 – 8
8	0.4920	0.4933	0.4913	0.4930	0.4920	0.502	0.482	1¼ – 4½	4¾ – 8
9	0.5910	0.5923	0.5903	0.5920	0.5910	0.601	0.581	1¼ – 5¼	5½ – 8
10	0.7060	0.7073	0.7053	0.7070	0.7060	0.716	0.696	1½ – 6	6¼ – 8
11	0.8600	0.8613	0.8593	—	—	0.870	0.850	—	2 – 8
12	1.0320	1.0333	1.0313	—	—	1.042	1.022	—	2 – 9
13	1.2410	1.2423	1.2403	—	—	1.251	1.231	—	3 – 11
14	1.5210	1.5223	1.5203	—	—	1.531	1.511	—	3 – 13

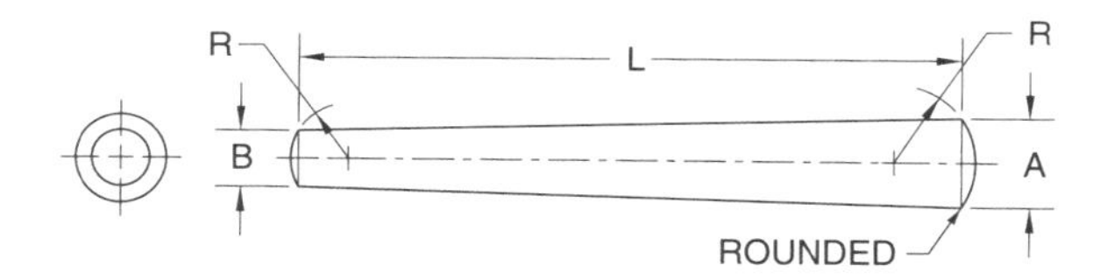

WOODRUFF KEYS AND KEYSEATS

Key No.	Nom Key Size A × B	Key Height C Max	Key Height D Max	Dist Below Center E	Keyseat Width W	Keyseat Depth H Max.
204	1/16 × 1/2	0.203	0.194	3/64	0.0630	0.1718
304	3/32 × 1/2	0.203	0.194	3/64	0.0943	0.1561
305	3/32 × 1/2	0.250	0.240	1/16	0.0943	0.2031
404	1/8 × 1/2	0.203	0.194	3/64	0.1255	0.1405
405	1/8 × 5/8	0.250	0.240	1/16	0.1255	0.1875
406	1/8 × 3/4	0.313	0.303	1/16	0.1255	0.2505
505	5/32 × 5/8	0.250	0.240	1/16	0.1568	0.1719
506	5/32 × 3/4	0.313	0.303	1/16	0.1568	0.2349
507	5/32 × 7/8	0.375	0.365	1/16	0.1568	0.2969
606	3/16 × 3/4	0.313	0.303	1/16	0.1880	0.2193
607	3/16 × 7/8	0.375	0.365	1/16	0.1880	0.2813
608	3/16 × 1	0.438	0.428	1/16	0.1880	0.3443
609	3/16 × 1 1/8	0.484	0.475	5/64	0.1880	0.3903
807	1/4 × 7/8	0.375	0.365	1/16	0.2505	0.2500
808	1/4 × 1	0.438	0.428	1/16	0.2505	0.3130
809	1/4 × 1 1/8	0.484	0.475	5/64	0.2505	0.3590
810	1/4 × 1 1/4	0.547	0.537	5/64	0.2505	0.4220
811	1/4 × 1 3/8	0.594	0.584	3/32	0.2505	0.4690
812	1/4 × 1 1/2	0.641	0.631	7/64	0.2505	0.5160
1008	5/16 × 1	0.438	0.428	1/16	0.3130	0.2818
1009	5/16 × 1 1/8	0.484	0.475	5/64	0.3130	0.3278
1010	5/16 × 1 1/4	0.547	0.537	5/64	0.3130	0.3908
1011	5/16 × 1 3/8	0.594	0.584	3/32	0.3130	0.4378
1012	5/16 × 1 1/2	0.641	0.631	7/64	0.3130	0.4848
1210	3/8 × 1 1/4	0.547	0.537	5/64	0.3755	0.3595
1211	3/8 × 1 3/8	0.594	0.584	3/32	0.3755	0.4065

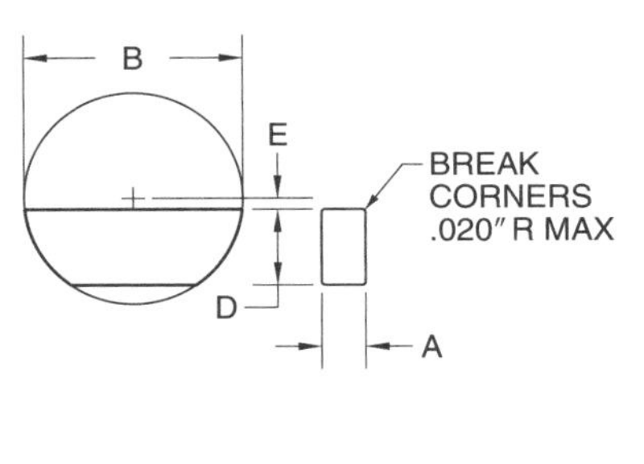

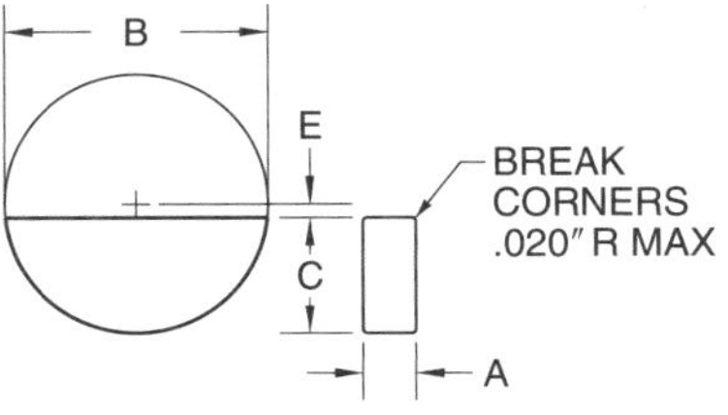

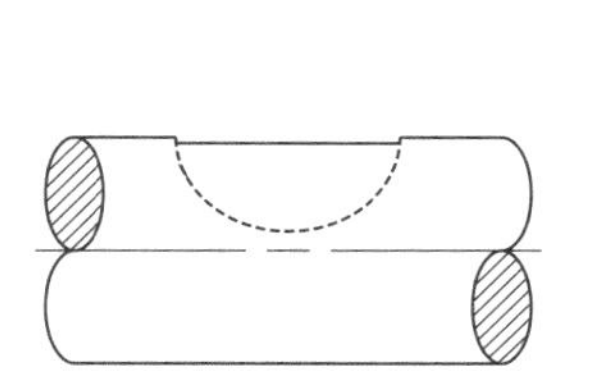

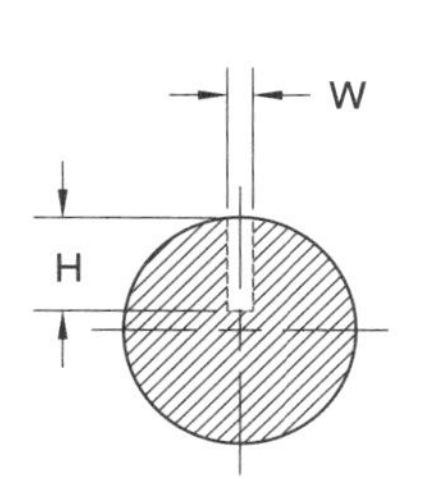

SQUARE AND FLAT KEYS

Shaft Dia	Square Key Sizes	Flat Key Sizes
1/2 - 9/16	1/8	1/8 × 3/32
5/8 - 7/8	3/16	3/16 × 1/8
15/16 - 1 1/4	1/4	1/4 × 3/16
1 5/16 - 1 3/8	5/16	5/16 × 1/4
1 7/16 - 1 3/4	3/8	3/8 × 1/4
1 13/16 - 2 1/4	1/2	1/2 × 3/8
2 5/16 - 2 3/4	5/8	5/8 × 7/16
2 7/8 - 3 1/4	3/4	3/4 × 1/2

A

abbreviation: A shortened version of the letters forming a word.

absolute programming: A method of generating NC commands where each point is specified in relation to the origin.

acute angle: Any angle smaller than 90°.

acute triangle: A triangle with each angle less than 90°.

addendum: The portion of the gear teeth between the pitch circle and the outside diameter circle.

addition: The process of uniting two or more numbers to make one number.

allowance: The amount of space between mating parts when at their actual size.

alloy: A metal that consists of two or more chemical elements.

altitude: The dimension of a triangle that is perpendicular to the base.

angularity: The condition of a feature's axis or surface being at a specified angle (other than 90°) from a datum.

annealing: The process of heating metal until the metal's crystalline structure changes, and then allowing it to cool very slowly.

arc: A portion of a circle's circumference.

area: A quantity that represents the size of a surface or two-dimensional shape.

arrowhead: A symbol that indicates direction or identifies the endpoint of a line.

assembly print: A print that illustrates how two or more parts fit together.

asymmetrical object: An object that cannot be divided in such a way that one half is the mirror image of the other half.

auxiliary section: A sectional view that is not one of the principal planes.

auxiliary view: A view that shows the true shape of an object surface that is not parallel to one of the six principal planes of projection.

axonometric drawing: A pictorial drawing that shows three sides of an object at the same scale, but contains no true view of any side.

B

back (transverse) pitch: The distance from the center of one row of rivets to the center of the adjacent row of rivets.

backlash: The small amount that one gear can rotate before a meshing gear begins to move.

barrel cam: A cylindrical cam with a groove in its surface that guides the follower.

base: **1.** The horizontal side of a triangle. **2.** A polygon at an end of a solid figure.

base circle: The circle formed at the radius of a cam drop.

base unit: A fundamental unit of measure that cannot be described using any other combination of units.

basic dimension: A numerical value used to describe a theoretical exact size, shape, or location of a feature or datum.

bevel: A sloped surface.

bevel gear: A gear with tapered teeth that is used in applications where shaft axes intersect.

blind hole: A drilled hole that does not completely pass through material.

blind rivet: A type of rivet that can be used when there is access from one side only.

blunt start: The removal of the partial thread at the starting end of the thread.

boring: The process of enlarging an existing hole or circular internal shape with a rotating cutting tool.

boss: A short projection with a finished surface that extends above the surface of a part.

break line: A line that indicates the omission of unnecessary portions of a drawing.

breakout side: The side of a stamped part that is opposite the die that breaks through the surface.

brittleness: The tendency of a material to fracture under pressure without significant deformation.

broken-out section: A sectional view in which a small portion designated by a freehand break line is removed.

C

cabinet drawing: An oblique drawing with receding lines drawn to one-half the scale of the true view.

caliper: A measuring tool for inside and outside dimensions.

cam: A machine part that transmits a pattern of motion using an irregular external or internal surface.

cam displacement: The maximum distance from the lowest to the highest point of the cam.

cam drop: The lowest point on a cam.

cam follower: A machine part in contact with a cam path that moves as the cam rotates.

cam profile: The shape of the cam perimeter or another path that actuates the cam follower.

cam rise: The highest point on a cam.

carburizing: A case hardening process in which carbon is introduced into a solid iron-base alloy heated above a certain temperature.

Cartesian coordinate system: A system of defining a position or motion in space by its distance in three perpendicular directions from an origin point.

case hardening: The process of increasing the hardness of a metal surface without changing the mechanical properties of the core.

casting: A manufacturing process where a shaped cavity is filled with molten metal, which takes on the shape of the cavity as it cools and solidifies.

cavalier drawing: An oblique drawing with receding lines drawn to the same scale as the true view.

centerline: A thin dashed line that locates axes or centerpoints of arcs and circles.

central processing unit (CPU): The control center of a computer.

chain line: A thick, dashed line that identifies a surface requiring special treatment or outlines a tolerance zone.

chamfer: A beveled edge.

characteristic: An aspect of a feature that can be toleranced.

characteristic symbol: A symbol that represents a feature characteristic.

chord: A line across a circle, but not through the centerpoint.

circle: A plane figure in which all points are equal distance from its center.

circularity: The condition of a circular feature where all points are equally distant from a centerpoint.

circular pitch: The distance between corresponding points of two adjacent gear teeth on the pitch circle.

circular runout: The maximum variation between high and low spots on the edge of a circular feature in relation to its centerpoint.

circumference: The perimeter of a circle.

clearance: The space between the bottom of a gear tooth space and the tip of a tooth fully meshed into that tooth space.

clearance fit: A fit design for two mating parts that always have space between them, as long as both parts are within their tolerance zones.

clearance requirement: A specification for the empty space needed around the outside surface of a piece of equipment.

code: An instruction that consists of a letter to designate the type of action required and a number to specify the command code, distance, or speed.

command: A line of the necessary codes to complete one step in an operation.

compass: A drafting instrument used to draw arcs and circles.

complete (full) thread: A thread having full form at both crest and root.

compound gearing: The use of several gears to achieve high gear ratios.

compressive strength: Strength under compressive forces, which squeeze the material.

computer-aided design (CAD): The creation of a technical drawing in computer software, which can then be output to printers or plotters.

concentric circles: Two or more circles that have different diameters but the same centerpoint.

concentricity: The condition in which a circular feature's axis exactly matches a datum axis.

cone: A solid figure with a circular base and a surface that tapers from the base to the vertex.

constant pitch series: A standard thread series with a set number of threads per inch for a range of diameters.

constant-velocity motion: Motion at a constant rate of speed.

conventional drafting: The manual creation of a technical drawing directly onto paper using pens, pencils, and special instruments.

conventional view: An exterior orthographic view used for assembly drawings.

core: A separate, internal part of a casting mold that is used to form a void or open area in a part.

corner: The angular space at the intersection of surfaces.

counterbored hole: An enlarged and recessed hole with square shoulders.

counterdrilled hole: A hole with a cone-shaped opening below the outer surface.

countersink: The tool that produces a countersunk hole.

countersunk hole: A hole with a cone-shaped recess at the outer surface.

crest: The surface joining the flanks that is farthest from the thread's axis.

crest clearance: The distance, measured perpendicular to the axis, between the crest of a thread and the root of its mating thread.

cutting plane line: A thick, dashed or broken line, terminating in right angle arrowheads, that indicates where an object is imagined to be cut in order to view internal features.

cutting speed: The speed of the surface of a cutting tool.

cylinder: A solid figure with circles as its bases.

cylindricity: The condition of a cylindrical feature where all points are equally distant from a common axis.

D

datum: A theoretically exact point, line, axis, or surface that serves as the origin for dimensions and a reference for tolerances.

datum feature: A datum that is an actual line or surface on a part.

datum target: A datum that is a point, line, or area that is designated on a part but may not be a physical feature.

decimal: A number expressed in base 10 notation.

decimal point: A period that separates a decimal's whole number from its fraction.

denominator: The number that shows the size of the parts in a fraction.

depth of cut: The penetration of a cutting tool for each pass.

derived unit: A unit of measure that can be defined as a combination of other units.

detailed representation: A method of thread representation in which the thread profiles are drawn realistically.

detail print: A supplemental print that provides certain types of additional information needed to produce a part.

detail view: A portion of a drawing that is enlarged to show intricate features more clearly.

diagonal pitch: The distance between the centers of rivets nearest each other in adjacent rows.

diameter: The distance across a circle through the centerpoint.

diametral pitch: The number of raised projections per unit of pitch diameter.

die: A very strong piece of metal with specially shaped cavities that is used to form shapes when pressed against softer materials.

digital readout (DRO): A device on a machine tool that provides a digital display of the cutting tool's position in space.

dimensioning: A method of identifying and quantifying the size of features.

dimension line: A thin line that indicates the extent and direction of dimensions.

dimetric drawing: An axonometric drawing with two axes drawn on equal angles to the third, but the angle between them is larger or smaller.

direct tolerancing: The practice of specifying a dimension's permissible range directly within the dimensioning lines.

dividend: A number to be divided.

divider: A drafting instrument used to transfer and compare dimensions.

division: The process of determining how many times one number contains the other number.

divisor: A number by which division is done.

draft angle: The small angle between a mold or die surface and an imaginary surface that is perpendicular to the parting line.

drawing: A manufacturing process in which material is pulled through a die in order to shape the material to final size and shape.

drilled hole: A round hole in a material produced by a twist drill.

drilling: The cutting of round holes in material with a rotating twist drill.

drive: The shape of the recess on the fastener head that fits the tool used to rotate the fastener.

dual dimensioning: The practice of showing both inch and millimeter dimensions together.

ductility: The ability of a material to deform under tensile forces without breaking or cracking.

dwell: The time during the cam revolution in which there is no motion of the cam follower.

E

edge: The intersection of two surfaces.

edge cam follower: A follower shaped like a chisel that contacts the cam with an edge.

elastomer: A flexible material that can be stretched up to twice its length and return to its original length when released.

engineering fit: The specification of the appropriate tolerances and allowances for the way two parts are intended to mate together.

equation: A statement of equality between two mathematical expressions.

equilateral triangle: A triangle that has three equal angles and three equal sides.

even number: A number that can be divided by 2 an exact number of times.

exploded view: A drawing that separates all the components of an assembly, but retains their alignment and orientation for reassembly.

extension line: A thin line that extends from a feature in order to facilitate dimensioning.

external thread: A thread on the external surface of a cylinder or cone.

extruding: A manufacturing process in which a material is pushed through a die in order to obtain the desired shape.

F

face cam: A cam that uses a groove cut into its face, instead of or in addition to the perimeter, to contact a follower.

feature: Any surface, angle, hole, round, or other characteristic on a part that can be dimensioned and controlled.

feature control: A tolerancing method that addresses special characteristics and can account for the interrelationships of multiple features.

feature control frame: A rectangular frame divided into sections that encloses characteristic symbols, numerical tolerances, datum references, and modifiers.

feature information: Installation information about the relationships between equipment components and equipment or facility features.

feed: The rate at which the cutting tool advances into the workpiece.

ferrous metal: A metal that has iron as the major alloying element.

field rivet: A rivet placed in the field.

fillet: A rounded interior corner.

first-angle projection: A projection system that places the object between the plane of projection and the observer.

flank: Either surface of a thread that connects its crest with its root.

flat cam follower: A follower that contacts the cam with a flat surface.

flatness: The condition of a surface feature with no surface variations.

flaw: An unintentional interruption in the characteristic texture of a surface.

foreshortening: The apparent shortening of inclined parts when shown in an orthographic view.

forging: A manufacturing process in which a workpiece is deformed into the desired shape with high compressive forces, usually between two dies.

forging print: A print that details the information needed to forge a part.

form: A thread's shape profile in an axial plane.

forming process: A process that shapes a part prior to final machining operations.

formula: An equation involving multiple variables that has been solved for one of the variables.

fraction: A portion of a whole number.

front auxiliary view: A primary auxiliary view with the inclined surface perpendicular to the frontal plane.

full section: A sectional view in which the cutting plane passes entirely across the object.

fundamental deviation: The difference between a basic dimension and the closest limit of the tolerance zone.

G

gauge: A standard that is used as a tool for checking sizes and pitches.

G-code: A programming language for NC machining that consists of a series of standardized letter and number codes for machining operations.

gear: A toothed wheel that is used with other gears in order to transmit rotational power from one shaft to another.

gear ratio: The relationship between the sizes of two meshing gears.

general tolerance: A specified tolerance that is common for all dimensions on a drawing that are not otherwise toleranced.

geometric dimensioning and tolerancing: A method of specifying the size, shape, and location of features on manufacturing prints, as well as the allowable variations.

graded pitch series: A standard thread series with a different number of threads per inch for most diameters.

graphics tablet: An electronic input device that consists of a drawing area and a menu.

grinding: The process of removing material with an abrasive.

grip: The effective holding length of a rivet.

groove: A shallow channel machined into a surface.

H

half section: A sectional view in which two cutting planes are passed at right angles to each other along centerlines or symmetrical axes.

hardening: The process of heating metal followed by quenching in oil, water, or another cooling medium to bring the temperature down quickly.

hardness: The ability of a material to resist permanent deformation, usually by indentation.

hardware: The physical components of a computer system.

harmonic motion: Motion in which acceleration changes smoothly, causing displacement to vary according to a sinusoidal curve.

heat treatment: The application of heat to change the properties of a metal without changing its size and shape.

helical gear: A gear with teeth that follow a helix (spiral) shape and are not parallel to the shaft axis.

herringbone gear: A gear that is composed of two rows of helical teeth.

hexahedron: A regular solid figure formed from six squares.

hidden line: A thin dashed line that represents an edge or contour that cannot be seen from the view of an object.

hypotenuse: The side of a right triangle opposite the right angle.

I

improper fraction: A fraction with a numerator larger than its denominator.

inclined surface: A plane surface perpendicular to one principal plane of projection and inclined to the others.

incomplete thread: A thread that is not fully formed.

incremental programming: A method of generating NC commands where each point is specified in relation to the last point where the machine stopped.

indicator: A measuring tool for very small linear displacements.

individual feature: A feature that is an independent characteristic of a part and does not relate to any datum.

input device: Hardware used to enter information into a computer system.

installation print: A print that illustrates the general configuration and information needed to install a specific piece of equipment.

interference fit: A fit design for two mating parts that always has some overlap between the parts, regardless of their actual sizes within their tolerance zones.

internal gear: A spur gear that meshes on its inside circumference.

internal thread: A thread on the internal surface of a hollow cylinder or cone.

intersecting surfaces: Surfaces that meet at an edge.

involute: The curve formed by the path of a point on a straight line as it unwinds from a round surface.

isometric drawing: An axonometric drawing with the axes drawn equally spaced at 120° apart.

isosceles triangle: A triangle that contains two equal angles and two equal sides.

K

key: A removable fastener that provides a positive means of transmitting torque between a shaft and a hub when mounted in a keyseat.

keyseat: A rectangular groove along the axis of a shaft or hub that mates with a key.

knurl: A raised pattern formed on a material for improving the grip.

L

lateral face: The side of a solid figure.

lay: The direction of the dominant surface texture pattern.

leader line: A thin line that connects a dimension, note, or specification with a particular feature.

least material condition (LMC): The condition of a feature having the minimum amount of material permitted by the tolerance zone.

left-hand thread: A thread that, when viewed axially, winds counterclockwise when receding.

limit dimensioning: The practice of including only the maximum and minimum values of a dimension.

lobe: A projecting part of a cam that causes the cam follower to be displaced.

lowest common denominator (LCD): The smallest number that can be used as a common denominator for a group of fractions.

M

machinability: The ease with which a material can be acceptably machined.

machining: The process of removing material from a workpiece with cutting tools in order to achieve a desired size and shape.

machining print: A print that details the information needed to machine a part.

major diameter: The diameter of the imaginary coaxial cylinder that bounds the crest of an external thread or the root of an internal thread.

malleability: The ability of a material to deform under compressive forces without developing defects.

margin: The distance from the edge of a plate to the centerline of the nearest row of rivets.

maximum material condition (MMC): The condition of a feature having the maximum amount of material permitted by the tolerance zone.

mechanical property: The response of a material under applied loads.

metal: A material consisting of one or more chemical elements having a crystalline structure, high thermal and electrical conductivity, the ability to be deformed when heated, and high reflectivity.

micrometer: A measuring tool for relatively small but extremely precise linear measurements.

milling: A cutting operation that combines the rotation of a cutting tool and the feeding of the workpiece into the path of the cutter.

minor diameter: The diameter of the imaginary coaxial cylinder that bounds the root of an external thread or the crest of an internal thread.

minuend: A number from which a subtraction is made.

miter gear: One of a pair of bevel gears having the same number of teeth.

mixed number: A combination of a whole number and a fraction.

modifier: A specification for a special condition of a feature or tolerance zone when a tolerance is applied.

modifier symbol: A symbol that represents a specific condition of the part or feature when the tolerance is applied.

monitor: A video display terminal.

mounting dimension: A dimension used to locate fastening points on equipment.

multiplicand: A number that is multiplied.

multiplication: The process of adding one number as many times as there are units in the other number.

multiplier: A number by which multiplication is done.

multiview drawing: A collection of two-dimensional drawings of a three-dimensional object that are shown in true view.

N

NC program: A collection of machining commands used to produce a specific feature or part.

neck: A groove cut into a cylindrical part to provide a space where the diameter on a cylinder changes.

nonferrous metal: A metal that does not contain iron.

nonthreaded fastener: A device that joins parts together without threads.

normal surface: A plane surface parallel to a principal plane of projection.

numerator: The number of parts in a fraction.

numerical control (NC): A process of automatically controlling the motion of machine tools using a set of programmed commands.

O

oblique drawing: A pictorial drawing that shows one surface of an object as a true view.

oblique surface: A plane surface not parallel to any principal plane of projection.

oblique triangle: A triangle that does not contain a right angle.

obtuse angle: Any angle greater than 90°.

obtuse triangle: A triangle with one angle greater than 90°.

odd number: A number that cannot be divided by 2 an exact number of times.

offset section: A sectional view in which the cutting plane line changes direction in order to include features that are not located in a straight line.

orthographic projection: A method of representing the true shape of one view of an object onto a single plane.

outline dimension: A dimension for the minimum space required to install the piece of equipment.

output device: Hardware that either displays or generates drawings.

outside diameter: The diameter of the circle formed by the tops of the gear teeth.

P

parabolic motion: Motion in which displacement follows a parabolic curve and the acceleration and deceleration is constant in absolute value.

parallelism: The condition of a feature's axis or surface where all the points are equidistant from a datum line or plane.

parallelogram: A four-sided plane figure with opposite sides that are parallel and equal.

parting line: A boundary dividing two parts of a mold.

parts list: An area located above the title block that contains specifications for any off-the-shelf components used in the assembly shown on the print.

patternmaking (casting) print: A print that details the information needed to make the mold for a cast part.

perpendicularity: The condition of a feature's surface or axis being at a right angle (exactly 90°) to one or two datums.

phantom line: A thin, dashed line that indicates additional reference information.

pin: A cylindrical, nonthreaded fastener that is placed into a hole to secure the position of two or more parts.

pinion gear: The smallest of meshing gears.

pitch: The distance between corresponding points on adjacent raised projections.

pitch circle: The imaginary circle located approximately halfway between the tops and the roots of raised projections.

pitch diameter: The diameter of the pitch circle.

plane figure: A two-dimensional figure.

plane of projection: An imaginary surface on which the shape of an object from that view is drawn.

plastic: A material made up of repeating groups of atoms or molecules linked in long chains called polymers.

plate cam: A cam made from a flat plate that contacts the follower at its perimeter.

plotter: A computer output device that generates finished drawings with pens.

plus and minus tolerancing: The practice of providing an ideal dimension along with its allowable deviations in the positive and negative directions.

pointed cam follower: A follower that ends in a conical tip.

point-to-point dimensioning: A dimensioning method that specifies sizes along an object from one feature to another.

polar coordinate dimensioning: A dimensioning method that determines location with angular and radius dimensions.

polygon: Any plane figure with three or more straight sides.

position: The condition of a feature's center, axis, or center plane being at a nominal distance from one or more datums.

primary auxiliary view: An auxiliary view that is projected to a plane that is perpendicular to one of the three principal planes and inclined to the other two.

prime number: A number greater than 1 that can be divided an exact number of times only by itself and the number 1.

print: A reproduction of a working drawing.

printer: A computer output device that generates finished drawings with liquid ink or powder toner.

prism: A solid figure with two bases that are identical, parallel polygons.

product: The result of multiplication.

profile: An outline of an object in a given plane.

profile of a line: A mathematically defined two-dimensional shape.

profile of a surface: A mathematically defined three-dimensional surface.

projection line: A line that connects the edges of an object in a principal view and an auxiliary view.

proper fraction: A fraction with a denominator larger than its numerator.

protractor: A measuring tool for angles.

pure metal: A metal that consists of one chemical element.

pyramid: A solid figure with a base that is a polygon and sides that are triangles.

Pythagorean theorem: A theorem that states that the square of the hypotenuse of a right triangle is equal to the sum of the squares of the other two sides.

Q

quadrant: One-fourth of a circle.

quadrilateral: A polygon with four sides.

quotient: The result of division.

R

rack: A spur gear that is flat rather than concentric.

radius: The distance from a circle's centerpoint to its circumference.

rapid prototyping: An NC process that builds parts in an additive process rather than the traditional subtractive (material removing) processes.

reaming: The process of enlarging an existing hole slightly in order to improve its dimensional accuracy and surface quality to tighter tolerances.

recess: A groove cut into the internal diameter of a cylinder.

rectangle: A quadrilateral with opposite sides that are equal and four right angles.

rectangular coordinate dimensioning: A dimensioning method that specifies all dimensions from a baseline or datum.

reference dimension: A dimension that is used for informational purposes only and is not intended to govern manufacturing operations.

reference line: A line representing the imaginary hinge on which an auxiliary view was rotated into the plane of the print.

regardless of feature size: The condition where the characteristic tolerances are not affected by the size tolerances of the feature.

regular polygon: A polygon with sides that are all of equal length.

regular solid figure: A solid figure with faces that are regular polygons.

related feature: A feature that is associated with one or more specific datums.

remainder: 1. The difference between a minuend and subtrahend. **2.** The part of a quotient left over when the quotient is not a whole number.

removed section: A sectional view that is detached from the projected view and located elsewhere on the sheet.

revision block: An area, located in the upper-right corner of a print, that contains information about changes made to the original drawing.

revolved section: A sectional view in which a cross-sectional shape is shown at the cutting plane location.

rhomboid: A quadrilateral with opposite sides that are equal and no right angles.

rhombus: A quadrilateral with sides that are all equal and no right angles.

right angle: An angle that is exactly 90°.

right-hand thread: A thread that, when viewed axially, winds clockwise when receding.

right triangle: A triangle with one right angle.

riser: An internal cavity in a casting mold that funnels molten metal into the part cavity and serves as a reservoir for extra molten metal after the part cavity is filled.

rivet: A cylindrical metal pin that is deformed in order to hold parts together.

rivet pitch: The distance from the center of one rivet to the center of the next rivet in the same row.

roller cam follower: A follower tipped with a rotating wheel that significantly reduces friction.

rolling: A manufacturing process in which material is squeezed between two revolving rolls to obtain the desired thickness.

root: The surface joining the flanks that is identical in position with, or immediately adjacent to, the cylinder or cone from which the thread projects.

root circle: A circle formed by the bottom of the tooth spaces.

root diameter: The diameter of the root circle.

roughness: The degree of irregularity in the smoothness of a surface.

round: A rounded exterior corner.

rule: A semiprecision measuring tool used for measuring length.

runout: 1. The curve produced by a plane surface tangent to a cylindrical surface. **2.** The measurement of the relationship of circular features to a datum axis.

S

scale: A drafting instrument used to measure lines and reduce or enlarge them proportionally.

scalene triangle: A triangle that has no equal angles or equal sides.

schematic assembly print: A print that illustrates in pictorial or plan view the relative locations and connections of equipment within a system.

schematic representation: A method of thread representation in which solid lines perpendicular to the axis represent roots and crests.

secant: A straight line touching a circle's circumference at two points.

secondary auxiliary view: An auxiliary view that is projected to a plane that is oblique to all of the principal planes.

sectional view: The view of a cross section of an object.

section line: A thin line used in a hatch pattern fill that identifies an area as being a cutting plane surface.

section lining: The pattern of section lines that fills the area of the surfaces formed by a cutting plane.

sector: A plane figure enclosed between two radii and an arc of a circle.

segment: A plane figure enclosed between the arc of a circle and a chord that connects the two ends of the arc.

semicircle: One-half of a circle.

shank: The cylindrical body of a rivet.

shaping: A cutting operation performed by the reciprocating motion of a cutting tool.

shop rivet: A rivet placed in a shop.

side auxiliary view: A primary auxiliary view with the inclined surface perpendicular to the side plane.

significant digit: A digit that indicates the precision of a measured value.

simplified representation: A method of thread representation in which hidden lines are drawn parallel to the axis at the approximate depth (minor diameter) of the thread.

sketching: Drawing without instruments.

slot: An elongated hole machined either through a part or to a specified depth.

software: A collection of programmed instructions in a computer.

solid figure: A three-dimensional figure.

special series: A screw thread series with combinations of diameter and pitch not in the standard screw thread series.

sphere: A solid figure generated by a circle revolving about an axis.

spherical cam follower: A follower with a rounded end.

spline: A series of teeth or parallel surfaces machined into a shaft (external splines) or hub (internal splines).

spotface: A flat surface machined at a right angle to a drilled hole.

spur gear: A gear with straight teeth that are parallel to the shaft axis.

square: **1.** A quadrilateral with sides that are all equal and four right angles. **2.** A tool for laying out and checking right angles.

stamping: A manufacturing process that involves forming sheet metal between dies with pressure while cutting the part from the sheet metal with the edge of the die.

stamping print: A print that details the information needed to stamp thin material parts into the desired final size and shape.

standard: **1.** A document, established by consensus, that provides rules, guidelines, or characteristics for activities or their results. **2.** An object that is created to exactly match a certain defined characteristic.

stitch line: A thin, dashed or dotted line that indicates a sewing path.

straight angle: An angle that is exactly 180°.

straightness: The condition of a line feature where all points on the feature are in perfect alignment.

strength: The pressure that a material can withstand before it fails.

subtraction: The process of removing a quantity from another quantity.

subtrahend: A number that is subtracted.

sum: The result obtained from adding two or more numbers.

supplementary block: An area located to the left of a title block that contains any additional information necessary for reading a print or manufacturing the part.

surface feature: An intentional deviation in an otherwise flat surface.

symbol: A simplified graphic representation of an object or idea.

symmetrical object: An object in which one half is the mirror image of the other half.

symmetry: The condition where the median points of elements on opposing features lie exactly on the part's centerline.

symmetry line: A centerline that defines a plane of symmetry for a partial view.

T

tabular dimensioning: A dimensioning method that assembles all the numerical feature dimensions in table form.

tangent: A straight line touching a circle's circumference at only one point.

taper: A solid or hollow cylinder in which the diameter changes uniformly from one end to the other.

taper thread: A thread formed on a cone.

technical pen: A drawing instrument that makes ink lines of a consistent width.

tempering: The process of heating metal followed by controlled cooling at a specific rate.

tensile strength: Strength under tensile forces, which pull on the material.

tetrahedron: A regular solid figure formed from four triangles.

thermal expansion: The elongation of a material when subjected to heat.

thermoplastic: A plastic that softens when heat is applied and reforms into a solid when cooled.

thermoset: A plastic that is chemically changed during initial processing and does not soften with subsequent application of heat.

thin section: A sectional view of a part that is too thin to be shown by the ordinary cross-sectioning convention.

third-angle projection: A projection system that places the plane of projection between the object and the observer.

thread: A ridge of uniform section that forms a spiral on the internal or external surface of a cylinder or cone.

thread angle: The angle between the flanks of a thread, measured in an axial plane.

threaded fastener: A device that uses threads to join parts together.

thread lock coating: A liquid coating applied to a threaded fastener to prevent the loosening of assembled parts from vibration, shock, and/or chemical leakage.

thread series: A group of diameter-pitch combinations for certain thread form.

through hole: A drilled hole passing completely through the material.

title block: An area located in the lower right-hand corner of a print that contains identifying information about the drawing.

tolerance stack: The accumulation of excessive tolerance due to the referencing of dimensions to features with their own tolerance.

tolerance zone: An area or volume that defines the space within which a feature may acceptably vary.

tolerancing: A method of specifying the allowable variations of a feature.

tooth face: The curved surface of a tooth located above the pitch circle.

tooth flank: The curved surface of a gear tooth located below the pitch circle.

top auxiliary view: A primary auxiliary view with the inclined surface perpendicular to the top plane.

torque: Rotational force.

total runout: The maximum variation between high and low spots on a cylindrical surface feature in relation to its axis.

toughness: The ability of a material to withstand fracture when stressed.

trace point: The point of contact between the cam and cam follower.

transitional fit: A fit design for two mating parts that may result in either clearance or interference, depending on the actual sizes of the parts within their tolerance zones.

translucent paper: Paper that allows some light to pass through.

trapezium: A quadrilateral with no sides that are parallel.

trapezoid: A quadrilateral with only two sides that are parallel.

triangle: 1. A drafting instrument used to draw vertical and inclined lines. **2.** A polygon with three sides.

trimetric drawing: An axonometric drawing with all axes drawn at different angles.

true view: A view in which the line of sight is perpendicular to the surface.

T-square: A drafting instrument used to draw horizontal lines and as a reference base for positioning triangles.

turning: A cutting operation performed with the workpiece rotating and a cutting tool fed into or across the workpiece.

U

undercut: A groove machined at the intersection of two perpendicular planes and runs the length of the part.

V

variable: A symbol used as a substitute for any real number.

vernier scale: An additional scale on a measuring tool that is used to precisely determine measurements that lie between the smallest marks on the main scale.

vertex: The common point of the triangular sides that form a pyramid.

viewing plane line: A thick, dashed or broken line, terminating in right angle arrowheads, that indicates the direction of an alternate external view of an object.

visible line: A thick line that represents an edge or contour that can be seen from the view of an object.

volume: A quantity that represents the size of a three-dimensional object.

W

waviness: A widely spaced surface texture pattern.

welding: The process of joining metal parts by heating them until molten and allowing the molten metals to merge, usually adding extra filler metal to the joint as well.

welding print: A print that details the information needed to weld assemblies together.

whole depth: The total height of the gear tooth from the root circle to the outside diameter circle.

whole number: A number with no fractional or decimal parts.

working depth: The depth a tooth extends into the tooth space when in full mesh with proper clearance.

working drawing: A set of plans that contains the information necessary to complete a job.

worm: A threaded rod that rotates a worm gear.

worm gear: A special type of spur gear that is driven by a worm.

Y

yoke cam: A cam that rotates within an enclosure in order to translate the rotary motion into both vertical and lateral movement.

Z

zone: An area in a print margin that is identified by a letter or number.

Page numbers in italic refer to figures.

A

B

C

H

I

J

K

L

M

N

T

U

V

W

X

Y

Z

USING THE *MACHINE TRADES PRINTREADING* INTERACTIVE CD-ROM

Before removing the Interactive CD-ROM from the protective sleeve, please note that the book cannot be returned for refund or credit if the CD-ROM sleeve seal is broken.

Windows System Requirements

To use this CD-ROM on a Windows® system, your computer must meet the following minimum system requirements:

- Microsoft® Windows® 7, Windows Vista®, or Windows® XP operating system
- Intel® 1.3 GHz processor (or equivalent)
- 128 MB of available RAM (256 MB recommended)
- 335 MB of available hard disk space
- 1024 × 768 monitor resolution
- CD-ROM drive (or equivalent optical drive)
- Sound output capability and speakers
- Microsoft® Internet Explorer® 6.0 or Firefox® 2.0 web browser
- Active Internet connection required for Internet links

Macintosh System Requirements

To use this CD-ROM on a Macintosh® system, your computer must meet the following minimum system requirements:

- Mac OS X 10.5 (Leopard) or 10.6 (Snow Leopard)
- PowerPC® G4, G5, or Intel® processor
- 128 MB of available RAM (256 MB recommended)
- 335 MB of available hard disk space
- 1024 × 768 monitor resolution
- CD-ROM drive (or equivalent optical drive)
- Sound output capability and speakers
- Apple® Safari® 2.0 web browser or later
- Active Internet connection required for Internet links

Opening Files

Insert the Interactive CD-ROM into the computer CD-ROM drive. Within a few seconds, the home screen will be displayed allowing access to all features of the CD-ROM. Information about the usage of the CD-ROM can be accessed by clicking on Using This Interactive CD-ROM. The Quick Quizzes®, Illustrated Glossary, Media Clips, Flash Cards, Tests, Prints, and ATPeResources.com can be accessed by clicking on the appropriate button on the home screen. Clicking on the ATP web site button (www.go2atp.com) accesses information on related educational products. Unauthorized reproduction of the material on this CD-ROM is strictly prohibited.

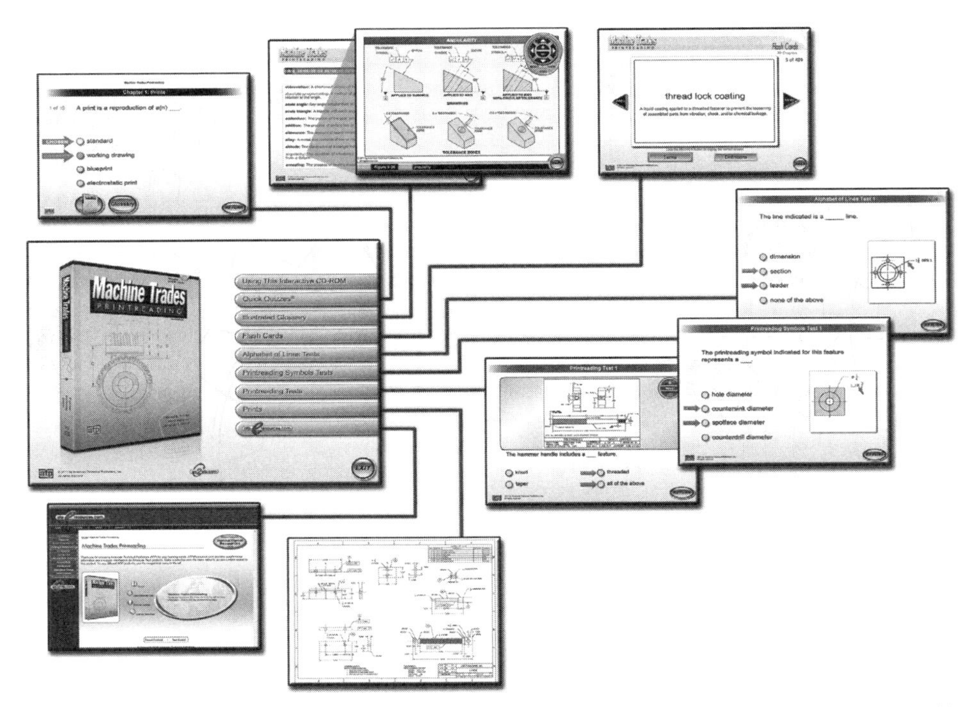

Punch Assembly (Parallel Clamp)

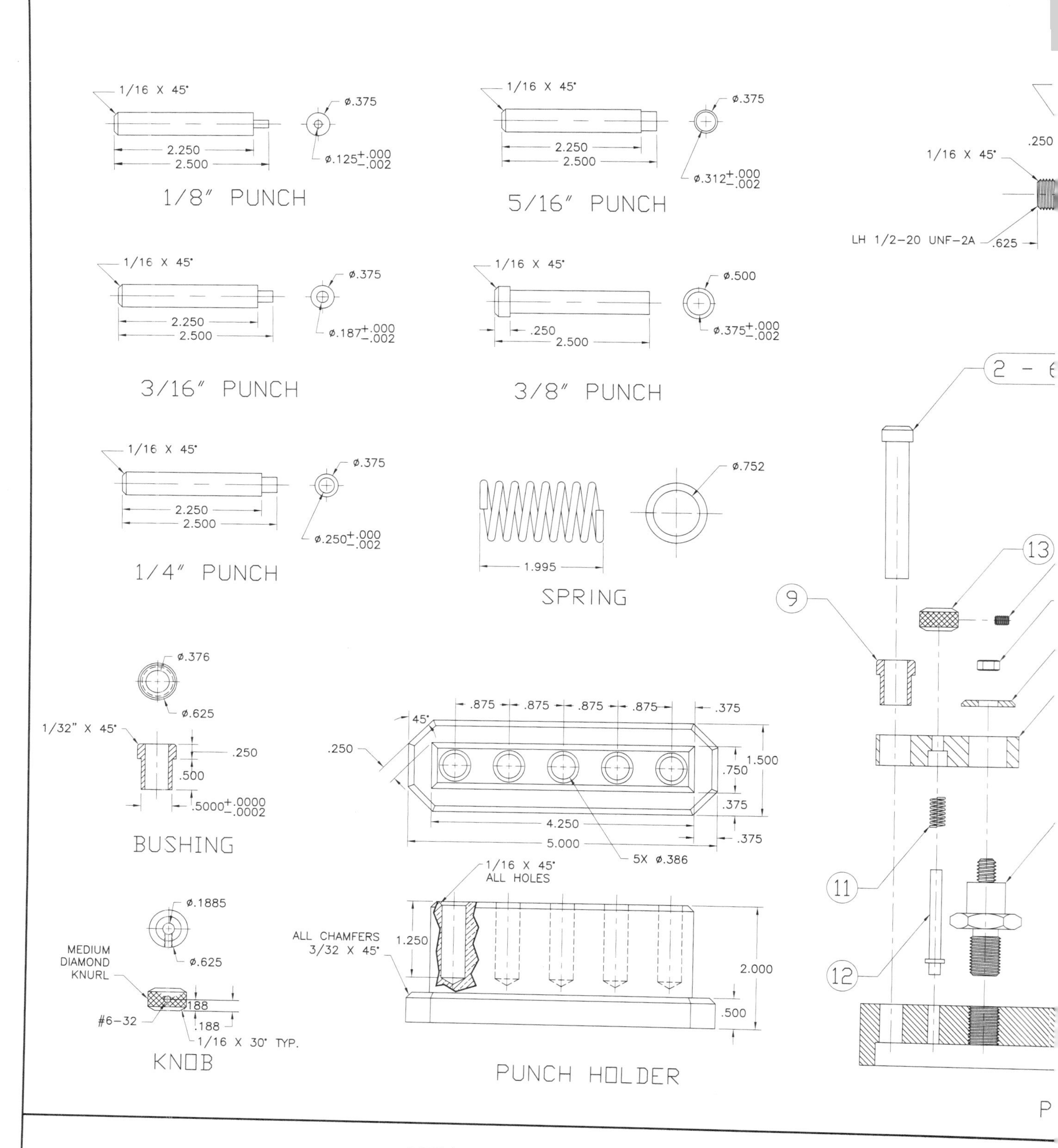

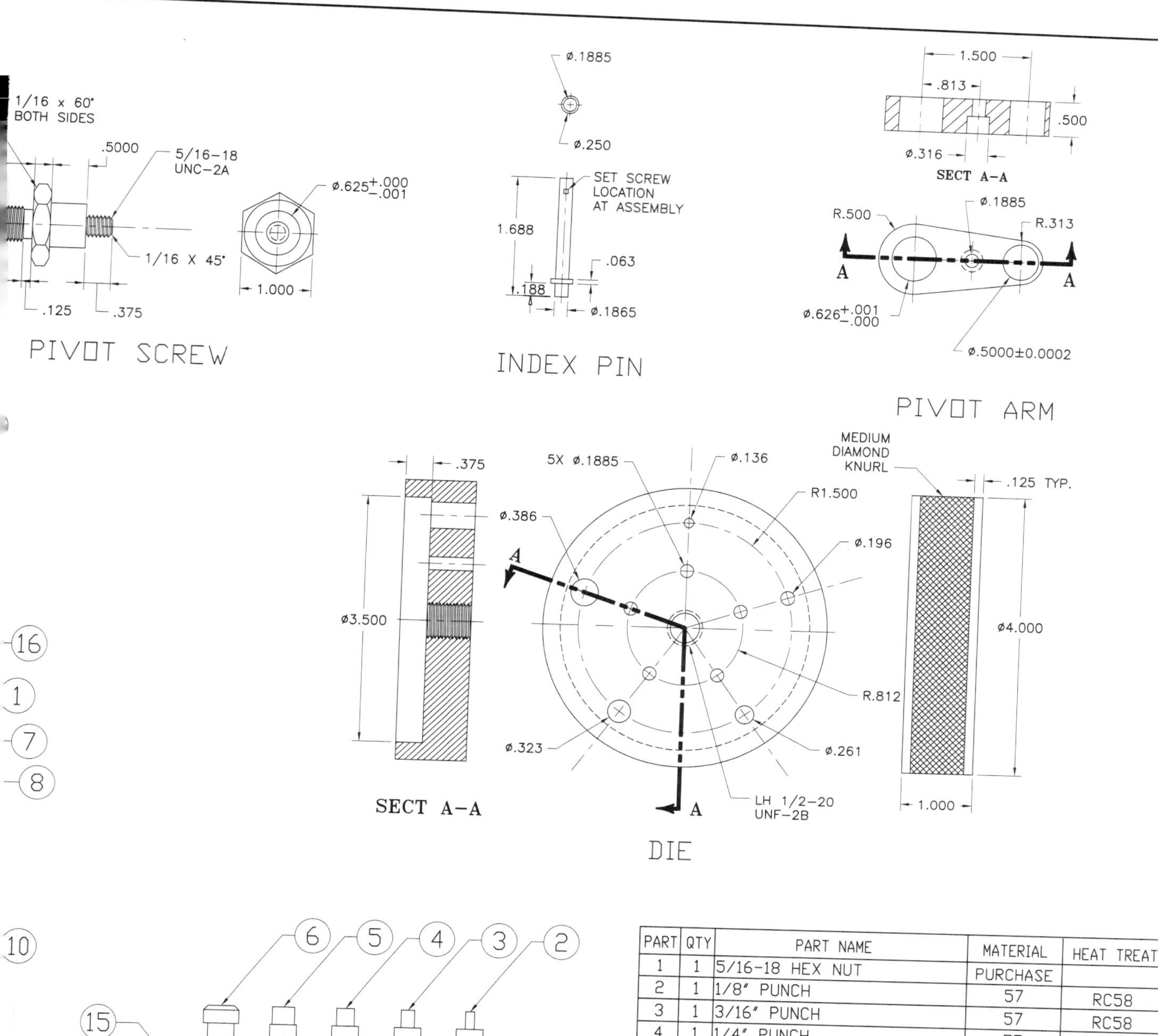

PART	QTY	PART NAME	MATERIAL	HEAT TREAT
1	1	5/16-18 HEX NUT	PURCHASE	
2	1	1/8" PUNCH	57	RC58
3	1	3/16" PUNCH	57	RC58
4	1	1/4" PUNCH	57	RC58
5	1	5/16" PUNCH	57	RC58
6	1	3/8" PUNCH	57	RC58
7	1	WASHER	PURCHASE	
8	1	PIVOT ARM	CRS	NONE
9	1	BUSHING	CRS	NONE
10	1	PIVOT SCREW	CRS	NONE
11	1	SPRING	ø.03 SPRING STEEL	NONE
12	1	INDEX PIN	CRS	NONE
13	1	KNOB	CRS	1
14	1	DIE	57 OR A2	RC58
15	1	PUNCH HOLDER	ALUM.	NONE
16	1	#6-32 X 7/32 SET SCREW	PURCHASE	

PUNCH ASSEMBLY

NO. REQD. 1	NOT TO SCALE	SHEET 1 OF 1
DRAWN BY: MB REVISED BY: JG		DATE: 4/19

Detail, Bottom Die Holder (Holder-Punch)

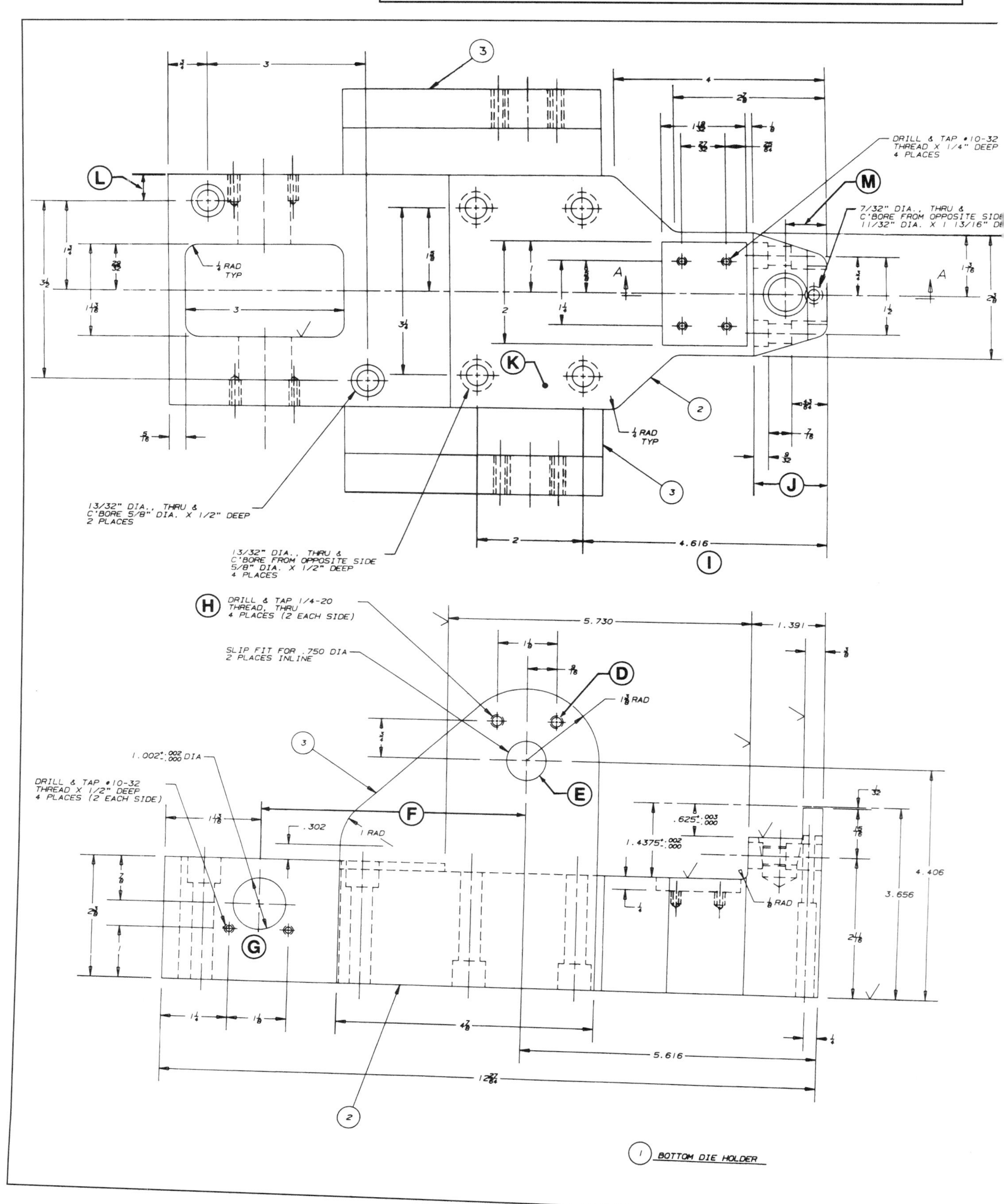

BILL OF MATERIAL

ITEM	MATERIAL	QTY.	DESCRIPTION
1	-	1	BOTTOM DIE HOLDER
2	4140	1	PLATE 3 3/4" X 4 1/2" X 12 7/16" LG.
3	4140	2	PLATE 1 21/32" X 4 7/8" X 5 13/16" LG.
4	-	2	BRONZE BUSHING 3/8" O.D. X 5/16" I.D. X 7/16" LG.

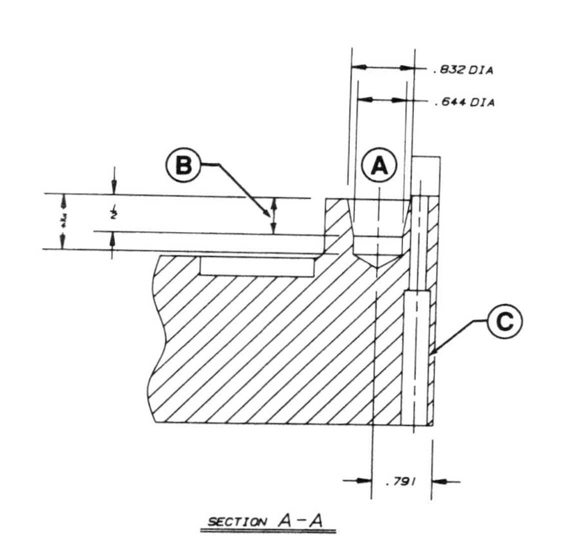

3/8" PRESS FIT FOR BRONZE BUSHING 2 PLACES

.5625 +.002 -.001 RAD

℄ VALVE

.500 DIA TYP

WELD ALL AROUND TYP

REV	DATE	BY	DESCRIPTION

REVISIONS

WORTHINGTON INDUSTRIES
1205 DEARBORN DR. , COLUMBUS OHIO

DETAIL, BOTTOM DIE HOLDER
360 CRIMP SCISSORS TYPE
NON-REFILLABLE VALVE

STANDARD TOLERANCES
FRACTION: ±1/64
DECIMAL : ±.005
ANGLES : ±1/2
UNLESS OTHERWISE SPECIFIED

DATE: 5-30 DRN. BY: MATY SCALE: NTS
SHEET 1 OF 1 DWG. NO.: D-ED-02-616-D0014